FORENSIC CHEMISTRY HANDBOOK

FORENSIC CHEMISTRY HANDBOOK

Edited by

Lawrence Kobilinsky

A JOHN WILEY & SONS, INC., PUBLICATION

Copyright © 2012 by John Wiley & Sons. All rights reserved.

Published by John Wiley & Sons, Inc., Hoboken, New Jersey
Published simultaneously in Canada

No part of this publication may be reproduced, stored in a retrieval system, or transmitted in any form or by any means, electronic, mechanical, photocopying, recording, scanning, or otherwise, except as permitted under Section 107 or 108 of the 1976 United States Copyright Act, without either the prior written permission of the Publisher, or authorization through payment of the appropriate per-copy fee to the Copyright Clearance Center, Inc., 222 Rosewood Drive, Danvers, MA 01923, 978-750-8400, fax 978-750-4470, or on the web at www.copyright.com. Requests to the Publisher for permission should be addressed to the Permissions Department, John Wiley & Sons, Inc., 111 River Street, Hoboken, NJ 07030, 201-748-6011, fax 201-748-6008, or online at http://www.wiley.com/go/permission.

Limit of Liability/Disclaimer of Warranty: While the publisher and author have used their best efforts in preparing this book, they make no representations or warranties with respect to the accuracy or completeness of the contents of this book and specifically disclaim any implied warranties of merchantability or fitness for a particular purpose. No warranty may be created or extended by sales representatives or written sales materials. The advice and strategies contained herein may not be suitable for your situation. You should consult with a professional where appropriate. Neither the publisher nor author shall be liable for any loss of profit or any other commercial damages, including but not limited to special, incidental, consequential, or other damages.

For general information on our other products and services or for technical support, please contact our Customer Care Department within the United States at 877-762-2974, outside the United States at 317-572-3993 or fax 317-572-4002.

Wiley also publishes its books in a variety of electronic formats. Some content that appears in print may not be available in electronic formats. For more information about Wiley products, visit our web site at www.wiley.com.

Library of Congress Cataloging-in-Publication Data:

Forensic chemistry handbook / edited by Lawrence Kobilinsky.
    p. cm.
  Includes index.
  ISBN 978-0-471-73954-8 (cloth)
  1. Chemistry, Forensic–Handbooks, manuals, etc. 2. Forensic sciences–Handbooks, manuals, etc. 3. Criminal investigation–Handbooks, manuals, etc. I. Kobilinsky, Lawrence.
    HV8073.F5595 2011
    363.25'62–dc22

                                                          2010053071

Printed in the United States of America

    ePDF ISBN: 978-1-118-06222-7
    oBook ISBN: 978-1-118-06224-1
    ePub ISBN: 978-1-118-06223-4

10 9 8 7 6 5 4 3 2 1

CONTENTS

Preface   xv

Contributors   xxi

1. Forensic Environmental Chemistry   1
Anthony Carpi and Andrew J. Schweighardt

    1.1   Introduction   2
    1.2   Chemical Fingerprinting   4
         1.2.1   Hydrocarbon Mixtures   4
         1.2.2   Polycyclic Aromatic Hydrocarbons   6
         1.2.3   Biomarkers   11
         1.2.4   Additives   11
         1.2.5   Isotopes   12
         1.2.6   Tracers   13
         1.2.7   Methods of Detection   16
         1.2.8   Weathering   18
    1.3   Spatial Association of Environmental Incidents   18
    References   20

2. Principles and Issues in Forensic Analysis of Explosives   23
Jimmie C. Oxley, Maurice Marshall, and Sarah L. Lancaster

    2.1   Introduction   24
    2.2   Sample Collection   25
    2.3   Packaging   29
    2.4   Sorting   30
    2.5   Documentation   31
    2.6   Environmental Control and Monitoring   31
    2.7   Storage   33
    2.8   Analysis   33
    2.9   Records   36
    2.10   Quality Assurance   36
    2.11   Safety and Other Issues   37
    Conclusion   37
    References   38

3. Analysis of Fire Debris   41
John J. Lentini

   3.1 Introduction   42
   3.2 Evolution of Separation Techniques   43
   3.3 Evolution of Analytical Techniques   47
   3.4 Evolution of Standard Methods   49
   3.5 Isolating the Residue   51
        3.5.1 Initial Sample Evaluation   51
        3.5.2 ILR Isolation Method Selection   51
        3.5.3 Solvent Selection   54
        3.5.4 Internal Standards   54
        3.5.5 Advantages and Disadvantages of Isolation Methods   56
   3.6 Analyzing the Isolated ILR   56
        3.6.1 Criteria for Identification   63
        3.6.2 Improving Sensitivity   90
        3.6.3 Estimating the Degree of Evaporation   95
        3.6.4 Identity of Source   98
   3.7 Reporting Procedures   101
   3.8 Record Keeping   102
   3.9 Quality Assurance   105
   Conclusion   105
   References   106

4. Forensic Examination of Soils   109
Raymond C. Murray

   4.1 Introduction   110
   4.2 Murder and the Pond   111
   4.3 Oil Slicks and Sands   113
   4.4 Medical Link   114
   4.5 Examination Methods   114
        4.5.1 Color   115
        4.5.2 Particle-Size Distribution   117
        4.5.3 Stereo Binocular Microscope   120
        4.5.4 Petrographic Microscope   122
        4.5.5 Refractive Index   124
        4.5.6 Cathodoluminescence   124
        4.5.7 Scanning Electron Microscope   125
        4.5.8 X-Ray Diffraction   126
   4.6 Chemical Methods   127
        4.6.1 FTIR and Raman Spectroscopy   128
   4.7 Looking Ahead   129
   References   130

5. Analysis of Paint Evidence — 131
Scott G. Ryland and Edward M. Suzuki

- 5.1 Introduction — 132
- 5.2 Paint Chemistry and Color Science — 134
  - 5.2.1 Binders — 134
  - 5.2.2 Pigments — 136
- 5.3 Types of Paint — 139
  - 5.3.1 Automotive Finish Systems — 139
  - 5.3.2 Architectural Coatings (Structural Paints or House Paints) — 140
  - 5.3.3 Other Coatings — 141
- 5.4 Paint Evidence Interpretation Considerations — 141
- 5.5 Analytical Methods — 142
  - 5.5.1 Microscopic Examinations — 143
  - 5.5.2 Physical Nature of the Transfer — 147
  - 5.5.3 Microscopy — 149
  - 5.5.4 Microspectrophotometry — 152
  - 5.5.5 Infrared Spectroscopy — 158
  - 5.5.6 Raman Spectroscopy — 175
  - 5.5.7 Pyrolysis Gas Chromatography and Pyrolysis Gas Chromatography–Mass Spectrometry — 178
  - 5.5.8 Elemental Analysis Methods — 188
  - 5.5.9 Other Methods — 205
- 5.6 Examples — 208
  - 5.6.1 Example 1 — 208
  - 5.6.2 Example 2 — 210
  - 5.6.3 Example 3 — 213
- References — 217

6. Analysis Techniques Used for the Forensic Examination of Writing and Printing Inks — 225
Gerald M. LaPorte and Joseph C. Stephens

- 6.1 Introduction — 226
- 6.2 Ink — 226
  - 6.2.1 Ink Composition — 227
- 6.3 Ink Analysis — 230
  - 6.3.1 Physical Examinations — 233
  - 6.3.2 Optical Examinations — 236
  - 6.3.3 Chemical Examinations — 238
  - 6.3.4 Ink Dating — 240
- 6.4 Office Machine Systems — 242
  - 6.4.1 Inkjet Ink — 242

		6.4.2	Inkjet Ink Analysis	243
		6.4.3	Toner Printing	245
		6.4.4	Toner Analysis	246
	Conclusion			247
	References			248

7. The Role of Vibrational Spectroscopy in Forensic Chemistry   251
Ali Koçak

7.1	Introduction to Vibrational Spectroscopy		252
7.2	Infrared Spectroscopy		253
7.3	Infrared Sampling Techniques		255
	7.3.1	Transmission Spectroscopy	255
	7.3.2	External Reflection Spectroscopy	255
	7.3.3	Attenuated Total Reflectance	256
	7.3.4	Diffuse Reflectance Spectroscopy	258
	7.3.5	Infrared Microspectroscopy	259
7.4	Raman Spectroscopy		260
7.5	Raman Spectroscopic Techniques		262
	7.5.1	Surface-Enhanced Raman Spectroscopy	262
	7.5.2	Resonance Raman Scattering	263
	7.5.3	Coherent anti-Stokes Raman Spectroscopy	263
	7.5.4	Confocal Raman Spectroscopy	263
7.6	Applications of Vibrational Spectroscopy in Forensic Analysis		264
	References		265

8. Forensic Serology   269
Richard Li

8.1	Introduction		270
8.2	Identification of Blood		271
	8.2.1	Oxidation–Reduction Reactions	272
	8.2.2	Microcrystal Assays	275
	8.2.3	Other Assays for Blood Identification	275
8.3	Species Identification		278
	8.3.1	Immunochromatographic Assays	278
	8.3.2	Ouchterlony Assay	280
	8.3.3	Crossed-Over Immunoelectrophoresis	281
8.4	Identification of Semen		282
	8.4.1	Visual Examination	282
	8.4.2	Acid Phosphatase Assays	283
	8.4.3	Microscopic Examination of Spermatozoa	284
	8.4.4	Immunochromatographic Assays	285

		8.4.5	RNA-Based Assays	286

 8.4.5 RNA-Based Assays 286
 8.5 Identification of Saliva 286
  8.5.1 Visual and Microscopic Examination 287
  8.5.2 Identification of Amylase 287
  8.5.3 RNA-Based Assays 289
 References 289

9. Forensic DNA Analysis 291
Henrietta Margolis Nunno

 9.1 Introduction 292
  9.1.1 Background on DNA Typing 292
  9.1.2 DNA Structure 294
  9.1.3 Nuclear and Mitochondrial DNA Organization 295
 9.2 Methodology 296
  9.2.1 Sample Collection and DNA Extraction 296
  9.2.2 DNA Quantification 297
  9.2.3 Polymerase Chain Reaction 298
  9.2.4 Short Tandem Repeats 298
  9.2.5 PCR of STRs 300
  9.2.6 Separation and Sizing of STR Alleles 301
  9.2.7 Combined DNA Index System (CODIS) Database 305
  9.2.8 Frequency and Probability 306
 9.3 Problems Encountered in STR Analysis 307
  9.3.1 Low-Copy-Number DNA 307
  9.3.2 Degraded DNA and Reduced-Size (mini) STR Primer Sets 308
  9.3.3 PCR Inhibition 310
  9.3.4 Interpretation of Mixtures of DNA 310
  9.3.5 Null Alleles and Allele Dropout 311
  9.3.6 Factors Causing Extra Peaks in Results Observed 312
  9.3.7 Stutter Product Peaks 312
  9.3.8 Nontemplate Addition (Incomplete Adenylation) 313
  9.3.9 Technological Artifacts 313
  9.3.10 Single-Nucleotide Polymorphism Analysis of Autosomal DNA SNPs 313
  9.3.11 Methods Used for SNP Analysis 314
  9.3.12 Mitochondrial DNA Analysis 315
 9.4 Methodology for mtDNA Analysis 316
  9.4.1 Preparation of Samples 316

	9.4.2 MtDNA Sequencing Methods	316
	9.4.3 Reference Sequences	317
	9.4.4 Screening Assays for mtDNA	318
	9.4.5 Interpretation of mtDNA Sequencing Results	319
	9.4.6 Statistics: The Meaning of a Match for mtDNA	320
	9.4.7 Heteroplasmy	320
	9.4.8 The Future of DNA Analysis	321
References		322

10. Current and Future Uses of DNA Microarrays in Forensic Science — 327
Nathan H. Lents

10.1	Introduction	328
10.2	What is a DNA Microarray?	328
	10.2.1 cDNA Microarray	329
	10.2.2 Other Types of DNA Arrays	330
	10.2.3 The Birth of "-omics"	331
10.3	DNA Microarrays in Toxicogenomics	332
	10.3.1 Sharing Information	333
	10.3.2 Forensic Application	333
10.4	Detection of Microorganisms Using Microarrays	334
	10.4.1 Historical Perspective	334
	10.4.2 DNA Fingerprinting	335
	10.4.3 DNA Fingerprinting by Microarrays	336
	10.4.4 DNA Sequence-Based Detection	337
	10.4.5 Where DNA Microarrays Come In	337
	10.4.6 Looking Forward: Genetic Virulence Signatures	338
10.5	Probing Human Genomes by DNA Microarrays	340
	10.5.1 STR Analysis	340
	10.5.2 SNP Analysis	343
	10.5.3 Exploring an Unknown Genome?	344
Conclusion		345
References		345

11. Date-Rape Drugs with Emphasis on GHB — 355
Stanley M. Parsons

11.1	Introduction	357
11.2	Molecular Mechanisms of Action	357
	11.2.1 Receptors and Transporters	357

	11.2.2	Real GHB Receptors	359
11.3	Societal Context of Date-Rape Agents		361
	11.3.1	Acute Effects of Date-Rape Agents on Cognition and Behavior	361
	11.3.2	Medicinal Uses of Date-Rape Drugs	361
	11.3.3	Self-Abuse	362
	11.3.4	Date Rape, Death, and Regulation	363
11.4	Metabolism Fundamentals		363
	11.4.1	Complexity in Unraveling Metabolism of GHB-Related Compounds	363
	11.4.2	Isozymes in GHB-Related Metabolism	364
	11.4.3	Subcellular Compartmentalization of Enzymes, Transporters, and Substrates	364
	11.4.4	Dynamics and Equilibria for Enzymes and Transporters	365
	11.4.5	Thermodynamics-Based Analysis of Metabolic Flux	366
	11.4.6	Metabolism of Endogenous GHB Versus Ingested GHB and Prodrugs	367
	11.4.7	Directionality of in Vivo and in Vitro Enzymatic Activity	367
	11.4.8	Transporters and Enzymes Mediating GHB-Related Metabolism	367
11.5	Biosynthesis of Endogenous GHB		368
	11.5.1	First Step for GHB Biosynthesis in the Known Pathway	368
	11.5.2	Second Step for GHB Biosynthesis in the known Pathway	368
	11.5.3	Third Step for GHB Biosynthesis in the known Pathway	371
	11.5.4	Which Step in GHB Biosynthesis is Rate Limiting?	373
	11.5.5	Are There Other Biosynthetic Pathways to Endogenous GHB?	374
11.6	Absorption and Distribution of Ingested GHB		376
	11.6.1	Gastrointestinal Tract	376
	11.6.2	Blood	377
11.7	Initial Catabolism of GHB		377
	11.7.1	Transport into Mitochondria	377
	11.7.2	Iron-Dependent Alcohol Dehydrogenase ADHFe1	377
	11.7.3	Poorly Characterized Catabolism of GHB	379

11.8	Chemistry of GHB and Related Metabolites not Requiring Enzymes	380
11.9	Experimental Equilibrium Constants for Redox Reactions of GHB	380
11.10	Estimated Equilibrium Constants for Redox Reactions of GHB in Vivo	381
11.11	Different Perspectives on Turnover of Endogenous GHB are Consistent	384
11.12	Disposition of Succinic Semialdehyde	385
11.13	Conversion of Prodrugs to GHB and Related Metabolites	386
	11.13.1 γ-Butyrolactone	386
	11.13.2 1,4-Butanediol	387
11.14	Subcellular Compartmentalization of GHB-Related Compounds	388
11.15	Comparative Catabolism of Ethanol, 1,4-Butanediol, Fatty Acids, and GHB	389
11.16	Catabolism of MDMA, Flunitrazepam, and Ketamine	390
11.17	Detection of Date-Rape Drugs	390
	11.17.1 Compounds Diagnostic for Dosing by Synthetic Date-Rape Drugs	390
	11.17.2 Compounds Diagnostic for Dosing by GHB	390
	11.17.3 Gold-Standard Testing	391
	11.17.4 Many Applications for Reliable Field Tests	392
	11.17.5 Hospital Emergency Department Example	392
	11.17.6 Preparation of a Sample for Delayed Analysis	393
	11.17.7 Time Window Available to Detect Dosing	393
	11.17.8 Extending the Time Window	394
11.18	Special Circumstances of GHB	395
	11.18.1 Industrial Connection	395
	11.18.2 Enzymes Acting on GHB in Bacteria, Yeast, and Plants	395
	11.18.3 Possible Accidental Intoxication by GHB in the Future	395
11.19	Considerations During Development of Field Tests	396
	11.19.1 Shortcomings of Antibody-Based Screens for Simple Analytes	396
	11.19.2 Advantages of Enzyme-Based Screens for Simple Natural Analytes	397
11.20	Development of an Enzymatic Test for GHB	399
	11.20.1 Sensitivity Required for the Hospital Emergency Department	399

		11.20.2	Choice of Enzyme	399
		11.20.3	Reliable Field Test for GHB	400
	Conclusion			402
	Notes			404
	References			406

12. Forensic and Clinical Issues in Alcohol Analysis   435
Richard Stripp

	12.1	Introduction		436
	12.2	Blood Alcohol Concentration		437
	12.3	Alcohol Impairment and Driving Skills		441
	12.4	Field Sobriety Tests		443
	12.5	Blood Alcohol Measurements		444
		12.5.1	Enzymatic Methods	444
		12.5.2	Headspace Gas Chromatography	445
		12.5.3	Breath Alcohol Testing	446
		12.5.4	Breath Alcohol Instrumentation	447
		12.5.5	Extrapolation from BrAC to BAC	449
		12.5.6	Urine and Saliva	450
		12.5.7	Ethyl Glucuronide	450
		12.5.8	Postmortem Determination of Alcohol	451
		12.5.9	Quality Assurance of Alcohol Testing	452
	References			453

13. Fundamental Issues of Postmortem Toxicology   457
Donald B. Hoffman, Beth E. Zedeck, and Morris S. Zedeck

	13.1	Introduction		458
	13.2	Tissue and Fluid Specimens		460
		13.2.1	Blood	460
		13.2.2	Urine	461
		13.2.3	Vitreous Humor and Cerebrospinal Fluid	461
		13.2.4	Gastric Contents	462
		13.2.5	Meconium	463
		13.2.6	Brain	464
		13.2.7	Liver and Bile	464
		13.2.8	Lung, Spleen, Kidney, and Skin	465
		13.2.9	Muscle	465
		13.2.10	Bone, Teeth, Nails, and Hair	465
		13.2.11	Other Materials for Analysis	466
	13.3	Specimen Collection and Storage		466
	13.4	Extraction Procedures		467
	13.5	Analytical Techniques		467
	13.6	Interpretation		470

13.6.1	Postmortem Redistribution	470
13.6.2	Pharmacogenomics	471
13.6.3	Drug Interactions	472
13.6.4	Drug Stability and Decomposed Tissue	473
13.6.5	Effects of Embalming Fluid	474
Conclusion		475
References		476

14. Entomotoxicology: Drugs, Toxins, and Insects — 483
Jason H. Byrd and Michelle R. Peace

14.1	Introduction	484
14.2	The Fly and Forensic Science	484
14.2.1	History of Forensic Entomology, Toxicology, and the Rise of Entomotoxicology	485
14.2.2	Drugs and the Fly Life Cycle	488
14.2.3	Why Use Insects as a Toxicological Specimen?	490
14.2.4	Drug Extraction Methods	492
14.2.5	Qualitative Versus Quantitative	493
14.2.6	Changes in Insect Development: Toxins and Drugs	494
14.2.7	The Future of Entomotoxicology	494
References		495

Index — **501**

PREFACE

In February 2009 a report entitled *Strengthening Forensic Science in the United States: A Path Forward* was issued by the National Research Council (NRC) of the National Academy of Sciences. The committee members who wrote the report included scientists, judges, lawyers, statisticians, and forensic scientists. The authors of the report recognized that there is an ongoing need to assure that evidence analysis is held to the highest standards and that what is reported in writing and in testimony must be reliable and credible. Forensic science is a large umbrella science consisting of many subdisciplines, including serology, forensic DNA analysis, toxicology, document examination, hair and fiber analysis, arson investigation, firearms and toolmarks, explosives analysis, blood spatter pattern analysis, digital evidence, impression evidence, forensic pathology, forensic anthropology, forensic odontology, and others. Crime scene personnel are trained to identify and collect biological and physical evidence for subsequent laboratory analysis. This evidence can shed light on events leading up to and during the commission of a crime. Often, evidence collected at a crime scene can be associated with either a victim or a suspect. Criminals can often be linked to the crime scene and/or to the victim. At the same time, those falsely accused can be excluded or exonerated based on reliable analysis of physical evidence. Following scientific analysis, the criminalist writes a report and will testify in a court of law. The testimony of an expert witness must be unbiased, accurate, and based on a sound scientific foundation. The admissibility of novel scientific evidence must be determined prior to expert testimony. The use of a novel scientific method to analyze evidence and the subsequent testimony that describes the results of such testing can be challenged for a variety of reasons.

The *Strengthening Forensic Science* (SFS) report recognized that not all forensic disciplines were at the same level with respect to standards and reliability. This problem is based in part on insufficient funding for forensic research and lack of oversight by any national organization. Although ASCLD-LAB accredits forensic laboratories, American Board of Criminalistics (ABC) certifies forensic analysts, and the National Institute of Justice provides funding for some forensic research projects, techniques and principles used in many forensic disciplines are not at the same level of reliability as that achieved by DNA scientists. The science of using DNA for human identification dates back to the work of Alec Jeffries, who applied restriction fragment length polymorphism analysis for this purpose in the mid-1980s. This major breakthrough was complemented by a second breakthrough technique, polymerase chain reaction, developed by Kary Mullis. He used this in vitro method to amplify a relatively small number of template nucleic acid molecules (derived from biological evidence) into billions of copies for subsequent identification of the DNA donor. These procedures are considered highly reliable and are admissible in all courts in the United States.

The SFS report spells out 13 recommendations that if accomplished would certainly expand funded research and establish the reliability of non-DNA forensic evidence analysis.

- Congress should establish a new National Institute of Forensic Science (NIFS), which would involve itself with research and education, the forensic science disciplines, physical and life sciences, measurements and standards, and testing and evaluation.
- NIFS should establish standard terminology and standards to be used in reporting and testifying about the results of forensic testing.
- NIFS should competitively fund research in forensic science to address the accuracy, reliability, and validity of the subdisciplines.
- NIFS, through congressional funding, should provide funding to remove public forensic labs from the administrative control of law enforcement agencies or prosecutors' offices.
- NIFS should encourage research on observer bias and sources of human error in forensic examinations.
- Congress should fund (through NIFS) the development of tools to advance measurement, validation, reliability, information sharing, and proficiency testing in forensic science and to establish protocols for forensic examinations, methods, and practices.
- Laboratory accreditation and analyst certification should be mandatory.
- Forensic laboratories should establish routine quality control and quality assurance programs to ensure the accuracy of forensic analyses.
- NIFS should establish a national code of ethics for all forensic disciplines.
- Congress should provide funding to NIFS to work with appropriate organizations and educational institutions to improve and develop graduate education programs and to provide scholarships and fellowships with emphasis on developing research methods and methodologies applicable to forensic science practice.
- NIFS should support replacing coroner systems with medical examiner systems to improve medicolegal death investigation.
- Congress should provide funds for NIFS to launch a new effort to achieve nationwide fingerprint data interoperability.
- Congress should provide funding for NIFS to help forensic scientists manage and analyze evidence from events that affect homeland security.

The full SFS report, which describes all of the above, is available through the National Academies Press, 500 Fifth Street N.W., Washington, DC 20001.

Some of the issues faced by forensic experts who write official reports and present courtroom testimony about their analysis are (1) whether or not the techniques of the discipline are founded on a sound scientific foundation, so that the experimental findings are accurate and reliable; (2) whether or not there is any significant possibility of human error or analyst bias that could potentially taint the results; and (3) whether or not rigorous standards have been established for interpreting the results.

The authors of the SFS report believe that with the exception of DNA, no subdiscipline of forensic science is sufficiently reliable that a unique identification can be made of physical evidence: for example, "This bullet was fired from this gun and from no other gun in the world"; or "This fingerprint was left by this person and by no other person"; or "This tire track impression was made by this tire and by no other tire". What is the statistical basis for such a statement? What is the measurement error rate? What is the human error rate? How should conclusions be expressed in a report or in testimony?

With all of this in mind, I invite you to read through the various chapters in this book and keep in mind that the subject matter is about science and technology and not about art. Keep in mind how the expert comes to a conclusion and how that conclusion is reported. What is the statistical basis for the expert's report and testimony? Can the expert testify to a high degree of scientific certainty that the questioned and known specimens have a common origin? I leave that to the reader to decide.

This handbook begins with a review of forensic environmental chemistry which involves the use of trace chemical techniques for investigating environmental spills in an effort to determine if there is any civil or criminal liability. The field can be broken down into two broad areas based on the techniques used to determine liability: chemical fingerprinting and spatial association. In chemical fingerprinting, complex mixtures of chemicals or chemical isotopes are used to associate a spill or environmental release with a specific source. In spatial association, a geographical information system and geochemical techniques are used to attribute the location of a contaminant with a possible source in physical space.

Chapter 2 addresses the principles and issues that exist in the forensic analysis of explosives. It lays out the foundation for proper handling of evidence, which is critical to identifying and convicting the criminal. Evidence at the scene of an explosion, especially a large explosion, offers some unique challenges. Basic principles of evidence collection, handling, storing, and identifying various explosives are discussed herein.

Chapter 3 is a review of arson and fire debris analysis. The isolation and identification of ignitable liquid residues (ILRs) from fire debris is a critically important aspect of arson investigation. This chapter covers common techniques for the isolation and identification of ILRs. Analytical procedures have become more sensitive, and results of testing play a very important role during litigation in a criminal or civil court. Quality control is an important component in fire debris analysis. Reports of findings should be written in a scientific manner describing the fire under investigation, evidence handling, a description of the evidence and where it was collected, the isolation procedure and what testing was done and with what kind of equipment, observations made, and conclusions, with a discussion of the meaning of the results.

Chapter 4 reviews the forensic examination of soils. Soils and sediment are excellent sources of trace evidence in both criminal and civil cases because there are an almost unlimited number of identifiable soil types based on the content of rocks, minerals, glasses, and human-made particles and chemicals. Forensic examination commonly identifies the original geographic location of soils associated with a crime, thus assisting an investigation. Studies of soil and related material samples associated with a suspect and crime scene can produce evidence that the samples had or did not have a common source, thus indicating whether or not a suspect was ever at a particular location. Gathering intelligence for criminal and civil investigations, as well as gem and art fraud studies, often use the methods of forensic geology

Chapter 5 deals with the analysis of paint evidence. Paint and coatings often appear in criminal, civil, and art-authenticity investigations. This chapter reviews the current methodologies and approaches used by forensic paint examiners to analyze this type of physical evidence as well as the problems that they may encounter. Fragments of multilayered in-service paint are one of the most complex types of materials encountered in the forensic science laboratory. They consist of both organic and inorganic components heterogeneously distributed in very small samples, often on the order of only 1 square millimeter. These characteristics dictate the requirements of the analytical chemistry approaches to be used, and they can present a formidable challenge to the forensic analyst responsible for classification of the materials and an evaluation of their evidential significance. Several case examples are presented to illustrate these concepts.

Chapter 6 describes analytical techniques used for the forensic examination of writing and printing inks. The analysis and identification of writing and printing inks and toners are generally very important in document examination, especially when used in conjunction with a reference library. Inks can be differentiated based on the chemistry of colorants, solvents, resins, and additives. Instrumental analysis, including GC–MS, HPLC, and FT–IR and Raman spectroscopy, can often be used following visual examination, microscopic observation, and thin-layer chromatography. Analysis of toners can be performed with XRF, SEM–EDS, or pyrolysis GC. Although chemical analysis of materials used to create documents can provide vast amounts of relevant information and strongly support associations between questioned and known materials, in nearly all cases, the data obtained will not support a conclusion that identifies a particular writing instrument of printing device.

Chapter 7 describes the role of vibrational spectroscopy in forensic chemistry. Spectroscopy is the study of the interaction of electromagnetic radiation with matter to determine the molecular structure of a solid sample or one dissolved in a specific solvent. This interaction depends on the intrinsic properties of the sample material and can be classified by the energy of the probing electromagnetic radiation. Energy can be in the form of ultraviolet, visible, or infrared light, as well as other forms of energy. Infrared spectroscopy is a good technique to use to identify such fibers as acrylics, nylons, or polyesters or paints or alkyds, acrylics, or nitrocellulose. The size of the sample may require the use of microscopic infrared spectroscopy, and the nature of the sample may indicate the use of external reflection spectroscopy or attenuated total reflectance spectroscopy. These techniques are reviewed as well as related methods of sample identification.

Chapter 8 discusses the important science of forensic serology, an important area of modern forensic science. The primary activity of the forensic serologist is the identification of bodily fluids, as these fluid stains are commonly associated with violent criminal cases. Proving the presence of blood, semen, saliva, and so on, can often confirm alleged violent acts.

Chapter 9 reviews the field of forensic DNA analysis. It describes how DNA became a valuable forensic tool in identifying the source of physical evidence left at a crime scene. The use of restriction fragment length polymorphism analysis in the mid-1980s was replaced by the use of the polymerase chain reaction (PCR) method, which is more sensitive, requiring far less high-molecular-weight DNA, uses less hazardous materials, and is faster and more economical. PCR-STR-based genetic profile typing methods have improved in sensitivity over the past 20 years and have become a basic tool in the crime lab. Where nuclear DNA is insufficient to generate a full genetic profile, mitochondrial DNA can sometimes be used to provide identifying information. Also described are low-copy-number procedures and the typing of single-nucleotide polymorphisms within the human genome.

Chapter 10 reviews current and future uses of DNA microarrays in forensic science. DNA microarrays have revolutionized basic research in molecular and cellular biology, biochemistry, and genetics. Through hybridization of labeled probes, this high-throughput technology allows the screening of tens or even hundreds of thousands of data points in a single run. The technology is most advanced with nucleic acids, but protein and antibody microarrays are coming of age as well. Because of the unique ability to screen for large numbers of molecules, such as DNA sequences, simultaneously, the potential utility to forensic investigations is tremendous. Indeed, progress has been made demonstrating that microarrays are powerful tools for use in the forensic laboratory. As the technology matures and associated costs come down, the day that microarray analysis becomes a routine part of the forensic toolkit draws nearer.

Chapter 11 reviews the problem of date-rape drugs such as MDMA, flunitrazepam, and ketamine, with an emphasis on GHB. Recreational, predatory, and lethal doses, metabolism, and diagnostic metabolites are described. Similarities to and differences from the effects and metabolism of ethanol are also discussed. The advantages of field tests to detect date-rape drugs, and limitations of antibodies and advantages of enzymes for field testing, are discussed. The development of a rapid enzymatic test for the detection of GHB is described.

Chapter 12 covers forensic and clinical issues in alcohol analysis. Ethanol, a clear volatile liquid that is soluble in water and has a characteristic taste and odor, is a central nervous system (CNS) depressant and causes most of its effects on the body by depressing brain function. CNS depression is correlated directly with the concentration of alcohol in the blood (BAC). The Estimation of a person's blood alcohol concentration is based on important parameters such as body weight, ethanol concentration of the beverage consumed and number of beverages consumed, and length of time and pattern of the drinking. Because men and women have different body water amounts (men average 68% and women 55%), there are differences between the ethanol concentration achieved in men and women of similar weight for the same amount of alcohol. Various methods are described that can help to determine BAC in the field as well as in the laboratory.

Chapter 13 discusses fundamental issues of postmortem toxicology. The basic principles of forensic postmortem toxicology are presented. This chapter covers the acquisition and usefulness of different specimens, current analytical techniques, and the interpretation of findings. Special problems associated with the interpretation of drug levels include the conditions of the specimens and the effects of postmortem redistribution, postmortem drug changes, pharmacogenomics, drug interactions, and embalming fluid.

Chapter 14 reviews a field of growing importance in forensic science, entomotoxicology: drugs, toxins, and insects. Forensic entomology is gaining widespread acceptance within the forensic sciences as one method of estimating a portion of the postmortem interval by utilizing the time of insect colonization of a body, also known as the period of insect activity. Additionally, insect evidence can be utilized as alternative toxicology samples in cases where no other viable specimens exist. This subfield, known as entomotoxicology, can provide useful qualitative information to investigators as to the presence of drugs in the tissues at the time of larval feeding. The presence of drugs can alter the developmental period of the insects and should always be taken into consideration by the forensic entomologist. The relationship between toxicology and forensic entomology is also examined.

LAWRENCE KOBILINSKY

CONTRIBUTORS

Jason H. Byrd, University of Florida, College of Medicine, Gainesville, Florida

Anthony Carpi, Environmental Toxicology, John Jay College of Criminal Justice, The City University of New York, New York

Donald B. Hoffman, Department of Sciences, John Jay College of Criminal Justice, The City University of New York, New York

Lawrence Kobilinsky, Chairman, Department of Sciences, John Jay College of Criminal Justice, The City University of New York, New York

Ali Koçak, Department of Sciences, John Jay College of Criminal Justice, The City University of New York, New York

Sarah L. Lancaster, Forensic Explosives Laboratory, Defence Science and Technology Laboratory, Fort Halstead, Sevenoaks, Kent, UK

Gerald M. LaPorte, National Institute of Justice, Office of Investigative and Forensic Sciences, Washington, DC

John J. Lentini, Scientific Fire Analysis, LLC, Big Pine Key, Florida

Nathan H. Lents, Department of Sciences, John Jay College of Criminal Justice, The City University of New York, New York

Richard Li, Department of Sciences, John Jay College of Criminal Justice, The City University of New York, New York

Henrietta Margolis Nunno, Department of Sciences, John Jay College of Criminal Justice, The City University of New York, New York

Maurice Marshall, (Formerly) Forensic Explosives Laboratory, Defence Science and Technology Laboratory, Fort Halstead, Sevenoaks, Kent, UK

Raymond C. Murray, University of Montana, Missoula, Montana

Jimmie C. Oxley, University of Rhode Island, Chemistry Department, DHS Center of Excellence for Explosives Detection, Mitigation, and Response, Kingston, Rhode Island

Stanley M. Parsons, Department of Chemistry and Biochemistry, Program in Biomolecular Science and Engineering, Neuroscience Research Institute, University of California, Santa Barbara, California

Michelle R. Peace, Department of Forensic Science. Virginia Commonwealth University, Richmond, Virginia

Scott G. Ryland, Florida Department of Law Enforcement, Orlando, Florida

Andrew J. Schweighardt, John Jay College of Criminal Justice, The City University of New York, New York

Joseph C. Stephens, United States Secret Service, Forensic Services Division, Questioned Document Branch, Instrumental Analysis Section, Washington, DC

Richard Stripp, Department of Sciences, John Jay College of Criminal Justice, The City University of New York, New York

Edward M. Suzuki, Washington State Crime Laboratory, Washington State Patrol, Seattle, Washington

Beth E. Zedeck, Pediatric Nurse Practitioner, New York

Morris S. Zedeck, (Retired) Department of Sciences, John Jay College of Criminal Justice, The City University of New York, New York

CHAPTER 1

Forensic Environmental Chemistry

ANTHONY CARPI and ANDREW J. SCHWEIGHARDT
John Jay College of Criminal Justice, The City University of New York, New York

Summary Forensic environmental chemistry involves the use of trace chemical techniques for investigating environmental spills in an effort to determine civil or criminal liability. The field can be broken down into two broad areas based on the techniques used to determine liability: chemical fingerprinting and spatial association. In chemical fingerprinting, complex mixtures of chemicals or chemical isotopes are used to associate a spill or environmental release with a source. In spatial association, geographical information systems and geochemical techniques are used to attribute the location of a contaminant with a possible source in physical space.

1.1	Introduction	2
1.2	Chemical fingerprinting	4
	1.2.1 Hydrocarbon mixtures	4
	1.2.2 Polycyclic aromatic hydrocarbons	6
	1.2.3 Biomarkers	11
	1.2.4 Additives	11
	1.2.5 Isotopes	12
	1.2.6 Tracers	13
	1.2.7 Methods of detection	16
	1.2.8 Weathering	18
1.3	Spatial association of environmental incidents	18
	References	20

Forensic Chemistry Handbook, First Edition. Edited by Lawrence Kobilinsky.
© 2012 John Wiley & Sons, Inc. Published 2012 by John Wiley & Sons, Inc.

1.1 INTRODUCTION

As technology for trace chemical analysis has expanded in recent decades, so has its application to criminal and civil casework. This has transformed traditional forensic investigations and has expanded their applicability to less traditional areas, such as those involving environmental crimes. Prior to 1950, environmental law in the United States was based on tort and property law and was applied to a very limited number of incidents. Driven by growing environmental awareness in the 1950s and 1960s, the U.S. Congress passed the first Clean Air Act in 1963. This was followed by a slow but steady string of further developments, including the founding of the Environmental Protection Agency (EPA) in 1970 and the passage of the Clean Water Act in 1972, the Endangered Species Act in 1973, and the Comprehensive Environmental Response, Compensation, and Liability Act (Superfund) in 1980. International law began to address environmental issues with the signing of the Convention on International Trade in Endangered Species of Wild Fauna and Flora (CITES) in 1975 and other international treaties. These early milestones have been bolstered by recent amendments, new agencies, and renewed funding, all of which make up a series of laws and regulations that define criminal practices and govern civil liability cases involving the environment. Increased legislation and improved enforcement have led to a significant decrease in easily identifiable environmental disasters, such as when the Cuyahoga River in Cleveland, Ohio burst into flames in 1969 as a result of industrial discharge. As these visible issues have diminished, environmental scientists have found themselves faced with questions that are more difficult to identify and are more intractable in nature. This has led, in turn, to advances in the investigative techniques used to investigate environmental crimes.

It is impossible to pinpoint the exact birth date of forensic environmental science. However, one source attributes the origin of the term *environmental forensics* to the scientific contractor Battelle in the late 1990s (Haddad, 2004). One of the company's specialties is forensic environmental chemistry, and the company provides services in hydrocarbon fingerprinting, contamination identification, and product identification. Regardless of when the field was named, most sources would agree that the field began gathering momentum about 30 years ago. Since that time, various subdivisions have emerged. Some of these divisions have their roots in diverse areas such as geology, toxicology, biology, physics, and chemistry. As such, the term *environmental forensics* might be considered a misnomer for two reasons. The first is the tendency of the word *forensics* to be semantically confusing, because it has no real meaning when used in this context. The second is the loss of the word *science*, for this serves as a necessary reminder of the field's vast and diverse capabilities, spanning across not just one but many sciences.

The term *environmental forensics* is often misapplied to what should rightfully be called *forensic environmental chemistry*. For example, environmental forensics has been defined as "the systematic investigation of a contaminated site or an event that has impacted the environment," a definition that is clearly biased toward the chemistry perspective (Stout et al., 1998). The broad capabilities of the field are unnecessarily simplified to the question: Who caused the contamination, and when did it occur? (Ram et al., 1999). Surely this is not the only question that environmental forensics is capable of answering. Nevertheless, this mindset has persisted because it is acknowledged and reaffirmed repeatedly. Many of the shortfalls of the earlier definitions of environmental forensics have been identified and amended in subsequent definitions. Many of these revisions offer a more generic, all-inclusive definition. One source defines forensic environmental science simply as "litigation science" (Murphy, 2000); another as "environmental 'detective work' ... operating at the interface junction points of several main sciences including chemistry and biochemistry, biology, geology and hydrogeology, physics, statistics, and modeling" (Petrisor, 2005). Vives-Rego (2004) defines it not just as the environmental application of chemistry, biology, and geology, but as "science and the art of deduction." Finally, Carpi and Mital (2000) define it as "the scientific investigation of a criminal or civil offense against the environment." These updated definitions more accurately reflect the capabilities of forensic environmental science beyond the chemistry realm. In particular, the definition provided by Carpi and Mital (2000) specifically includes the use of DNA to solve crimes perpetrated against wildlife and plant life. In this chapter we focus on the specific subarea of forensic environmental chemistry and leave to another source the broader description of the methods and techniques that apply to environmental forensics.

However one chooses to define this growing field, one thing is certain: Forensic environmental science is filling the significant niche left void by forensic science and environmental science. Due in large part to its close association with the core sciences, forensic environmental science has experienced significant growth since its inception, especially in recent years. Aside from technological achievements in the past 30 years, several important advances have helped propel forensic environmental science from a burgeoning offshoot of forensic science to a scientific discipline in its own right. One such advancement was the founding of the journal *Environmental Forensics* in 2000 (Taylor & Francis, London). Although research pertaining to forensic environmental science occurred before the journal existed, the journal can be credited with offering a place for environmental research that falls under the forensic science umbrella. Thus, *Environmental Forensics* provides a forum to facilitate the exchange of information, ideas, and investigations unique to forensic environmental science (Wenning and Simmons, 2000).

Forensic environmental science has become such a diverse field that it is difficult to find a single work that adequately covers all its subdisciplines. The literature on the subject that enjoys the most success does so because it focuses on a specific area of forensic environmental science. As such, in this chapter we focus on forensic environmental chemistry. Our aim is to elaborate on several key areas of forensic environmental chemistry, perhaps where other resources have been unable to or have failed to do so. In particular, we focus on chemical fingerprinting and its subsidiaries, such as hydrocarbon fingerprinting, isotope fingerprinting, and complex mixture fingerprinting. Chemical fingerprinting attempts to individualize a chemical and trace it back to its origin. This technique has become increasingly important not only to identify that a chemical spill has indeed occurred, but also to identify the party responsible. We also focus on spatial analysis for the purpose of source attribution. Several cases are discussed that are illustrative of the capabilities of spatial analysis and chemical fingerprinting as they pertain to forensic environmental chemistry.

1.2 CHEMICAL FINGERPRINTING

Chemical fingerprinting is a subsidiary of forensic environmental chemistry that examines the constituents of a mixture for the purpose of creating a unique chemical signature that can be used to attribute the chemicals to their source. At one time it was sufficient to arrive at a generic classification and quantitation of the chemical mixture so that appropriate remediation measures could be designed and implemented. However, modern analytical techniques that are focused on individualizing and associating a mixture with a source have become increasingly popular, both for liability reasons and because of the recognition and attempt to apportion liability when multiple and/or temporally distant parties may be responsible for chemical contamination. The main objectives of chemical fingerprinting are to characterize, quantitate, and individualize a chemical mixture (Alimi et al., 2003). In this section we provide the reader with a review of some of the constituents of a mixture that are useful for assembling a chemical fingerprint as well as the techniques used to screen for these constituents. The efficacy of these analytes and of detection techniques are evaluated by illustrating their application in several cases.

1.2.1 Hydrocarbon Mixtures

The majority of chemical spills involve hydrocarbon mixtures; as a result, many techniques are tailored for these mixtures (Sauer and Uhler, 1994). Early techniques were used simply to quantify the total petroleum hydrocarbon concentration, but modern techniques must be capable of quantification as well as identification and individualization (Zemo et al., 1995). The latter two are especially important for litigation purposes. However, identification and individualization may also provide for the design of a more effective remediation plan that accounts for dispersal, weathering, and degradation of the chemical mixture (Zemo et al., 1995).

Petroleum hydrocarbon mixtures may be broadly classified into three general groups. *Petrogenic hydrocarbons* are present in crude oil or its refined products. *Pyrogenic hydrocarbons* are the combusted remnants of petrogenic hydrocarbons and other by-products. *Biogenic hydrocarbons* are those that arise from more recent natural processes: for example, swamp gas or the volatile hydrocarbon mixtures released by decaying plant or animal tissue exposed to anaerobic conditions. Within each of these three broad groups, hydrocarbons are generally separated into three types: saturated aliphatics (alkanes), unsaturated aliphatics (alkenes, etc.), and aromatic hydrocarbons. Aromatic hydrocarbons include both light petroleum products [e.g., benzene, toluene, ethylbenzene, and xylenes (BTEX)] and heavier products such as polycyclic aromatic hydrocarbons.

Analytical techniques such as gas chromatography are usually adequate for differentiating among petrogenic, pyrogenic, and biogenic hydrocarbon mixtures because of the unique ratios of alkanes, alkenes, and aromatic structures that can be expected in these mixtures. Furthermore, gas chromatography can be used to differentiate different grades of petrogenic hydrocarbons because crude mixtures have a variety of hydrocarbon components (i.e., unresolved complex mixtures), which often present themselves as a "hump" on a chromatogram, whereas more refined mixtures have less variety in their components. Retention time for various petrogenic compounds is affected by the structure of these compounds: for example, gasoline elutes first (C_4 to C_{12}), along with Stoddard solvents (C_7 to C_{12}), which are followed by middle distillate fuels (C_{10} to C_{24}), and crude mixtures (up to C_{40}) (Zemo et al., 1995). Crude oil mixtures contain a diverse array of hydrocarbons and, on average, are comprised of 15 to 60% paraffins and isoparaffins, 30 to 60% naphthenes, and 3 to 30% aromatics, with the remainder of the mixture being composed of asphaltenes and various trace compounds (Bruce and Schmidt, 1994). Pyrogenic hydrocarbon mixtures can be recognized on the chromatogram because large molecules undergo combustion first, leaving behind a disproportion of smaller molecules. However, pyrogenic compounds are more difficult to attribute to a source because the chemical signature is further removed from the original petrogenic source (Bruce and Schmidt, 1994). Steranes and hopanes are often used as target analytes when the focus of a study is biogenic hydrocarbons, because these analytes are more resistant to many more forms of weathering than are other biogenic components (Alimi et al., 2003).

Although it is worthwhile to classify a mixture as petrogenic, pyrogenic, or biogenic in origin, this is commonly not enough. To arrive at a unique chemical signature, the analysis must extend beyond identifying the class characteristics of a mixture. Modern methods often involve the examination of ancillary components of a mixture, such as dyes, additives, stable isotopes, radioactive isotopes, biomarkers, polycyclic aromatic hydrocarbons (PAHs). PAH homologs, and metabolized PAHs. It is customary to screen for many of these analytes with the intent of providing the most comprehensive chemical signature possible. Before selecting a suite of analytes, it is wise first to consider if these analytes may already have been present at a location (due to a prior contamination or natural processes) and if these analytes are highly susceptible to degradation. Indeed, the characteristics of a good chemical marker are that it is resistant to degradation and that it can uniquely identify the hydrocarbons released from other sources (Sauer and Uhler, 1994).

The first step in confirming hydrocarbon contamination is accomplished by screening for saturated hydrocarbon molecules such as pristane and phytane. These are isoparaffins that are resistant to degradation and are highly indicative of hydrocarbon contamination (Sauer and Uhler, 1994). Pristane and phytane usually represent themselves to the right of C_{17} and C_{18} peaks on a chromatogram. Fresh hydrocarbon mixtures have prominent C_{17} and C_{18} peaks in relation to pristane and phytane peaks, whereas the converse is true for degraded mixtures (Bruce and Schmidt, 1994; Morrison, 2000b). Due to the proportionality between the ratios of these compounds and the extent of degradation, the ratios of pristane and phytane to the C_{17} and C_{18} peaks are often used to estimate the degree of weathering.

1.2.2 Polycyclic Aromatic Hydrocarbons

Polycyclic aromatic hydrocarbons (PAHs) are hydrocarbon compounds with two to six rings. Homologs of the PAH compounds may be similar to the parent compound except that they are substituted for by one or more alkyl groups. The ratio of two PAHs to two other PAHs is sometimes expressed in double-ratio plots, in which certain regions of the plot are diagnostic for one source or another. It is also becoming increasingly common to screen for metabolized PAHs (as well as BTEX compounds), whose structures differ predictably from the original PAH. PAHs are often very useful in studies involving weathered mixtures, because the complex structure of PAHs makes them more resistant to degradation. The rate of degradation is proportional to the complexity of the ring structure, with the compounds having the fewest number of rings degrading first (Alimi et al., 2003). Some target parent PAH compounds and their alkyl homologs are shown in Table 1.1.

One of the most prominent applications of PAH analysis has been the study of the *Exxon Valdez* oil spill. The spill occurred when the tanker hull was punctured as it ran aground on March 24, 1989, releasing some 10.8 million gallons of oil into Prince William Sound, Alaska. The oil released was dispersed by water currents and a windstorm that followed the spill a few days later, and concern

TABLE 1.1 Target Parent PAH Compounds and Their Alkyl Homologs

Naphthalenes	Chrysenes
C_0-naphthalene (N)	C_0-chrysene (C)
C_1-naphthalenes (N1)	C_1-chrysenes (C1)
C_2-naphthalenes (N2)	C_2-chrysenes (C2)
C_3-naphthalenes (N3)	C_3-chrysenes (C3)
C_4-naphthalenes (N4)	C_4-chrysenes (C4)
Phenanthrenes	EPA priority pollutant
C_0-phenanthrene (P)	Biphenyl (Bph)
C_1-phenanthrenes (P1)	Acenaphthylene (Acl)
C_2-phenanthrenes (P2)	Acenaphthiene (Ace)
C_3-phenanthrenes (P3)	Anthracene (An)
C_4-phenanthrenes (P4)	Fluoranthene (Fl)
Dibenzothiophenes	Pyrene (Py)
C_0-dibenzothiophene (D)	Benzo[*a*]anthracene (BaA)
C_1-dibenzothiophenes (D1)	Benzo[*b*]fluoranthene (BbF)
C_2-dibenzothiophenes (D2)	Benzo[*k*]fluoranthene (BkF)
C_3-dibenzothiophenes (D3)	Benzo[*e*]pyrene (BeP)
Fluorenes	Benzo[*a*]pyrene (BaP)
C_0-fluorene (F)	Perylene (Pe)
C_1-fluorenes (F1)	Indeno[1,2,3-*cd*]pyrene (IP)
C_2-fluorenes (F2)	Dibenz[*a,h*]anthracene (DA)
C_3-fluorenes (F3)	Benzo[*ghi*]perylene (BP)

Source: Alimi et al. (2003).

was raised over the dispersal of the oil into adjacent bodies of water (Galt et al., 1991). Many of these concerns were seemingly corroborated by the detection of oil in neighboring bays. However, it was speculated that some of the oil detected in these neighboring waters may have been from biogenic sources, petrogenic sources from previous spills, or pyrogenic sources from hydrocarbons that had previously undergone combustion. A massive effort was mounted to identify the extent to which Exxon was responsible for oil detected in these adjacent waters.

The study immediately focused on components of oil that were the most resistant to degradation, such as PAHs and biomarkers. A substantial part of the investigation focused on evaluating the effects of weathering on the *Exxon Valdez* cargo if it was to be accurately differentiated from other sources (Figure 1.1). As expected, lighter components of the oil matrix were preferentially lost to weathering. With the effects of weathering understood, the investigation then turned to PAH analysis. Two PAHs that were focused on for distinguishing different crude mixtures were phenanthrenes and dibenzothiophenes; chrysenes were used to differentiate crude from refined mixtures because chrysenes are removed during the refining process (Boehm et al., 1997). The ratios of the PAH compounds to one another were particularly useful

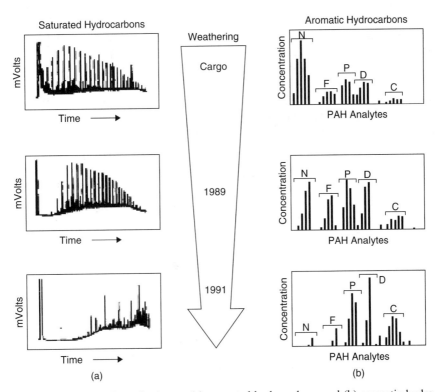

Figure 1.1 Effects of weathering on (a) saturated hydrocarbons and (b) aromatic hydrocarbons from the *Exxon Valdez* spill. N, naphthalenes; F, fluorenes; P, phenanthrenes; D, dibenzothiophenes; C, chrysenes (Boehm et al., 1997).

because the concentrations of the PAHs will change with weathering; however, the ratio of one PAH to another generally remains constant (Boehm et al., 1997). In this case, researchers created a double-ratio plot comparing dibenzothiophenes to phenanthrenes in order to distinguish PAHs of the *Exxon Valdez* spill from PAHs of other sources (Figure 1.2). As seen in the figure, the double-ratio plot showed distinct clustering of oil samples from different sources, allowing a differentiation to be made. When the PAHs in neighboring bays were analyzed, some were attributed to the *Exxon Valdez* spill, but many were found to have originated from other sources, both natural and anthropogenic (Boehm et al., 1998).

PAHs have also been used to study contamination at former manufactured gas plant (MGP) facilities. Prior to the use of natural gas, MGPs made coal gas to use as fuel. Former MGP sites are evaluated for contamination by screening for PAHs that would have been introduced to the environment as coal tar, which is a by-product of the coal gas manufacturing process. This can sometimes be a difficult task because the sites often contain PAHs that may be unrelated to the MGP, having been introduced via other natural and anthropogenic avenues. The investigations are further complicated because similar PAH signatures are obtained for MGP coal tar residues and background residues. Although the composition of the PAHs contained in MGPs and background sources may be similar, the PAH ratios and patterns (i.e., petrogenic or pyrogenic) can be used to differentiate PAHs from different sources. One study examined the ratios and patterns of PAHs for the purpose of distinguishing MGP PAHs from background PAHs in soil samples collected in and around a stream near an MGP (Costa et al., 2004).

PAHs may be present either as unsubstituted parent compounds or as a substituted alkyl homolog (see Table 1.1 for examples). Petrogenic patterns of PAHs are recognized because they contain a bell-shaped distribution of the parent PAH and its homologs where concentration of the single- or double-substituted homologs

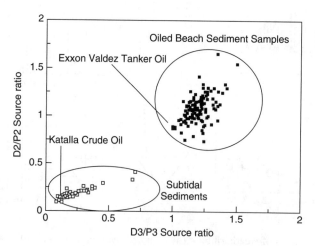

Figure 1.2 Double-ratio plot showing how the ratio of PAHs (dibenzothiophenes to phenanthrenes) can be diagnostic for one source or another (Boehm et al., 1997).

are highest, and concentrations decrease as one moves in either direction toward the unsubstituted parent or toward the complex, multisubstituted homolog (see Figure 1.4 for an example). Pyrogenic patterns of PAHs are recognized because they contain a distribution in which the parent PAH is more abundant, due to preferential combustion of the substituted homologs. Researchers observed a pyrogenic pattern in PAH residues derived from the MGP site in question (Figure 1.3), but samples collected from an adjacent stream indicated a mix of petrogenic and pyrogenic PAHs (Figure 1.4).

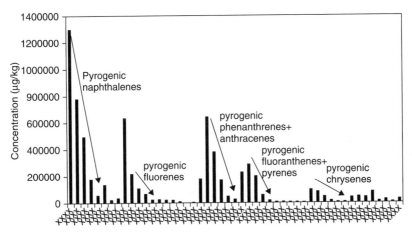

Figure 1.3 PAH composition of residues derived from an MGP site. Decreasing concentrations of substituted homologs of the parent compounds indicate the pyrogenic origin of the sample (Costa et al., 2004).

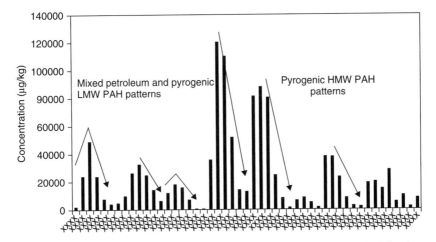

Figure 1.4 PAH composition of residues derived from a streambed. Mixed patterns indicate the presence of a mixture of pyrogenic and petrogenic hydrocarbons (Costa et al., 2004).

The researchers then turned to high-molecular-weight PAH ratios to determine if the pyrogenic pattern observed in the stream was from weathered MGP residues or from recent background contamination. Several PAHs were chosen to create double-ratio plots in which certain sections of the plot were diagnostic for either the MGP, background sources, or a mix of the two. A comparison of samples from the streambed surface (Figure 1.5) and samples from the streambed subsurface (Figure 1.6) indicated that most of the surface (i.e., newer) PAHs were derived

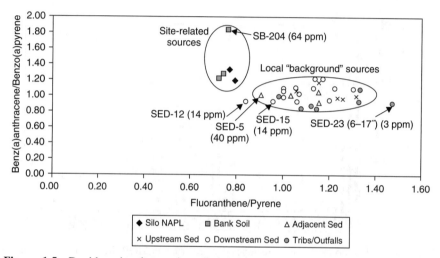

Figure 1.5 Double-ratio plot used to distinguish site-related, background, and mixed PAH signatures in streambed surface samples (Costa et al., 2004).

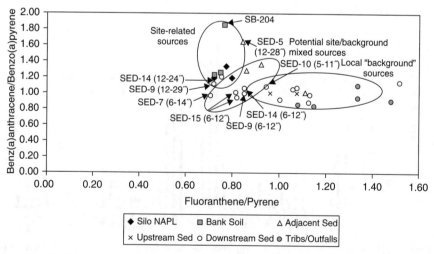

Figure 1.6 Double-ratio plot used to distinguish site-related, background, and mixed PAH signatures in streambed subsurface samples (Costa et al., 2004).

from background sources, whereas most of the subsurface (i.e., older) PAHs were derived from the MGP site. Results of studies such as this can help to draw attention to other potential sources of contamination in order achieve the most efficacious remediation effort.

1.2.3 Biomarkers

Biomarkers such as steranes and hopanes are hydrocarbon remnants of deceased organisms that are useful in chemical fingerprinting because they are extremely resistant to weathering (Alimi et al., 2003). Thus, biomarkers can often be useful to individualize a hydrocarbon mixture when saturated hydrocarbons and PAHs have already been degraded (Sauer and Uhler, 1994). A comprehensive list of biomarkers that are useful in hydrocarbon mixture studies is provided by Alimi et al. (2003).

1.2.4 Additives

Inorganic compounds are often added to hydrocarbon mixtures to serve as anti-knock agents, octane boosters, corrosion inhibitors, and anti-icers (Kaplan, 2003). Additives are not present in crude mixtures, of course, so their presence is indicative of a refined mixture (Bruce and Schmidt, 1994). Because refining practices and additives change over time, and since these changes have been well documented, the presence of particular additives in a hydrocarbon mixture can be highly indicative of a certain time frame during which a sample was produced. For example, lead was first added to gasoline in 1923, and its concentration in gasoline decreased steadily until it was phased out in U.S. automobile fuels in 1995 (Kaplan, 2003). Other gasoline additives that have predictably appeared and disappeared throughout history are methylcyclopentadienyl manganese tricarbonyl (MMT) and methyl *tert*-butyl ether (MTBE). The chronology of some popular additives has been thoroughly documented in several sources (Morrison, 2000a,b; Kaplan, 2003). Some additives, such as lead, have been used over large time frames, but the concentration of lead in gasoline has varied predictably over the years. Although this can be used to arrive at a reasonable estimate of time of manufacture of the hydrocarbon mixture in question, it is not an infallible method because additive concentrations are often reported based on a pooled standard, which ignores batch-to-batch variation (Morrison, 2000a).

The use of additives for dating a release can be complicated by the fact that additives may be discontinued in certain countries or for certain applications, but may still be used in others. Additives that are supposedly absent in a mixture may also be present in very dilute amounts. The utility of additives in dating manufacture or release dates is greater than their capacity to individualize a mixture. This is because many companies often purchase additives from the same manufacturer. These additives are then added unaltered into various hydrocarbon mixtures, so many different mixtures may have the same additives present (Morrison, 2000a). Further complications when screening for old additives may be encountered because these compounds often contain oxygen, which contributes to their rapid weathering over time (Morrison, 2000b).

1.2.5 Isotopes

When complex hydrocarbon mixtures cannot be identified by analysis of stable components, the mixtures can be identified by analysis of stable isotopes within the mixture. Stable isotopes are often analyzed with respect to one another. In other words, the ratio of one stable isotope to another stable isotope within the same mixture can often be unique, thereby allowing for the creation of an isotope signature. In contrast to stable isotopes, unstable isotopes decay predictably such that the degree of decay can be correlated with the age of the mixture. Analysis of unstable isotope decay is often referred to as a *long-term method* because it is capable of estimating release dates thousands of years prior. Unstable isotopes are also useful because their decay is independent of environmental factors such as weathering (Kaplan, 2003).

Isotopes can be useful in chemical fingerprinting in two ways. The ratio of two isotopes can be compared as a means of individualization because no two mixtures will have exactly the same ratio of two isotopes. Carbon and lead isotope ratios are commonly used for source identification. Radioactive isotopes are also useful for dating a release because these isotopes have known rates of decay that are independent of environmental conditions.

Carbon isotopes were used in one study to determine the origin of soil gas methane near the site of a prior gasoline spill (Lundegard et al., 2000). The investigation was triggered by the detection of high methane levels near a service station where approximately 80,000 gallons of gasoline had been spilled 20 years earlier. Initially, it was speculated that the methane was due to the bacterial degradation of the gasoline, but the investigators were considering other possibilities. Suspicion was raised because high levels of methane were detected outside the original gasoline plume, and in some cases the levels detected outside the plume were higher than those within the plume (Figure 1.7).

The initial hypothesis of methane generation by bacterial degradation of the gasoline was also challenged because this is not a common degradation pathway. For gasoline to be fermented to methane, it would first have to be converted to the necessary precursor compounds for methanogenesis by fermentation (Lundegard et al., 2000). Although the generation of methane via this pathway is possible, the investigators were considering more plausible origins of the methane that, coincidentally, were unrelated to the gasoline spill. One of the potential origins considered was the biodegradation of organic matter.

The methanogenesis of petrogenic compounds can be distinguished from that of organic compounds from biogenic origins through the use of ^{13}C, which is a stable carbon isotope. Differentiating the methanogenesis of petrogenic and organic compounds is accomplished based on the idea that older, petrogenic compounds have lower quantities of ^{13}C isotopes than does newer organic matter. The process of methanogenesis significantly reduces the amount of ^{13}C present in the original

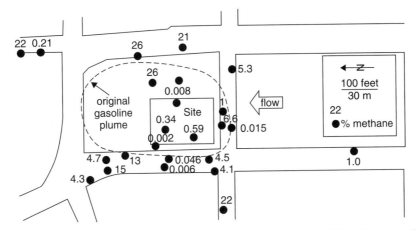

Figure 1.7 Service station map showing methane concentrations within and surrounding the original plume (Lundegard et al., 2000).

organic matter, but the ^{13}C in the nascent methane remains stable regardless of environmental conditions (Lundegard et al., 2000). The study indicated that wood fill from beneath the service station site and gasoline from within the original plume had indistinguishable ^{13}C quantities.

Another way of differentiating methane from petrogenic and biogenic sources is through the use of ^{14}C, which is a naturally occurring radioactive isotope of carbon taken up by all living organisms. The age of the source from which the methane was formed can be predicted because ^{14}C has a half-life of about 5700 years, and therefore it will still be detectable in methane formed from biogenic organic matter less than 50,000 years old. The hypothesis that methane originated from the degradation of biogenic organic matter was corroborated by the ^{14}C analysis, which indicated that the highest ^{14}C levels were detected outside the original plume (Figure 1.8). The level of ^{14}C in petrogenic hydrocarbons is zero, so the researchers concluded that the methanogenesis must be of biogenic origins. This hypothesis was further supported because a review of the site history indicated that the area consisted of organic fill, including wood and sawdust.

1.2.6 Tracers

When none of the analytes discussed previously are amenable to the case at hand, techniques that rely on tracers can sometimes be used for forensic tracking of environmental chemicals. A tracer can be any molecule that is diagnostic of one source but not others. Sometimes, multiple tracers are used to augment the significance of the results. One study used organic tracers to determine the origin of gas- and particle-phase air pollutants in two California cities (Schauer and Cass, 2000). The objective of the study was to determine the primary source(s) of air pollutants in Fresno and Bakersfield, California. The results and chemical composition of samples collected at the two locations were compared to those collected at a remote

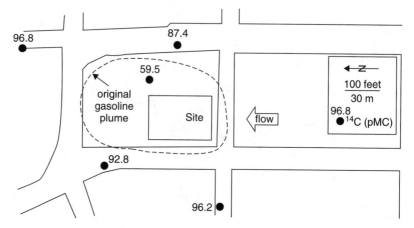

Figure 1.8 ^{14}C levels detected within and surrounding the original plume of a gasoline spill exhibiting high methane concentrations. High ^{14}C concentrations indicated methanogenesis of biogenic hydrocarbons and countraindicated gasoline as the source of origin (Lundegard et al., 2000).

site located at the Kern Wildlife Refuge and distant from anthropogenic sources of air pollutants. Previous tests for air pollutants have used generic compounds to draw connections between air pollutants and their sources (Harley et al., 1992). However, some of the analytes used in these other studies are not exclusive to a particular source. The researchers in the California study aimed to develop a more accurate method for tracing the origin of the air pollutants. Atmospheric samples in the two cities were collected, as well as single-source control samples that consisted of combustion emissions from gasoline-powered motor vehicles, diesel engines, hardwood combustion, softwood combustion, and meat-cooking operations. Tracers that were unique to sources and those that were common between multiple sources were chosen both to fingerprint and then apportion emissions in particulate samples with mixed origins. Further criteria used to choose tracers were (1) that they were not selectively removed from the environment, and (2) that they were not formed by atmospheric reactions to any significant extent. The researchers used direct measurements of these tracer compounds to draw conclusions about the source of particulate pollutants in the areas indicated.

Specific tracers were used to apportion the results obtained from specific sources. For example, the compound levoglucosan was found to be specific to wood combustion, so the concentration of levoglucosan in proportion to other constituents in a mixed sample could be used to apportion the contribution of wood combustion to particulate loading in an area (see Figure 1.9). Based on the concentrations of the other tracers in the samples, the relative contributions of each source (e.g., automobiles, wood combustion) were apportioned (Figure 1.10). Low levels of pollutants derived from anthropogenic sources at the remote site were noteworthy. The researchers concluded that local anthropogenic emissions (particularly automobile exhaust) were responsible for the majority of air pollutants in the two urban

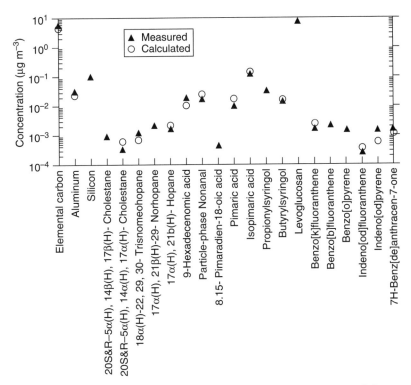

Figure 1.9 Ambient concentrations of various air pollution tracer compounds in samples (Schauer and Cass, 2000).

environments, whereas naturally occurring dusts were primary contributors at the remote site.

Inorganic compounds can also be used for forensic chemistry purposes. One study used metal tracers to identify dust that resulted from the collapse of the World Trade Center (WTC) buildings (Scott et al., 2007). The analysis of particulate matter arising from this catastrophic event has been an area of great interest because there are significant health implications associated with inhalation of the dust. Substantial amounts of the dust were transferred from the collapse to nearby buildings, so the objective of the research was to develop a method based on metal tracer detection to determine which buildings were most affected, and for those that were severely affected, to determine if appropriate remediation efforts had been undertaken.

Techniques employed included screening for human-made vitreous fibers, as well as trace metals, including As, Cd, Cr, Cu, Pb, Mn, Ni, V, and Zn. Trace metal detection was found to be more applicable to the identification of WTC dust because the atmosphere and buildings around the WTC were probed routinely for these metals after the collapse (Scott et al., 2007). Although these metals can originate from other sources as well, the researchers expected trace metals to be detected in quantities and ratios that were unique to WTC dust.

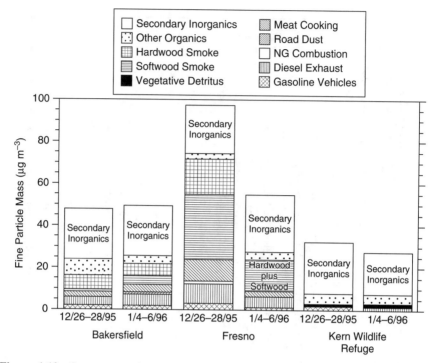

Figure 1.10 Source apportionment to particulate air pollution in two urban areas and one remote site in California (Schauer and Cass, 2000).

Concentrations of the nine metals in WTC dust as reported by four studies were compared to concentrations in background dust collected from Arizona. A discriminant analysis model was used to classify each sample as having originated from WTC or background dust based on the relative concentrations of the nine metals (Figure 1.11). The analysis indicated that WTC dust had elevated levels of Cr and Mn and low levels of As, Cd, and Cu compared to background dust. The researchers were able to demonstrate that trace metals could be used to distinguish pure WTC dust from background dust with 94% accuracy; however, mixed dust samples had lower levels of accuracy (Scott et al., 2007).

1.2.7 Methods of Detection

One of the most widely used techniques in chemical fingerprinting for hydrocarbons is gas chromatography. This is based on the concept that each compound has a unique structure and will therefore be retained differentially in the gas chromatograph before being eluted. As long as other parameters (e.g., temperature, column length, column packing) are held constant, any differences in retention time can be attributed to the structure of the compound (Bruce and Schmidt, 1994). Mixtures contain many different compounds, so a gas chromatogram represents a chemical fingerprint of all the chemical constituents in a mixture. Gas chromatography

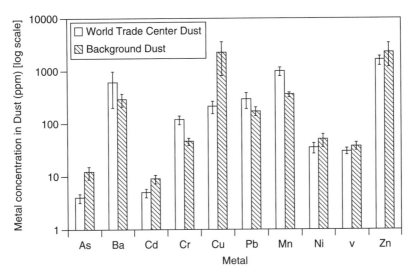

Figure 1.11 Comparison of trace metal fingerprints in dust from the WTC (white bars) and from background (striped) (Scott et al., 2007).

is often combined with other techniques to achieve a more detailed analysis. For example, a gas chromatograph is commonly used as a preliminary separation technique that is followed by detection using mass spectroscopy. Some researchers even use two-dimensional gas chromatography (GC × GC) to achieve superior resolution (Gaines et al., 1999). A good review of the literature focusing on these techniques is provided by Suggs et al. (2002). The potential weaknesses and vulnerabilities of these techniques are discussed by Morrison (2000b).

Although many of the aforementioned techniques are highly effective, they often have a deleterious impact on the sample. That is, substantial portions of the sample are often destroyed in the course of the analysis. Sample destruction may not be a major concern in other disciplines, but evidence is sometimes limited in forensic investigations, and what little sample may be available often attains the status of a precious and rare commodity. Another example of when the destruction of a sample is avoided is when the sample itself is, quite literally, a rare commodity, such as an archaeological treasure or artifact. For samples of limited quantity or prized value, less invasive methods of analysis are often sought.

One study that warranted the use of a minimally invasive technique involved the analysis of ancient tools made of obsidian (Tykot, 2002). The purpose of the analysis was to evaluate the Mediterranean sources and trade routes of obsidian tools without damaging them. To bolster the results and to compensate for the potential weaknesses of certain techniques, this study relied on a series of methods, including scanning electron microscopy (SEM), x-ray fluorescence (XRF), neutron activation analysis (NAA), and inductively coupled plasma mass spectroscopy (ICP-MS). The elemental compositions indicated by the four techniques were used to construct possible sources and distributions of obsidian. The results helped lend credence to the theory of a vast distribution network for the tools rather than the lone source theory that was once promulgated.

1.2.8 Weathering

When screening for various analytes, one variable that must be kept in mind is weathering. This is the process by which the chemical signature of a mixture is altered due to evaporation, dispersal, biodegradation, or oxidation of certain components of the mixture. Short-chain hydrocarbons are most vulnerable to weathering mainly because their simple structure makes them susceptible to degradation, particularly to biodegradation (Alimi et al., 2003). Compounds with more complex structures are generally more resistant to weathering. Weathering generally occurs at predictable rates such that the age of the mixture can be estimated accurately based on the relative amount of weathering of short-chain hydrocarbons to larger, more resistant molecules. However, weathering can vary because of site-specific differences in environmental conditions (Morrison, 2000b). The compounds that exhibit the greatest longevity are generally the most useful for estimating the degree of weathering and therefore the age of a particular compound. Isotopes, BTEX compounds, PAHs, and biomarkers have all been used with varying degrees of success for determining the extent of weathering of hydrocarbon samples.

Certain studies have focused on families of constituents of oil that are resistant to degradation, specifically paraffins, isoparaffins, aromatics, naphthenes, and olefins, all of which generally range from three to 13 carbons (Kaplan et al., 1997). These compounds, commonly referred to as PIANO compounds, are useful because their ratios vary among different hydrocarbon mixtures. For example, fuels of different grades and octane levels are composed of unique ratios of PIANO compounds. PIANO analysis is particularly useful to spatial analysis because the concentration of the various PIANO constituents and additives has changed over time, due to evolving oil refining practices, varying octane levels, and increasingly stringent legal regulations (Davis et al., 2005).

1.3 SPATIAL ASSOCIATION OF ENVIRONMENTAL INCIDENTS

Spatial analysis is used to associate a pollutant release or plume with a source by tracing back its geographic point of origin to a particular place in space or time. This approach is often thought of as one of the more esoteric areas of forensic environmental chemistry; however, this reputation is, for the most part, undeserved because spatial analysis is inherently simple and straightforward. Spatial analysis applies all of the usual tools of forensic environmental chemistry to a spatial problem. For example, spatial analysis relies on many of the analyses with which we are familiar, such as those that screen for various hydrocarbons, fuel additives, isotopes, and biomarkers. However, spatial analysis attempts to determine more than the source and content of a particular contamination. Spatial analysis goes further by elucidating not only the "what" and "who" of a chemical contamination incident, but also the "where" and "for how long."

Spatial analysis of chemical transport involves integrating the results of chemical analyses with spatial and sometimes historical information about a site. For example, Carpi et al. (1994) examined the spatial distribution of airborne mercury pollution around a municipal solid-waste incinerator. The study used transplanted and prepared samples of sphagnum moss as biological monitors of air pollutants. Clean moss samples were distributed to 16 sites within a 5-km (3.1-mi) radius of the waste incinerator, plus one remote site about 20 km (12.4 mi) away. Samples of the moss at each of these stations were collected in duplicate every 2 weeks for about three months. The samples were analyzed by cold vapor atomic absorption spectroscopy for mercury contamination in two ways: Each sample was split and half was analyzed as received and the other half was first oven-dried at 105°C for 24 h before analysis.

Higher concentrations of mercury were correlated with sites closest to the incinerator, which then led the researchers to use meteorological data from the nearest weather service station [38 km (23.6 mi) away] to determine if mercury near the incinerator could indeed be traced back to the incinerator. It was determined that plants from sites with the highest levels of mercury were downwind of the incinerator. A locally weighted spatial statistics technique called *kriging* was used to develop regression surfaces for the pollutants over the area and these regression surfaces showed that proximity to the incinerator accounted for a high degree of the variability in mercury concentrations with location (Opsomer et al., 1995). The study benefited by incorporating topographical data to support the conclusions.

Interestingly, the mercury concentration in moss from the remote site was approximately equivalent to the mercury detected in some of the samples within the 5-km radius of the incinerator. However, comparison of the undried to dried moss samples demonstrated that the volatility of mercury at the remote site, and thus chemical species of mercury accumulating in samples at the remote site, was significantly different than it was near the incinerator. Samples collected near the incinerator demonstrated relatively low volatility, and indeed incinerators are known to emit high levels of $HgCl_2$, which has a low volatility (Carpi, 1997). In contrast, mercury at the remote site demonstrated high volatility, indicating that the form of mercury in the samples was primarily volatile elemental mercury. The authors conducted a site history at the remote site that revealed that whereas the site was distant from any anthropogenic source of mercury, it was close to a recently flooded reservoir system, and flooding of land is known to release naturally occurring elemental mercury from soil.

Spatial analysis can sometimes be complicated if samples have been collected sporadically or randomly, or if the data are otherwise incomplete. Such was the case in a study that investigated unusually high radiocesium (^{137}Cs) levels in a river basin near the Chernobyl nuclear power plant (Burrough et al., 1999). An explosion occurred at the power plant in April 1986 and released radioactive materials to large areas surrounding the plant. After the explosion, ^{137}Cs levels in the contaminated areas generally decreased due to radioactive decay and various environmental factors that resulted in dispersal and dilution. However, some locations near a river basin exhibited high or increasing levels of ^{137}Cs after the explosion.

The research attempted to find a correlation between hydrological events (e.g., flooding) and spatial and temporal variations in ^{137}Cs contamination by using a method of statistical analysis involving *geographical information systems* (GISs) (Burrough, 2001).

GIS was used to construct various maps showing soil types, land cover, and proximity of the flooded areas to main rivers. Using the maps to inspect the spatial distribution of ^{137}Cs over the contaminated area was complicated because samples were collected only in 1988, 1993, and 1994. The sparsely collected samples necessitated data interpolation, in which new data points were created to augment existing data points. However, interpolation of GIS data can sometimes be unrealistic unless the propagation of errors is understood through the use of geostatistics. Ultimately, the use of a GIS and geostatistics in this study helped to establish that there was a relationship between flood events in the river basin and high concentrations of ^{137}Cs.

GISs have been used for many years, but it has only recently been suggested that a GIS be combined with geostatistics, for the precise reasons illustrated in the case above. The GIS technique was developed to automate the mapmaking process by aiding in storage, retrieval, analysis, and display of spatial data (Burrough, 2001). The flaw in the GIS method is that it analyzes the attributes of an object or surface but does not consider spatial variation. Geostatistics has been proposed as the perfect complement to GISs because geostatistics is more realistic, in that it considers chance, uncertainty, and incompleteness in a data set (Burrough, 2001). Unfortunately, the fundamental differences between the two fields, combined with the fact that GISs were not designed with geostatistics in mind, have often caused some recalcitrance when the combination of GISs and geostatistics is suggested (Wise et al., 2001). The benefits of using both methods have only begun to be appreciated, mostly because the advantages can no longer be ignored. Geostatistics stands to profit from the union primarily because a visual interpretation of the data is made available. GISs will benefit because they often rely on data sets that are incomplete and uncertain, and geostatistics offers a method for interpolating such data and understanding the error (Burrough, 2001).

Spatial analysis is a blend of many techniques that are not necessarily exclusive to spatial analysis. At times, its boundaries may even seem amorphous. Spatial analysis may be regarded as superior to other disciplines that share these techniques because only spatial analysis attempts to discern both the spatial and temporal extent of a contamination. Other disciplines merely apply these techniques to determine the identity of the contaminant. Spatial analysis is likely to be relied on increasingly in the future as a supplement to more conventional techniques when there is a need to identify the source and dispersal pattern of widespread contamination.

REFERENCES

Alimi, H., Ertel, T., et al. (2003). Fingerprinting of hydrocarbon fuel contaminants: literature review. *Environ. Forensics*, 4(1):25–38.

Boehm, P. D., Douglas, G. S., et al. (1997). Application of petroleum hydrocarbon chemical fingerprinting and allocation techniques after the *Exxon Valdez* oil spill. *Mar. Pollut. Bull.*, **34**(8):599–613.

Boehm, P. D., Page, D. S., et al. (1998). Study of the fates and effects of the *Exxon Valdez* oil spill on benthic sediments in two bays in Prince William Sound, Alaska: 1. Study design, chemistry, and source fingerprinting. *Environ. Sci. Technol.*, **32**(5):567–576.

Bruce, L. G., and Schmidt, G. W. (1994). Hydrocarbon fingerprinting for application in forensic geology; review with case studies. *AAPG Bull.*, **78**(11):1692–1710.

Burrough, P. A. (2001). GIS and geostatistics: essential partners for spatial analysis. *Environ. Ecol. Stat.*, **8**(4):361–377.

Burrough, P. A., van der Perk, M., et al. (1999). Environmental mobility of radiocaesium in the Pripyat Catchment, Ukraine/Belarus. *Water Air Soil Pollut.*, **110**(1):35–55.

Carpi, A. (1997). Mercury from combustion sources: a review of the chemical species emitted and their transport in the atmosphere. *Water Air Soil Pollut.*, **98**(3–4): 241–254.

Carpi, A., and Mital, J. (2000). Expanding use of forensics in environmental science. *Environ. Sci. Technol.*, **34**(11):A262–A266.

Carpi, A., Weinstein, L. H., et al. (1994). Bioaccumulation of mercury by sphagnum moss near a municipal solid waste incinerator. *J. Air Waste Manag. Assoc.*, **44**(5):669–672.

Costa, H. J., White, K. A., et al. (2004). Distinguishing PAH background and MGP residues in sediments of a freshwater creek. *Environ. Forensics*, **5**(3):171–182.

Davis, A., Howe, B., et al. (2005). Use of geochemical forensics to determine release eras of petrochemicals to groundwater, Whitehorse, Yukon. *Environ. Forensics*, **6**(3): 253–271.

Gaines, R. B., Frysinger, G. S., et al. (1999). Oil spill source identification by comprehensive two-dimensional gas chromatography. *Environ. Sci. Technol.*, **33**(12):2106–2112.

Galt, J. A., Lehr, W. J., et al. (1991). Fate and transport of the *Exxon Valdez* oil spill: 4. *Environ. Sci. Technol.* **25**(2):202–209.

Haddad, R. I. (2004). Invited editorial: What is environmental forensics? *Environ. Forensics*, **5**(1):3.

Harley, R. A., Hannigan, M. P., et al. (1992). Respeciation of organic gas emissions and the detection of excess unburned gasoline in the atmosphere. *Environ. Sci. Technol.*, **26**(12):2395–2408.

Kaplan, I. R. (2003). Age dating of environmental organic residues. *Environ. Forensics*, **4**(2):95–141.

Kaplan, I. R., Galperin, Y., et al. (1997). Forensic environmental geochemistry: differentiation of fuel-types, their sources and release time. *Org. Geochem.*, **27**(5–6):289–317.

Lundegard, P. D., Sweeney, R. E., et al. (2000). Soil gas methane at petroleum contaminated sites: forensic determination of origin and source. *Environ. Forensics*, **1**(1):3–10.

Morrison, R. D. (2000a). Application of forensic techniques for age dating and source identification in environmental litigation. *Environ. Forensics*, **1**(3):131–153.

Morrison, R. D. (2000b). Critical review of environmental forensic techniques: II. *Environ. Forensics*, **1**(4):175–195.

Murphy, B. L. (2000). Editorial. *Environ. Forensics*, **1**(4):155.

Opsomer, J. D., Agras, J., et al. (1995). An application of locally weighted regression to airborne mercury deposition around an incinerator site. *Environmetrics*, **6**(2):205–219.

Petrisor, I. (2005). Sampling and analyses: key steps of a forensics investigation. *Environ. Forensics*, **6**(1):1.

Ram, N. M., Leahy, M., et al. (1999). Environmental sleuth at work. *Environ. Sci. Technol.*, **33**(21):464–469.

Sauer, T. C., and Uhler, A. D. (1994). Pollutant source identification and allocation: advances in hydrocarbon fingerprinting. *Remediation*, **5**(1):25–46.

Schauer, J. J., and Cass, G. R. (2000). Source apportionment of wintertime gas-phase and particle-phase air pollutants using organic compounds as tracers. *Environ. Sci. Technol.*, **34**(9):1821–1832.

Scott, P. K., Unice, K. M., et al. (2007). Statistical evaluation of metal concentrations as a method for identifying World Trade Center dust in buildings. *Environ. Forensics*, **8**(4):301–311.

Stout, S. A., Uhler, A. D., et al. (1998). Environmental forensics unraveling site liability: an interdisciplinary analytical approach can unravel environmental liability at contaminated sites. *Environ. Sci. Technol.*, **32**(11):260A–264A.

Suggs, J. A., Beam, E. W., et al. (2002). Guidelines and resources for conducting an environmental crime investigation in the United States. *Environ. Forensics*, **3**(2):91–113.

Tykot, R. H. (2002). Chemical fingerprinting and source tracing of obsidian: the central Mediterranean trade in black gold. *Acc. Chem. Res.*, **35**(8):618–627.

Vives-Rego, J. (2004). Environmental forensics: a scientific service at the service of justice and society. *Environ. Forensics*, **5**(3):123–124.

Wenning, R. J., and Simmons, K. (2000). Editorial. *Environ. Forensics*, **1**(1):1.

Wise, S., Haining, R., et al. (2001). Providing spatial statistical data analysis functionality for the GIS user: the SAGE project. *Int. J. Geogr. Inf. Sci.*, **15**:239–254.

Zemo, D. A., Bruya, J. E., et al. (1995). The application of petroleum hydrocarbon fingerprint characterization in site investigation and remediation. *Ground Water Monit. Remediation*, **15**(2):147–156.

CHAPTER 2

Principles and Issues in Forensic Analysis of Explosives

JIMMIE C. OXLEY

University of Rhode Island, Chemistry Department, DHS Center of Excellence for Explosives Detection, Mitigation, and Response, Kingston, Rhode Island

MAURICE MARSHALL

(Formerly) Forensic Explosives Laboratory, Defence Science and Technology Laboratory, Fort Halstead, Sevenoaks, Kent, UK

SARAH L. LANCASTER

Forensic Explosives Laboratory, Defence Science and Technology Laboratory, Fort Halstead, Sevenoaks, Kent, UK

Summary Proper handling of evidence is critical to identifying and convicting a criminal. Evidence at the scene of an explosion, especially a large explosion, offers some unique challenges. Basic principles of evidence collection, handling, storing, and identifying are discussed herein.

2.1	Introduction	24
2.2	Sample collection	25
2.3	Packaging	29
2.4	Sorting	30
2.5	Documentation	31
2.6	Environmental control and monitoring	31
2.7	Storage	33
2.8	Analysis	33
2.9	Records	36
2.10	Quality assurance	36
2.11	Safety and other issues	37
	Conclusion	37
	References	38

Forensic Chemistry Handbook, First Edition. Edited by Lawrence Kobilinsky.
© 2012 John Wiley & Sons, Inc. Published 2012 by John Wiley & Sons, Inc.

2.1 INTRODUCTION

Many types of laboratories engage in chemical analysis. In forensic science laboratories, a wide variety of chemical, biological, and physical analyses are undertaken. The principal difference between forensic analysis and more general analysis is the degree of certainty required of the results. The technical issues depend on the nature of the sample: in particular, on whether bulk samples or invisible traces are being sought. Bulk samples are considered to be anything that is visible to the naked eye and can range from micrograms to several kilograms of material.

The most obvious difference between analyses of bulk versus trace samples is the relationship between the sample and the environment. Sometimes the desired analyte may be in the environment, and sometimes a species in the environment (e.g., water, oxygen, iron particles) may degrade the sample or affect the results. The analyte in the environment is not generally an issue in bulk samples. In trace samples, the amount of contamination may be large enough to distort results. Thus, if the sample contains a tiny amount of the analyte sought and the environment contains a large concentration of that species, extreme precautions will need to be taken to protect the sample and exclude contact with the environment. Conversely, if the sample contains a large concentration of the analyte and the environment a tiny amount, the issue is trivial. An understanding of the composition of the background environment is therefore highly desirable, but not always possible. This needs to be considered when reporting results.

A robust and well-designed trace analysis protocol is likely to involve (1) physical separation between the analyst and the sample; (2) the use of disposable items for handling, packaging, and containment; (3) appropriate blank and control samples; and (4) environmental monitoring. The precise detail of the measures will depend on the environmental challenge to the integrity of the analysis, and it is often possible to strengthen one protective technique to counter the weakness or absence of another.

Although explosives are, of course, simply chemicals or chemical mixtures, in some respects their analysis is easier because many explosives (e.g., the organic explosives) are rarely found in the general public environment. In response to the Provisional Irish Republican Army (PIRA) bombing campaign on the mainland (1970s to 1990s), the United Kingdom (UK) led the way in protocols pertaining to explosive evidence. Over the course of a decade four studies were produced documenting explosives in the environment: two on background levels of military explosives and two on levels of inorganic ions (Crowson et al., 1996; Walker et al., 2001; Cullum et al., 2004; Sykes and Salt, 2004). Of 670 samples collected on the British mainland, only eight showed traces of organic explosives. A recent repeat of this study in the United States showed that only three out of 333 samples had traces of high explosives (Laboda et al., 2008). Both the UK and U.S. studies showed nitrates at the microgram level in 20 to 30% of the samples.

Much attention has quite correctly been paid to issues of cross-contamination in the analysis of explosive traces because of the generally very serious nature of the criminal offences involved. However, in reality, all forensic trace analyses need to be protected against the risk of compromise by ill-founded suspicions of all types of cross-contamination. There should, of course, be no such suspicion that is well founded! Trace analysis procedures need to be designed, tested, and validated to ensure that positive evidence is produced showing the integrity of the results. This applies whether the sample is a few nanograms of explosive or a few nanograms of DNA.

This is a field where contamination of evidence can easily occur due to the wide range of vapor pressures exhibited by explosive formulations. Did the Madrid bombing use Goma 2 ECO or Titadyn? The answer is critical because it would point to one or another terrorist group. Dinitrotoluene (DNT) and nitroglycerin were found as part of the evidence. Was it in the explosives used to make the terrorist devices, or was it a result of cross-contamination during storage since DNT is highly volatile?

2.2 SAMPLE COLLECTION

Unfortunately, forensic chemists do not always have control over the vital aspect of collection and packaging of the materials they must examine. There is a world of difference between the effort and preplanning required for dealing with a large bombing attack and that required for the investigation of a bombing of a mailbox or single residence. Most forensic scientists will only deal with the latter. Nonetheless, preplanning will be worthwhile. Clean containers and packaging materials should be procured and stockpiled ready for use. Examples of such items are disposable scoops, scrapers, dustpans, and brushes, as well as metal cans and nylon bags of various sizes. Similarly, collection devices such as brushes, scoops, scrapers, and vacuum pumps and filters should be obtained. Minivacuums can be constructed from disposable plastic tubing, syringe filters, and plastic syringes.

Preferably all items used for collection should be subjected to quality assurance tests before use. The easiest way to ensure the cleanliness of tools used for collection of trace explosive evidence is to use disposable items from a known supplier which have just come from the box. If possible, a statistical sample of each item should be prescreened before operational employment. However, if a prescreen is not possible, a more rigorous regime of analysis of blank and control samples can be substituted. It is to be understood that this may entail the risk of loss of evidence if a control is analyzed as being positive. It should also be noted that suitable control samples should still be obtained, even with the use of prescreened materials.

Swabbing to collect trace explosive evidence is a common practice. Swabs may be pre-prepared using solvent-washed cotton balls that are either dry or have been wetted with a solvent (Jenkins and Yallop, 1970). Although ideally the swabs should be premade and preanalyzed, necessity may drive the investigator to use improvised swabs (e.g., alcohol-wetted hand-wipes, facial tissues, paper towels). Although swabbing is a superficially simple technique, in fact a plethora of interacting variables and issues need to be considered. A key issue is to identify the type of explosive being sought: for example, inorganic or organic? Another issue is practicality and generality. Although it is arguably possible to design swabbing protocols that are optimized for the collection of particular explosives, our experience is that it is better to design for the widest possible application. Bomb scenes are generally places of chaos where decision making is handicapped by lack of information, and it is operationally much better to avoid the need to make choices by providing sample collection kits of general application.

In the UK in the early 1970s, dry swabs, water-wetted swabs, and solvent-wetted swabs were all used for collecting different types of residue. Subsequently, it was realized that the choice of dry or water- or solvent-wetted swabs was usually little better than guesswork prior to laboratory analysis. Furthermore, the choice of solvent or water was less significant than it first appeared. Although the solubility of diverse inorganic and organic explosive species varies dramatically in water and organic solvents, the small amounts present in trace samples means that the actual concentration that has to be dissolved in the swabbing solvent is rather low. Moreover, too strong a solvent will not necessarily recover more explosive residue; rather, it will merely pick up more unwanted background material, thereby complicating the subsequent preanalytical cleanup and concentration in the laboratory. Another very important consideration in the choice of a swabbing solvent is toxicity. These various issues led the UK to adopt a 50:50 ethanol–water mixture in their swab kits. This was found to provide recovery of a broad spectrum of both inorganic and organic explosive traces and to be compatible with subsequent laboratory protocols (Douse, 1985; Warren et al., 1999).

Very large samples do not lend themselves to solvent washing. Large containers for solvent extraction may be available, but plastics may allow interferents to elute. Generally, borosilicate glass rather than plastic containers are preferred, but extraction of ions from soda glass can also be a problem. Since large containers are generally not available, large samples should be swabbed. The swabs found in premise kits could be used. (Premise kits are useful if prepared ahead of time. They contain clean, validated swabs; disposable gloves; tweezers; and a solvent such as ethanol–water 50:50. A control in a premise kit would be a solvent-wetted but unused swab. Like hand kits, premise kits are heat sealed under positive pressure.)

Driven by the pressures of dealing with the long-term PIRA campaign, the UK has pioneered the use of premade, screened kits for both hand testing and premise screening. In addition, in a pilot program hair kits have been prepared specifically for explosives, and some laboratories include materials for hair sampling in kits used for the recovery of gunshot residue.

Every effort should be made to ensure that the investigator does not contaminate the scene. This includes using fresh, disposable tweezers to handle and manipulate swabbing materials and donning disposable outerwear before entering the scene. Ideally, different investigators should obtain samples from the crime scene and a suspect's premise to eliminate the possibility of cross contamination (of explosives traces and other types of evidence, such as DNA or fiber). If this is not possible and the same investigators are screening the scene and the suspect's premises, they need to take steps to ensure that (1) they do not inadvertently transfer contamination between scenes, and (2) they provide objective documented evidence to prove the efficacy of their measures to prevent cross-contamination. Such measures might include, for example, use of disposable overalls, gloves, hats, hoods or hairnets, and bootees or overshoes; handwashing, hairwashing, or whole-body showers, and possibly the taking of personal control swabs or personal checking with airport-style explosive screening instruments. In addition, attention needs to be paid to the cleaning or overwrapping of personal items such as spectacles, hair bands, wristwatches, and jewelry. These are excellent potential sources of trace evidence from suspects and present a risk of cross contamination for the scene examiner and the laboratory scientist. The simplest approach is, of course, the best; do not wear such items to a scene.

Searching a scene for bulk and trace evidence involves somewhat different approaches. The investigator generally is called upon by law enforcement to identify bulk explosives based on visual identification. This is likely to come about because police have entered a suspect's premises and found something that seems suspicious, have intercepted an attempted bombing and disrupted the device, or have discovered a device that fails to function or functions inefficiently, such as a poorly constructed pipe bomb. Trace explosive evidence is not evident; that is, it is invisible to the investigator or the collector of evidence. Both the collector and the investigator must rely on good training, sense, and good luck. It is vital that collection material and people be shown to be free of contamination prior to entering a scene.

At the scene of an explosion, the investigator will start by attempting to identify the seat of the explosion. Generally, this will be indicated by the presence of a crater. A crater is not always formed, such as in the case of a low level of explosives; if a disperse-phase (gas-phase) explosion has occurred; or if a high-explosive condensed-phase explosion has occurred either in midair or on a frangible surface that has been destroyed in the process. The location of the explosion seat may also be identified by tracking back from the patterns of physical damage at the scene, such as the marks made by objects that have been moved, or from patterns of fragment penetration in nearby items.

Once the explosion seat has been located, the investigator can begin to look for likely places to find explosive residue. The best candidates will be substrates that were close to the explosion when it happened. The obvious first step is to take samples from the crater left by the explosion. However, most of the residue is expected to be found "near," not "in," the crater where it has been thrown, spalled from the exploding charge, or where it has condensed out of the fireball onto relatively colder surfaces. Nearby metallic items such as door frames and furniture, window frames, and light fittings are such surfaces and should be examined. In general, if a surface shows evidence of explosive damage or disruption [i.e., microcratering, pitting, or sooting (either visible to the naked eye or through a low-power lens)], it may be a good source of explosive residue. In the search for explosives at a bomb scene, impermeable surfaces offer the best chance of finding explosive residue for the majority of explosives. However, for very volatile explosives, such as triacetone triperoxide (TATP), the issue is the rapid loss of the explosive from surfaces due to evaporation or sublimation. In these circumstances, an absorbent, porous substrate (e.g., carpet or wood) is likely to retain the volatile explosive much longer than is a hard surface such as metal. In any case, sealed packaging is critical for retaining this evidence. As discussed below, for most explosives evidence nylon bags are preferable. Because one does not know a priori which type of explosive has been used, the on-scene investigator must take samples that cover either possibility.

In the case of a car bombing, such as a booby trap attack intended to kill the occupants, the preferred option would be to package the vehicle and take it to the forensic laboratory. Whether the vehicle is examined at the scene or in the lab, the first step is to identify the seat of the explosion and then take samples from both the nonporous surfaces, such as plastic and metal, and the porous surfaces, such as carpet and upholstery. If a vehicle, a car or truck, was used to contain the bomb, explosive residue will probably be found on those scattered vehicle parts that were in close proximity or contact with the explosive before detonation; thus, depending on the initial location of the bomb, residues are likely on parts from the trunk or passenger compartment, whereas they are generally not to be found on the engine block.

When searching for evidence that explosives may have been stored or hidden in a location or a suspect's premises, the investigator will look for explosive traces in likely storage places. Apart from cupboards and shelving, other locations, such as hidden spaces behind removable paneling in bathrooms, should also be sampled. In some instances, dusty surfaces may reveal the outline of containers that have been removed. If it is suspected that a premises has been used for explosives manufacture or processing, it will also be worthwhile to remove the pipes and waste traps from sinks and washbasins to see if any explosive residue has collected there.

If manufacture or chemical synthesis is suspected, clearly any obvious items of laboratory equipment will be seized for examination. However, amateur experimenters and makers of improvised explosives are also quite likely to press ordinary domestic equipment into service: for example, kitchen mixing bowls, cooking spatulas, measuring cylinders or jugs, food processors and stirrers, and scales or balances. All of these should be examined for explosive residue. In the absence of any obvious laboratory work area at a suspect's premises, it is often worthwhile to swab the kitchen work surfaces for traces of explosives residue, paying particular attention to cracks or crevices where residues might have been missed by any cleaning efforts. Discovery of clandestine laboratories presents a number of difficult issues for an investigator. First and foremost is the likelihood that unknown hazardous materials are present, often in unlabeled containers. Second is the very real risk of booby traps and antihandling devices. The investigator will wish at an early stage to assess whether a clandestine laboratory is likely to be drug or explosives related. The presence of technical literature dealing with one or the other subject can help. In addition, the chemical synthesis of most illicit drugs is chemically more complex than that of the common organic explosives. Hence, as an initial guide, an elaborate laboratory with complex equipment is more likely to be drug than explosive related. It should be noted that there have been examples of drug laboratories being booby trapped with explosive devices precisely to guard against the arrival of law enforcement. For example, packets wrapped to appear as illegal drugs were in reality Armstrong's mixture, an extremely sensitive formulation used in toy caps and party poppers. Moreover, the first encounter with TATP in a criminal context in the United States was in booby traps in a drug laboratory (Christian, 2005).

Frequently, the investigator needs to collect evidence from a suspect's person. Typically, the options are hands, hair, clothing, jewelry, and glasses. The law enforcement investigator and forensic examiner will be familiar with the use of pre-prepared kits for examining hands for gunshot residue. The principle can be extended to explosive residues, but the design details must be modified. Hair is washed much less frequently than hands; furthermore, it has been shown that explosive handlers can be allowed to wash their hair and in some instances still exhibit detectable levels of contamination. Examination of hair for trace explosives can be performed simply by combing the hair with a comb on which clean gauze has been threaded. After multiple passes through hair, the gauze can be disengaged from the comb and extracted with solvent for chemical analysis (Sanders et al., 2002; Oxley et al., 2005,2007a,b). Chemical digestion of hair is performed when looking for evidence of drug use, but such procedures have not yet been investigated for explosives.

Clothing evidence has been pivotal in a number of high-profile cases. The Oklahoma City bomber, Timothy McVeigh, had traces of pentaerythritol tetranitrate (PETN) on his clothing; and Ressam Amad, the would-be millennium bomber, had acid burns and other traces of explosives on his trousers. In high-profile cases it is usual not only to confiscate and examine the clothing worn when the suspect was apprehended, but also the entire contents of his wardrobe as well as bags, backpacks, and holdalls, which may have been used to transport devices. It is commonplace when a person is handling items for him to wipe his hands on the seat of his trousers. Other common hand movements that potentially transfer residue from hands to clothing are adjusting the knees of one's trousers when sitting down or putting hands in one's pockets. All these spots are good places to look for explosives traces.

As mentioned previously, jewelry, rings, bracelets, watch straps, hair bands, and glasses are likely to be worn for long periods of time with no rigorous attempts at cleaning. These readily become contaminated with any explosive handled by a suspect and are likely to retain traces in crevices over long periods of time. This is a positive feature in terms of finding evidence of malfeasance and a negative feature in preventing the investigator from contaminating the evidence. Generally, these items are seized and solvent washed, preferably with the aid of an ultrasonic bath, although items such as leather watch straps can be wiped with a solvent-wetted swab rather than being solvent washed.

2.3 PACKAGING

Without the proper packaging, all the effort made in the collection of evidence is for naught. Selection of container material should be done well in advance of need and should consider the size of the object to be collected and the degree of isolation required between the sample and the environment. For example, does the sample need to be protected against sorption of external vapors? Or is it a potential source of vapors that might be lost as evidence or contaminate other samples? Some samples, of course, should not be sealed in nonpermeable bags; these included dried or wet blood, which may be subject to attack by bioorganisms, as well as items likely to undergo chemical decomposition and generate gaseous decomposition products.

Small samples are usually sealed in nylon bags with the neck twisted and taped, doubled over twice, and taped again. Of the plastics commonly available, nylon is the most impervious to vapors; commercial evidence bags are usually polyethylene. Large samples, such as cars, light fixtures, or window and door frames, which may have trace residue and might show evidence of explosive cratering, should be sealed in polymer sheeting (tarpaulins) and taped together with adhesive tape. Medium-sized samples that will not be analyzed for trace explosives can be packed in plastic tubs or crates with lids, evidence seals, and labels.

It is likely that the collection of evidence at the scene will result in contamination of the outside of the collection container. For this reason, after the initial bagging at the scene, a second bagging should occur well away from the scene at an established workstation. Some laboratories actually perform a third bagging upon receipt at reception to contain any external contamination picked up in transit (or at the scene) and to prevent any contamination from the nontrace areas of the laboratory itself. An on-site workstation outside the target premises can be produced simply by laying out clean disposable paper at a convenient spot protected from the elements [e.g., the back of a police van (assuming it is known and can be shown not to have contained explosives or firearms)]. If it is known that evidence of trace explosives is to be sought and operational constraints allow time for preplanning, it may be worthwhile considering hiring clean vehicles and prescreening them to ensure the absence of explosive contamination before the operation.

Opening the evidence at the forensic laboratory requires as much thought as does packaging it. If the sample has been triple-wrapped, its wrappings need to be removed sequentially as it moves to progressively cleaner areas. The multiple layers of packaging are pealed off sequentially with gloved hands in a manner that ensures no transfer of contamination from the outside world into the trace area. This means that fresh gloves are used at each step. The innermost layer of packaging is not opened until the item is in a location guaranteed to be free from contamination.

2.4 SORTING

The Forensic Explosive Laboratory in the UK, which has had a great deal of experience processing debris from large bombing events, has found it useful to use large, lidded plastic bins for large pieces of evidence. At the height of the PIRA bombing campaign, stashes of these bins were maintained at various locations in England. With huge volumes of debris it was found much easier to locate the anomalies if the debris were sorted into bits of similar size. Therefore, in the field or directly on returning to the analytical facility, they presorted what was considered the most promising debris. The goal of such sorting was to find pieces derived from the bomb or its container.

2.5 DOCUMENTATION

A great deal of material is required to fully document a crime scene: for example, photographs and drawings as well as notes and logbooks. Every item of evidence that is collected needs to be uniquely labeled and controlled. Each person involved in evidence collection and labeling must have a notebook to keep track of the exhibit references he or she has generated. For forensic purposes the collection and packaging procedure needs to ensure evidential continuity and integrity, with a properly documented chain of custody. This is usually accomplished by attaching a signature seal and receipt that is countersigned as the item passes from hand to hand. As one might expect, modern electronic solutions have been tested in some jurisdictions with the aim of supplanting the traditional paper-and-ink systems. Although barcode labels and electronic scanners are widely used in retailing, this technique is more easily implemented in the laboratory than in the field.

Adequate documentation is essential to the satisfactory progress of a case. Conversely, it is all too easy to generate superfluous paperwork, with accompanying nugatory effort. In the late 1980s it was normal in the UK to divide a bomb scene into numerous zones, often on a grid pattern, and expend much effort recording the fact that a particular item was found in a specific zone. Eventually, the pressure of work generated by a massive terrorist bombing campaign forced an evaluation of such practices. It was realized that it was rarely, if ever, possible to draw any useful evidential conclusion from such detailed information about the location of bomb debris. The key piece of information for a court was that a specific item had been found at the bomb scene. Consequently, the practice of zoning at bomb scenes was curtailed dramatically. Typically, a scene might use the area around the crater and other natural boundaries for zoning, or if the scene was a building, perhaps divided according to rooms and whether the evidence was recovered from inside or outside the building.

2.6 ENVIRONMENTAL CONTROL AND MONITORING

As stated in Section 2.1, the precautions necessary to protect sample integrity depend on the challenges presented by the environment. For example, does the environment contain significant amounts of the target analyte so that contamination of the sample becomes a real rather than a theoretical risk? This issue is particularly relevant when handling trace rather than bulk evidence. To protect the integrity of the sample, a rigorous protocol is needed to ensure laboratory cleanliness, complemented by a measuring regime to demonstrate that the requisite standards of cleanliness have actually been achieved and maintained. The process begins with the initial design of the laboratory. The laboratory layout should impose barriers between the unknown outside environment and the controlled trace environment. Physical features include airlocks, filtered air with positive pressure, and a separate antechamber through which personnel must enter or exit, in the process changing into clean, disposable overgarments, removing watches, jewelry, and eyeglasses.

The antechamber. In the antechamber, personnel will wash hands and other exposed skin and don hair covers, disposable overalls, shoe covers, and gloves. They should also ensure cleanliness by a quick screen using explosive detection instrumentation [e.g. ion mobility spectrometry (IMS)]. Pens, papers, watches or other jewelry, and eyeglasses should be left here. Duplicates should be waiting on the other side of the airlock in the trace examination room. Administrative controls include access control to ensure that only appropriately trained or supervised personsonal can enter the trace laboratory. In fact, since every entry and egress represents a theoretical contamination risk, the number of people entering the analysis area should be minimized. A log of visits should be maintained. It is from this antechamber that the forensic chemist and the sample, still wrapped in its innermost layer of packaging, enter the trace laboratory.

The trace laboratory. A regular cleaning schedule is necessary. The primary defense is a filtered air system, whose performance is monitored both by differential pressure gauges and atmospheric particulate monitors. Floors should be washed at least once a week with strong alkaline detergent. One approach to ensuring cleanliness is to predefine a number of squares (e.g., 50 cm × 50 cm) on the floor which are regularly swabbed for contamination (e.g., weekly). The tops and insides of cupboards and drawers and air intakes need a less frequent scheduled cleaning and screening. The goal is to see that every surface is cleaned and tested for contamination at a known interval. It is important that the forensic chemist hypothesizes where contamination might accumulate. Analytical instruments may be elevated on small blocks to facilitate cleaning under each instrument. It is useful if instruments and materials can be dedicated to a particular stage in the analysis. As far as possible, disposable tools should be used. For example, the trace laboratory should have its own paper and pens, which are used only once. Wrenches, screwdrivers, and beakers should be cleaned by heating in an oven ($\sim 250°C$) and either storing the materials in the oven or removing then and wrapping them individually in nylon bags until use.

At the beginning of the work cycle, the analytical bench should be cleaned: for example, with alkaline detergent solution and ethanol, and wiped down with paper towels. A fresh disposable benchtop surface, such as butcher paper, aluminum foil, or Benchkote, should be used for each sample. The forensic chemist should change gloves frequently, at least between samples. To provide evidence that no contamination is present, the forensic chemist should take control swabs of the work surface, the front of her suit and gloved hands, and the disposable benchtop surface. These are considered control samples; they are necessary to show that contamination has not occurred (Hiley, 1998; Crowson et al., 2001; Beardah et al., 2007).

2.7 STORAGE

Like most evidence, aging can result in loss or contamination of explosives evidence. A rapid turnaround time is important. In the UK during the PIRA terrorist campaign, the Forensic Explosives Laboratory, responsible for analysis of trace and other evidence, pledged a 6 hour turnaround time for a single hand-test kit and two weeks for a preliminary report, regardless of the type of case.

Storage conditions are also critical; for example, the first bombings using TATP were in Israel. Even in cases where intelligence pointed to the use of TATP, forensic examiners found no residue. Subsequent studies showed that TATP is so highly volatile that evidence should be thoroughly sealed and, ideally, examined immediately, to maximize the chances of recovery and detection. Sealing and storing at subambient temperatures are not always necessary but can hardly hurt explosive evidence. (*Note:* Samples with moist blood or other biological media present may be subject to bacterial growth if sealed without the additional precaution of subambient temperature. However, consideration should then be given to the effect of thawing on the other materials.)

2.8 ANALYSIS

Bulk analysis may be of visible particles of an explosive and/or its combustion products, but it may be of bomb components, such as timers, fuses, wiring, or tape. The analysis is usually performed to identify the material and to make comparisons, for example, between a material recovered from a crime scene and one recovered from a suspect's premises. At best, it is only possible to say that two materials are different or to list those aspects in which they are similar. However, to make such a statement, it is necessary to have a well-founded understanding of the likely variations in a material and the practical significance of such variations. This highlights the value of reference collections and surveys of items and materials likely to be encountered in case work (e.g., carpets, car paints, accelerants, and explosives). Comprehensive databases or reference collections are useful for all sorts of bomb-making gear (e.g., pipes, timers, tapes, and batteries).

In some cases the explosive material used in the bomb can be seen. Inefficient devices such as those using smokeless powder or black powder often leave visible residue. High explosives usually leave residue, too, but since it is not visible to the human eye, it may not make the multiple down-selects to end up in the final analytical protocol. In efficient bombs, the residual explosive has probably been thrown out from the surface of the bomb—spalled off the surface (Kelleher, 2002). Thus, looking just outside the crater may be more productive than looking at evidence inside the crater.

Visible explosive particles should first be examined intact (i.e., with nondestructive techniques, as clues are available at this point). Although prilled ammonium nitrate is made at several hundred locations around the world, there are wide variations in prill shape (to be determined visually) and prill coating (to be determined analytically). With the proper reference collection, such variations may make it possible to trace the material back to its original manufacturer.

For bomb scene materials with no visible residue, prescreening for trace explosives may be possible using an explosive detector. In practice, this approach is less labor-saving than it first appears and, more important, runs the risk of inadvertent contamination of valuable trace evidence or the inadvertent exclusion of evidence with traces of material not detected by the specific detector. Otherwise, the more accurate but laborious approach is a multistep extraction of the debris to separate organic from polar species. One approach is to extract the debris with the aid of an ultrasonic bath using a solvent for both polar and nonpolar species. If one uses a strong solvent, such as acetone or acetonitrile, all of the explosive residue will dissolve, but so will many unwanted species, which serve to obscure the analysis and clog the system. The Forensic Explosives Laboratory in the UK has settled on a technique using a relatively weak solvent system.

The debris is extracted with ethanol–water (50:50) and run through a glass pipette containing a few grams of a solid adsorbent such as Chromosorb 104. The packing retains the common organic explosives (e.g., TNT, PETN) and allows the polar species to elute. This polar fraction is analyzed by ion chromatography looking for species such as nitrates (ammonium or uronium) or chlorates. The Chromosorb column is then eluted with ethyl acetate to extract the organic species. These are analyzed by gas chromatograph (GC) or liquid chromatography (LC), depending on the anticipated volatility of organic explosives suspected (see Figure 2.1). For example, the thermal stability of hexamethylene triperoxide diamine (HMTD) is so poor that LC is the preferred technique; otherwise, samples break down in the inlet port of the GC. Obviously, the GC or LC can be attached to a variety of detectors. For GC, an electron capture detector (ECD) is the most sensitive detector, and a thermal energy analyzer (TEA), which analyzes for nitro groups, is the most specific. If LC is used, a mass selective detector (MSD) is usually employed. This has the added feature of providing orthogonal confirmation: not only retention time but also fragmentation patterns. While the field of mass spectroscopy (MS) is developing rapidly, for the common benchtop GC–electron ionization–MS, most explosives rarely exhibit a parent peak (P). Identification is based on fragmentation patterns, where P–OH is often observed. Yinon has discussed this extensively (Yinon and Zitrin, 1981, 1993; Yinon, 2007). An MS database of explosives is available online at http://expdb.chm.uri.edu, but it is essential that forensic scientists also create their own libraries using their own experimental conditions, as mass spectral fragmentation patterns are highly dependent on the exact conditions under which they are obtained.

Orthogonal techniques are defined as two techniques for which sample identification is based on completely different principles (e.g., infrared spectroscopy and mass spectroscopy). In practice, finding two absolutely orthogonal techniques is rarely achievable; rather, in practice, significantly different techniques are used (e.g., using two different GC columns which yield significantly different retention times). The more orthogonal the techniques, the higher the level of confidence in the assignment of identity.

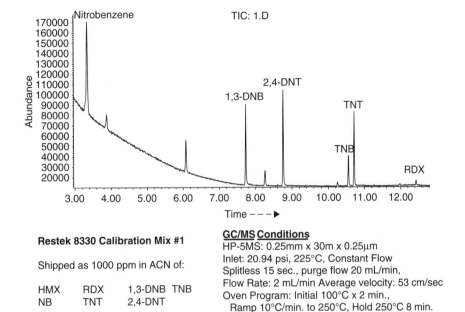

Figure 2.1 Chromatogram of explosive calibration standard.

In developing the analytical protocols discussed above, it is important to use standard solutions with known concentrations of explosives. Mixed standards are useful. Not only do they demonstrate the system suitability and performance across the entire range of explosive species being sought, but the mixed standard can never be confused as the real result. For high precision in retention results, a retention time marker [e.g., musk tibetine or fluoronitrotoluene (for GC)] should be added to both sample and standard solutions.

Although forensic samples are rarely quantified, if accurate quantification is desired, explosive recoveries should be determined. A good approach is to use the method of standard additions. This involves dividing the sample into two parts, spiking one part of the sample with a reasonable amount of explosive, typically between one-third and two-thirds of the amount estimated to be present in the sample, then performing the standard workup on the original and spiked samples and determining what percentage of the spike was recovered.

Isotope ratio mass spectroscopy (IRMs) applied to forensic evidence is a vibrant research area at the time of this writing. Much work is needed to establish reproducibility of results and degree of variability among "identical" samples. In principle it is easy to say that if two samples differ in some significant respect, they are in fact different. However, the converse is not true. One can never state with absolute forensic certainty that two items are identical. All one can do is to list all those features in which the two samples match. In practice there comes a point at which a reasonable person will conclude that a match has been established beyond a reasonable doubt. Thus, it may be possible, in a few years, that with the aid of foreseeable improvements in isotope ratio mass spectrometry, one could testify that a sample of explosive recovered from a pipe bomb at a scene matches a sample of explosive recovered from a mixing bowl at a suspect's premises and that it is highly probable that the samples came from the same original batch of explosive. This issue of the definition of a match is a more general one for forensic science, and would benefit from wider consideration, and preferably agreement within the judicial system. The obvious parallel is the long-established UK criterion that for two fingerprints to match, at least 16 points of agreement must be identified. Other jurisdictions accept a lower figure; for example, in the United States, 12 points are considered sufficient. For this to become possible, large reference collections and databases will become essential, and possibly the very expensive IRMS instrument will become so widely purchased that the price will become affordable to every forensic lab (much as has happened with GC–MS). Rapid improvements in electronic instrumentation suggest that the limitation in precision currently encountered in this technique may be attributed to the chemical methods used in the sample preparation.

2.9 RECORDS

While keeping good records of work is always useful, it is absolutely essential in a forensic laboratory. The written record becomes a legal document. Whether the employee leaves the laboratory or is not called to testify for five years, every step of the procedure and every aspect of the results, including negative results, are extremely important to document. Given the fugacity of electronic records, it is wise to archive all materials in the form of printed paper. Such paper records should, of course, be signed and dated by the scientist producing them. Records should include the printouts from various instruments: chromatograms with standards, mass spectral fragmentation patterns, infrared spectra.

ISO 9001 requires that all quality-related records be retained for a minimum of seven years. Although this would apply to forensic evidence, some jurisdictions require much longer document retention (e.g., 20 years in the UK for explosive offenses).

2.10 QUALITY ASSURANCE

For forensic work only techniques and methods of proven validity should be used. These methods should be documented in the form of standard methods or standard operating procedures (SOPs) to which the laboratory staff can refer readily. Similarly, only persons of proven competence should be allowed to report forensic results, and these people should be subjected to periodic proficiency tests. Requirements for personnel training and method validation are found, for example, in ASCLD (American Society of Crime Laboratory Directors) protocols and ISO 17025 (for laboratory work) and ISO 17020 (for scenes).

To create a new laboratory protocol, the developer should undertake any necessary research and precision and validation studies, producing both a draft method and a report documenting his or her findings. Once the method is accepted, the originator would be expected to become the first trainer in this protocol; it would be sensible to bear the latter requirement in mind when selecting a scientist to undertake the method research and development. Elements that should be part of the new method are a clear statement of precision based on a statistical study of replicate analyses; procedures for calibration, preparation of standards, traceability and uncertainty of measurements; and guidance as to the interpretation of results for forensic purposes.

2.11 SAFETY AND OTHER ISSUES

For a newly set up explosive laboratory, a number of issues will arise. These include acquisition of pure explosive standards, explosive storage, licensing and regulatory compliance, and disposal of hazardous materials.

Generally, licensing and approved storage are required before any explosive can be acquired or stored. Once storage is available, it will be necessary to find an entity that can provide pure explosive standards. Most commercial explosives have been formulated with plasticizers, stabilizers, phlegmatizers, and so on. In consequence, commercial explosives will contain significantly less than 100% of the desired explosive. It may be necessary to synthesize the desired explosive or recrystallize a commercial explosive. The physical properties of this explosive must be characterized and recorded. Differential scanning calorimetry (DSC) allows the determination of material purity, provided that the materials have observable melting points (see Figure 2.2). It should be noted that the purified explosive, freed from the additives present in commercial products, may be significantly more hazardous.

Waste from the explosive laboratory will represent various types of hazards: explosive; flammable; toxic; biological; sharps. Proper arrangements need to be made for disposal for each of these waste streams. In general this will require the services of specialist contractors. Explosive laboratories attached to explosive firing ranges may be able to send their explosive waste for disposal. In large facilities, this is typically achieved by incineration. However, for small amounts of explosive and in most laboratories, the best option is chemical destruction. Suitable methods may be found in the literature.

One point that has yet to be mentioned is the fact that many explosives are toxic. This should be obvious from a historic point of view since nitrate esters are used to treat angina, and cyclonite (RDX) was patented for use as a rat poison.

CONCLUSION

Forensic analysis differs from routine chemical analysis in the critical nature of the results and the fact that the results and documentation of it may come under severe

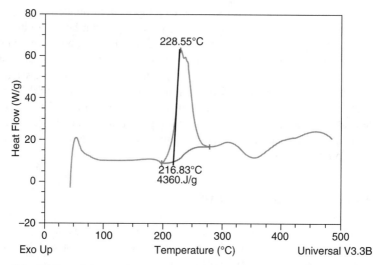

Figure 2.2 Differential scanning calorimetric scan of unknown black solid (suspected of being an explosive). *(See insert for color representation.)*

scrutiny. For trace analysis, every pathway for contamination must be anticipated and mitigated. The differences in analyte concentration between sample and environment must be characterized and proper barriers put in place. There are only two standards in forensic analysis: perfect and useless.

REFERENCES

Beardah, M. S., Doyle, S. P., and Hendey, C. E. (2007). Effectiveness of contamination prevention procedures in a trace explosives laboratory. *Sci. Justice*, **47**:120–124.

Christian, D. R. (2005). The multifaceted demands and dangers you face when investigating clandestine laboratories. *Evidence Technol. Mag.*, May–June, pp. 18–21.

Crowson, C. A., Cullum, H. E., Hiley, R. W., and Lowe A. M. (1996). A survey of high explosives traces in public places. *J. Forensic Sci.*, **41(6)**:980–989.

Crowson, A., Hiley, R. W., and Todd, C. C. (2001). Quality assurance testing of an explosive trace analysis laboratory. *J. Forensic Sci.*, **46(1)**:53–56.

Cullum, H. E., McGavigan, C., Uttley, C. Z., Stroud, M. A. M., and Warren, D. C. (2004). A second survey of high explosives traces in public places. *J. Forensic Sci.*, **49(4)**:684–690.

Douse, J. M. F. (1985). Trace analysis of explosives at the low nanogram level in handswab extracts using columns of Amberlite XAD-7 porous polymer beads and silica capillary column gas chromatorgraphy with thermal energy analysis and electron capture detection. *J. Chromatogr.*, **328**:155–165.

Hiley, R. W. (1998). Quality control in the detection and identification of traces of organic high explosives. In: Beveridge, A. (Ed.), *Forensic Investigation of Explosions*. London: Taylor & Francis, pp. 315–342.

Jenkins, R., and Yallop, H. J. (1970). The identification of explosives in trace quantities on objects near an explosion. *Explosivstoffe*, **6**:139–141.

Kelleher, J. D. (2002). Explosives residue: origin and distribution. *Forensic Sci. Commun.*, 4.

Laboda, K. G., Collin, O. L., Mathis, J. A., LeClair, H. E., Wise, S. H., and McCord, B. R. (2008). A survey of background levels of explosives and related compounds in the environment. *J. Forensic Sci.*, **53(4)**:1556–4029.

Oxley, J. C., Smith, J. L., Kirschenbaum, L., Shinde, K. P., and Marimganti, S. (2005). Accumulation of explosives in hair. *J. Forensic Sci.*, **50(4)**:826–831.

Oxley, J. C., Smith, J. L., Kirschenbaum, L., Shinde, K. P., and Marimganti, S. (2007a). Accumulation of explosives in hair: II. Factors affecting sorption. *J. Forensic Sci.*, **52(6)**:1291–1296.

Oxley, J. C., Smith, J. L., Bernier, E., Moran, J. S., and Luongo, J. (2007b). Hair as forensic evidence of explosives handling. *Propellants Explosives Pyrotech.*, Nov.

Sanders, K. P., Marshall, M., Oxley, J. C., Smith, J. L., and Egee, L. (2002). Preliminary investigation into the recovery of explosives from hair. *Sci. Justice*, **42(3)**:137–142.

Sykes, A.-M., and Salt, I. (2004). Survey of inorganic traces in the environment. *Proceedings of the 8th International Symposium on Analysis and Detection of Explosives*, Ottawa, Ontario, Canada, June 6–10.

Walker, C. J., Cullum, H. E., and Hiley, R. W. (2001). An environmental survey relating to improvised and emulsion/gel explosives. *J. Forensic Sci.*, **46(2)**:254–267.

Warren, D., Hiley, R. W., Phillips, S. A., and Ritchie, K. (1999). Novel technique for the combined recovery, extraction and clean-up of forensic organic and inorganic trace explosives samples. *Sci. Justice*, **39**:11–18.

Yinon, J. (Ed.) (2007). *Counterterrorist Detection Techniques of Explosives*. Amsterdam: Elsevier.

Yinon, J., and Zitrin, S. (1981). *The Analysis of Explosives*. Oxford, UK: Pergamon Press.

Yinon, J., and Zitrin, S. (1993). *Modern Methods and Applications in Analysis of Explosives*. New York: Wiley.

CHAPTER 3

Analysis of Fire Debris[†]

JOHN J. LENTINI

Scientific Fire Analysis, LLC, Big Pine Key, Florida

Summary The isolation and identification of ignitable liquid residues (ILRs) from fire debris are critically important aspects of arson investigation. In this chapter we describe common techniques for the isolation and identification of ILRs. Analytical procedures have become more sensitive, and results of testing play a very important role during litigation in a criminal or civil court. Quality control is an important component in fire debris analysis. Reports of findings should be written in a scientific manner, describing the fire under investigation, evidence handling, a description of the evidence and where it was collected, the isolation procedure and what testing was done with what kind of equipment, observations made, and conclusions, with a discussion of the meaning of the results.

3.1	Introduction	42
3.2	Evolution of separation techniques	43
3.3	Evolution of analytical techniques	47
3.4	Evolution of standard methods	49
3.5	Isolating the residue	51
	3.5.1 Initial sample evaluation	51
	3.5.2 ILR isolation method selection	51
	3.5.3 Solvent selection	54
	3.5.4 Internal standards	54
	3.5.5 Advantages and disadvantages of isolation methods	56
3.6	Analyzing the isolated ILR	56
	3.6.1 Criteria for identification	63
	3.6.2 Improving sensitivity	90

[†]This chapter is essentially an updated version of the chapter entitled "Analysis of Ignitable Liquid Residues" that appears in the author's textbook, *Scientific Protocols for Fire Investigation* (CRC Press, 2006), and is used here with permission.

Forensic Chemistry Handbook, First Edition. Edited by Lawrence Kobilinsky.
© 2012 John Wiley & Sons, Inc. Published 2012 by John Wiley & Sons, Inc.

	3.6.3 Estimating the degree of evaporation	95
	3.6.4 Identity of source	98
3.7	Reporting procedures	101
3.8	Record keeping	102
3.9	Quality assurance	105
	Conclusion	105
	References	106

3.1 INTRODUCTION

The laboratory analysis of fire debris is one of the most important hypothesis tests that can be performed in an investigation, especially when the investigator forms a hypothesis that the fire was set using ignitable liquids. It has been widely acknowledged that making a determination that ignitable liquids were used in a fully involved compartment is not valid based on visual observation alone (NFPA 921, 2011). A laboratory analysis is necessary. Even when the compartment is not fully involved, and there is an exceptional burn pattern similar to the one shown in Figure 3.1, the investigator still needs to determine the identity of the ignitable liquid.

In the past, the laboratory analysis was referred to as "the icing on the cake," because by the time samples were collected, the fire investigator had already decided what caused the fire, and the stated purpose of the laboratory analysis was merely to help determine the identity of the flammable or combustible liquid used to start the fire. Investigators were accustomed to receiving negative reports from their laboratory, even when they "knew" that a fire had been intentionally set with ignitable liquids.

In those days, the term *ignitable liquid* had not yet been coined. *Flammable or combustible liquids* were generally referred to as *accelerants*, even by people in the laboratory who had no idea how such liquids may have been used. Many of the findings were, in fact, false negatives, because laboratory methods were not sensitive enough to detect ignitable liquid residues at low levels. The sensitivity has improved dramatically since this author began distilling fire debris in 1974, so much so that it is now possible to detect the petroleum products that are a natural part of the background. Many fire investigators, however, still distrust negative reports from laboratories, based on this earlier experience.

The development of analytical procedures has paralleled the development of standards for fire debris analysis. The first vague outline of a standard was not published until 1982 (AA Notes, 1982), and prior to that, analysts would report that a sample contained "an oily liquid" that exhibited "sufficient similarities" to a known sample of gasoline or kerosene or diesel fuel. What was sufficient, however, was not clearly defined until there were standards.

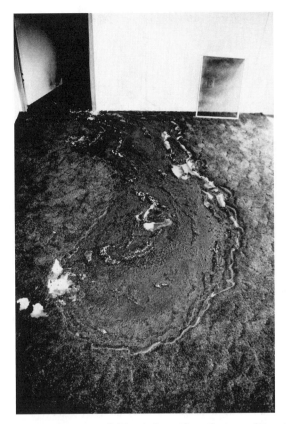

Figure 3.1 "Obvious pour pattern." The only surface that was burned in this mobile home was the floor. There were no furnishings in the house. The carpet was tested and found to be positive for the presence of a medium petroleum distillate such as mineral spirits or charcoal lighter fluid. This is a unique case where visual observation alone can lead to valid conclusions about what caused the pattern. *(See insert for color representation.)*

3.2 EVOLUTION OF SEPARATION TECHNIQUES

Paul Kirk reported in 1969 that the normal manner of isolating a liquid accelerant from other materials is "to distill the liquid from a solid residue in a current of steam." Kirk reported that the distillate would then be subjected to fractional distillation, flash point, refractive index, or density determination, but that a better procedure was to employ a gas–liquid chromatograph. Kirk stated, "From all of these laboratory procedures, the most important single piece of information that is made available is that a foreign flammable liquid was present at the fire scene. This alone is strong evidence for arson, at least after the possibility of accidental placing of the liquid is eliminated" (Kirk, 1969). The operative word here is *foreign*.

Figure 3.2 shows a steam distillation apparatus for the isolation of ignitable liquid residue from fire debris. In this classical separation technique, the debris is

44 ANALYSIS OF FIRE DEBRIS

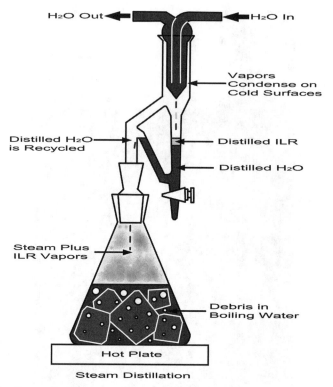

Figure 3.2 Steam distillation apparatus. The sample is boiled. Vapors condense on a "cold finger" and fall into the trap, which allows the water to recycle while the immiscible oil layer builds up on top of the water column.

covered with water and boiled, and the steam and other vapors are condensed in a trap that recycles the water and allows any nonmiscible oily liquids to float on top. There is actually a visible layer of liquid isolated from the sample. More often than not, this layer consisted of a drop or two, or simply a rainbow film on top of the column of water. This could be extracted with a solvent and analyzed, but even so, steam distillation was not a very sensitive technique. If the sample did not exhibit a detectable petroleum odor, steam distillation was almost always ineffective at isolating any ignitable liquid. Despite its lack of sensitivity, steam distillation did have the advantage, given a sufficiently concentrated sample, to produce a visible layer of liquid that could be shown to a jury.

In 1969, Kirk suggested that the debris could be heated in a closed container and the internal gaseous phase could be sampled and analyzed by gas chromatography, but indicated that he was unaware of its use in routine analyses (Kirk, 1969). By the mid-1970s, this technique, known as *heated headspace* (Figure 3.3), was used routinely but had sensitivity limitations similar to those of steam distillation and, because of the low volatility of higher-molecular-weight compounds, was ineffective at isolating the heavier components of common combustible liquids

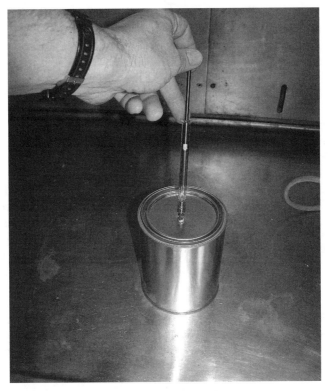

Figure 3.3 Using a gastight syringe to withdraw a sample of headspace from a fire debris sample. The headspace sample, about 500 µL, is injected directly into the GC–MS injection port.

such as diesel fuel. Heated headspace (Kirk called it a "shortcut") is still used as a screening tool in some laboratories today.

The first quantum jump in sensitivity took place in 1979, when two chemists from the Bureau of Alcohol and Firearms Philadelphia laboratory reported in the *Arson Analysis Newsletter* on the separation process known as *dynamic headspace concentration*. They used a dry nitrogen purge and a vacuum pump to draw ignitable liquid vapors from a heated sample through a Pasteur pipette filled with activated coconut charcoal (Chrostowski and Holmes, 1979). The vapors were rinsed off the charcoal using carbon disulfide and analyzed by gas chromatography. Over the next decade, the apparatus for conducting this type of analysis resulted in the publication of many articles describing newer and more wonderful apparatuses. Dynamic headspace concentration is still a recognized analytical technique, but because it is both destructive and complicated, it is not used by many laboratories.

In 1982, Juhala made the case for using *passive headspace concentration* on fire debris, wherein an adsorbent package is placed in the sample container and heated up. He used charcoal-coated copper wires and Plexiglas beads, and reported an increase in sensitivity of two orders of magnitude over distillation and heated headspace analysis (Juhala, 1982). Many laboratories had just completed setting up their dynamic systems; however, adoption of passive headspace concentration took some time, but gradually its advantages made it the dominant method of separation. Dietz reported on an improved package for the adsorbent called C-bags (Dietz, 1993), but these quickly gave way to activated carbon strips (ACSs), which required much less preparation. In 1993, Waters and Palmer reported on the essentially nondestructive nature of ACS analysis, performing up to five consecutive analyses on the same sample with little discernible change and no change in the ultimate classification of the residue (Waters and Palmer, 1991). This separation technique is the method of choice in most laboratories today. Figure 3.4 is a conceptual drawing of the procedure for passive headspace concentration, and Figure 3.5 is a photograph of a typical adsorption device, which consists of a 10 mm × 10 mm square of finely divided activated charcoal impregnated on a polytetrafluoroethylene (PTFE) strip. The technique was actually adapted from the industrial hygiene industry. Charcoal disk badges are worn by employees to determine their exposure to hazardous chemicals. The charcoal adsorbs a wide variety of organic compounds.

Solid-phase microextraction (SPME) represents yet another kind of passive headspace concentration technique. The SPME fiber is a more active adsorber of most ignitable liquid residues than is an activated carbon strip. Exposing an ACS to the headspace of a sample at elevated temperatures for 16 h allows for the isolation of less than 0.1 μL of ignitable liquid residue if there is no competition from

Figure 3.4 Schematic drawing of passive headspace concentration using an activated carbon strip. Vapors are produced by heating the container with debris to 80°C. The ACS adsorbs the vapors for 16 h, then is rinsed with diethyl ether spiked with 100 ppm, and the resulting solution is analyzed by GC–MS.

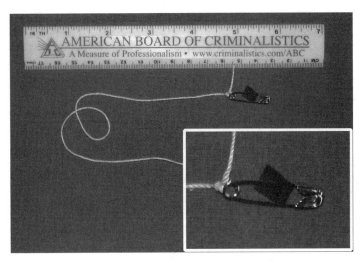

Figure 3.5 Close-up view of an activated carbon strip.

the substrate. A SPME fiber can accomplish the same task in 20 min or less. The advantages and disadvantages of the various separation techniques are discussed later in the chapter.

3.3 EVOLUTION OF ANALYTICAL TECHNIQUES

As the ability of the separation step to isolate smaller quantities of ignitable liquid residue improved, the sensitivity of the analytical instruments improved as well. In the 1950s and 1960s, extracts were analyzed by infrared (IR) or ultraviolet (UV) spectroscopy (Midkiff, 1982), but because most extracts were mixtures, these techniques were neither very sensitive nor very specific. The IR spectrum of gasoline looks very much like the IR spectrum of kerosene. Gas chromatography (GC) using pattern recognition techniques became the analytical method of choice beginning in the late 1960s. GC is actually a separation technique rather than an identification technique, but unlike separating the ignitable liquid residue from the sample matrix, GC works by separating similar compounds in an extract from each other.

In the 1970s, GC columns, the engine that makes the technique work, were glass or metal tubes, $\frac{1}{4}$ in. in diameter by 6 to 10 ft long. Chemists typically purchased empty columns and packed the columns themselves, using a coated powdery substance (stationary phase). It was known that $\frac{1}{8}$-in. columns provided better resolution than $\frac{1}{4}$-in. columns, but these had to be made from metal, and the chemist could not see inside the tube to check for gaps. These columns were usually purchased already packed. As the column manufacturers experimented with narrower and narrower columns, they went back to drawn-glass tubes coated on the inside with the oily stationary phase. There were problems with these early capillary columns, not the least of which was the forensic science community's resistance to change. As the bugs were worked out, capillary columns became the standard choice, but as late as 1990, packed columns were contemplated in ASTM E1387, *Standard Test Method for Flammable or Combustible Liquid Residues in Extracts from Fire Debris Samples by Gas Chromatography*. In the 1995 edition the standard "recommended," but did not require, capillary columns. The 2001 edition requires "A capillary, bonded phase, methylsilicone or phenylmethylsilicone column or equivalent. Any column length or temperature program conditions may be used provided that each component of the test mixture is adequately separated" (ASTM E1387-01, 2001). Today, very few laboratories use packed columns.

Gas chromatography detectors originally measured the change in thermal conductivity (TC) of the effluent from the column. Flame ionization detection (FID) improved the sensitivity by a couple of orders of magnitude over TC detectors. In a flame ionization detector, there is a hydrogen flame burning between two charged plates. As indicated by the flow of current between the two plates, the electrical conductivity changes when a hydrocarbon compound comes through the hydrogen flame and is burned.

Some laboratories were using gas chromatography–mass spectrometry as early as 1976 (GC–MS) (Stone, 1976). At that time, mass spectrometers were expensive, not terribly reliable, and required a computer (this was before the days of the personal computer, when computers took up half of the room), and GC–MS was the exception rather than the rule. With the earlier instruments, the operator watched a stripchart recorder and pushed a button to collect a mass spectrum when a peak indicated that a compound was coming off. This was a very labor-intensive process. There were people who argued that chemists had an obligation to use the best technology available, and the advantages of GC–MS over GC–FID required that MS be used. One of the leading proponents of GC–MS was Jack Nowicki, who also noted that GC–MS would make the previous accelerant classification system obsolete (Nowicki, 1990). He was eventually proved correct. Most laboratories stayed with the FID methods because of the difficulties with implementing GC–MS and because they felt comfortable with their ability to read patterns using FID alone. By the early 1980s, mass spectrometry was still expensive, but its use had become more widespread in fire debris analysis, particularly in the better-funded laboratories. Public laboratories acquired GC–MS instruments for use in drug identification, and this was another reason that they became available for fire debris analysis. The instrumentation of the 1980s was more automated and could collect a mass spectrum several times per second, even if no peak was eluting. This resulted in a much more efficient process, but the data files were very large. Today's GC–MS, collecting data every tenth of a second, uses sophisticated software to keep the file size to around a megabyte, an amazing feat considering that 18,000 spectra may be collected during a single run.

In 1982, Martin Smith published an article about a technique he had developed, called *mass chromatography*, which utilized a computer to separate the mass spectral signals according to the functional groups of the compounds that produced them. This technique allowed chemists to view many simple and easy-to-recognize patterns as opposed to looking at one large complicated pattern (Smith, 1982).

Today this powerful analytical tool is known as *extracted ion profiling* or *extracted ion chromatography* and forms the basis of most identifications.

The development of the personal computer made it possible for average laboratories to control a mass spectrometer, and the instrument manufacturers responded to the demand by producing benchtop models with increasing sensitivity and extraordinary robustness. The *quadrupole mass filter*, which is at the heart of the most popular mass selective detectors, has no moving parts. Only those parts of the instrument required to maintain a vacuum are subject to mechanical problems.

3.4 EVOLUTION OF STANDARD METHODS

As fire debris analysis technology improved, so did the approach of the forensic science community to the problems encountered by analysts. *Arson Analysis Newsletter* continued publication through 1984, and much valuable information was exchanged. Through this informal journal, forensic scientists analyzing fire debris had a means of communicating with each other that was unavailable to scientists in many other disciplines. In 1982, based on work conducted at the Center for Fire Research at the National Bureau of Standards (NBS) (now the NIST) and the ATF National Laboratory, an *accelerant classification system* was published (AA Notes, 1982). Not only was there a description of five classes recognized as "usually identifiable by GC–FID patterning alone when recovered from fire debris," but the authors (who were not named in the publication but are believed to include Philip Wineman and Mary Lou Fultz) also published minimum requirements for class identification. This was the first time that anyone explained what "sufficient similarities" should mean and was a watershed moment in the history of fire debris analysis. Although the original publication of the classification system stated that the final report was not yet available and that the results of the evaluation would be printed in a future issue of the *Arson Analysis Newsletter*, that never happened. The classification system was used informally for the next six years.

The International Association of Arson Investigators (IAAI) has had, almost since its beginning, a standing committee of forensic scientists and engineers that it called upon to advise fire investigators about laboratory analysis issues. In 1987, IAAI President John Primrose approached the Forensic Science Committee and requested that it produce a position paper on what should appear in a laboratory report. It quickly became apparent to members of the committee that in order to prescribe the contents of a report, it would first be necessary to set down an acceptable method of analysis. Four sample preparation techniques (steam distillation, headspace analysis, solvent extraction, and dynamic headspace–purge and trap) were the separation techniques described. Gas chromatography with flame ionization detection, photo ionization, or mass spectral detection was required. Although the publication was called a "guideline," it contained the following sentence: "Unless a petroleum distillate has been identified by the pattern recognition techniques described below, it has not been sufficiently identified" (IAAI Forensic Science Committee, 1988). The guidelines then reproduced the NBS–ATF classification and identification scheme and described how some materials would not fall within the guidelines. Isoparaffinic hydrocarbons were cited as one example of an ignitable liquid not described in the guidelines. At that time, the only place that a fire debris chemist would be likely to encounter isoparaffinic hydrocarbons was in Gulflite charcoal starter fluid. With the introduction of the IAAI guidelines, fire debris chemists became accustomed to the idea that they needed to follow standard methods. In the private sector, laboratories advertised to the membership of the IAAI, their main source of clients, that they followed the guidelines. Laboratories that did not follow the guidelines lost business.

In 1990, ASTM Committee E30 on Forensic Sciences took the IAAI guidelines and formulated them into six different standards for the preparation, cleanup, and analysis of fire debris extracts. ASTM E1387, the *Standard Test Method for Flammable or Combustible Liquid Residues in Extracts from Fire Debris Samples by Gas Chromatography*, was published originally in 1990. In 1995, the cumbersome phrase "flammable or combustible liquid" was changed to "ignitable liquid." In 1993, Committee E30 enlisted the aid of Martin Smith, Jack Nowicki, and several other prominent chemists to draft a *Guide for Fire Debris Analysis by Gas Chromatography–Mass Spectrometry*. The guide was revised in 1997, and in 2001 it was "promoted" to a standard test method.

The ignitable liquid classification scheme from 1982 was updated in an attempt to keep up with the ever-changing output of the petrochemical industry. New products were constantly being brought on line, including "environmentally friendly" alternatives to solvents such as mineral spirits and fuels such as charcoal lighter fluid. When ASTM E1387 was first published, a "Class 0" was added to the original classification scheme to account for the liquids that did not fit into one of the five original classes. Further classification within Class 0 was, however, possible, so Classes 0.1 through 0.5 appeared in ASTM E1387-95. By the time the next revision was due, Class 0.6 had been created for de-aromatized distillates, but the committee realized it was time for a change, as the miscellaneous classes now outnumbered the original classes. In 2000, the system was completely redesigned, resulting in nine differently named but no longer numbered classes, with subdivisions of light (C_4-C_9), medium (C_8-C_{13}), and heavy (C_8-C_{20+}) in eight of those nine classes (ASTM E1618-10, 2010).

In 2009, after much discussion, Committee E30 on Forensic Sciences made the decision that analysis of ILRs by GC alone no longer represented the "best practice." ASTM E1387 was allowed to expire and become a "historical" standard. Most fire debris analysts had long ago made the transition to GS-MS. As of this writing, gas chromatography-mass spectrometry is the only generally accepted method for analyzing ignitable liquid residues.

The last 30 years have seen dramatic improvements in separation technology, in analytical technology, and in the scientific community's approach to fire debris analysis. In 1999, the U.S. Department of Justice Office of Law Enforcement Standards produced a report entitled *Forensic Sciences: Review of Status and Needs*, compiled by more than 40 eminent forensic scientists. In reviewing the state of the art, fire debris analysis was described as "a sub discipline of trace analysis that is in good standing because there is sufficient published work on the analysis and interpretation of the material involved. Standard guides for the examination and interpretation of chemical residues in fire debris have been published through the consensus process of ASTM Committee E30 on Forensic Science. These standardization documents are often quoted in the scientific literature, helping to meet the requirements of the legal community" (U.S. Department of Justice, 1999). Most of the other forensic disciplines discussed in this review were reported as still needing standardization and/or validation of standard methods. Fire debris analysts can point to a history of standardization that existed even before the *Daubert* court made it a necessity.

3.5 ISOLATING THE RESIDUE

3.5.1 Initial Sample Evaluation

Once the receipt of a sample has been documented, and the chain of custody protected (as described in ASTM E1492-05, 2005) the first critical step in the analysis of fire debris is the selection of a separation technique. Choosing an inappropriate technique could result in a false negative, a misidentification, or the destruction of evidence. The first step in this selection process (and the first step in any chemical analysis process) is to *look at the sample*. One purpose of looking at the sample is to ensure that it is what it purports to be. Sample characteristics will determine the most appropriate method for isolating any ignitable liquid residues (ILRs) that may be present. Once the visual examination has taken place, the next step is a "nasal appraisal." Occupational safety experts will no doubt frown on this recommendation, but it can be done carefully. There is no need to put one's nose in the can, even though the analyst can be reasonably certain that the fire investigator who collected the sample has already done exactly that. Unless the sample is a liquid sample for comparison purposes, it can be safely appraised by removing the lid, and waving the hand gently over the top of the sample to see if it exhibits any obvious odors. If there is an odor present, it becomes possible to do a rapid and accurate analysis by removing a small piece of the sample and extracting it with solvent. If the odor is very strong, it is advisable to remove a small piece of the sample and place it into a separate container for analysis.

3.5.2 ILR Isolation Method Selection

Solvent extraction according to ASTM E1386 is an appropriate method for rinsing out empty containers, for extracting small aliquots of samples with a high concentration of ILR, and for isolating residues from very small samples. Not every investigator who comes to the laboratory will have the experience and knowledge to know how to find the best samples, and it is not unusual for an inexperienced investigator to bring in a sample in a film canister.

The vast majority of samples, however, are likely to be samples of burned building materials, floor coverings, and furnishings that do not exhibit a strong odor and are best analyzed by passive headspace concentration, as described in ASTM E1412 (ASTM E1412-07, 2007). This technique is essentially nondestructive. If the analyst decides to use another technique later, running passive headspace concentration will not interfere with that.

Other methods of isolation have been studied thoroughly over the last 30 years, and although they have some utility, none match the advantages of passive headspace concentration using an activated carbon strip. Headspace sampling, described in ASTM E1388, which involves warming the container and sampling the vapors in the headspace directly and then injecting those vapors into the gas chromatograph, is a good screening technique, but it does not result in the production of an archiveable extract, nor does it detect compounds much heavier than C_{15} (ASTM E1388-05, 2005). Dynamic headspace concentration, described in ASTM E1413, was useful in demonstrating the effectiveness of adsorption or elution as a valid approach to ILR isolation, but it is destructive, requires far more attention than passive headspace concentration, and the apparatus can be cumbersome and finicky. It is no more sensitive than passive headspace concentration, and it is possible for *breakthrough*, loss of analyte out the effluent end of the tube, to occur (ASTM E1413-07, 2007). The only advantage that it offers is speed. If the results can be reported in 24 h, there is no advantage at all. Most laboratories (unfortunately) take several days or weeks to report on a fire debris analysis, due to sample backlogs. Solid-phase microextraction is another alternative, but like dynamic headspace concentration, it is very labor intensive and, like headspace sampling, does not have the potential to produce an archiveable sample that can be analyzed again (ASTM E2154-01, 2008) Using passive headspace concentration results in a solution that can be injected many times, and when the carbon strip is left in the solution, it will gradually readsorb ignitable liquid residues as the eluting solvent evaporates. (Juhala reported the readsorption by small portions of the activated charcoal that fell off his Plexiglas beads in 1982.) The solution can be reconstituted years later if a second look at the sample is desired. Because of the transient nature of many fire debris sample containers, the archived activated carbon strip is often the best evidence after a few years have passed.

The only equipment required for passive headspace concentration is a convection oven, vials, ACS strips, and a solvent dispenser. Caseload will determine the required oven size. In our laboratory, the convection oven can hold up to ten 1-gallon cans and more than twenty 1-quart cans.

Every laboratory should optimize the parameters in its ACS procedure to make sure that they are getting the best results possible. "Good" results from ACS are those where the chromatogram of the concentrated headspace vapors of a standard closely matches the chromatogram of that same standard in the eluting solvent.

A 10 mm × 10 mm carbon strip is the minimum size recommended. This 100-mm^2 strip can easily accommodate the headspace vapors from 10 μL of any ignitable liquid placed on a Kimwipe in a 1-quart can. It is possible to overload carbon strips, and this results in the preferential adsorption of heavier hydrocarbons over light hydrocarbons and of aromatics over aliphatics, but this effect is generally not large enough to affect the identification. Usually, samples that are capable of overloading the carbon strip will exhibit a strong odor, and the analyst can take an aliquot of the sample or reduce the analysis time.

Figure 3.6 One-quart can equipped with a pressure relief device, a short strip of cellophane tape over a small hole pierced in the lid.

For samples with a high water content, there exists a danger that the vapor pressure in the sample container will cause the lid to pop off. This has the potential to contaminate the sample oven. A pressure relief device is easy to construct for such samples. Puncture a small hole in the lid, and cover with cellophane tape. Figure 3.6 shows a can so equipped.

The adsorption time of a typical ACS procedure is 16 h. The analyst is encouraged to experiment with different adsorption times and temperatures, with the goal of finding a balance between the maximum recovery and the minimum time necessary. Sixteen hours is convenient because the samples can be put in the oven at 4:00 P.M. and taken out at 8:00 A.M. the next day. One of the major advantages of ACS is that it requires very little attention from the analyst. Once the strip is in the can and the can is in the oven, nothing is going to happen until it is time to take the strip out of the can, put it in the vial, and add the eluting solvent. At this point in the analysis, however, the analyst must be extremely attentive to the procedure. Once the carbon strips are taken out of the sample container, they have an identical appearance. It is at the point of placing the strips into their vials that an unrecoverable error can occur—a strip can be misidentified. The analyst should not allow himself or herself to be distracted by phone calls or other people in the lab. The operative word here is *focus*.

Many laboratories use a pre-concentration step, wherein they add approximately 500 μL of solvent to the strip in a vial; then after the strip has had a chance to equilibrate, it is removed and the solvent is evaporated down to 100 μL or so. This will result in a fivefold increase in concentration, but that increase can usually be achieved electronically with very little loss of signal or increase in noise. The first analysis, in this analyst's view, does not require the pre-concentration step. That can be accomplished at a later time, if necessary, but pre-concentration runs the risk of skewing the results if not done very carefully. The only safe way to evaporate the solvent is to blow a stream of dry nitrogen over it. Heating the solvent is a bad idea, because lower-molecular-weight compounds are likely to be lost.

3.5.3 Solvent Selection

The solvent used for the elution is another critical choice, not so much in terms of the quality of results but in terms of the analyst's quality of life. The most popular eluting solvent is carbon disulfide, a highly toxic, carcinogenic, teratogenic, smelly, nasty liquid that will ignite upon exposure to boiling water. It does work very well to elute aromatics and aliphatics approximately equally from ACSs, but so does diethyl ether. Studies indicate that carbon disulfide is superior to diethyl ether, or to pentane, the other solvent recommended by ASTM Committee E30 (Newman and Dolan, 2001), but if standards are prepared using diethyl ether, the slight change in the chromatographic profile does not affect the identification. Armstrong and Lentini (1997) found only marginal differences between diethyl ether and carbon disulfide when applied to carbon strips exposed to 10-μL samples of ignitable liquid residues. Comparing the relative health risks makes diethyl ether an obvious choice for this analysis.

Carbon disulfide was originally selected as a fire debris solvent because of its high desorption efficiency and its relatively quiet signal when passing through a flame ionization detector. When using a mass spectral detector, the advantage of its low signal disappears because the detector is turned off while the solvent is passing through.

Some concerns have been expressed about the capability of diethyl ether to form explosive peroxides, but that will not occur if the ether is kept in a refrigerator and used on a regular basis. Explosions of cans of ether have only been reported when those cans have been allowed to sit for years, unused, in the back of an unrefrigerated stockroom. Carbon disulfide, diethyl ether, and pentane are all highly flammable; however, with respect to fire, carbon disulfide poses the greatest risk, as it has the lowest ignition temperature and the broadest flammable limits (NFPA 325, 1994). Finally, carbon disulfide costs almost 10 times what the other solvents cost. A comparison of the properties of the three solvents recommended by ASTM E1412 is shown in Table 3.1.

3.5.4 Internal Standards

There are two places in the analysis of fire debris where the use of internal standards is appropriate. Addition of an internal standard to the sample itself allows the analyst to develop at least a qualitative feel for the "tenacity" of the sample and for

TABLE 3.1 Elution Solvent Comparison[a]

	Carbon Disulfide	Diethyl Ether	Pentane
Flash point	−30°C (−22°F)	−45°C (−49°F)	−40°C (−40°F)
Lower elution limit (LEL) (vol% in air)	1.0	1.9	1.5
Upper elution limit (UEL) (vol% in air)	50.0	48.0	7.8
Specific gravity	1.3	0.7	0.9
Boiling point	40°C (115°F)	35°C (95°F)	36°C (97°F)
Autoignition temp.	90°C (194°F)	180°C (356°F)	260°C (500°F)
Exposure limit, TWA[b]	4 ppm	400 ppm	600 ppm
Exposure limit, STEL[c]	12 ppm	500 ppm	750 ppm
Carcinogenic	yes	no	no
Teratogenic	yes	no	no
IDLH	500 ppm	19,000 ppm (LEL)	15,000 ppm (LEL)
FID signal	small	large	very large
Cost per liter[d]	$399.7.70	$44.25	$38.10

[a] Refer to your laboratory's hazard communication literature or MSDS for more complete information.
[b] Time-weighted average for an 8-h exposure.
[c] Short-term exposure limit (up to 15 min).
[d] Source: J.T. Baker, online catalogue, Aug. 18, 2007. All prices are for case quantities of Ultra Resi–analyzed grade.

the effectiveness of the isolation procedure. In our laboratory, this is accomplished by the addition of 0.5 μL of 3-phenyltoluene (actually, 20 μL of a 2.5% solution of 3-phenyltoluene in ether). In the eluting solvent, we use a second internal standard consisting of 0.1% (100 ppm) perchloroethylene. If the 3-phenyltoluene does not appear in the chromatogram, this means that we have an exceedingly tenacious sample (some samples come equipped with their own active sites, which can compete very effectively with the ACS for analyte molecules) and suggests that the tenacity of the sample might be the reason that the chromatogram appears so flat. If the perchloroethylene peak does not appear, or is significantly reduced in height, something has gone wrong with the injection.

Comparison of the signal from the sample to the perchloroethylene signal allows for a semiquantitative determination of the amount of ILR present. For 10-μL standards of known ignitable liquids isolated according to ASTM E1412, the two internal standard peaks are roughly on the same order of magnitude as the sample peaks. Essential blanks include the 3-phenyltoluene applied to filter paper and the blank strip eluted with the spiked solvent. Some analysts may perceive the danger of being accused of "contaminating" a sample, but this is easily overcome by having a proper blank in the file. The advantages of using internal standards far outweigh this disadvantage.

3.5.5 Advantages and Disadvantages of Isolation Methods

Two drawbacks have been cited for use of the ACS method, the time required to perform the adsorption, and the relative lack of sensitivity compared to SPME. If laboratories were in the habit of providing same-day service, the first argument might have merit. Turnaround times generally range from two days to two months. In that context, a 16-h versus a 15-min analysis time is meaningless. With respect to sensitivity, the ACS method is capable of routinely detecting 0.1 μL of ignitable liquid from a nontenacious background, which should be low enough. Our goal is to help the fire investigator understand whether a foreign ignitable liquid was present at the fire scene. We now have the ability to detect the solvent in polyurethane finish five years or more after it has been applied to a hardwood floor. We have no need to be more sensitive than that. A technique that is capable of adsorbing significant quantities of ignitable liquid residue in 15 min is also capable of becoming contaminated much more easily than a carbon strip that might take 16 h to come to equilibrium with a dilute sample.

The use of a "screening" technique for fire debris samples is an issue that each laboratory should address. There is usually not a need to screen samples, since an ACS separation is likely to be more sensitive. A sample that tests negative on a quick headspace analysis (per ASTM E1388) needs to be tested further anyway. If a request is made to look for light oxygenates (alcohols, acetone) or to get a ballpark estimate of analyte concentration (other than through a "nasal appraisal"), running a headspace can be useful. Screening techniques also allow for a swift, if less than definitive, result. For routine analyses, however, passive headspace concentration, conducted according to ASTM E1412, should be the norm. Solvent extraction, as described in ASTM E1386, is appropriate for sampling aliquots of very strong samples or for extracting very small samples or empty containers (ASTM E1386-10, 2010).

Steam distillation may be selected in the odd case where it is desired to produce a neat liquid extract of the fire debris. The benefit of this is that a vial of the liquid can be brought into a courtroom, shown to a jury, lit on fire on a Q-tip, and passed around. Because steam distillation is appropriate only on very concentrated samples, however, it is preferable to make sure that the sample is preserved, and the sample itself can be passed around for the jury to smell. For the most part, steam distillation is a technique whose time has come and gone. The last reapproval of an ASTM steam distillation standard took place in 2001, and the Committee on Forensic Sciences in 2006 took the decision to allow the standard to expire. A comparison of the advantages and disadvantages of the various isolation techniques is shown in Table 3.2.

3.6 ANALYZING THE ISOLATED ILR

Despite the improvements in separation and detection technology, the overall approach to identification of ignitable liquid residues is the same as it was in the early 1970s. A chromatogram from the sample is compared with chromatograms

TABLE 3.2 Comparison of ILR Isolation Techniques

Method	Advantages	Disadvantages
E1385, steam distillation	Produces a visible liquid, simple to explain	Labor intensive, destructive, not sensitive, requires expensive glassware
E1386, solvent extraction	Useful for small samples and empty containers, does not cause significant fractionation, useful for distinguishing heavy petroleum distillates from each other	Labor intensive, expensive, co-extracts nonvolatile substances, increased risk of fire, solvent exposure, destructive
E1388, headspace sampling	Rapid, more sensitive to lower alcohols, nondestructive	No archiveable sample, not sensitive to heavier compounds, poor reproducibility
E1412, passive headspace	Requires little analyst attention, sensitive, nondestructive, produces archiveable sample, inexpensive	Requires overnight sampling time
E1413, dynamic headspace using activated charcoal	Rapid, sensitive, produces archiveable sample, inexpensive	Labor intensive, subject to breakthrough, destructive
E1413, dynamic headspace using Tenax	Rapid, sensitive	Labor intensive, requires thermal desorption, no archiveable sample, destructive
E2154, solid-phase microextraction	Rapid, highly sensitive, useful for field sampling with portable GC–MS	Labor intensive, expensive, requires special injection port, reuse of fibers, no archiveable sample

from known standards, and the analyst determines whether there are "sufficient similarities" to make an identification. What has changed is the quantity of information available because of the increased resolution provided by capillary columns and the ability to obtain a mass spectrum up to 10 times per second, as well as reaching a consensus on the meaning of "sufficient." So, although there is more information to compare, the technique is still one of pattern recognition and pattern matching.

An argument can be made that when one looks at a mass spectrum, one is looking at structural details rather than simply matching patterns, but the patterns still have to match. The same argument has been made about structural elucidation in the use of FTIR for drug identification. Although it would be nice to think that analysts routinely consider molecular structure, the day-to-day operation is one of pattern matching.

There exists a specific skill set that is required to compare chromatographic patterns. This skill set must be learned carefully and used routinely if it is to remain sharp. Pattern recognition is one "scientific" skill that has historically been problematic. Although fire debris analysts use the same equipment as drug analysts, the drug analyst typically looks for a single peak with a particular retention time. The fire debris analyst must compare an entire pattern of peaks produced by a sample to the pattern of peaks produced by a standard. This exercise includes a comparison of inter- and intragroup peak heights and can be quite complex. Although it is true that many drug analysts also conduct ignitable liquid residue analyses, a different set of skills is involved.

Fire debris analysis also involves a set of skills different from those employed by environmental scientists, who are typically trying to quantify the components of an oil spill or contaminants at a Superfund site. Environmental methods typically assume the presence of gasoline or other petroleum products, then look for benzene, toluene, ethylbenzene, and xylene (BTEX) to quantify the amount present. Unless they are trying to identify the source of the spill, environmental analysts are usually not employing the same skill set used by fire debris analysts.

If one thinks about the nature of many petroleum products and the processes that are going on when those products are isolated from debris samples, one can begin to understand why chromatographic patterns look the way they do. Many petroleum products are straight-run distillates, particularly the medium and heavy petroleum distillates. The overall pattern of these products is a Gaussian (bell-shaped) distribution of peaks, dominated by the normal alkanes. The patterns produced by kerosene and diesel fuel have been likened to a stegosaurus, because the chromatograms bear a passing resemblance to the dinosaur's dorsal fin. (Actually, a spinosaurus' sail back fin more closely resembles the pattern, but stegosaurus is the term of art.). A medium petroleum distillate produces the same pattern, shrunken and coming off early.

A fractionation process similar to distillation occurs in the isolation of an ignitable liquid residue from a sample. This is caused by the very low vapor pressure of ILR components above C_{18} and by the selective adsorption of the heavier hydrocarbons on complex substrates. Hydrocarbons up to C_{23} can be captured from diesel fuel placed on a noncompeting substrate such as filter paper. That same fuel placed on charred wood, however, may exhibit a pattern that ends at C_{18}. If compounds do not get into the air in the headspace, they will not be adsorbed onto the carbon strip. The tenacity of the sample needs to be considered, particularly when trying to distinguish between kerosene and diesel fuel.

This author learned pattern recognition the old-fashioned way: running dozens, then hundreds of standards, and learning what the patterns looked like. Today's analysts are fortunate in that the ASTM standards give examples of many of the patterns that an analyst is likely to see in positive samples, and there exists a detailed compilation of literally hundreds of patterns available in a standard text (Newman et al., 1998). Both the ASTM standards and the *GC–MS Guide to Ignitable Liquids* provide sufficient information to allow the analyst to set up an instrument to provide patterns that look very much like the ones in the texts. Although these texts are an important resource, it is imperative that every fire debris analysis laboratory have its own library of ignitable liquid residues. This provides for patterns with exactly the

TABLE 3.3 Ions Used for Extracted Ion Chromatography

Ion	Compounds	Ion	Compounds
57	Alkanes	119	C_4 alkylbenzenes
83	Cycloalkanes	131	Methyl, dimethylindans
91	Toluene, xylenes	142	Methylnaphthalenes
105	C_3 alkylbenzenes	156	Dimethylnaphthalenes
117	Indan, methylindans		

same retention times and mass spectra with exactly the same fragmentation patterns. The in-house library is also a quality assurance tool, which lets the analyst know when there has been some drift in the instrument, when a sensitivity loss has taken place, and when it is time to run a new set of standards. Whatever approach is taken, it will take time to develop the ability to recognize ILR patterns.

The mass spectrometer provides the ability to simplify what can be very complex and confusing patterns. This is a result of the ability of the data analysis software to separate out only those peaks having particular ions present in the pattern. For example, if one wants to look at the alkanes, one only needs to obtain an extracted ion chromatogram for *m/z* (mass/charge ratio) 57, hereafter referred to as ion 57, and most of the balance of the components will disappear. It is thus possible to break down a total ion chromatogram (TIC) into its component parts—see Table 3.3.

Although there are more patterns to learn, they are simpler patterns and easier to remember. This approach to data analysis, known as *mass chromatography*, was first proposed by Smith in 1982. There are basically two ways to approach mass chromatography: the single-ion approach and the multiple-ion approach. Dolan (2004) has proposed referring to the single-ion chromatograms as *extracted ion chromatograms* (EICs) and multiple-ion chromatograms as *extracted ion profiles* (EIPs).

When working with software that scales the extracted ion chromatogram or profile to the tallest peak, some caution is advised when using multiple ions. To the extent that the second, third, and fourth ions contribute to a pattern, they tend to make it more complicated, thus defeating some of the purpose of extracting the ions in the first place. To the extent that these additional ions do not change the pattern, they may convince the analyst that he or she is seeing more than is actually present. Finally, to the extent that these additional ions are present at substantially lower concentrations, they would be better observed on their own rather than in the profile. Figure 3.7 is a comparison of ion 57 from a kerosene standard versus a profile based on ions 57, 71, 85, and 99. The profile is slightly more complicated, but since ion 57 is the base peak for almost every component in the chromatogram, the ion 57 profile is the tallest.

Another example is shown in Figure 3.8, which presents ions 128, 142, and 156, the naphthalenes, from a gasoline sample. In the top chart, the ions are combined into a profile, which although it gives the analyst an idea about the relative abundance of the three ions, shows very little detail for the dimethylnaphthalenes represented by ion 156. When the extracted ions are presented separately, the analyst still gets the quantitative data by reading the abundance numbers next to the

60 ANALYSIS OF FIRE DEBRIS

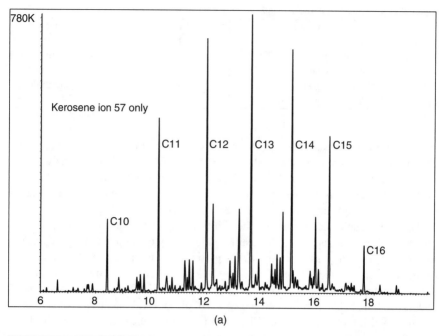

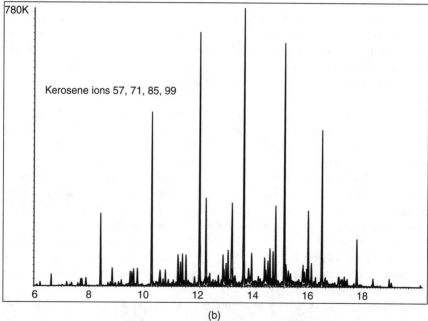

Figure 3.7 Ion profiling versus ion chromatography. Comparison of (a) the ion 57 chromatogram with (b) the ion profile combining ions 57, 71, 85, and 99 from a kerosene standard. The four ions in the profile are plotted in the merged format. Plotted individually, they are all very similar, with ion 57 presenting the tallest peaks.

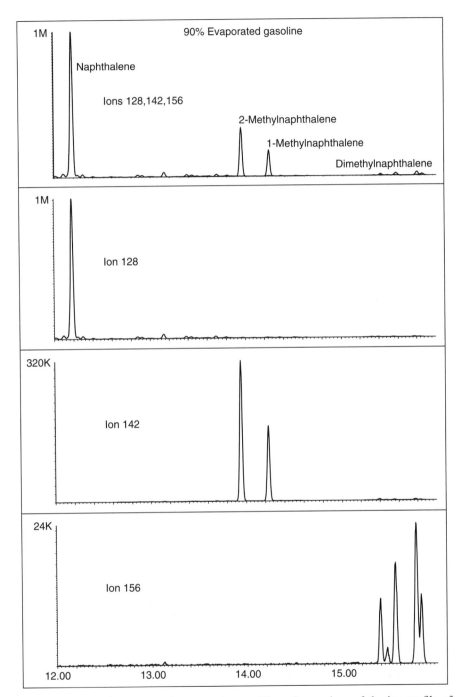

Figure 3.8 Ion chromatography versus ion profiling. Comparison of the ion profile of the naphthalenes from a gasoline standard (top chart). Compare with the detail provided by presenting the three ions (128, 142, and 156) separately.

Y-axis, but also gets to see the fine details in the shape of the peaks at the right side of the chart.

A similar example is shown in Figure 3.9(a). These charts show ions 91, 105, and 119 from a 75% evaporated gasoline standard. The effect is even more pronounced at 90% evaporation, shown in Figure 3.9(b), when the ion 91 peaks are smaller. Whether an analyst chooses extracted ion chromatography or extracted ion profiling is largely a matter of taste. It is easier for this author to use the simpler patterns

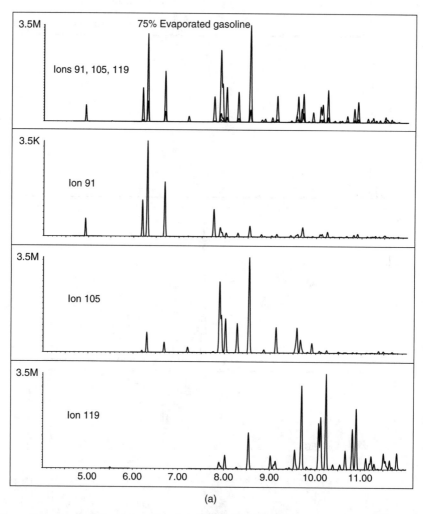

Figure 3.9 Ion profiling versus ion chromatography. (a) Comparison of ion profiles from 75%-evaporated gasoline. The top chart shows a combination of ions 91, 105, and 119. The next three charts show those ion chromatograms plotted independently. (b) Comparison of ion profiles from 90%-evaporated gasoline. The top chart shows a combination of ions 91, 105, and 119. The next three charts show those ion chromatograms plotted independently. Note the improvement in the level of detail presented for the xylenes.

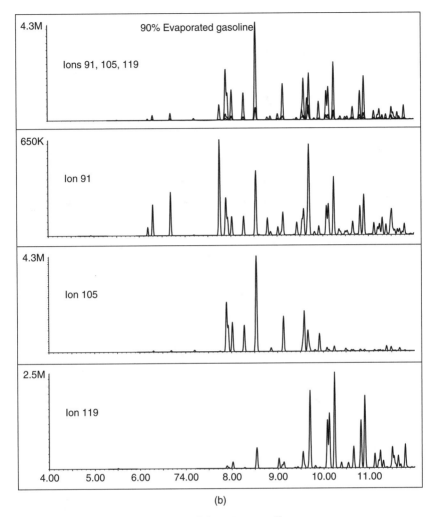

Figure 3.9 (*Continued*)

and avoid the eyestrain required to see the smaller peaks in the profiles. The typical set of six ion chromatograms used in our laboratory to document the presence of gasoline is shown in Figure 3.10, and the set of three ion chromatograms used for distillates is shown in Figure 3.11. Details of each are discussed below.

3.6.1 Criteria for Identification

Most of the chromatograms that an analyst uses to make a positive identification of an ignitable liquid residue will match (exhibit "sufficient similarities" to the standard) at the level of the TIC. Samples with low concentrations of ILR, high backgrounds, or both, can sometimes yield a positive identification for ILR if the

64 ANALYSIS OF FIRE DEBRIS

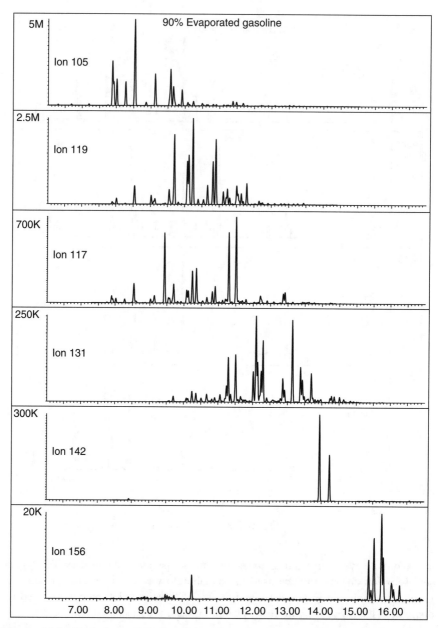

Figure 3.10 Typical set of six ion chromatograms used in our laboratory to document the presence of gasoline. Ions we use are 105 for C_3 alkylbenzenes, 119 for C_4 alkylbenzenes, 117 for indan and methyl indans, 131 for dimethyl indans, 142 for 2- and 1-methylnaphthalene, and 156 for dimethylnaphthalenes.

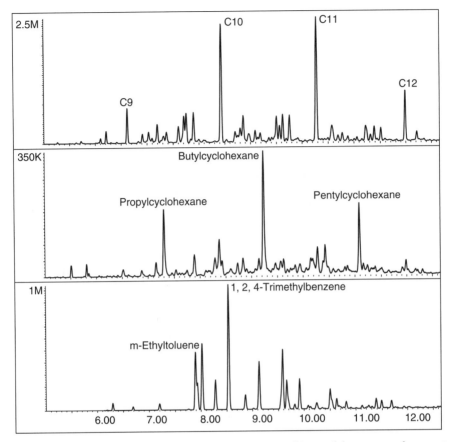

Figure 3.11 Typical set of three ion chromatograms used in our laboratory to document the presence of distillates. These are the ion 57, 83, and 105 chromatograms from Smokey Bear charcoal lighter, a medium petroleum distillate.

analyst is very careful. Usually, but not always, if the TIC from the sample does not match the TIC from the standard, the sample is likely to be negative (determined to contain no detectible ignitable liquid residue). Examples of cases where this is not true are presented following this general discussion of criteria for identification of an ILR in routine cases.

ASTM E1618 identifies eight classes of ignitable liquids identifiable by GC–MS:

1. Gasoline
2. Petroleum distillates
3. Isoparaffinic products
4. Aromatic products, including dearomatized distillates
5. Naphthenic–paraffinic products

6. N-Alkane products
7. Oxygenated solvents
8. Miscellaneous

With the exception of gasoline and oxygenated solvents, each of the classes above can be placed into one of three ranges: light (C_4–C_9), medium (C_8–C_{13}), and heavy (C_8–C_{20+}). This standard presents criteria for identifying the various kinds of compounds (alkanes, cycloalkanes, aromatics, and condensed ring aromatics) found in each one of the nine classes. The alkanes include both straight-chain and branched hydrocarbons, and can be extracted using ion 57. The cycloalkanes are mostly substituted cyclohexanes, which can be seen by extracting ion 83. Cycloalkanes also have a strong ion at 55, and as the length of the substituted alkane increases, ion 57 begins to dominate the mass spectra. *Aromatics* means alkyl-substituted benzenes with a single ring. These can be extracted using ions 91, 105, and 119. Ion 91 will show toluene and xylenes, ion 105 will show xylenes and C_3 alkylbenzenes, and ion 119 will show C_4 alkylbenzenes. There will be some overlap in the extracted ion chromatograms, as shown in Figure 3.9. *Condensed ring aromatics* refers to indans and naphthalenes. Indans have a five-membered ring attached to a benzene ring, and may be substituted. The naphthalenes that we usually see are naphthalene itself, 2- and 1-methyl naphthalene (written in that order because that is the order of elution from a nonpolar column), and the dimethyl naphthalenes. Naphthalene presents a single peak when ion 128 is extracted. There are two peaks for the methyl naphthalenes seen when ion 142 is extracted, and there are eight peaks in the dimethyl naphthalene chromatogram seen when ion 156 is extracted. The indans can be visualized by extracting ions 117 and 131.

Identification of Gasoline The composition of petroleum products as found in fire debris is influenced by three factors: crude oil parentage, the effects of petroleum refining processes, and the effects of weathering. Gasoline is the way it is largely because of the second of these influences, petroleum refining. Although all crude stocks contain aromatics, aliphatic hydrocarbons comprise the bulk of most crude oils. Because the aliphatics cause knocking in gasoline engines, the value of the crude stock is enhanced by *cracking*, making small molecules out of larger ones, and by *reformation* through a process of dehydrogenation. Toluene and xylenes are the most abundant compounds produced in these processes. Gasoline does contain numerous light aliphatics (as light as butane) present when gasoline is first pumped, but by the time it has been through a fire, most of the lighter ($<C_7$) aliphatics have evaporated. Consequently, toluene is usually one of the first tall peaks seen in a sample of gasoline, and once the gasoline has weathered to 50% or more, the toluene peak is much shorter than the C_3 alkyl benzene or xylene peaks. Gasoline changes considerably as it evaporates, which is what makes it one of the more difficult classes to identify. When a Gaussian distribution of normal alkanes changes because of evaporation, it is still a Gaussian distribution of normal alkanes, just a heavier one. Most of the samples of gasoline that this author has

seen in samples of fire debris have been more than 75% evaporated. (Estimating the degree of evaporation is discussed later.)

Gasoline is, in this author's experience, the most frequently misidentified ignitable liquid residue. That is because many of the compounds present in gasoline as it comes from the pump are also produced when polymers, such as poly(vinyl chloride) (PVC) and polystyrene, degrade as a result of exposure to heat. The key to avoiding misidentifications is making sure that the ratios between groups of compounds and *within* groups of compounds are consistent with the standard. Toluene is a very common pyrolysis product. Note that the polymer need not be an aromatic one to produce aromatic pyrolysis products. In fact, the pyrolysis products of PVC, because of the way it responds to heat, consist almost entirely of aromatics (Stauffer, 2003). It is an unusual fire debris sample that does not contain toluene at some level. Since it is one of the first gasoline compounds to evaporate, one does not expect to see a tall toluene peak in the absence of equally tall xylene peaks in a sample that is positive for gasoline. Figure 3.12 shows the chromatogram of a 10-μL standard of gasoline adsorbed using an activated charcoal strip and eluted with diethyl ether spiked with 100-ppm perchloroethylene. The toluene and xylene peaks are almost equally tall. If a fire debris sample contains toluene from gasoline, it will be accompanied by xylenes

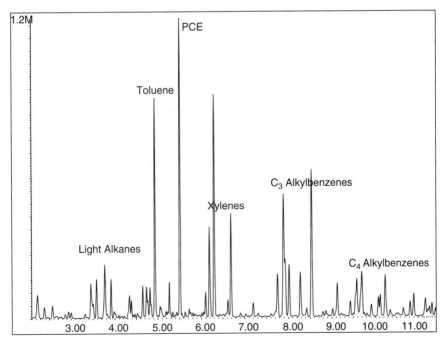

Figure 3.12 Total ion chromatogram of fresh gasoline; 10 μL was spotted on a piece of filter paper and the headspace was concentrated using ASTM E1412. The eluting solvent was diethyl ether spiked with 100 ppm of perchloroethylene, shown here eluting between toluene and the xylenes.

and the higher peak groupings of gasoline. Toluene that is not so accompanied comes from something other than gasoline.

Xylenes are also produced by the decomposition of plastics, but unlike toluene, the xylenes can be examined for correct intergroup ratios. Figure 3.13 shows ion 91, the base ion for xylenes, from gasoline in three different stages of evaporation, kerosene, and a medium petroleum distillate. The relative ratios for the three peaks are almost indistinguishable. Note that the ethylbenzene peak in the 50% evaporated gasoline is slightly lower. This trend continues as the degree of evaporation increases, but generally, one will find that if the xylenes in a sample are from an ignitable liquid, they will exhibit this characteristic ratio. Note that there are three isomers of xylene, *ortho-*, *meta-*, and *para-*, but that the three peaks seen in the chromatograms actually represent four compounds, because *meta-* and *para-*xylene cannot be resolved except in the longest columns (ethyl benzene followed by the three xylenes). If xylenes are found in ratios other than the one shown in Figure 3.13, particularly if ethyl benzene is the tallest peak in the group, one can safely conclude that neither gasoline nor any other petroleum product was the sole source of the xylene. If styrene is present at a higher concentration than the ethylbenzene, one is certainly looking at polymer decomposition products.

The next group to consider is the C_3 alkyl benzenes. This is by far the most important of the patterns in any sample of gasoline. Like the xylenes, it is also found in all petroleum products from which the aromatics have not been removed, and unless evaporation has decreased the concentration of the lighter compounds, the peak ratios will always be the same. This is the group identified in the 1982 guidelines (and in every standard since) as "the m-ethyltoluene pseudocumene five-peak group." This group appears in many samples that contain no gasoline, and it is the failure to exhibit the proper ratio of peaks that is sometimes the analyst's only clue that he or she is looking at data that can easily be misinterpreted. The group of peaks must be present to identify gasoline, but sometimes it is overemphasized, and sometimes it is "seen" where it really is not present. Figure 3.14 shows one such sample, where there was a misidentification of gasoline. The figure shows the suspect sample at the top, as well as four standards that exhibit the proper ratios of components in the C_3 alkyl benzene group. As with the xylenes, *meta-* and *para-*ethyl toluene cannot be resolved, although the resolution is somewhat better here than in the xylenes—we see a shoulder. In nearly all cases where the C_3 alkyl benzene group is, in fact, present as a component of a petroleum product, the pseudocumene peak will be the tallest. Because most of the residues encountered in gasoline will be highly evaporated, the pseudocumene peak will be the tallest peak on the chart—both the TIC and the EIC. Note that in the questioned sample at the top of Figure 3.14(a), it is the second peak that is the tallest, not the pseudocumene peak. The first peak has the same retention time as m- and p-ethyltoluene, but the shape is wrong, and the mass spectrum indicates that the peak actually represents benzaldehyde. The mass spectrometer unequivocally identified the second peak in the questioned sample as 1,3,5-trimethylbenzene, but the third peak, which may have represented o-ethyltoluene, also contained a significant 118 ion, indicating the

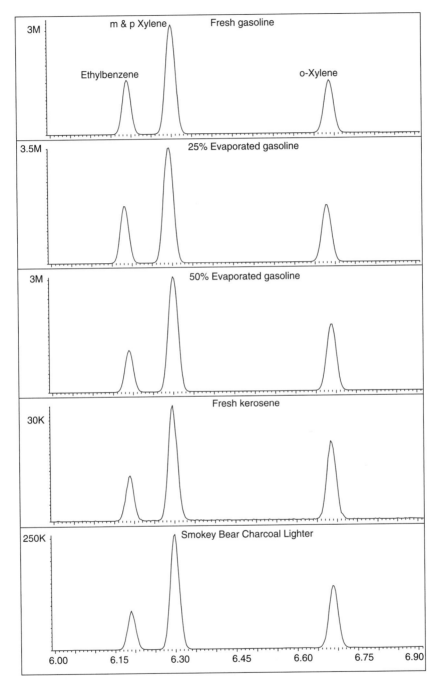

Figure 3.13 Ion chromatogram for ion 91 from gasoline in three different stages of evaporation, kerosene, and a medium petroleum distillate. Note that the relative ratios for the three peaks in the xylene group are almost indistinguishable. Evaporation causes the ethylbenzene peak in the 50%-evaporated gasoline to be slightly shorter.

70   ANALYSIS OF FIRE DEBRIS

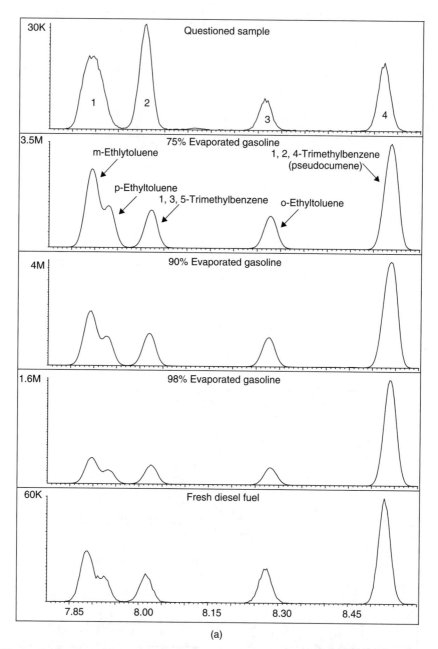

Figure 3.14 Study of the C_3 alkylbenzene group. (a) The top chromatogram shows an unknown sample. This is followed by gasoline in three stages of evaporation, and fresh diesel fuel. The TIC of the unknown did not look like gasoline, but based on this EIP and a few other ion chromatography comparisons, an analyst called the sample positive for gasoline.

ANALYZING THE ISOLATED ILR 71

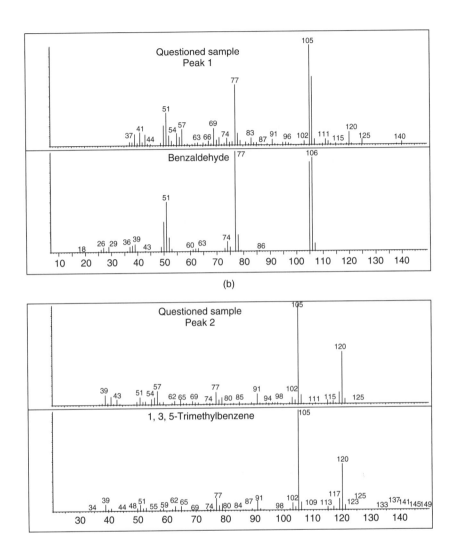

Figure 3.14 (*Continued*) The mass spectra shown in parts (b) to (e) show that two of the peaks, 1 and 3, were not gasoline components (b). Mass spectrum and library match of the first peak in the top chart shown in part (a). This peak actually represents benzaldehyde, not *m*-ethyltoluene. (c) Mass spectrum of the second fully resolved peak in the top chart shown in part (a). As with gasoline, this peak is identified as 1,3,5-trimethylbenzene (d). Mass spectrum of the third major peak in the chart shown at the top of part (a). This mass spectrum changed across the peak but clearly contained α-methylstyrene. 1-Decene, a common decomposition product, was also suggested by the spectrum. There may be some *o*-ethyltoluene coeluting as well, but that cannot be demonstrated (e). Comparison of the mass spectrum fourth peak in the chart shown in part (a) with 1,2,4-trimethyl benzene (pseudocumene). This is a reasonably straightforward match.

72 ANALYSIS OF FIRE DEBRIS

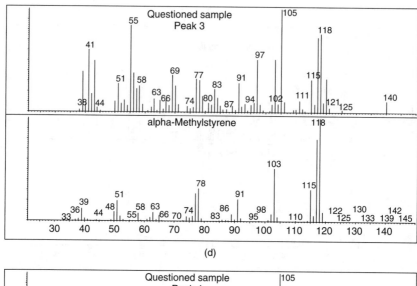

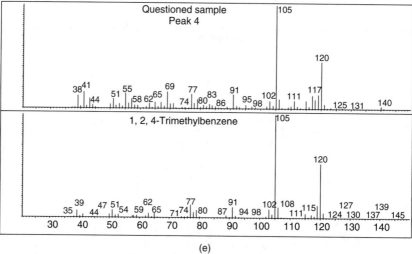

Figure 3.14 (*Continued*)

presence of α-methylstyrene, a common pyrolysis product. Phenol, which exhibits a base peak at 94, is also a confounding compound that elutes in the range of the five-peak group. *Simply matching the components when the ratios are not right can lead to misidentifications*. The odds against three background components coeluting in exactly the right concentrations to skew the peaks in this group are pretty high. Three other samples from the same fire scene were similarly misidentified, and exhibited similar peak ratios and mass spectral characteristics. As with almost all misidentifications, there was not a good pattern match with the TIC. Extracted ion chromatography or extracted ion profiling can be very useful; however, the analyst should remember that it is a spectral as well as a chromatographic technique. The individual mass spectra should be examined, especially when the peak ratios are "off."

A word about the use of the mass spectrometer beyond generating mass chromatograms is in order. Most analysts will use the mass selective detector or mass spectrometer as a tool for generating extracted ion chromatograms and extracted ion profiles. Obviously, if one is looking at very simple mixtures or single components, the mass spectrum is necessary in order to make an identification. A sometimes-overlooked function of the mass spectrometer is the evaluation of extracted ion chromatograms and profiles. As we saw in Figure 3.14(a), an extracted ion chromatogram with the intragroup ratios just slightly "off" has the potential to be misleading. Figure 3.14(b) to (e) show the mass spectra of each of those five peaks, as well as the library's best match. The first peak, which is coincident with m-ethyl toluene, has a large peak at m/z 77, but no such ion is present in the spectrum of m-ethyl toluene. There may be some m-ethyl toluene hidden under this peak, but there is definitely some benzaldehyde as well.

One way to determine whether there are coeluting compounds under a chromatographic peak is to examine the mass spectra at different points across the peak. If the peak represents a pure compound, the mass spectrum will change little, if at all. In Figure 3.14(c), the second peak in the five-peak group is a nearly-perfect match for 1,3,5-trimethylbenzene, which is the third peak in the five-peak group required to identify gasoline. The third peak, whose mass spectrum is shown in Figure 3.14(d), clearly contains more than one substance, but the strong peak at 118, as well as the retention time, indicates that α-methylstyrene is coeluting. Like the second peak, the fourth peak is a pure compound and a nearly perfect match for pseudocumene. (The mass spectra of all of the aromatics in this group are quite similar. In fact, using a library search usually results in a list of match candidates that includes all four members of the group.) Because the peak ratios are off, and especially because two of the four peaks in the ion 105 chromatogram represent compounds not found in gasoline, it must be concluded that the first identification was in error.

Background subtraction is a tool found in most data analysis software and can frequently resolve questions as to whether a particular compound is really there. A detailed evaluation of the quality of the spectra underlying a mass chromatogram, whether it is a single-ion extraction or a multiple-ion profile, should be carried out periodically just to keep the analyst in practice, but it should be carried out routinely on any sample where either the peak ratios or retention times are "just a little off." Note in Figure 3.14(a) that when a sample is sufficiently evaporated, the peaks at the left side of the chart begin to diminish. This result is expected. The initial guidelines for identifying gasoline stated that the m-ethyl toluene/pseudocumene five-peak group was still present in gasolines that had lost 90% of their fresh weight. In fact, this group does not disappear until the gasoline is more than 98% evaporated. If this group cannot be positively identified, an analyst is on very thin ice indeed when identifying a sample as containing gasoline. In such a case, there must be a peak-for-peak match of all of the higher peak groupings. As with many of the other components of gasoline, this five-peak group is present in almost all petroleum products from which the aromatics have not been removed. The bottom chart in Figure 3.14(a) is from unevaporated diesel fuel.

The C_4 alkyl benzenes are best viewed by extracting ion 119. Because there are more ways to build a C_4 alkyl benzene than a C_3, this is a more complex pattern, but it is present in almost all petroleum products. Figure 3.15 shows the patterns found in highly evaporated gasoline, as well as in kerosene and diesel fuel, which are produced by the presence of the C_4 alkyl benzenes. Note that in kerosene and diesel fuel, this pattern is more complex than in gasoline.

The next group of compounds that must be present in order to make a solid identification of gasoline is the indans. These can be extracted using ions 117 and 131. The doublet at 11.3 and 11.5 min in the ion 117 chart is, in this analyst's experience, always present. It has traditionally not been an absolute requirement of the ASTM standards, but it probably should be. The standard states, "Indan (dihydroindene) and methylindans are usually present." The first tall peak in the 117 chart is indan, appearing at 9.4 min. Figure 3.16 shows the peak groupings characteristic of gasoline for the indans, methylindans, and dimethyl indans. Like the C_3 alkyl benzenes, the indans and alkyl-substituted indans can be found in roughly the same proportions in gasoline that is 98% evaporated. Other petroleum products, particularly the distillates, contain indans and methylindans, but their mass chromatographic patterns are more complex than the ones for gasoline.

The next group of compounds that should be present in a sample identified as containing gasoline is the methyl- and dimethylnaphthalenes. Naphthalene is also present, but naphthalene is so common that its presence in a sample is meaningless. 2-Methylnaphthalene elutes before 1-methylnaphthalene and is almost always more abundant than the 1-methylnaphthalene. In any case where this ratio is reversed, the extract should be considered suspect. The boiling point of the methylnaphthalenes is high enough that the ratio, unlike xylenes or C_3 alkyl benzenes, is unlikely to be affected by evaporation. All petroleum products are likely to contain both the naphthalenes and the dimethylnaphthalenes, and the ratios should be highly comparable to each other.

There are some aromatic products in the marketplace that meet almost all of the ASTM criteria for the identification of gasoline. Such products are used as solvents for stains, insecticides, adhesives, and industrial and commercial products. The way to differentiate these aromatic products from gasoline is to look for the presence of alkanes. All gasolines contain high percentages of alkanes when fresh, but even highly evaporated gasolines will contain some branched alkanes. Straight-chain hydrocarbons are an unusual finding in gasoline, because they are undesirable components that cause knocking when burned in gasoline engines. [Knocking is

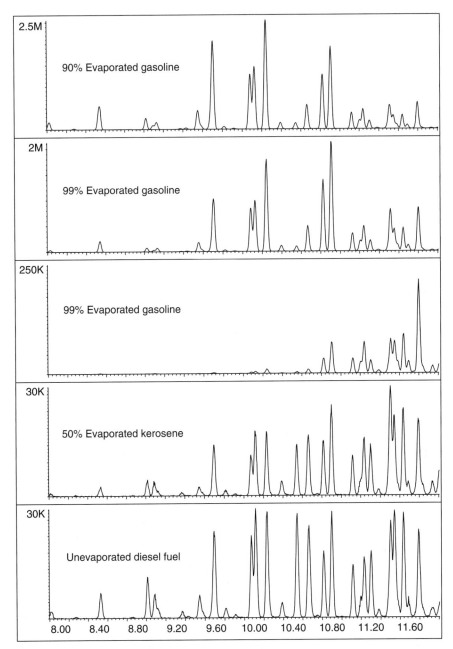

Figure 3.15 C$_4$ alkylbenzenes represented by ion 119 in gasoline in three different stages of evaporation, 50%-evaporated kerosene, and unevaporated diesel fuel. All samples were prepared by spotting 10 µL on a piece of filter paper, and processed using ACS adsorption–elution.

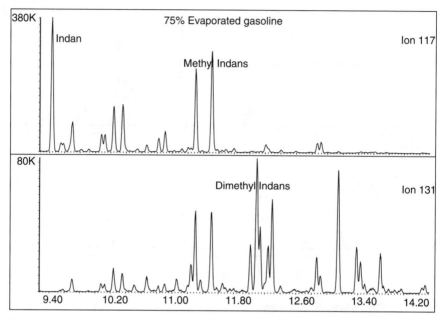

Figure 3.16 Ion chromatograms for indan, the methylindans, and dimethylindans found in 75%-evaporated gasoline.

the premature detonation of a fuel and occurs more readily with straight-chain hydrocarbons than with branched-chain hydrocarbons or aromatics. Resistance to knock is known as the *octane rating*. Octane rating is based on a scale of 0 to 100, where 0 is the resistance to knock of 100% n-heptane and 100 is the resistance to knock of 100% iso-octane (2,2,4-trimethylpentane). N-Octane actually has an octane rating below zero.]

Identification of Distillates   Most distillates likely to be encountered in fire debris are straight-run distillates from crude oil. They have not been subjected to cracking or reforming, so they lack the high aromatic content found in gasoline, but aromatics are still present. The distillates are characterized by an abundance of normal alkanes, along with branched alkanes and cycloalkanes.

Distillates are usually easy to recognize, except for light petroleum distillates, which might be missed unless the analyst checks for them in every sample before calling it negative for ILR. Because of their high volatility, they tend to be present at low concentrations in fire debris samples and are typically found on the left side of the chart, where they may be mistaken for decomposition products. This is a particular hazard when the higher-boiling components are decomposition products. Light petroleum distillates (LPDs) do not generally exhibit the Gaussian distribution seen in medium and heavy petroleum distillates (MPDs and HPDs). When the peaks that elute prior to 8 min are examined, and the analyst sees a mixture of cycloalkanes, branched alkanes, and perhaps some normal alkanes in the C_7

through C_9 range, an LPD should be suspected. The analyst should have in the library as many LPDs as possible, because unlike the heavier distillates, the patterns tend to vary from one to the other,. Figure 3.17(a) shows three different brands of cigarette lighter fluid, each exhibiting a different pattern of peaks. The recommended extracted ion profiles for LPDs are 57, 55, 83, and 91, which can be seen in Figure 3.17(b). This particular LPD is unusual in that it contains xylenes but no toluene is present. Because of their high volatility, LPDs are not persistent in the environment and are not usually expected to be found as background material. There are some cleaning agents and automotive products that contain LPDs, but if the sample was collected from somewhere other than the garage, the workshop, or under the kitchen sink, a finding of LPD generally indicates the presence of a foreign ignitable liquid.

This is not the case with medium or heavy petroleum distillates, which are far more common in our environment (Lentini et al., 2000). A typical medium petroleum distillate, Sparky charcoal lighter fluid, is shown in Figure 3.18(a). The ions necessary to make a valid identification of a medium petroleum distillate are 57 and 83, shown in Figure 3.18(b). The analyst should also check for ion 105 or 91, to be certain that the aromatics have not been removed, in which case the ILR can be identified as a dearomatized distillate. One does not notice much difference between the TIC and the ion 57 EIC, because distillates are dominated by the normal and branched alkanes. The alkylcyclohexanes, which elute about midway between the normal alkanes, present an overall appearance similar to that of the alkane chromatogram. If the normal alkanes are not present, or if they are present

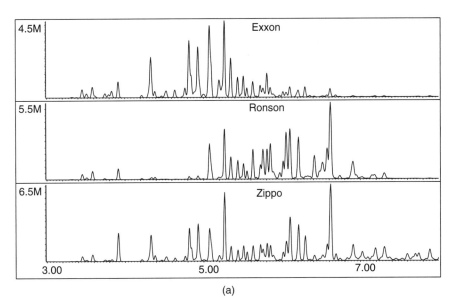

Figure 3.17 (a) Total ion chromatograms of three different brands of cigarette lighter fluid. (b) Ion chromatograms of ions 57, 55, 83, and 91 from a standard of Ronson lighter fluid.

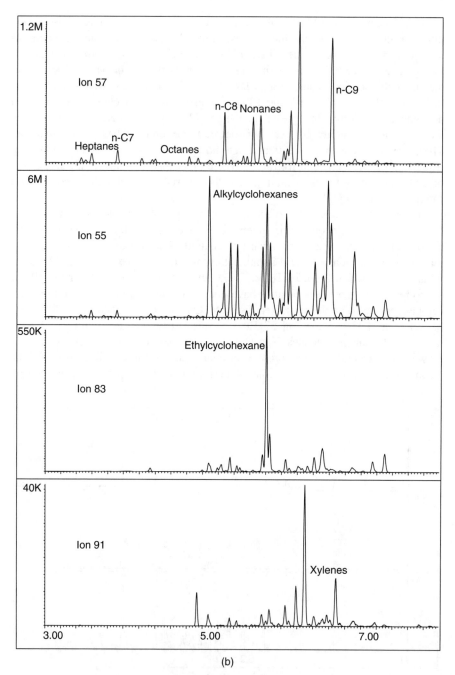

Figure 3.17 (*Continued*)

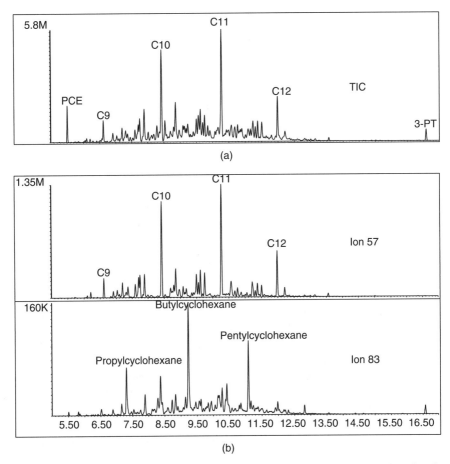

Figure 3.18 Total ion chromatogram of (a) a typical medium petroleum distillate, Sparky charcoal lighter fluid, and (b) extracted ion profiles for ions 57 and 83.

at approximately the same concentration as the branched alkanes, an isoparaffinic product should be suspected. When the cycloalkanes are present at an abundance greater than about 20% of the abundance of the alkanes, a naphthenic–paraffinic source is indicated. In most distillates, the cycloalkanes will be present at about 5 to 10% of the concentration of the normal and branched alkanes. Medium petroleum distillates cover a wide range of products and may be used as fuels, such as lamp oil or charcoal starter, or as solvents, such as mineral spirits or insecticide carriers. This wide range of products and formulations requires that the library of reference materials include numerous MPDs.

Heavy petroleum distillates (HPD) include kerosene and diesel fuel. Except for HPDs that are formulated for specific applications, such as jet fuel, the carbon number range of heavy petroleum distillates can vary. Thus, unless one has a sample of the unevaporated liquid, it is difficult to determine the degree of evaporation

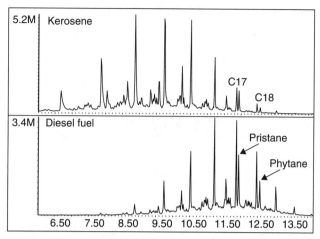

Figure 3.19 Total ion chromatograms showing a comparison of kerosene and diesel fuel.

of the residue isolated from a sample. In northern parts of the country, diesel fuel sold in the winter may actually be kerosene.

Figure 3.19 shows a comparison of kerosene and diesel fuel. These distillates come with internal carbon number markers, in the form of pristane and phytane. Pristane is 2,6,10,14-tetramethyl pentadecane ($C_{19}H_{40}$) and it elutes immediately after normal heptadecane. Phytane is 2,6,10,14-tetramethyl hexadecane ($C_{20}H_{42}$) and it elutes immediately after normal octadecane. (Although pristane and phytane elute immediately after two normal hydrocarbons, they come from different sources. The n-alkanes started out as fatty acids, but pristane and phytane started out as parts of the chlorophyll molecule.) Thus, one can identify the two doublets on the high side of the bell-shaped curve and can count carbons up and down from there. In Figure 3.19 these doublets occur at 12 and 12.5 min. Kerosene, which is a known standard that has been evaporated to 50% of its original volume, covers the range C_{11} to C_{19}, while the diesel fuel range is C_{12} to C_{21}. If the kerosene is evaporated further, it will look more like the diesel fuel. Had the diesel fuel been evaporated less, it would look more like the kerosene.

As with the medium petroleum distillates, interpretation of a finding of heavy petroleum distillates should be approached with caution. There are numerous household products that contain HPDs, including many of the same kinds of products in which MPDs are found. The safer charcoal lighters are made from kerosene rather than mineral spirits. They are safer because of their higher flash point. Figure 3.20 shows a total ion chromatogram of lemon oil furniture polish. This polish has a carbon number range from C_{12} to C_{22}, with C_{17} being the tallest peak on the chart. The limonene peak at the left side of the chart could easily be attributed to a pine substrate. Pinenes and limonene are very common in samples containing structural (coniferous) wood. If found in a fire debris sample, this particular furniture polish could easily be misreported as diesel fuel.

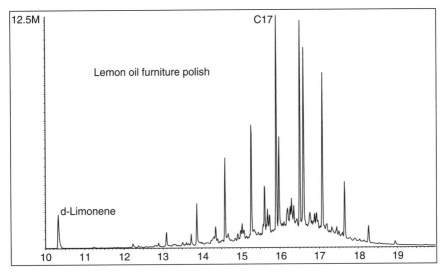

Figure 3.20 Total ion chromatogram of lemon oil furniture polish. This sample could easily be reported out as diesel fuel.

It is not necessary for a liquid to be present in order for a distillate to be detected. Figure 3.21(a) shows the total ion chromatogram of a piece of pine wood that was stained with Minwax finish 10 months before it was subjected to headspace concentration. The naturally occurring terpenes, α- and β-pinene and d-limonene, are the dominant peaks on the chart, but the mineral spirits solvent is still clearly visible. When ions 57 and 83 are extracted, the terpenes disappear, to yield the charts shown in Figure 3.21(b). These results show the critical necessity of asking for comparison samples, particularly when samples of flooring are submitted for analysis. The flooring does not need to be recently painted to exhibit the distillate solvent used to apply the floor coating. This author has reported finding distillates 24 months after application (Lentini, 2001) and has detected distillates in samples of finished flooring and furniture up to 10 years old. There is no reason to believe that these solvents do not persist indefinitely, trapped in either the wood matrix or in the polymer coating matrix.

In addition to true distillates, there are some distillate-like residues produced as the result of decomposition of other products. Asphalt is what is left at the bottom of the distillation pot after all of the volatiles have been distilled from crude oil. It contains hydrocarbons ranging from C_{30} to C_{60}. When these long-chain hydrocarbons undergo pyrolysis, they do so in much the same way as that of the long-chain hydrocarbons in polyethylene, via random scission. This results in the production of normal alkanes in the range C_9 to C_{18} (Lentini, 1998). When these pyrolysis products are present in fire debris and subjected to headspace concentration, they produce a chromatogram that can be and has been mistaken for the chromatogram of a heavy petroleum distillate. Such a chromatogram is shown in Figure 3.22(a). In 1982 it was reported that roof shingles could produce "accelerant-like residues,"

82 ANALYSIS OF FIRE DEBRIS

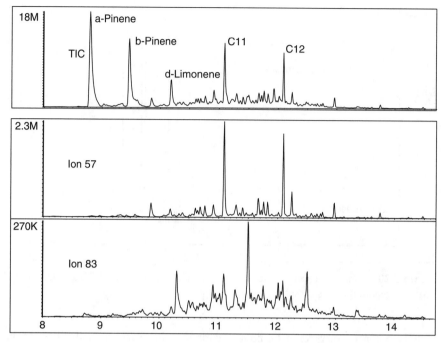

Figure 3.21 Total ion chromatogram of (a) a piece of pine wood stained with Minwax Finish 10 months prior to its analysis, and (b) extracted ion profiles for ions 57 and 83.

and at that time, there existed no reliable methods for distinguishing HPDs from asphalt shingle residues (Lentini and Waters, 1982). The increased use of capillary columns allowed for the occasional visualization of a "double-peak kerosene," sometimes called "pseudokerosene," but even with capillary columns, it was not uncommon for asphalt residue to be misidentified as a liquid petroleum distillate. This author learned how to make the differentiation in 1995 in connection with the investigation of an insurance claim that had been erroneously denied because of a finding of HPDs where none should have been present. Figure 3.22(b) shows how the distinction is made. One compares ion 57 with ion 55. If a second peak appears in front of the n-alkane peak, or if a small peak grows larger, we can conclude (particularly after we collect the mass spectrum) that the second peak is the 1-olefin. A sample such as this must be classified as asphalt smoke condensate or asphalt decomposition residue.

The amount by which the olefin peak "grows" will vary depending on the sample. When one looks at kerosene and diesel fuel, the only difference in appearance between the ion 55 chart and the ion 57 chart is that the abundance of ion 55 is lower. The relative abundances of the individual peaks with respect to each other do not change.

Another way to distinguish asphalt decomposition products from HPDs is to look for the cycloalkanes. They are not present in asphalt decomposition products. A Gaussian pattern will be observed when ion 83 is extracted, but this is due

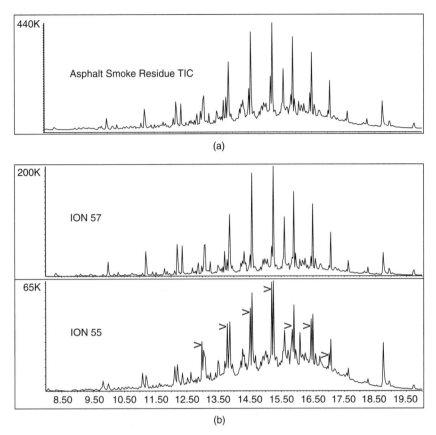

Figure 3.22 Total ion chromatogram of (a) asphalt smoke residue of the type that can be mistaken for a petroleum distillate, and (b) extracted ion profiles for ions 57 and 55. The growth of the peak in front of the normal alkane when comparing 55 to 57 demonstrates the presence of alkenes, which allows the identification of the residue as a decomposition product rather than a foreign petroleum distillate.

entirely to the presence of the olefins. The ion 83 and ion 55 ion chromatograms show the same peaks. Asphalt smoke condensates also contain little if any pristane and phytane.

Polyethylene is another substance that produces a distillate-like appearance in the chart, but if the capillary column has any resolution at all, it will be obvious that one is looking at polyethylene residue rather than at a distillate. Figure 3.23(a) shows the total ion chromatogram of polyethylene smoke condensate. All of the peaks are doublets and most are actually triplets, owing to the presence of the $1,(n-1)$-diene in the mixture. When one looks at ion 57 and compares it with ion 55, as shown in Figure 3.23(b), the growth in the olefin peak is obvious relative to the alkane peak. The diene peak also grows, because the dienes contain more ion 55 than ion 57. Lubricating oils are subject to the same decomposition processes as asphalt and polyethylene, and the smoke condensates of lubricating oils appear similar.

84 ANALYSIS OF FIRE DEBRIS

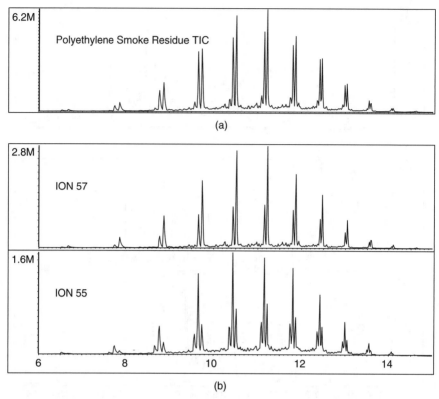

Figure 3.23 Total ion chromatogram of (a) polyethylene smoke condensate. The first peak in each doublet is the alkene; the second is the alkane. Pristane, phytane cycloalkanes, and aromatics are all absent. (b) Ion 57 EIC, and (c) ion 55 EIC. In the ion 55 chromatogram, a third peak, representing the $n, (n-1)$-diene, appears.

A careful examination of the chromatographic and mass spectral data will prevent the analyst from misidentifying decomposition products as distillates, but automation of pattern recognition is coming and presents some dangers if not handled properly. As with all computer "answers," one should always do a "reality check." The same pattern recognition software that allows a mass spectrum from an unknown compound to be matched against the spectra of 100,000 compounds can be applied to chromatograms. The chromatogram is converted using Microsoft Excel into a bar graph that has the same general appearance as a mass spectrum. This graph can then be compared against a library of known ignitable liquids that have been transformed similarly. As with the mass spectral libraries, "extra" peaks do not necessarily keep the database from recognizing a "match." The polyethylene smoke condensate chart shown in Figure 3.23(a), matched up against a database that contains only ignitable liquids, will yield a match of exceptionally high quality for diesel fuel. This is why it is necessary to populate ignitable liquid libraries used for this purpose with known background chromatograms. The chart would make an even better match for polyethylene smoke condensate, but only if it is available in the database.

Identifying Other Classes of Products The remaining classes of ignitable liquid residues can frequently be identified by what is and is not present. A dearomatized distillate, for example, will have a signal from the aromatics that is less than 1% of the signal from the aliphatics. Otherwise, it will look the same with respect to both the alkanes and the cycloalkanes. Because of environmental regulations governing the aromatic content of distillates, there has been, since 1973, an ASTM standard for this determination. ASTM D3257-07 is entitled *Standard Test Methods for Aromatics in Mineral Spirits by Gas Chromatography* (ASTM D3257-07, 2007) and uses a specified test blend for calibration. If the calculation of aromatic content is an issue, a forensic scientist might avoid a *Daubert* challenge by using this established method rather than by devising a new one.

The normal alkane products are very easy to recognize, as long as one makes sure that one is looking at a homologous series of normal alkanes as opposed to a homologous series of aldehydes or some other group of homologues that are pyrolysis or decomposition products. Groups of compounds that differ from each other only in that they have one additional CH_2 group look pretty much the same as a series of normal alkanes, and require some caution. Figure 3.24 shows a series of normal alkane products marketed by Exxon, as well as a sample of Lamplight Farms ultrapure lamp oil, which also consists of normal alkanes. One common source of normal alkanes is carbonless forms (NCR paper). The bottom chromatogram in Figure 3.24 shows the chromatogram of concentrated headspace vapors from a 2-in. × 2-in. square of a carbonless form. These forms contain microspheres filled with normal alkane solvent that, when broken, causes color to develop in the ink. There are only a few microliters per square foot, but this concentration is easily detectable. Figure 3.25 is a scanning electron micrograph, showing the bottom side of a carbonless form. It should be noted that the alkanes found in carbonless forms are usually accompanied by a pair of substituted biphenyls. Normal alkane products are also found in some brands of linoleum floor covering. Any time that a sample is collected from a floor likely to have linoleum in its structure (kitchen, bathroom, laundry room), a finding of normal alkanes is probably not meaningful. A comparison sample is an absolute necessity in such cases.

Isoparaffinic hydrocarbons are made by removing the normal hydrocarbons with a molecular sieve. These liquids are becoming more common as petrochemical manufacturers move to more environmentally friendly and odor-free replacements for the straight-run distillates, such as mineral spirits. Similar to the isoparaffinic hydrocarbons are the naphthenic–paraffinic products, which are characterized by an abundance of cycloalkanes. Whereas one might expect to find cycloalkanes present at less than 5% in the isoparaffinic hydrocarbons, they may be present at up to 30% in naphthenic–paraffinic products. The distinction requires looking at the ion profiles for the cycloalkanes versus the alkanes, as well as looking at the abundance numbers on the left side of the chart. Figure 3.26 shows a comparison

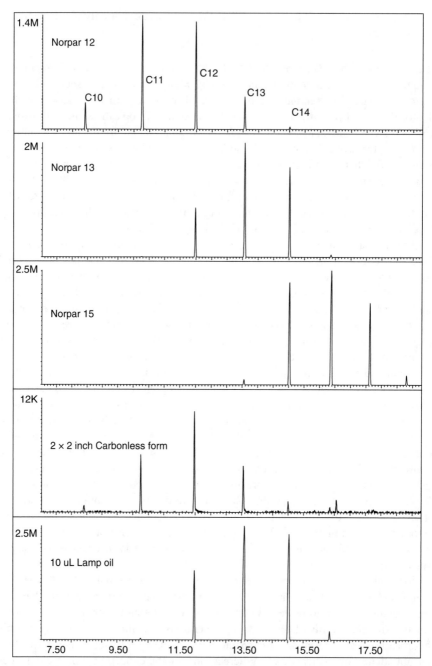

Figure 3.24 Ion 57 chromatogram of three Exxon Isopar products, compared with an ACS extract from a carbonless form, and the ACS extract of a 10-μL sample of Lamplight Farms ultrapure lamp and candle oil.

ANALYZING THE ISOLATED ILR 87

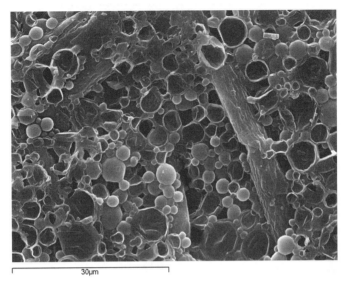

Figure 3.25 Scanning electron micrograph of the underside of a carbonless form. The microspheres range in size from 2 to 20 μm and are filled with normal alkanes.

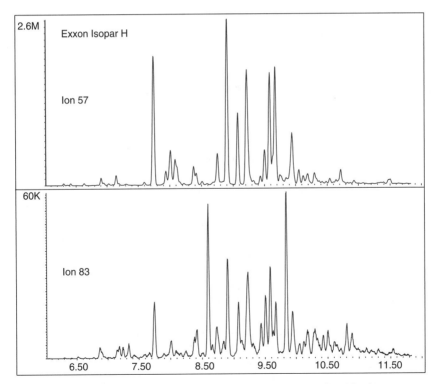

Figure 3.26 Comparison of the alkane (ion 57) and cycloalkane (ion 83) chromatograms from Exxon Isopar H. The cycloalkane peaks are only about 2% of the height of the branched alkane peaks.

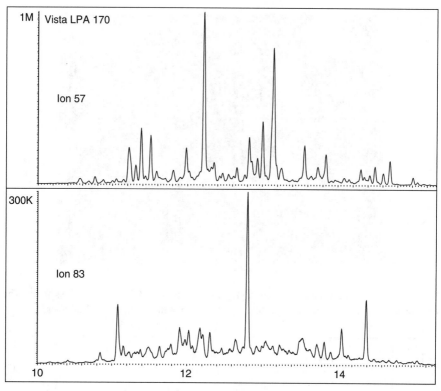

Figure 3.27 Comparison of the alkane (ion 57) and cycloalkane (ion 83) chromatograms from Vista LPA 170, a naphthenic–paraffinic product. The height of the cycloakane peaks is about 30% of the height of the branched alkane peaks.

of the ion 57 and ion 83 chromatograms from Isopar H, a common solvent used in cosmetics and other household products. Compare this to Figure 3.27, which compares those same ion chromatograms from a naphthenic–paraffinic solvent, Vista LPA 170. Instead of having an ion 83 chromatogram that is 2% the height of the ion 57 chromatogram, the ion 83 is more than 30% of the height of the ion 57 chromatogram in the naphthenic–paraffinic solvent. Also, note that the tall peaks in the ion 57 chromatogram from the naphthenic–paraffinic product are not normal alkanes, but are branched alkanes.

Another way to make the distinction between the isoparaffinic products and the naphthenic–paraffinic products is to take an average mass spectrum. Representative isoparaffinic products are shown in Figure 3.28. A average mass spectrum of any of these looks very much like the mass spectrum of a single alkane. Ion 57 is the base peak, and the other fragments are spread out in a Gaussian distribution with a spacing of 14 mass units between them. A typical isoparaffin average mass spectrum, that of Isopar L, taken over a 3-min portion of the chromatogram, is shown in Figure 3.29.

ANALYZING THE ISOLATED ILR **89**

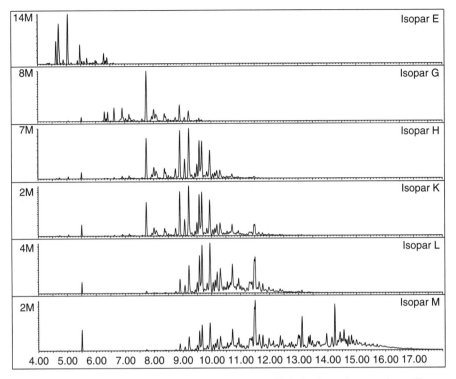

Figure 3.28 Total ion chromatograms of six isoparaffinic hydrocarbon products, Exxon Isopars E, G, H, K, L, and M.

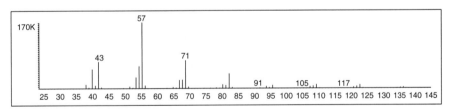

Figure 3.29 Average mass spectrum taken from Isopar L, over the range of 9 to 12 min. This spectrum has an appearance very similar to that of a branched alkane.

The same exercise can be performed on the naphthenic–paraffinic products, several of which are shown in Figure 3.30. The average mass spectrum of a typical naphthenic–paraffinic product, shown in Figure 3.31, looks quite different from that of the isoparaffinic product. While it still exhibits a base peak of 57, there will be a greater abundance of ions 55, 69, and 83 than in the isoparaffinic product.

The identification of other ignitable liquid residues requires an individual examination of the peaks in the chromatogram. It is possible to find just about any single compound in just about any sample. Findings of alcohols, turpentines,

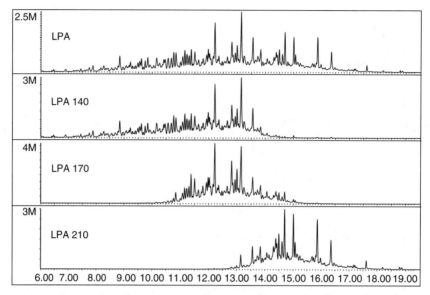

Figure 3.30 Total ion chromatograms of four naphthenic–paraffinic products: Vista LPA, LPA 140, LPA 170, and LPA 210. The numbers in the product names correspond approximately to flash points.

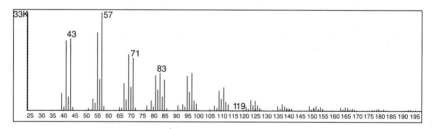

Figure 3.31 Average mass spectrum of Vista LPA taken from 8 to 17 min..

aromatic solvents, and other "flammable" liquids need to be undertaken with great caution. ASTM E1618 recommends not making an identification of these single compounds unless they are present in such concentrations that the signal is *at least two orders of magnitude greater than the background*. Just about every fire debris sample will contain methanol. Just about every fire debris sample will contain toluene. Unless these substances are present at concentrations sufficiently high for the analyst to feel comfortable saying they are not native to the background, they should not be reported.

3.6.2 Improving Sensitivity

The techniques used routinely to isolate and analyze ignitable liquid residues are quite sensitive. In samples with nontenacious substrates, it is possible to detect as little as one-tenth of 1 µL of ignitable liquid residue such as gasoline in 1 kg or more of debris. The task gets a little more difficult when there are active surfaces in the fire debris, such as charcoal, but an activated carbon strip is more tenacious than most debris that one is likely to encounter.

Before attempting to improve the sensitivity of the detection method, it is useful to have some idea about how much sensitivity one already has. This is best accomplished by preparing serial dilutions of an ignitable liquid with an internal standard, and comparing the signals to those generated by extracts from known amounts of ignitable liquid run through the laboratory's adsorption or elution process. In our laboratory we typically use 10-µL standards. The liquid is applied to a piece of filter paper in a 1-quart can, and the filter paper is subjected to ACS headspace concentration, just like any debris sample. When exposed to 10 µL of 75% evaporated gasoline, a 10 mm × 10 mm carbon strip eluted with 500 µL of diethyl ether spiked with 100 ppm of perchloroethylene yields a concentration of approximately 1000 ppm. This is easily detectable. (This translates to a recovery rate of about 5%; 500 µL at 1000 ppm equals 0.5 µL recovered out of 10 µL placed in the sample can.) One microliter in the sample can yields a concentration of approximately 100 ppm in the eluate: again, very easily detectable. At 0.1 µL (10 ppm in the eluate), the signal is noisy and it becomes difficult to extract acceptable ion chromatograms except for ion 105, showing the C_3 alkyl benzenes. To obtain better ion chromatograms, a number of different strategies can be followed. The lower limit of detection is formally defined as that concentration of analyte that will give a signal that is twice the background level; for this analyst, the detection limit is that concentration of analyte that produces a signal large enough to feel comfortable about a determination. One can lower the detection limit either by increasing the signal or by decreasing the background noise. Increasing the signal can be accomplished in a number of ways. One obvious way is to increase the concentration of the analyte. In the case of fire debris eluates from C-strips, this is accomplished by evaporating off some of the solvent. This strategy is not without its costs. It is time consuming and must be done very carefully if one is to avoid losing the analyte along with the solvent. No heat should be used in this process. At some point, the volume of the solution becomes too small to be handled by an autosampler, and this requires manual injection. Although it can be done, easier ways exist to increase the signal. Another obvious strategy is simply to increase the size of the injection. Normally, we inject 1 µL. When we are looking at samples of low concentration, we inject 2 µL.

One way to increase the signal dramatically is to run the gas chromatograph in the *splitless mode*. (Running splitless is one reason that SPME techniques are so sensitive.) Most laboratories use a split ratio of 20:1 to 50:1. This dilution of the sample improves resolution. Turning off the split will increase the amount of analyte presented to the detector, and the signal will increase by up to a factor of 10. The only cost is likely to be a small decrease in resolution.

Another way to increase the signal from the ions of interest in order to produce better ion chromatograms is to increase the dwell time, the time that the detector spends looking at a particular ion. This is accomplished by using the selected ion monitoring (SIM) mode on a quadrupole instrument. In the full-scan mode, we typically look from 33 to 300 amu, resulting in a dwell time of less than 1 ms per ion. If we ask the instrument to look only at the ions in which we are interested, we can increase the dwell time dramatically. The detector response will increase in proportion to the square root of the dwell time (i.e., if one increases the dwell time by a factor of 4, the signal strength will increase by a factor of 2. Our instrument performs 5.24 scans/s in the full-scan mode, resulting in a dwell time of 0.7 ms. In the SIM mode, there is only about one scan per second, resulting in somewhat rougher peaks, but with a dwell time of 50 ms (an increase of about 70-fold), the signal will increase by about a factor of 8. It is not the increasing signal that is the most attractive feature of SIM, however. It is the reduction in noise. "Changing channels" 267 times per scan (1400 times per second) generates an abundance of noise in the instrument. This is known as *housekeeping noise*. If one looks only at the 23 ions listed in Table 3.4, the noise is reduced by a factor of 60 or more.

Figure 3.32 shows a comparison of the noise generated in the full-scan mode versus the noise generated in the SIM mode. It is easy to see that even with a much smaller signal, the peaks of interest will be plainly visible.

The price to be paid for this increase in sensitivity is a decrease in specificity. (There is no such thing as a free lunch.) The mass spectra produced in the SIM mode do not contain enough ions for a library to match. The analyst can, however, make certain that the ion of interest is the base peak for a particular compound, and make sure that a "qualifier ion" is present. The abbreviated spectrum from a SIM peak can be compared with the SIM spectrum from a known compound, but that is as far as it can go. For this reason, the analyst should apply more rigid criteria when trying to determine whether a SIM pattern exhibits "sufficient similarities" to a standard pattern.

Figure 3.33 shows a comparison of a 10-ppm standard of 75% evaporated gasoline run in the full-scan mode, in the full-scan splitless mode, and in the SIM mode. Looking at just the total ion chromatograms, one might conclude that the SIM mode does not add value compared to the splitless mode, but when one generates the

TABLE 3.4 Ions Used for SIM

Ion	Compounds	Ion	Compounds
31	Methanol	117	Indan, methylindans
43	Alkanes	118	Methylstyrene
45	Ethanol	119	C_4 alkylbenzenes
55	Alkenes, cycloalkanes	120	C_3 alkylbenzenes
57	Alkanes	128	Naphthalene
69	Alkenes	131	Methyl, dimethylindans
71	Alkanes	133	C_5 alkylbenzenes
83	Cycloalkanes, alkenes	134	C_4 alkylbenzenes
85	Alkanes	142	Methylnaphthalenes
91	C_1, C_2 alkylbenzenes	156	Dimethylnaphthalenes
104	Styrene	168	3-phenyltoluene
105	C_3 alkylbenzenes		

ANALYZING THE ISOLATED ILR 93

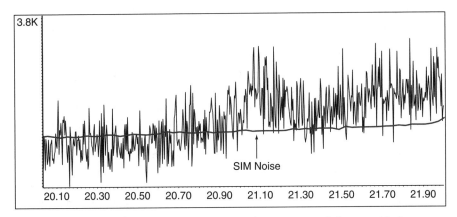

Figure 3.32 Comparison of the baseline noise generated by a full scan with that generated in the SIM mode. *(See insert for color representation.)*

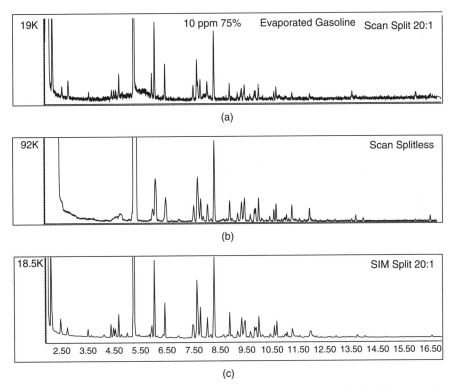

Figure 3.33 10-ppm solution of 75%-evaporated gasoline. (a) Run in the full-scan mode with a 20:1 split ratio. In each of these three charts the off-scale peak at 5.5 min is the perchloroethylene internal standard. (b) Run in the full-scan mode with a splitless injection. The abundance is increased by about a factor of 5 over the split injection. (c) Run in the SIM mode with a 20:1 split ratio. The abundance is not increased compared to the scan mode, but the noise is reduced significantly.

94 ANALYSIS OF FIRE DEBRIS

six extracted ion profiles for gasoline, the SIM chromatograms clearly outshine those produced in the full-scan mode, even with no split. This is demonstrated in Figure 3.34.

This process of increasing sensitivity can be taken even one step further, by using a splitless injection and selected ion monitoring, although in doing so, we are approaching territory where findings of lower and lower amounts may not be meaningful. Figure 3.35 shows the six gasoline ions from a solution of 0.1-ppm standard of 75% evaporated gasoline. This is roughly equivalent to a sample containing 0.001 µL of gasoline (1 nL!). Even at a concentration in the eluate of 0.01 ppm (10 ppb), the baseline does not go completely flat in the SIM splitless mode.

By simply using the tools available, we can see that the sensitivity of the analysis can be increased by four orders of magnitude. It does not seem to this analyst that lowering the detection limit further than this would be worthwhile. Table 3.5 shows

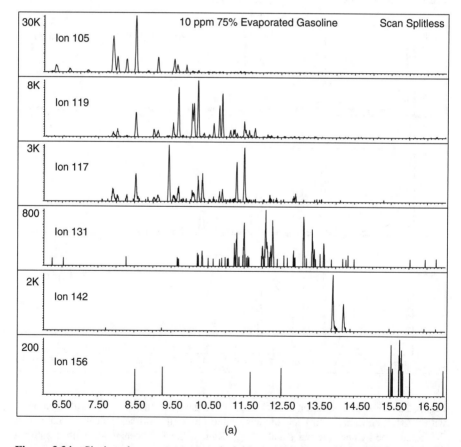

Figure 3.34 Six ion chromatograms produced by a 10-ppm solution of 75% evaporated gasoline run in the full-scan mode, with splitless injection; (b) run in the SIM mode, with a 20:1 split ratio..

ANALYZING THE ISOLATED ILR 95

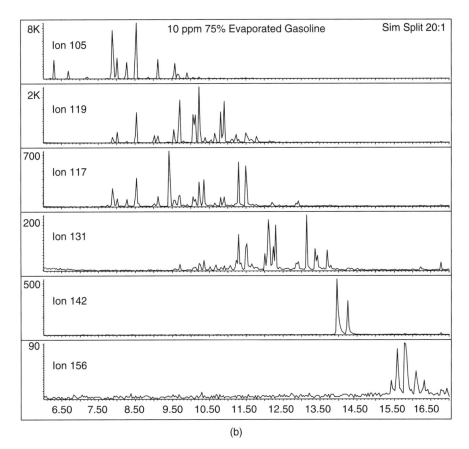

(b)

Figure 3.34 (*Continued*)

a rough order-of-magnitude calculation of the detection limits available using the techniques described above.

3.6.3 Estimating the Degree of Evaporation

Estimating the degree of evaporation is a task that an analyst is occasionally asked to perform. Usually, evaporation is not an issue, and frequently, even if it is an issue, estimating how much of the original volume of a sample may have evaporated is difficult. Comparing degrees of evaporation may be helpful in source or common source determinations, as will be discussed later.

Unless we know what the ignitable liquid was to start with, it is next to impossible to determine how much it has evaporated. This is particularly true for distillates, which may range from C_7 through C_{12} in one batch and C_9 through C_{13} in the next batch of the same product. Similarly, kerosene and diesel fuel formulations change with latitudes and seasons. So, unless there is a container that is suspected of being

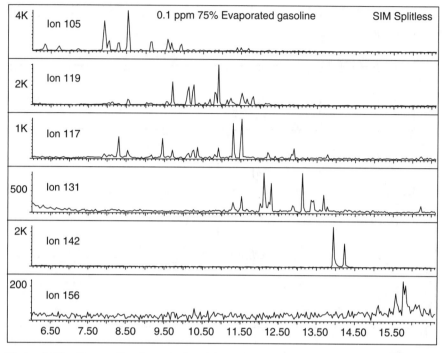

Figure 3.35 Six gasoline ions produced by a 2-μL splitless injection of a solution of 0.1 ppm 75%-evaporated gasoline.

TABLE 3.5 Detection Limits for Four Instrumental Configurations

Method	LLD (Solution)[a]	LLD (Sample)[b]
Full scan, split 20 : 1, 1-μL injection	100 (ppm)	1.0(μL)
Full scan, splitless, 1-μL injection	10	0.1
SIM, split 20 : 1, 2-μL injection	1	0.01
SIM, splitless, 2-μL injection	0.1	0.001

[a] Concentration at which this analyst feels comfortable calling a sample positive.
[b] Assumes little or no competition from the sample, and a 5% recovery rate.

the source of the residue, this is a task that simply cannot be accomplished, with the possible exception of gasoline.

Fresh from the pump, gasolines look pretty much alike, particularly with respect to major components, and they change in a reasonably predictable fashion as a result of exposure to fire. The issue of evaporation frequently arises in the context of trying to decide whether gasoline was present in a sample before the fire or introduced during or after the fire. It sometimes happens that firefighters bring gasoline-powered equipment to fire scenes, and it is sometimes necessary to eliminate such equipment as the source of residue found in a debris sample. A number of different techniques can be brought to bear on this question. The simplest way to make the estimation is to compare the total ion chromatograms. In fresh gasoline, toluene will be the tallest peak, and there will be numerous light (C_5 to C_8) alkane peaks to the left of the toluene peak. By the time that gasoline is 25% evaporated, many of these light alkane peaks have disappeared. Also, in a 25%-evaporated gasoline sample, the tallest peak is the *m*- and *p*-xylene peak.

The analyst's library of gasolines should include gasoline in many stages of evaporation, as well as fresh gasoline. In our laboratory, we keep gasoline standards on file at 0-, 25-, 50-, 75-, 90-, 98- and 99%-evaporated. Performing a rough estimation of the degree of evaporation happens every time we identify gasoline, simply because it is necessary to put a hard copy of at least one standard in the file (ASTM E1618), and it is desirable for the reference chromatogram to match the sample chromatogram as closely as possible.

By the time gasoline is 50% evaporated, the toluene peak is significantly shorter than the *m*- and *p*-xylene peak, and in fact, is shorter than the ethylbenzene peak. The pseudocumene peak, meanwhile, has grown so that it is almost as tall as the *m*- and *p*-xylene peak. At 75% evaporation, the pseudocumene peak is the tallest peak in the chart, and the toluene peak has almost disappeared. At 90% evaporation, the toluene peak is gone and the xylenes are far shorter than the C_3 alkylbenzenes. The C_4 alkylbenzenes, meanwhile, have grown as tall as most of the C_3 alkylbenzenes, but pseudocumene is still the tallest peak in the chart. By the time the sample is 90% evaporated, the C_4 alkylbenzenes are taller than the C_3 alkylbenzenes.

A more quantitative technique for estimating evaporation involves use of the mass spectrometer. Collecting the average mass spectrum from C_6 to C_{13} (about 3 to 13 min in our laboratory) yields an average that changes predictably from one stage of evaporation to the next. In fresh gasoline, ions 91 and 105 are the tallest, followed by 119, 43, 57, and 71. The 43 fragment is produced by the light branched alkanes to the left of toluene. At 25% evaporation, 91 and 105 still dominate the mass spectrum, but the 57 ion is now taller than the 43. At 50% evaporation, ion 91 is no longer equal to 105, and 119 and 120 exceed the height of ion 57.

At 90% evaporation, the 119 ion is almost equal in size to the 105 ion, and ion 134 is beginning to catch up with ion 91. At 98% evaporation, ion 119 is the tallest, followed by 105, 134, and 91. Figure 3.36 shows these six average mass spectra. Keep in mind that the substrate can contribute to the spectra, thus making this approach more difficult in some cases.

Any attempt to estimate evaporation should be done by comparing "apples to apples." Although it is hoped that the analyst has optimized the isolation process so that an eluate from an ACS looks like the liquid in solution that has not gone through the ACS process, standards should be eluates rather than solutions prepared simply by diluting the liquids. Further, samples and standards should have approximately equal concentrations. More concentrated samples tend to appear more evaporated, due to the effects of displacement (Newman et al., 1996).

There are certain substrates that hold certain compounds preferentially, and substrate effects need to be considered before reporting a sample as having a problematic evaporation level. Concrete, for example, tends to retain lighter hydrocarbons, making residues appear less evaporated than the fire scene conditions

98 ANALYSIS OF FIRE DEBRIS

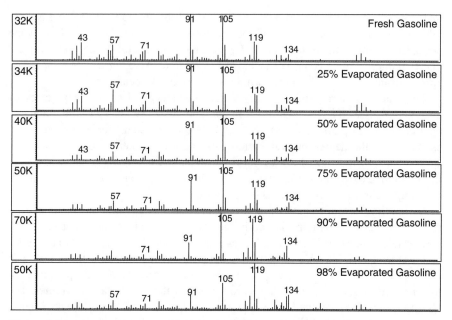

Figure 3.36 Average mass spectra from 3 to 13 min of fresh gasoline, followed by the same spectrum taken for gasoline evaporated 25, 50 75, 90, and 98%.

would indicate. One of the more important reasons for estimating the degree of evaporation is to allow for an "apples to apples" comparison when attempting to match a questioned sample to a known source.

3.6.4 Identity of Source

The ability to match a weathered ignitable liquid residue with a proposed source of unburned ignitable liquid has eluded fire debris analysts until just recently. The environmental forensics community has developed some tools to identify the sources of petroleum spills, but they have the advantage of having huge quantities of material available for characterization. Petroleum hydrocarbons contain trace quantities of biomarkers, substances that have changed little since they were first synthesized inside a living organism. The relative amounts of these biomarkers frequently allows for the identification of the source of a major spill. Polycyclic aliphatic and aromatic hydrocarbons are also useful in "fingerprinting" hydrocarbons. Examination of additives, such as oxygenates, or in the case of old gasolines, the alkyl lead compounds, can also provide clues as to the identity of the source of a spill.

There are three levels of "control" on the composition of petroleum. The primary control is the genesis of the petroleum: its "genetic" features, including its geographic location, the original source of the fuel (terrestrial or marine), and the conditions in the rock where the oil resided for millions of years.

The second level of control affecting the hydrocarbon fingerprint is processes imposed on the hydrocarbon by the refiner. Activities such as distillation, hydrocarbon cracking, isomerization, alkylation, and blending change the composition of the fuel. The tertiary controls are those that occur after the petroleum product leaves the refiner. The most important tertiary control of interest to fire debris analysts is, of course, weathering, or evaporation. Mixing in the service station tank is also critical. Because the tanks are seldom empty, each time a new delivery is made, the fuel service station tank assumes a new temporary identity. For a more extended discussion on chemical fingerprinting in the environmental forensic science arena, see *Introduction to Environmental Forensics* (Murphy and Morrison, 2002).

Some of the early work to identify the source of ignitable liquid residues from fire debris was done by Dale Mann (1987a,b). Using a 60-m column, Mann learned that it was possible to make comparisons on samples of fresh gasoline and to correctly identify the source by examining the relative ratios of the light hydrocarbons, ranging from n-pentane to n-octane. It is, unfortunately, exactly these hydrocarbons that are lost before a gasoline is even 25% evaporated. Mann's second paper describes the limitations of making comparisons between ignitable liquid residues and fresh samples, due to the contamination of residues with pyrolysis products, changes introduced by the isolation method, and the loss of volatiles resulting from weathering. Comparison of residues with fresh sources was, therefore, not studied frequently after Mann's extensive work. More recently, however, Dolan and Ritacco reported that they had devised a way to measure the relative abundances of 20 peak pairs in gasolines and were able to identify the source of 30 samples that were evaporated to 25% and 50%. This process uses the relative abundances of branched alkanes that occur generally in the center of the gasoline chromatogram. Because the peaks are sequential, eluting only a few seconds apart, the relative ratios are not affected significantly by evaporation. The peaks examined are all minor components. The major components of gasoline are similar enough to each other that they are not useful in discriminating between sources (Dolan and Ritacco, 2002).

Following up on this work, Wintz and Rankin applied principal components analysis only to learn that while the ratios within the pairs show little if any change between unevaporated and 50%-evaporated gasoline, only a few of those pairs were actually useful in distinguishing between one gasoline source and another, and some of those pairs occurred in the light petroleum distillate range, and so are unlikely to be found in gasolines evaporated to more than 50% (Wintz and Rankin, 2004).

Further work by Barnes and Dolan identified six ratios of sequentially eluting minor components in 50% evaporated gasoline and four ratios in 75% evaporated gasoline. Using these ratios, they were able to discriminate successfully among 16 gasoline samples evaporated 50% and 10 gasoline samples evaporated 75%. They used a blind study to confirm the ability of the technique to identify the source of three gasolines correctly matched against a library of 10 candidate sources. The technique uses a 60-m column, and is automated using a target compound program. The program takes the abundance of the base peak from the mass spectrum of each compound of interest and divides the base peak abundance of the later eluting compound by the base peak abundance of the earlier eluting compound. As well as this technique appears to work, the authors state that it is most useful for *eliminating* a source, and caution that the analyst's statement of conclusions reflect that a common origin is indicated but is not a certainty. (Barnes et al., 2004).

A somewhat simpler approach, using an analysis of naphthalene, methyl, and dimethylnaphthalenes has been reported by Sandercock and Du Pasquier in a three-part series published in 2003 and 2004 (Sandercock and Du Pasquier, 2003, 2004a,b). The column used was an everyday 30-m HP-5 column of the type used for routine fire debris analysis. A SIM method was used, and in order to have all of the peaks capable of being integrated at the same approximate level of precision, the signal for naphthalene and the methylnaphthalenes was selectively reduced by monitoring ions 127 and 129 rather than 128 for naphthalene, and 141 and 143 rather than 142 for methylnaphthalene. The base peak of 156 was used for the dimethylnaphthalenes. Sandercock and Du Pasquier started their research by pretreating the gasoline with an alumina column to separate out both polynuclear aromatic hydrocarbons and the polar alkylphenols, but learned that the alkylphenols did not provide much help in discriminating among different samples. The PAH ratios do not change significantly upon evaporation and seem to be a characteristic imparted by the refinery.

One way to compare any type of ILR to a suspected source, or to see if two ILRs within the same class came from the same source is to compare the average molecular weights. When exposed to a fire, ignitable liquid evaporates, and its average molecular weight necessarily increases. No predictable experience will cause the average molecular weight of the fire-exposed liquid to decrease. If the suspected source of an ignitable liquid, which was not exposed to the fire, exhibits a higher average molecular weight than the residue extracted from samples collected at the scene, the suspected source can be conclusively eliminated.

In doing comparisons of gasolines or any ignitable liquid, the best approach takes the proposed source liquid, evaporates some of it so that it matches the residue in terms of carbon number range, and then subjects it to the same separation procedure that was applied to the questioned sample. Once this has been accomplished, a detailed examination of the finer points in the chromatogram can be carried out.

A really good match between chromatograms not only matches the peaks but also matches the valleys. Examining multiple extracted ion chromatograms for both quality and quantity is a necessary step when doing comparisons. This practice has been codified with respect to the analysis of distillates heavier than kerosene in ASTM D5739-06, *Standard Practice for Oil Spill Source Identification by Gas Chromatography and Positive Ion Electron Impact Low Resolution Mass Spectrometry* (ASTM D5739-06, 2006). The approach in this method compares 24 extracted ion chromatograms of polynuclear aromatics and certain biomarkers. Although this method is not applicable to kerosene and lighter mixtures such as gasoline, the general approach seems workable. Given the state of the art, it is probably a good thing that comparisons are not frequently requested.

3.7 REPORTING PROCEDURES

The analyst's report can be one of the most important documents generated during the investigation of a fire. The report should therefore be written carefully enough so that readers are not misled. The objective of fire debris analysis is to determine whether any foreign ignitable liquid residue is present in a sample.

The report from a forensic science laboratory is supposed to be a scientific report. As such, it should include an introduction, a section detailing the test methods and results, and a discussion and conclusion section, if necessary. The laboratory report should also state what was done with the evidence. There is an unfortunate tendency in some agencies to provide checklists and merely state "positive for gasoline" or "negative for accelerants" rather than preparing a real laboratory report. Reading such a report, a reviewer cannot tell whether the analyst used a GC–MS or a Ouija board!

Given that it is now possible to store templates with all possible results in them, the excuse that it is "too time-consuming" to prepare an understandable narrative report is unacceptable. Although laboratory reports are intended to assist investigators, they should also be understandable to people asked to review the work. At a minimum, a report should include the following:

- Identification of the fire in question.
- A description of the how the sample was delivered to the laboratory, when, and by whom.
- A description of the samples, including container size, substrate material, and a reported location from where the sample was collected.
- A description of the isolation procedure used to separate the ILR from the sample substrate.
- A description of the analytical technique applied to the sample extract.
- The results of the analysis of the data.
- A discussion of the meaning of the results if there is any chance of misinterpretation or misuse.

In the discussion section, the analyst can provide examples of potential sources for whatever ignitable liquid residues may have been identified. This is also an appropriate place to put in a disclaimer about the possibility that the ILR may not be foreign to the background.

- A conclusion or bottom line understandable even to an attorney is a helpful thing.
- A sentence stating what has happened to the sample.

If a substance is found that is natural or incidental to the background, it is the analyst's job to say so. ASTM E1618 allows for the inclusion of disclaimers on both positive and negative reports. With respect to negative reports, a disclaimer to the effect that negative results do not preclude the possibility that ignitable liquids were present at the fire scene is suggested as an aid to help readers avoid misunderstanding the report. Similarly, in the case of a positive report, the standard states: "It may be appropriate to add a disclaimer to the effect that the identification of an ignitable liquid residue in a fire scene does not necessarily lead to the conclusion that a fire was incendiary in nature. Further investigation may reveal a legitimate reason for the presence of ignitable liquid residues." Certainly, a finding of gasoline in the living room is noteworthy. It is not so noteworthy to find gasoline in the basement near the chain saw. On occasion, it seems that the fire debris analyst is too eager to "help" with an arson investigation. Reporting a sample as being "positive" when all the analyst has identified is background compounds is neither helpful nor morally defensible. Karen and Paul Stanley of Akron, Ohio, were falsely accused of setting a fire that killed their infant son. The fire debris samples were submitted to a laboratory that reported finding turpentine in a sample consisting of charred Douglas fir. The laboratory analyst had absolutely no business reporting turpentine, or at the very least, should have reported the finding of turpentine with a huge disclaimer stating that the turpentine was indistinguishable from naturally occurring turpentine found in coniferous woods. The case was dismissed after the prosecutor understood the significance (or lack thereof) of the chemist's findings (Trexler, 2002).

A similar case of an analyst being too "helpful" occurred in a homicide case in Georgia, where the laboratory analyst reported finding "toluene, a flammable liquid" on a suspect's "clothing." Actually, the "clothing" included a pair of tennis shoes, and the examination of a new pair of tennis shoes directly from the shoestore revealed the presence of a high concentration of toluene. A closer examination of the first analysis revealed that in addition to toluene, diethylene glycol and butylated hydroxytoluene (BHT) were also present in the suspect's shoes and the exemplar shoes in identical relative proportions. Less than a year earlier, there had been a presentation on the analysis of suspects' shoes in arson cases at an AAFS seminar (Cherry, 1996). Perhaps if the analyst had attended that presentation (or read the proceedings), the toluene would have been reported properly as being native to the sample background, or more appropriately, not reported at all. Typical positive and negative reports from our laboratory are shown as Figure 3.37.

3.8 RECORD KEEPING

Each case file that includes a positive identification of an ignitable liquid residue should contain not only the sample charts but also a standard to which that sample can be compared. This means that both the sample and the standard should be printed with the same time scale, so that the data can easily be reviewed. Even if there is no criminal prosecution, the analyst should be aware that many fire cases involve civil litigation, so the case file should be kept for a reasonable period of time.

Electronic data files should also be protected and stored. There are only a few versions of GC–MS software in widespread use, so it is possible for one analyst

APPLIED TECHNICAL SERVICES, INCORPORATED

1190 Atlanta Industrial Drive, Marietta, Georgia 30066
• (770)423-1400 Fax (770) 424-6415 e-mail fire@atslab.com website www.atslab.com
This report may not be reproduced except in full.

CHEMICAL ANALYSIS REPORT

| **File No.** Lab File # | **Date** | **Page** 1 **of** 1 |

Fire Investigator
Agency
Address
City, State ZIP

Subject

Incident # (Claim # 78910)
Your File # 23456.
Victim (Insured): Name.
Date Of Loss: July 28, 2004.
Analysis of Fire Evidence.

Background

On August 3, 2004, John Lentini of ATS received from client via UPS the following:

    Item 1. A one-quart can containing debris identified as removed from the north end of the laundry room.

    Item 2. A one-quart can containing debris identified as removed from the center of the laundry room.

ATS was requested to analyze the samples for ignitable liquid residues.

Test Methods and Results

The samples were separated according to ASTM Practices E1412-07 and E1386-05, and analyzed according to ASTM Standard Method E1618-10.

Gas chromatographic/mass spectrometric (GC/MS) analysis of concentrated headspace vapors from Item 1 reveals the presence of components having retention times and mass spectra characteristic of components of known weathered gasoline.

GC/MS analysis of a solvent extract from Item 2 reveals the presence of components having retention times and mass spectra characteristic of components of known weathered gasoline.

Conclusion

Gasoline was present in both samples.

Sample Disposition

Samples have been returned to client via UPS.

Prepared by_____ John J. Lentini, F-ABC

Reviewed by_____ P. E. Rogers, Senior Chemist

Figure 3.37 (a) Typical positive report from our laboratory; (b) typical negative report from our laboratory.

 APPLIED TECHNICAL SERVICES, INCORPORATED

1190 Atlanta Industrial Drive, Marietta, Georgia 30066
• (770)423-1400 Fax (770) 424-6415 e-mail fire@atslab.com website www.atslab.com
This report may not be reproduced except in full.

CHEMICAL ANALYSIS REPORT

File No. Lab File # **Date** **Page** 1 of 1

Fire Investigator
Agency
Address
City, State ZIP

Subject

Incident # (Claim # 78910)
Your File # 23456.
Victim (Insured): Name.
Date Of Loss: July 21, 2004.
Analysis of Fire Evidence.

Background

On August 4, 2004, John Lentini of ATS received from client via UPS the following:

> Item 1. A one-quart can containing burned carpet and pad identified as removed from the north end of the south bedroom.
>
> Item 2. A one-quart can containing debris identified as removed from the center of the living room.

ATS was requested to analyze the samples for ignitable liquid residues.

Test Methods and Results

The samples were separated according to ASTM Practices E1412-07 and E1386-05, and analyzed according to ASTM Standard Method E1618-10.

Gas chromatographic/mass spectrometric (GC/MS) analysis of concentrated headspace vapors from Items 1 and 2 fails to reveal the presence of any ignitable liquid residues.

These results do not eliminate the possibility that ignitable liquids were present at the fire scene.

Sample Disposition

Samples have been returned to client via UPS.

Prepared by_____ John J. Lentini, F-ABC

Reviewed by_____ P. E. Rogers, Senior Chemist

Figure 3.37 (*Continued*)

to review another's raw data. Such reviews should be facilitated by preservation of the data. Instrumental data such as tune reports or spectrum scans should be kept, if only to keep track of the performance of the instrument. Certainly, if a case file has matching standards and samples, there is no need to review the tune report for that particular week. No matter what happened, the instrument tune parameters would not cause a false positive. Blanks, on the other hand, should be stored somewhere, particularly for those cases where the detection limit of the procedure is being reduced.

3.9 QUALITY ASSURANCE

The factors affecting the overall quality of a laboratory's work product are numerous. For this reason, accreditation programs have been set up so that an outside agency can verify that a laboratory is performing appropriately. Each laboratory should have a detailed written procedure for the examination of fire debris. As part of the quality assurance program, fire debris analysts should participate in proficiency testing at least once a year. This can be internal or external, but certainly, external proficiency testing carries more credibility. Proficiency tests are available from a number of commercial suppliers, but it is not necessary to actually purchase proficiency tests. Groups of analysts can form round-robin groups and take turns preparing samples for each other. Because it is not necessary to demonstrate one's ability to place a strip in a can, samples can be prepared by exposing 10 strips to the same sample, then just mailing the strips. By doing regular proficiency tests or round robins, analysts can monitor their own performance, and laboratory directors can be assured that their analysts are performing adequately.

Individual professional development and continuing education is an important component of any quality assurance program. Analysts should take the time to keep up with the literature and should attend professional meetings when possible. Although the science of fire debris analysis is reasonably settled, this is an interesting field with a large cadre of professionals who perform and publish research on a continuing basis. Keeping up with this research is a professional responsibility.

The ultimate sign that an analyst is keeping up with the profession is the decision to become certified. The American Board of Criminalistics offers a certification in fire debris analysis. Certification can be maintained only by continuing education and annual participation in external proficiency testing. Becoming certified is a way for an analyst to demonstrate that he or she cares about professional development. Supporting individual certification is a way that a laboratory director can assure his or her employees that the agency also cares.

CONCLUSION

The isolation and identification of ignitable liquid residues from fire debris samples is an important part of the fire investigation process. The fire debris analyst should be familiar with the common techniques of fire investigation and understand the language used by investigators. The analyst must also understand the stakes involved. If an ignitable liquid residue is identified, a hypothesis that a fire was incendiary may be supported, and a long litigation process may ensue in either civil or criminal court. The laboratory results are often the deciding factor in a prosecutor's decision to indict or an insurance company's decision to resist a claim.

The technology for fire debris analysis has improved dramatically over the last 20 years, to the point where our methods are now as sensitive as they need to be, and possibly more sensitive than they need to be. The analytical procedure requires focus and creativity, but also adherence to the generally accepted criteria for making an identification. The analysis should be conducted and reported in such a way that it is capable of being reviewed by another analyst. There is no reason that a reviewing analyst should reach a different conclusion, and such reviews should not only be expected but be welcomed. Science is based on "multiple witnessing." Fire debris analysts need to be acutely aware of the stakes involved in what they are doing and of the need to communicate effectively about the meaning of their findings.

REFERENCES

AA Notes (1982). *Arson Anal. Newsl.*, **6(3)**:57.

Armstrong, A., and Lentini, J. (1997). Comparison of the eluting efficiency of carbon disulfide with diethyl ether: the case for laboratory safety. *J. Forensic Sci.*, **42(2)**:307.

ASTM D3257-07 (2007). *Standard Test Methods for Aromatics in Mineral Spirits by Gas Chromatography*. Annual Book of Standards, Vol. 6.03. West Conshohocken, PA: American Society for Testing and Materials.

ASTM D5739-06 (2006). *Standard Practice for Oil Spill Source Identification by Gas Chromatography and Positive Ion Electron Impact Low Resolution Mass Spectrometry*. Annual Book of Standards, Vol. **11.02**. West Conshohocken, PA: American Society for Testing and Materials.

ASTM E1386-10 (2010). *Standard Practice for Separation and Concentration of Ignitable Liquid Residues from Fire Debris Samples by Solvent Extraction*. Annual Book of Standards, Vol. **14.02**. West Conshohocken, PA: American Society for Testing and Materials.

ASTM E1387-01 (2001). *Historical Standard Test Method for Ignitable Liquid Residues in Extracts from Fire Debris Samples by Gas Chromatography*. Annual Book of Standards, Vol. **14.02**. West Conshohocken, PA: American Society for Testing and Materials.

ASTM E1388-05 (2005). *Standard Practice for Sampling of Headspace Vapors from Fire Debris Samples*. Annual Book of Standards, Vol. **14.02**. West Conshohocken, PA: American Society for Testing and Materials.

ASTM E1412 (2007). *Standard Practice for Separation of Ignitable Liquid Residues from Fire Debris Samples by Passive Headspace Concentration with Activated Charcoal*. Annual Book of Standards, Vol. **14.02**. West Conshohocken, PA: American Society for Testing and Materials.

ASTM E1413-07 (2007). *Standard Practice for Separation and Concentration of Ignitable Liquid Residues from Fire Debris Samples by Dynamic Headspace Concentration.* Annual Book of Standards, Vol. **14.02**. West Conshohocken, PA: American Society for Testing and Materials.

ASTM E1492-05 (2005). *Standard Practice for Receiving, Documenting, Storing, and Retrieving Evidence in a Forensic Science Laboratory.* Annual Book of Standards, Vol. **14.02**. West Conshohocken, PA: American Society for Testing and Materials.

ASTM E1618-10 (2010). *Standard Test Method for Ignitable Liquid Residues in Extracts from Fire Debris Samples by Gas Chromatography–Mass Spectrometry.* Annual Book of Standards, Vol. **14.02**. West Conshohocken, PA: American Society for Testing and Materials.

ASTM E2154-01 (2008). *Standard Practice for Separation and Concentration of Ignitable Liquid Residues from Fire Debris Samples by Passive Headspace Concentration with Solid Phase Microextraction (SPME).* Annual Book of Standards, Vol. **14.02**. West Conshohocken, PA: American Society for Testing and Materials.

Barnes, A., Dolan, J., Kuk, R., and Siegel, J., (2004). Comparison of gasolines using gas chromatography–mass spectrometry and target ion response. *J. Forensic Sci.*, **49(5)**:1018.

Cherry, C. (1996). Arsonist's shoes: clue or confusion. *Proceedings of the American Academy of Forensic Sciences*, Nashville, TN, p. 20.

Chrostowski, J., and Holmes, R. (1979). Collection and determination of accelerant vapors. *Arson Anal. Newsl.*, **3(5)**.

Dietz, W. R. (1993). Improved charcoal packaging for accelerant recovery by passive diffusion. *J. Forensic Sci.*, **38(1)**:165.

Dolan, J. (2004). Analytical methods for the detection and characterization of ignitable liquid residues from fire debris. In: Almirall, J., and Furton, K. (Eds.), *Analysis and Interpretation of Fire Scene Evidence.* Boca Raton, FL: CRC Press, p. 152.

Dolan, J., and Ritacco, C. (2002). Gasoline comparisons by gas chromatography–mass spectrometry utilizing an automated approach to data analysis. *Proceedings of the American Academy of Forensic Sciences Annual Meeting*, Atlanta, GA, Feb. 16, p. 62.

IAAI Forensic Science Committee (1988). Guidelines for laboratories performing chemical and instrumental analyses of fire debris samples. *Fire Arson Investig.*, **38(4)**:45.

Juhala, J. A. (1982). A method for adsorption of flammable vapors by direct insertion of activated charcoal into the debris samples. *Arson Anal. Newsl.*, **6(2)**:32.

Kirk, P. (1969). *Fire Investigation.* New York: Wiley, p. 153.

Lentini, J. (1998). Differentiation of asphalt and smoke condensates from liquid petroleum distillates using GC–MS. *J. Forensic Sci.*, **43(1)**:97.

Lentini, J. (2001). Persistence of floor coating solvents. *J. Forensic Sci.*, **46(6)**:1470.

Lentini, J., and Waters, L. (1982). Isolation of accelerant-like residues from roof shingles using headspace concentration. *Arson Anal. Newsl.*, **6(3)**:48.

Lentini, J., Dolan, J., and Cherry, C. (2000). The petroleum-laced background. *J. Forensic Sci.*, **45(5)**:968.

Mann, D. C. (1987a). Comparison of automotive gasolines using capillary gas chromatography: I. Comparison methodology. *J. Forensic Sci.*, **32(3)**:606.

Mann, D. C. (1987b). Comparison of automotive gasolines using capillary gas chromatography: II. Limitations of automotive gasoline comparisons in casework. *J. Forensic Sci.*, **32(3)**:616.

Midkiff, C. (1982). Arson and explosive investigation. In: Saferstein, R. (Ed.), *Forensic Science Handbook*. Englewood Cliffs, NJ: Prentice Hall, p. 225.

Murphy, B., and Morrison, R. (Eds.), (2002). *Introduction to Environmental Forensics*. San Diego, CA: Academic Press.

Newman, R., and Dolan, J. (2001). Solvent options for the desorption of activated charcoal in fire debris analysis. *Proceedings of the American Academy of Forensic Sciences*, Seattle, WA, Feb., p. 63.

Newman, R. T., Dietz, W. R., and Lothridge, K. (1996). The use of activated charcoal strips for fire debris extractions by passive diffusion: I. The effects of time, temperature, strips size, and sample concentration. *J. Forensic Sci.*, **41(3)**:361.

Newman, R., Gilbert, M., and Lothridge, K. (1998). *GC–MS Guide to Ignitable Liquids*. Boca Raton, FL: CRC Press.

NFPA 325 (1994). *Guide to Fire Hazard Properties of Flammable Liquids, Gases and Volatile Solids*. Quincy, MA: National Fire Protection Association.

NFPA 921 (2004). *Guide for Fire and Explosion Investigations*. Quincy, MA: National Fire Protection Association, pp. 45–49.

Nowicki, J. (1990). An accelerant classification scheme based on analysis by gas chromatography–mass spectrometry (GC–MS). *J. Forensic Sci.*, **35(5)**:1064.

Sandercock, M., and Du Pasquier, E. (2003). Chemical fingerprinting of unevaporated automotive gasoline samples. *Forensic Sci. Int.*, **134**:1–10.

Sandercock, M., and Du Pasquier, E. (2004a). Chemical fingerprinting of gasoline: 2. Comparison of unevaporated and evaporated automotive gasoline samples. *Forensic Sci. Int.*, **140**:43–59.

Sandercock, M., and Du Pasquier, E. (2004b). Chemical fingerprinting of gasoline: 3. Comparison of unevaporated automotive gasoline samples from Australia and New Zealand. *Forensic Sci. Int.*, **140**:71–77.

Smith, R. M. (1982). Arson analysis by mass chromatography. *Anal. Chem.*, **54(13)**:1399.

Stauffer, E. (2003). Concept of pyrolysis for fire debris analysts. *Sci. Justice*, **43**:29–40.

Stone, I. C. (1976). Communication to *Arson Anal. Newsl.*, **1(1)**:5.

Trexler, P. (2002). Prosecution expert rejects short as cause. *Akron Beacon Journal*, Akron, OH, Feb. 8.

U.S. Department of Justice (1999). *Forensic Sciences: Review of Status and Needs*. Washington, DC: US DOJ, Office of Law Enforcement Standards.

Waters, L., and Palmer, L. (1991). Multiple analysis of fire debris using passive headspace concentration. *J. Forensic Sci.*, **36(1)**:111.

Wintz, J., and Rankin, J. (2004). Application of principal components analysis in the individualization of gasolines by GC–MS. *Proceedings of the American Academy of Forensic Sciences*, Dallas, TX, Feb., p. 48.

CHAPTER 4

Forensic Examination of Soils

RAYMOND C. MURRAY

University of Montana, Missoula, Montana

Summary Soils and sediment constitute excellent trace evidence in both criminal and civil cases because there are an almost unlimited number of identifiable soil types based on the content of rocks, minerals, glasses, human-made particles, and chemicals. Forensic examination commonly identifies the original geographic location of soils associated with a crime, thus assisting an investigation. Studies of soil and related material samples associated with a suspect and crime scene can produce evidence that the samples had or did not have a common source, thus indicating whether or not the suspect was ever at a particular place. Gathering intelligence and mine, gem, and art fraud studies often use the methods of forensic geology.

4.1	Introduction	110
4.2	Murder and the pond	111
4.3	Oil slicks and sands	113
4.4	Medical link	114
4.5	Examination methods	114
	4.5.1 Color	115
	4.5.2 Particle-size distribution	117
	4.5.3 Stereo binocular microscope	120
	4.5.4 Petrographic microscope	122
	4.5.5 Refractive index	124
	4.5.6 Cathodoluminescence	124
	4.5.7 Scanning electron microscope	125
	4.5.8 X-Ray diffraction	126
4.6	Chemical methods	127
	4.6.1 FTIR and Raman spectroscopy	128
4.7	Looking ahead	129
	References	130

Forensic Chemistry Handbook, First Edition. Edited by Lawrence Kobilinsky.
© 2012 John Wiley & Sons, Inc. Published 2012 by John Wiley & Sons, Inc.

4.1 INTRODUCTION

As with so many other types of physical evidence, forensic geology began with the writings of Sir Arthur Conan Doyle, who wrote the Sherlock Holmes series between 1887 and 1927. He was a physician who apparently had two motives: writing salable literature and using his scientific expertise to encourage the use of science as evidence.

In 1893, Hans Gross, an Austrian forensic scientist, wrote a handbook for examining magistrates (Gross, 1893) in which he suggested that "perhaps the dirt on someone's shoes could tell more about where a person had last been than toilsome inquiries." It was only a matter of time before these ideas from an author of fiction and a criminalists' handbook would appear in a courtroom.

In 1908, a German chemist, Georg Popp, undertook the study of soil evidence in the homicide case of Margarethe Filbert. In this case Popp was able to demonstrate that soil samples from two locations associated with the crime had characteristics similar to the sample collected from the suspect's shoes. In addition, using soil from the suspect's shoes he was able to show a sequence of events in soil accumulation that was consistent with the theory of the crime, and he found no soil evidence that supported the suspect's alibi (Thorwald, 1967).

A century later, the use of geologic materials in criminal and civil cases is commonplace (Pye and Croft, 2004). Public and private laboratories around the world have trace evidence examiners qualified to examine soil, glass, and related material (Murray, 2011; Pye, 2007; Ruffell and McKinley, 2008).

Forensic geology studies vary in scope. A common type of investigation involves identifying a material that is key to a case: for example, examining pigments in a painted picture or material in a sculpture when authenticity or value is at issue. Identification is also important in questions of mining, mineral, or gem fraud to determine if the material is what its sellers claim it to be. Identification of fire-resistant safe insulation on a person or individual's property may provide probable cause for further investigation.

Beyond identification, forensic geologists can also look at the source of particular material. Here the examiner needs a broad knowledge of the geology and the best geologic and soil maps to answer questions. For example, if the soil on a body does not match the location where the body is found, from where was the body moved? Similarly, examiners can compare two samples, one associated with the suspect and the other collected from the crime scene, to see if they had a common source. For example, does the soil on the suspect's shoe have characteristics similar to those of the soil type collected at the crime scene?

Another new developing area of forensic geology is its use in intelligence work. For example, a person may claim never to have been to a particular location, but is then found with rocks from that spot, thus linking the person to the location. Remember the outcrop you saw behind Osama bin Laden on your television screen after 9/11? What was the location? A geologist who has done field work in the area would be able to locate that outcrop, and that actually happened: A geologist, John Shroder, was able to identify the region where bin Laden had been sighted in

Afghanistan in 2001. As with all class evidence, geologic evidence rarely provides a unique solution for which the geologic mind cannot imagine another possibility (Murray, 2005). But there are some exceptions, as illustrated in the following two cases.

4.2 MURDER AND THE POND

The murder of John Bruce Dodson produced one of the most interesting cases in the entire history of forensic geology. Here, the geologic evidence is unequivocal, in that it tied the suspect directly to the crime and eliminated the suspect's alibi. Most important, the investigator of the crime recognized the potential importance of the geologic evidence and arranged for an examination of that evidence. The testimony of the forensic geologist was critical to the prosecution of the case. The case began on October 15, 1995, when John Dodson was found dead while on a hunting trip with his wife of three months, Janice. The scene was high in the Uncompahgre Mountains of western Colorado.

At first glance it appeared to be a hunting accident. However, the autopsy revealed two bullet wounds to the body and one bullet hole through John's orange vest. The investigation showed that the Dodsons were camped near other hunters, one of whom was a Texas law enforcement officer. He responded to Janice's frantic call that her husband had been shot. She was standing about 200 yards from the camp in a grassy field along a fence line. The officer determined that John was dead and started the process of getting help. Prior to calling for help, Janice had returned to her camp and removed her hunting coveralls, which were covered with mud from the knees down. She later told investigators that she had stepped into a mud bog along the fence near the camp. Investigators found a. 308-caliber shell case approximately 60 yards from the body. In addition, they found a. 308-caliber bullet in the ground on the other side of the fence, which created a direct line from the location of the case to the body to the bullet.

Janice's ex-husband, J. C. Lee, was camped three-fourths of a mile from the Dodsons. Janice knew that the site was his favorite camp location. He naturally came under suspicion. However, Lee was hunting far away from camp with his boss at the time of the shooting. Most important, Lee reported to investigators that while he was out hunting, someone had stolen his. 308 rifle and a box of. 308 cartridges from his tent. Winter comes early at 9000 ft in the Umcompahgre, and little more could be done at the scene. However, investigators Bill Booth, Dave Martinez and Wayne Bryant returned during the summers of 1996, 1997, and 1998 and searched for the rifle and other evidence. They tried to search every place a weapon could have been hidden. They combed the entire area, including ponds, with metal detectors in the hope of finding the rifle; it has never been found. During the final search of the pond near Janice's ex-husband's camp, Al Bieber of NecroSearch International commented that the mud in and around a cattle pond near Lee's camp was bentonite, a clay that someone brought to the pond to stop the water from seeping out of the bottom. That evening, Booth and Martinez were camped near the crime scene. They were discussing the evidence in the

Figure 4.1 Pond lined with bentonite clay where Janice Dodson got clay on her shoes and pants when she went to steal the rifle that she used to kill her husband, John. *(See insert for color representation.)*

case when investigator Booth said, "The mud." He was referring to the dried mud that was found on Janice Dodson's clothing. If Janice had obtained the rifle from Lee's camp, she would probably have stepped or fallen into the bentonite clay that drained across the road from the cattle pond (see Figure 4.1). Remembering Janice's statement that she was returning to camp on the morning of the crime and stepped into a mud bog near her camp, Booth and Martinez decided they needed to obtain dried mud samples from the bog near the Dodsons' camp, the area around a pond nearby the camp, and the human-made pond and runoff near Lee's camp.

Booth and Martinez packaged the dried mud from each location and sent the samples along with the dried mud that had been recovered from Janice's overalls to the laboratory section of the Colorado Bureau of Investigation in Denver, where it was examined by Jacqueline Battles, a forensic scientist and lab agent. She concluded and later testified to the fact that the dried mud found on Janice Dodson's clothing was consistent with the dried mud recovered from the pond near Lee's camp. The dried mud that had been recovered from Janice's overalls was found not to be consistent with the mud bog or the pond near her camp. This was a breaking point in the case that allowed Booth and Martinez to put Janice Dodson in her ex-husband's camp around the time that his rifle had been stolen. There are no other bentonite-lined ponds in the area and no bentonite deposits.

Booth and Martinez went to Texas and served an arrest warrant on Janice. She was extradited to Colorado, tried in court, and convicted in the murder of John Bruce Dodson. The jury understood the results that followed Booth's insightful "the mud" exclamation. Janice is now serving a life sentence without the possibility of parole in Colorado's state prison for women. An appreciation of the value of soil evidence and the collection of that evidence was critical in this case.

4.3 OIL SLICKS AND SANDS

A case that illustrates many of the issues comparing soil and related material occurred in Canada a few years ago. The body of 8-year-old Gupta Rajesh was found alongside a road outside Scarboro, Ontario. The back of his shirt had a smear of oily material, and the preliminary conclusion was that he was the victim of a hit-and-run accident, with the oily material coming from the undercarriage of a vehicle. But examination of the oily material and the particles suspended in it by forensic geologist William Graves of the Centre of Forensic Sciences in Toronto told a different story.

Investigators had collected samples of oily material on the floor of an indoor concrete parking garage where a suspect, Sarbjit Minhas, parked her Honda automobile. Analysis of the samples showed that the sand and other particles within the oil from the victim's clothes and the parking garage were similar. Analysis of the oil from the victim's shirt and garage floor showed them to be both similar and different from oil collected on the floor of 10 other garages in the area.

Particles in samples from the victim's clothes and the suspect's parking place provided considerable information. The sand from both samples was sieved and subsamples produced of the various size grades for the two samples. When compared after the oil had been removed, the color of each pair of subsamples was identical.

Additionally, the heavy minerals in both samples were similar, and three distinct types of glass were found in the two samples: amber glass, tempered glass, and lightbulb glass. Each of the different glasses was identical in refractive index value (the amount that a ray of light bends when passing through the glass into another medium). Small particles of yellow paint with attached glass beads were found in both samples. This type of paint is often found on center stripes of highways and reflects light.

Graves concluded that there was a high probability that the body of Gupta Rajesh had been in contact with the concrete floor of the garage at the place where the suspect parked her car. Interestingly, the same oil and particles were found in the suspect's Honda. Whether the oil and particles on the victim came from inside the vehicle or the floor of the garage, the presence and distinctiveness of the samples still strongly associated those two areas with the victim. Minhas was tried in the Superior Court of the Province of Ontario in November 1983 and convicted, with help from testimony by Graves.

This case illustrates an important concept in the presentation of soil evidence and perhaps all physical evidence, except DNA. We have become awed and impressed by the high probabilities that result from DNA evidence. Some people expect that other types of evidence should have similar statistical information. But in the Minhas case, we see a conclusion based on at least 10 different materials and observations. Because we do not know the probability of a tempered glass fragment, a particular group of heavy minerals, or sand of the same color being on a particular parking place in a concrete garage in Scarboro, Ontario—and in all likelihood we will never know—a frequency statistic cannot be generated. A useful database of sands, particles, glass, oils, and heavy minerals would be too difficult to generate. Additionally, it may not apply to any one specific case because of the variability of mineral particles—the very distinctiveness that makes geologic materials such good evidence. Thus, we rely on the skilled and honest examiner to reach a conclusion expressed in words rather than in numbers to inform the jury or judge so that they can reach a verdict. In this way the expert is a teacher, instructing the judge, attorneys, and jury in the basic concepts and premises that allow them to do the work they do. The triers of fact must be schooled in the methods of production of the evidence (how light bulb glass is made, for example), the procedures used to analyze it, and what makes the evidence significant. That understanding will lead the courts to an appreciation of unquantifiable evidence and give the jury a basis for weighing its significance (Houck, 1999).

4.4 MEDICAL LINK

A recent case does not fit the pattern of most soil evidence but clearly illustrates the contribution being made by forensic geologists. Washington State Patrol forensic geologist Bill Schneck became involved in the investigation into the serious illness of a small child caused by arsenic poisoning. The suspected person was absolved when an examination of the child's house revealed a number of mineral specimens left in the house and the yard by a former occupant who was a mineral collector. Many of those specimens were arsenopyrite, an iron arsenic sulfide. The child had been eating and chewing on the material. This case is a good reminder that lead is not the only material that can cause health problems in children.

4.5 EXAMINATION METHODS

The value of any type of physical evidence depends in large part on how many different types that can be characterized exist on this Earth and how they are distributed. The value of soil evidence rests on the fact that there are an almost unlimited number of rock types, mineral types, fossils, and artificial rock material, including glass and brick and concrete. In addition, it is not uncommon for soil material to have incorporated human-made particles such as plastics, metals, and chemicals such as fertilizer and pesticides.

Various instruments, methods, and procedures are used to study minerals, rocks, soils, and related materials for forensic purposes. These methods are used to collect data from the various questioned and known samples, which are then used to make a judgment as to comparison or lack of comparison. The evidential value of some of these methods is greater than others. However, they are all standard methods long accepted for use in the identification of material. Because of the extreme diversity of soils and related material, the examiner may decide that in a particular case some other method will provide information that will assist in reaching a conclusion. Because soils often have added particles, such as fibers, hair, and/or paint chips, these can be collected for examination by experts in those forensic science fields (Junger, 1996). One of the reasons that microscopic examination is so important is that it is only way that particles of other forms of evidence can be found and unusual mineral particles discovered. We must remember that examination is focused on determining whether two samples have the same properties and thus have the possibility of having a common source. That determination is greatly strengthened when rare and unusual minerals and particles are found. The experienced forensic geologist can express an opinion based on experience and education (Stam, 2002). The strength of that opinion is often based on knowledge of the rarity of particular minerals, rocks, particles, and related properties.

4.5.1 Color

Color is one of the most important identifying characteristics of minerals and soils (Dudley, 1975; Sugita and Marumo, 1996; Guedes et al., 2009). Minerals form a mosaic of grays, yellows, browns, reds, blacks, and even greens and brilliant purples. Virtually all possible colors of the visible light spectrum are represented. With most geologic materials and soils, the native minerals contribute directly to the soil color. This is particularly true with stream deposits, windblown silts, and other recent formations that have been in place a comparatively short period of time. If sands along a river channel are examined, the color of each sand grain can generally be recognized individually; however, after a deposit has weathered for a long period of time, there is a degree of leaching, accumulation, and/or movement of substances within the soil. Soil particles become stained, coated, and impregnated with mineral and organic substances, giving the soil an appearance different from its original one. The mineral grains, especially the larger ones, are generally coated. In most situations the coatings on the soil particles consist of iron, aluminum, organic matter, clay, and other substances. The coloring of the coatings alone can give some indication as to the history of the sample.

The "redness" of a soil depends not only on the amount of iron present but also on its state of oxidation, with a highly oxidized condition tending to have a more reddish color. The iron on the coatings of the particles probably is in the form of hematite, limonite, goethite, lepidocrocite, and other iron-rich mineral forms. Black mineral colors in the soil are generally related to manganese or various iron and manganese combinations. Green colors are generally due to concentrations of specific minerals rather than of the mineral coatings. For example, some copper minerals, chlorite, and glauconite are usually green. Deep blue to purple coloration in the soil is generally due to the mineral vivianite, an iron phosphate.

Apart from the mineral colors in the soil are those that result from organic matter. The organic litter on the soil surface is generally black. Humus percolates through the mineral horizons, giving various dark colors. In some instances the iron and humic acids combine to form a dark reddish brown to nearly black color. To have some uniformity in descriptions of the color of geologic materials and soils, certain standards have been established. The color standards most frequently used in the United States are those of the Munsell Color Company. The color standards are established on three factors: hue, value, and chroma. *Hue* is the dominant spectral color, *value* is the lightness color, and *chroma* is the relative purity of the spectral color. Soil and rock colors are generally recorded as, for example, 7.5YR5/2 (brown). The 7.5YR refers to the hue, 5 the value, and 2 the chroma. This standardization of colors offers some degree of uniformity, but moisture content will also affect the color of the soil, as will light intensity and wavelength. Soil color is different in natural light than in fluorescent and different still in incandescent light. If a soil is air dry, it may be recorded as yellow, but if moist the recording may be yellowish brown. Moisture added to a dry soil will usually result in a more brilliant appearance. It is therefore important to record not only the color of the soil but also an estimate of the "wetness factor" at the time of the recording.

In the 1970s, investigators at the UK Home Office Forensic Science Service studied the use of color as an examination method and made many contributions to establishing the study of color as an important first step. Soil, being a mixture of materials of various sizes and compositions, contains individual minerals of different colors. If soil is fractionated into various sizes—coarse sand, medium sand, fine sand, silt, and clay—there is a tendency for the finer-sized particles to exhibit more red or reddish-brown colors as opposed to grays and yellows in the coarser fractions. In considering coarser sand particles, the matrix will commonly have a speckled appearance, with the quartz and feldspar particles being gray or yellowish; but the heavy minerals such as ilmenite and magnetite will generally be black. Sand particles from soils of recent origin, such as recent glacial or stream deposits, usually retain their original mineral appearance and we can usually detect a mosaic of colors. But sand fractions from the old landscapes commonly have coatings of clay, and the sand grains may be iron stained, which results in a more uniform color of the entire matrix. Soil grains veneered with organic matter give the particles a dark gray appearance. It is important first to record the color of the untreated soil sample and then to remove the organics so that the true color and appearance of the sand grains can be studied.

In studying soil samples for forensic purposes, the sample is normally dried at approximately 100°C and viewed with natural light, preferably coming from a northerly direction. A north-facing window is a good location for such observations. Such studies should be made on samples that have the same general size distribution of particles. The color of samples prepared from the individual sieved-out particle-size ranges gives important additional data. Two or more samples, collected for study, can be compared directly by the observer (see Figure 4.2). It is then possible to use a color chart with the samples to determine the Munsell color numbers for precise description of the color.

Instrumental determination of color provides more quantitative information and can now be done with a precision which exceeds that the human eye. Pye Associates of Great Britain have found that rapid, reliable, and highly reproducible results are achieved using the Minolta CM-2002 photospectrometer. This instrument covers the visible wavelength range 400 to 700 nm, providing a number of light source types and observer angles, and producing a range of color indices, including Munsell hue,

EXAMINATION METHODS   **117**

Figure 4.2 Two soil samples available for examination for color comparison. *(See insert for color representation.)*

value, and chroma; L*a*b* indices; and reflectance data and curves. Calibration of the instrument is carried out for every analysis to black, white, and CERAM II color standard materials.

4.5.2 Particle-Size Distribution

The determination of the distribution of particle sizes in a sample can often provide significant evidence. This determination is often produced for a variety of reasons: (1) to produce samples for comparison studies that are similar, in which case the control sample may contain some larger or smaller particles that are not present in the sample being questioned or an associated sample, and they must be removed; (2) the samples may be broken down into subsamples in which all the particles are in the same size range for mineral or color studies; or (3) a determination of the distribution of particle sizes may be produced as a method of comparison. A diagram showing the distribution of grain sizes can be used as a presentation and in many cases may be of evidential value (see Figure 4.3). For example, when abrasive particles have been introduced into machinery for the purpose of sabotage, the size distribution of the particles may be diagnostic of the material, assuming that changes in particle size have not taken place in the machinery.

The basic methods used for the separation of sizes are (1) passing the sample through a nest of wire sieves, with the size of the openings decreasing from top to bottom; (2) determining the rate of settling of the grains in a fluid, which is a measure of the size of the particles; and (3) instruments that measure the size of particles in a microscopic view and record the number of particles of each size. The distribution of particle sizes is then plotted on a diagram.

118   FORENSIC EXAMINATION OF SOILS

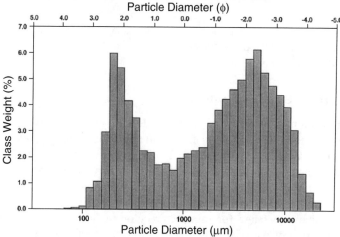

Figure 4.3 Nest of sieves used to separate particles into different size grades and a diagram showing the size distribution of particles. *(See insert for color representation.)*

Before making a mechanical analysis to determine the size distribution of particles, it is necessary to disperse the soil. Individual soil particles tend to stick together in the form of aggregates. Cementing agents of the aggregates must be removed; otherwise, a cluster of silt and clay particles would have the physical dimensions of sand or gravel. Cementing agents consist of organic matter, accumulated carbonates, and iron oxide coatings, and in some situations there is a mutual attraction of particles by physicochemical forces.

If carbonates have cemented the particles together, it is desirable to pretreat the sample with dilute hydrochloric acid to remove the carbonates. The sample is then treated with hydrogen peroxide to remove the organic cementing agents. All samples must be treated in the same way, and it must be determined before treatment that important information will not be lost, such as dissolving carbonate cement from grains that should be treated as single grains. It is almost always desirable to determine the size distribution of soil by sieving in a liquid, usually water. Dry sieving of the entire sample is generally unsatisfactory because the small particles tend to cluster together, and clay tends to adhere to larger particles. Sometimes a dispersing agent is added to the water.

A number of methods can then be used to determine the size distribution of the finer particles in a dispersed suspension. The hydrometer method is a rapid method for determining the percentage of sand, silt, and clay in a sample. It is based on the principle of a decreasing density of the suspension as the solid particles settle out. This method, although rapid and accurate, is unsatisfactory if we want to make a subsequent examination of the various size ranges, because there is actually no physical separation of the various-sized particles.

One of the most accurate and satisfactory procedures for fractionating soil samples is by the pipette method. This consists of pretreating the sample as is done in the hydrometer method, dispersing the soil in water, and calculating the time required for various-sized particles to settle out from the suspension. The principle is based on the fact that the rate of settling depends on the size of the mineral matter, with larger particles settling at a more rapid rate. The procedure is based on Stokes' law: $V = (2/9)gr^2(d - d')/n$, where V is the velocity of fall in centimeters per second g the acceleration due to gravity, r the radius of the particle in centimeters, d the density of the particle in grams per cubic centimeter, d' is the density of the fluid in grams per cubic centimeter, and n the viscosity of the fluid in poise. Although this method is generally considered the most satisfactory in regard to accuracy, it is not infallible. Several assumptions are made: that all particles have the same shape and that all the soil particles have the same density, neither of which is usually the case. Nevertheless, the pipette method is generally considered to be the best available mechanical method.

In making a mineral analysis of a sample, one fact becomes apparent: the sample contains different groups of minerals within various size ranges. Sands are made up of a set of minerals that are usually completely different from those within the clay size range. Therefore, in comparative analysis it is important to make comparisons within the same size ranges. It can be quite deceptive to compare the minerals found in one size range in one sample with a different size range in another sample.

When a transfer is made from a soil it is unlikely that the transferred sample is truly representative of the distribution of sizes that existed at that place. For this reason two samples that are to be compared using grain size distribution may not show exactly the same distribution of grain sizes. The method can be useful in coming to a conclusion of comparison or lack of comparison but is seldom definitive in itself.

Figure 4.4 Forensic scientist using a binocular microscope to identify particles in a soil sample.

4.5.3 Stereo Binocular Microscope

The stereo binocular microscope can be a most useful tool to the forensic geologist (see Figure 4.4). After an analysis of color has been made, the information that can be obtained by examination of soil samples with this instrument makes it the logical first step in the direct examination of particles.

Light microscopes are generally of two types, transmitted light and reflected light. In transmitted light microscopes, the light source is placed beneath the specimen, which must be transparent. Biological microscopes that are used in studying tissue are of this type. Reflected light microscopes have a light source above the object, and the surface features of the particle are viewed. Such a microscope is essentially a stationary, higher-power magnifying glass. Most of these microscopes have two sets of lenses, and thus the object is viewed in three dimensions, that is, in stereo. The magnification of a microscope is determined by multiplying the magnification of the ocular lens, commonly 10×, by the magnification of the objective lens, which differs from microscope to microscope but seldom exceeds 10×. This gives a maximum magnification of approximately 100×. The objectives may be individual lenses of fixed magnification, or in some microscopes a zoom objective is used. Such lenses can change magnification continuously from less than 1× to about 5×. Most viewing with stereo binocular microscopes is done at magnifications between 10× and 40×. Some of these microscopes are seated on a base that contains a second light source so that objects can be viewed in both transmitted and reflected light. When the transmitted light is polarized, the microscope may be used for both stereo reflected light viewing and low-power transmitted polarizing light studies.

Objects as small as approximately 10 μm in diameter can be viewed with a stereo binocular microscope. The upper limit is determined by how large a sample can fit under the instrument so that the surface of pebbles and cobbles can easily be viewed. The sample is normally placed on a tray having a dull black finish for ease of viewing light-colored minerals or a white finish for viewing dark-colored minerals. Various inserts are available for these microscopes that permit measurement of the size of objects or provide grids that aid in counting the various particles. The sample tray may have an etched grid that serves a similar purpose. Trays are available with various gummed surfaces that hold the grains in place for ease of counting.

In examining a soil sample or similar material, the scientist will commonly first examine the entire sample as it is received and observe the types of grains and particles. Recording a general impression of the material is normally done at this time. It is not uncommon to observe in soil samples nonmineral materials such as fibers, metals, paint, glass, and plastics. These objects can have important evidential value. In some situations they can be the most important parts of the sample. Materials such as metals, hair, fibers, paint, and plastic are removed for further examination by specialists. Plant particles can be of great value. The total amount of plant material is in general relatively useless for forensic purposes. However, identification of the individual grasses, seeds, leaves, and the like can be most useful.

Preliminary examination of the entire sample with the binocular microscope is normally very difficult. The mixture of particles of all sizes commonly obscures the grains and makes identification difficult. The presence of organic material contributes to this problem. The sample must be cleaned for the study of minerals and rock. Sieving of the samples removes the larger particles and most of the larger organic fragments. If the sample is washed carefully in water, the lighter organic particles will generally float and can be removed and saved for study. Treatment with hydrogen peroxide will remove the fine organic matter and clean the sample. The use of ultrasonic cleaners is dangerous in many cases. Many times the scientist has taken a sample that contained chips of red shale before ultrasonic cleaning and found that the sample contained thousands of silt-sized quartz grains and clay minerals after cleaning and that the shale chips had been broken up into its components. If the rocks and minerals are of a type that would not be disaggregated by ultrasonic cleaning, the method can be most useful.

With a clean sample, the experienced scientist can identify the rocks and minerals at sight or by using simple tests. In addition, it is possible to observe the texture and coatings on the surface of the grains. Properties of grains such as shape, rounding, weathering, inclusions, color, and polish can be observed and recorded. The counting of different types of grains is especially important. When we record our information in numbers, it is normally more useful than qualitative impressions. However, the samples may be so different on first examination that further work is not useful because a determination of comparison could never be made. The number or percentage of different types of grains is an extremely important tool in determining comparison or lack of comparison. In counting grains of different types, it is important that the sample counted be representative of the whole sample, that the identification be consistent and accurate, and that the same grain not be counted twice because the sample moved. The latter problem can usually be avoided by placing the sample on a gummed surface or by removing the grains as they are counted and placing each grain, as it is removed, in a container or gummed individual tray. Most important is the judgment and caution used by the scientist, whatever method is used.

4.5.4 Petrographic Microscope

A petrographic microscope differs in detail from an ordinary compound microscope (see Figure 4.5). However, its primary function is the same: to produce an enlarged image of an object placed on the stage. The magnification is produced by a combination of two sets of lenses, the objective and the ocular. The function of the objective lens, at the lower end of the microscope tube, is to produce an image that is sharp and clear. The ocular lens merely enlarges this image. For mineralogical work, three objectives—low, medium, and high power—are normally used. The magnification produced by objectives is usually $2\times$ (low), $10\times$ (medium), and $50\times$ (high). Oculars have different magnifications, usually $5\times, 7\times, 10\times, 15\times$, and $20\times$. The total magnification of the image is determined by multiplying the magnification of the objective by that of the ocular as follows: $50\times$ times $10\times = 500\times$ (see Figure 4.2).

Oculars normally contain a crosshair that is useful for locating grains under high power when changing objectives. A condensing lens system is normally provided under the stage for use with high magnification and for assisting the viewing of the various optical effects produced by minerals. Petrographic microscopes have a rotating stage and a polarizing filter under the stage that transmits light usually

Figure 4.5 Forensic scientist using a petrographic microscope to identify particles.

vibrating in a N-S (front-to-back) direction. Above the stage a second, removable polarizing filter is placed in the tube of the microscope. It transmits light usually in an E-W (left-to-right) direction. When the upper filter is inserted, light is blocked out from passing through the microscope. In this case the filters, which are called *polars*, are said to be crossed. Only when an *anisotropic* material, that is, a material that is not isotropic (meaning that it forms in the isometric crystal system or is amorphous), is placed on the microscope stage under crossed polars can it be seen. The effect of the anisotropic mineral is to rotate the N-S vibrating light from the lower polarizing filter, thus permitting some of it to pass the upper E-W polarizing filter. When the stage is rotated, there will be four positions when the vibration directions in an anisotropic crystal will line up with the N-S and E-W direction. At these positions the crystal is said to be at extinction and will appear black and thus not be seen.

In identifying mineral grains under the petrographic microscope, it is common to use the immersion method. Mineral grains are placed on a microscope slide in a liquid of known refractive index. These liquids are available commercially. The range 1.46 to 1.62, with a difference of 0.02 between adjacent liquids, serves most purposes. When the grain is viewed, a narrow line of light is commonly seen surrounding the grain. If the distance between the objective and the sample is increased slightly, usually by raising the tube of the microscope, the line of light, called a *Becke line*, will move in the direction of the higher refractive index. If the mineral has a higher refractive index, the Becke line will move into the grain. If the liquid of known refractive index is higher, the Becke line will move away from the grain into the liquid.

By trial and error with different liquids, a match is found. At this point the grain will be almost invisible in the liquid. In most cases the refractive index of the grain is found to fall between the refractive indexes of two liquids, and the value can be estimated by an experienced observer.

The petrographic microscope is an important tool in many aspects of forensic work and is the best method for a study of the optical properties of rocks and minerals. A study of individual mineral grains or thin sections of rocks and related material is easily accomplished by anyone trained in the use of the instrument. A thin section is a thin slice of rock mounted on a glass slide. The slice is normally 30 μm in thickness and may be prepared from a solid rock or loose material impregnated with plastic. The rock is cut with a diamond saw and the surface polished. This polished surface is cemented to a glass microscope slide with an adhesive of known refractive index such as epoxy or Canada balsam. A saw cut is then made parallel to the glass, leaving a wafer of rock cemented on the slide. Grinding of the wafer proceeds until it is thinned to approximately 30 μm. A thin class cover is then glued onto the polished rock surface to protect the rock and improve viewing with the microscope. Most rocks are transparent at this thickness and can be viewed in transmitted light (see Figure 4.6). Loose mineral grains of the same general size, also commonly mounted in epoxy or Canada balsam on a microscope slide, are covered with a thin platelet of glass (cover glass) and studied. This is the method used when the heavy minerals, that is, those minerals with high

Figure 4.6 Photomicrograph of a rock thin section as viewed through a petrographic microscope, Note the different minerals with different sizes and shapes. *(See insert for color representation.)*

specific gravity (such as rutile, garnet, zircon, and tourmaline), are separated from the more common lighter minerals (such as quartz and feldspar) by settling in a heavy liquid such as sodium polytungstate or one of the other tungsten-based heavy liquids and studied.

4.5.5 Refractive Index

The index of refraction of a transparent material is the ratio of the velocity of light in a vacuum, normally considered to be 1, to the velocity of light in the material being analyzed. Thus, a refractive index of 2.4553 means that light travels 2.4553 times as fast in a vacuum than in the transparent material. The measurement of refractive index, which is one of the most important methods for the comparison of glass, may be made using the Becke line method discussed for minerals. Most forensic laboratories today use the semiautomated refractive index instrument GRIM 11 for glass identification (see Figure 4.7).

4.5.6 Cathodoluminescence

The instrument used for cathodoluminescence is a luminoscope that is attached as a stage on a microscope or a scanning electron microscope. The specimen—for example, mineral grains or a thin section—is bombarded with a beam of electrons generated by the instrument. When the electrons strike the surface of the specimen, an optical luminescence is produced, which is seen as a display of colors. The colors and their intensity depend in large part on very small changes in the concentration of

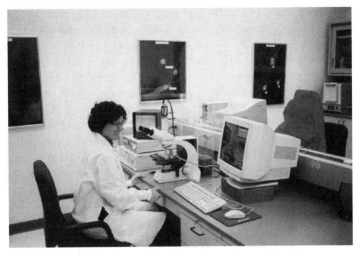

Figure 4.7 Forensic geologist using GRIM 11 instrumentation to determine the refractive index of a glass.

trace impurities, the minerals present, and where the trace impurities are located in the structure of the minerals. Thus, the method has wide application in determining or observing a variety of differences in mineral grains that otherwise appear similar.

4.5.7 Scanning Electron Microscope

The scanning electron microscope (SEM) has a wide range of magnifications, generally from $25\times$ to over $650,000\times$ and can record something as small as 1.5 nm. Needless to say, most of the work of the forensic geologist falls well within these limits. This instrument became commercially available in the mid-1960s and has been introduced rapidly in forensic work, especially in the study of gunshot residues and various other very small particles. It has the advantage that the surface of a sample may be viewed directly. However, an ultrathin coating of carbon or gold plated on the specimen improves the quality of the picture. The depth of field is very large, and most SEM pictures have an excellent three-dimensional appearance. In using the instrument it is possible to change magnification easily and thus study the appearance of the surface from very low to very high magnification. Differences in very small fossils that were not previously known can now be seen during routine examination. Surface features of individual grains of minerals such as quartz can be seen and shown to have many different types of scratches, pitting, and mineral growth. Some of these features may be useful in telling us the past history of the individual grain. It is not uncommon to observe other minerals, such as clay flakes, filling the scratches and thus adding another characteristic that can be useful in comparing the minerals.

When using these powerful instruments in forensic work, it is well to keep in mind that no two objects are ever exactly the same. No two sand grains are ever exactly alike when studied under the high magnifications of an electron microscope. This is true even if they have been side by side for the past million years. Observations made with these instruments can be very useful for establishing similarity or dissimilarity between samples. However, the very power of the instruments permits the possibility of their abuse in the hands of the unscrupulous. If we were to do a complete chemical analysis of a total person by the most modern methods in the morning and repeat the analysis on the same person in the afternoon, we would find chemical differences. However, this would not demonstrate that we had analyzed two different people. Similarly, the demonstration of small differences in soil does not prove in itself that they do not compare. It is equally true that showing a common similarity among soil samples, such as their both containing quartz, the most common mineral in soil and sediment, is poor evidence on which to base a comparison. The professional judgment of the scientist thus becomes increasingly important when these powerful instruments are used.

Scanning electron microscopes have the ability to determine the elemental composition of the particles being examined. This is possible because x-rays are produced when the electron beam of the microscope strikes a target. The scanning electron microscope can be coupled to an x-ray analyzer. The emitted x-rays are sorted by their energy or wavelength values, which are related to specific elements, and the analyzer produces information that identifies the elements present in the material being viewed. The amount of each element present is determined by the intensity of the emitted x-rays. Thus, the examiner can determine the chemical composition of the individual particle or particles being viewed.

4.5.8 X-Ray Diffraction

X-ray diffraction is one of the most important and reliable methods of identifying the composition of geologic, soil, and other crystalline substances (see Figure 4.8). The method is based on the arrangement of atoms, ions, and molecules within the specimen. The sample is analyzed by passing x-rays through a crystal and measuring the angle of the diffracted x-rays. Each crystalline material has its own distinctive x-ray pattern. The x-ray diffraction pattern of a sample is controlled by the internal structure of the specimen. The diffraction pattern can be collected on film, on an image plate, or by using an electronic detector. The interpretation of x-ray patterns under normal situations is a comparatively simple matter. Two factors are of prime importance in the interpretation: the d-spacing from d in Bragg's law, which is expressed in angstroms (Å; 1×10^{-8} centimeter or 0.0039 millionths of an inch), and the intensity.

There are at least two avenues for interpreting x-ray diffraction data: The first involves measuring the d values and intensities and comparing this information with published lists of data on minerals. The second involves comparing the x-ray pattern directly with the pattern produced by a known mineral. If a comparison between samples is to be made, there are situations in which the x-ray diffractograms themselves may be compared without actual identification of the substance. This is obviously less useful as evidence than actual identification.

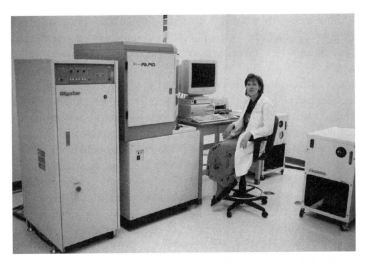

Figure 4.8 Forensic geologist using an x-ray diffractometer to identify minerals.

Among the strong points favoring the use of x-ray diffraction patterns is that they record the crystal structure. If two substances, diamond and graphite, were analyzed chemically, they would be identical because both are composed of pure carbon, but x-ray diffractograms of the two minerals would be quite different. Many samples are mixtures of two or more substances. If substances are analyzed chemically, some difficulties may be encountered because the actual chemical form of the substances at times cannot be established. As an example, we can use a mixture of two salts, sodium chloride and potassium nitrate. If the composition of the sample were determined by the usual chemical methods, it would reveal sodium, potassium, chloride, and nitrate; but what were the original compounds? Were they sodium chloride and potassium nitrate, sodium nitrate and potassium chloride, or a mixture of four salts? An x-ray diffractogram of such a salt mixture would tell us specifically the form of the salt.

X-ray diffraction is used as the principal tool in the modern identification of clay minerals. The chemical composition of clays generally tells us very little as to the nature of clay substances, but the possibilities of identification of clays by x-ray diffraction are almost unlimited. Clays, as well as other crystalline substances, can be x-rayed, and if identification is desired, the composition of the sample can be ascertained by measuring the diffraction patterns from a standard reference book or from reference cards.

4.6 CHEMICAL METHODS

There are a large number of instruments and methods that measure the chemical composition of organic and inorganic materials. There are times where these methods can be very useful and provide valuable information. For example, the chemical composition of glass is often a valuable addition to the optical and physical properties. Identification of organic compounds in a soil such a fertilizer adds an entirely new dimension to the list of properties used in determining comparison.

Methods for determining the amounts and types of elements in a sample may rely on the fact that as one of its properties, an element selectively emits or adsorbs light. This is the basis for emission spectroscopy and atomic adsorption spectrophotometry. Neutron activation analyses is a nondestructive method that produces a detection sensitivity of one-billionth of a gram. This method requires a nuclear reactor to bombard the sample with neutrons. The resulting gamma-ray radioactivity is measured to identify the elements present and their amounts. Needless to say, this method is expensive to operate and maintain and leaves the samples radioactive. Organic compounds contain carbon, and their identification requires different methods. There are several techniques for separating out the various compounds in a mixture. Generally, these methods depend on the relative amounts of the gas phase of a compound and the liquid phase under fixed conditions. That amount is a characteristic property of each compound. Because those compounds that have a higher tendency to go to the gas phase will move faster away in a given time, the distance they move in that time identifies and separates the compounds. These methods are generally called chromatography. Identification is also done using spectrophotometers, which measure the light-adsorbing properties of a compound. Mass spectrometers have the ability to uniquely identify compounds if analyzed under proper laboratory procedures. This instrument bombards the sample with high-energy electrons, causing the molecules to lose electrons and become positively charged. This is an unstable state and the molecules immediately break up into fragments. The instrument then passes the fragments through an electric or magnetic field where they are separated according to their masses. This permits the specific identification of the compounds because the distribution of masses is a unique property.

4.6.1 FTIR and Raman Spectroscopy

Fourier transform infrared spectroscocopy (FTIR) and Raman spectroscopy are nondestructive analytical tools that are commonly applied to the identification of minerals and material of organic origin. They are often combined with a microscope, thus facilitating the identification of a very small object or part of a sample. The nondestructive aspect is important because most owners do not appreciate removing part of their art object or gems for analyses. In addition, in criminal cases the sample may be very small and removal of material for analysis might leave little remaining for additional study or verification. In applying these methods, a source generates light across the spectrum of interest. The sample absorbs light according to its chemical properties. A detector collects the radiation that passes through the sample. Computer software analyzes the data collected and the results are compared with known spectra of inorganic and inorganic materials.

4.7 LOOKING AHEAD

The last several years have witnessed a tremendous increase in both the quantity and quality of forensic examination of soil and related material. There are now three comprehensive books on the subject, and the reader is encouraged to use them as source material (Murray, 2011; Pye, 2007; Ruffell and McKinley, 2008). The future will depend on how well we address a series of issues that include the following:

1. New methods are being developed that take advantage of the discriminating power inherent in earth materials. The opportunities appear to lie primarily in quantitative mineral identification. For this reason, the development of quantitative x-ray diffraction techniques that will provide quantitative data on the mineral composition of a sample would seem to be an important direction for research.

2. Considerable effort must be devoted to defining appropriate sampling methods and the training of those who collect samples. Communicating to law enforcement personnel the potential evidential value of soil and related material is particularly important, because unless the evidence is collected there will never be an opportunity for forensic examination.

3. There is a tremendous need for studies that attempt to demonstrate the diversity of soils. Such studies provide the opportunity for establishing worthwhile information on the distribution of the various soil types. The information should be data based and widely distributed.

4. There must be a continuing effort at all levels to improve the qualifications of examiners in forensic geology.

5. A problem has developed that result from the availability of new instruments that provide increasingly detailed measurements or observations, such as the scanning electron microscope. These instruments are capable of discriminating between individual grains. When you discriminate between individual grains, you lose the ability to say that the sample being examined was once part of, and thus had a common source with, another sample. When this is done, the entire concept of comparison is eliminated and the evidential value is lost. The forensic geologist must choose methods that provide the maximum discriminating power between samples without falsely excluding samples that are in fact associated.

The real future of the science lies with education: the education of the investigator and evidence collector that earth materials can make a major contribution to justice if they are both collected and studied properly. The real challenge of yesterday, today, and tomorrow is the production of the best evidence by the most skillful and objective people to serve the cause of justice. Our system of justice is still run by people, people who are human and in many cases people who are trained advocates. Those who the courts honor by allowing them to express an opinion, the expert scientific witnesses, must rise to the highest standard. If they don't, the advocates may find ways to remove the privilege and return to a legal world populated only by human witnesses reciting their stories from memory.

REFERENCES

Dudley, R. J. (1975). The use of color in the discrimination between soils. *J. Forensic Sci. Soc.*, **15**:209–218.

Gross, H. (1893). *Handbuch für Untersuchungsrichter*. Munich, Germany: N.P.

Guedes, A., et al. (2009). Quantative colour analysis of beach and dune sediments for forensic purposes: a Portuguese example. *Forensic Sci. Int.*, **190**:42–51.

Houck, M. M. (1999). Statistics and trace evidence: the tyranny of numbers, Oct., Vol. 1, No. 3, http://www.fbi.gov/hq/lab/fsc/current/index.htm.

Junger, E. P. (1996). Assessing the unique characteristics of close-proximity soil samples: Just how useful is soil evidence? *J. Forensic Sci.*, **41**:27–34.

Murray, R. C. (2005). Collecting crime evidence from soil. *Geotimes*, Jan., pp. 18–22.

Murray, R. C. (2011). *Evidence from the Earth* 2nd edition, Missoula, MT: Mountain Press.

Pye, K. (2007). *Geological and Soil Evidence*. Boca Raton, FL: CRC Press.

Pye, K., and Croft, D. J. (Eds.) (2004). *Forensic Geoscience*. Special Publication 232. London: Geological Society of London.

Ruffell, A., and McKinley, J. (2008). *Geoforensics*. Hoboken, NJ: Wiley-Blackwell.

Stam, M. (2002). The dirt on you. *Calif. Assoc. Criminalist's Newsl.*, 2nd quarter, pp. 8–11.

Sugita, R., and Marumo, Y. (1996). Validity of color examination for forensic soil identification. *Forensic Sci. Int.*, **83**:201–210.

Thorwald, J. (1967). *Crime and Science: The New Frontier in Criminology*. New York: Harcourt Brace Jovanovich.

CHAPTER 5

Analysis of Paint Evidence

SCOTT G. RYLAND
Florida Department of Law Enforcement, Orlando, Florida

EDWARD M. SUZUKI
Washington State Crime Laboratory, Washington State Patrol, Seattle, Washington

Summary Paint and coatings often appear in criminal, civil, and art authenticity investigations. In this chapter we review the current methodologies and approaches used by forensic paint examiners to analyze this type of physical evidence as well as the problems that they may encounter. Fragments of multilayered in-service paint are one of the most complex types of materials encountered in the forensic science laboratory. They consist of both organic and inorganic components heterogeneously distributed in very small samples, often on the order of only 1 mm^2 in size. Each layer is an entity onto itself and its autonomy must be maintained throughout the analysis. These characteristics dictate the requirements of the analytical chemistry approaches to be used, and they can present a formidable challenge to the forensic analyst responsible for the classification of the materials and an evaluation of their evidential significance. Several case examples are presented to illustrate these concepts.

5.1	Introduction	132
5.2	Paint chemistry and color science	134
	5.2.1 Binders	134
	5.2.2 Pigments	136
5.3	Types of paint	139
	5.3.1 Automotive finish systems	139
	5.3.2 Architectural coatings (structural paints or house paints)	140
	5.3.3 Other coatings	141
5.4	Paint evidence interpretation considerations	141
5.5	Analytical methods	142
	5.5.1 Microscopic examinations	143
	5.5.2 Physical nature of the transfer	147

Forensic Chemistry Handbook, First Edition. Edited by Lawrence Kobilinsky.
© 2012 John Wiley & Sons, Inc. Published 2012 by John Wiley & Sons, Inc.

	5.5.3	Microscopy	149
	5.5.4	Microspectrophotometry	152
	5.5.5	Infrared spectroscopy	158
	5.5.6	Raman spectroscopy	175
	5.5.7	Pyrolysis gas chromatography and pyrolysis gas chromatography–mass spectrometry	178
	5.5.8	Elemental analysis methods	188
	5.5.9	Other methods	205
5.6	Examples		208
	5.6.1	Example 1	208
	5.6.2	Example 2	210
	5.6.3	Example 3	213
	References		217

5.1 INTRODUCTION

The body of Wendy Coffield, 16, was discovered in the Green River, south of Seattle, on July 15, 1982. In the months to follow, the bodies of several more young women were found in or near the Green River. By the early 1990s, the total number of victims believed to be associated with this unsolved case had reached 60 or more, and investigators were faced with the very real possibility that the most prolific serial killer in U.S. history would never be apprehended. In 2001, based on recently developed DNA technology, semen samples collected from three of the victims were found to be those of Gary L. Ridgway, a paint detailer at a local truck manufacturing plant. Ridgway pleaded not guilty when charged with homicide, claiming that he only had sex with the victims. Although usable DNA evidence was not recovered from any of the other bodies, hundreds of very small colored spherical particles were found on the clothing of four of them as well as on the clothing of two of the victims associated with Ridgway's DNA (Palenik, 2007). Similar particles were found on Ridgway's work clothing and in his home and work environments. These particles were identified as dried droplets produced from the plume of spray paint, and the paint was found to have an acrylic urethane composition. In the early 1980s, this formulation was not used at all on North American original automotive finishes and it would also have been quite unusual for architectural coatings, with the possible exception of certain varnishes, but these would not have been pigmented as were the samples found on the bodies. The paint was identified as Imron, a DuPont high-end specialty product not sold to the general public; more specifically, it was found to be Imron that had been manufactured before 1984. This product had been used at the truck plant where Ridgway was employed, and the paint evidence was the basis for charging him with four additional counts of homicide. In 2003, Gary L. Ridgway pleaded guilty to 48 counts of homicide, prompted by his desire to avoid the death penalty and his fear of a conviction in the face of more physical evidence linking him to the victims. Ridgway's attorney was quoted in the press as stating that the paint evidence was crucial in his client's decision to change his plea.

This example illustrates the value of paint as physical evidence and as an investigative tool, and although the majority of cases involving paint may not be in the national limelight, this important category of trace evidence provides crucial links in criminal investigations and prosecutions on a daily basis. Typically, forensic paint examiners are asked to perform a comparison of two or more samples to determine if they could share a common origin. The samples may result from transfer of paint from one item to another, as may occur when a vehicle strikes another vehicle, a building or road structure (such as a guardrail), or an individual. If the object that is struck also has a painted surface, a double transfer may result, with each item now having paint from the other. Other circumstances may also facilitate a transfer, as when a crowbar is used to pry open a window, safe, or some other item with a painted surface; a bullet or other projectile strikes a painted surface; a person or object comes into contact with wet paint; or when paint particles are transferred from the clothing of one person to another due to contact. The latter two transfer scenarios are believed to have occurred in the Green River cases.

In other instances, an unknown paint chip is recovered from the scene of a hit-and-run accident or from clothing of a victim struck by a vehicle, and the forensic paint examiner is asked to try to identify the type of vehicle involved. On a few occasions, the paint examiner may be asked to determine if a material *is* paint, and if so, the type of the paint and its possible relevance to the circumstances of the investigation.

In the Green River cases, elements of all three types of examinations (comparison, vehicle identification, and determination if paint) were involved. After the unknown colored particles were removed from the victims' clothing, the trace analyst first had to determine what they were. Finding them to be dried droplets of spray paint, the analyst next had to determine their chemical composition and whether the specimens were similar to those associated with the suspect. Based on chemical composition, an original automotive finish—which usually serves as the basis for identifying the type of vehicle involved in a hit-and-run incident—appeared very unlikely. The nature of the paint, dried droplets of spray paint, also suggested that an original automotive finish was probably not involved, as the nearest automotive assembly plant was thousands of miles away. Nonetheless, an identification of the paint was sought to assess the significance of the match that was found between the specimens from the suspect and the victims. In this case, this resulted in the determination not only of the specific brand of paint that was involved, but also of a time frame for when it was manufactured.

One other type of examination that the forensic paint examiner may be asked to perform concerns whether a painting or some other artifact is a forgery. Although authenticity in such cases cannot be established by paint composition alone, in certain instances, forgeries can. If a painting or other artifact that was supposed to have been made on a certain date or during a certain time period contains pigments or other paint components that were first available only after this date or period, the painting or artifact could not have been produced when purported (assuming that no modern restorations were performed).

134 ANALYSIS OF PAINT EVIDENCE

How the forensic paint examiner goes about accomplishing these varied tasks is described in this chapter. We begin with an overview of paint chemistry and color science, briefly describe the types of coatings encountered, and discuss some issues and concerns specific to paint evidence of which analysts must be cognizant. The various analytical methods that are used to examine paint evidence and the types of information that these methods provide are then described. Since correct interpretation of analytical data is the cornerstone of formulating appropriate conclusions, we discuss analytical methods in some depth. Because data collected for two or more samples are frequently compared, differences that may occur and *their significance* are central issues in forensic science. Aspects of a particular analysis that can significantly affect data—such as sample preparation, instrument parameters used, nature of the samples, and possible alterations of samples caused by the analysis process—are therefore emphasized. We conclude by presenting three more case examples to illustrate the concepts discussed and to demonstrate further the wide variety of analyses that an examination of paint evidence may entail.

5.2 PAINT CHEMISTRY AND COLOR SCIENCE

Paint is a film or coating that serves both decorative and protective functions. In the coatings industry, *paint* usually refers to a pigmented coating, but in this chapter it is synonymous with the term *coating*. Dried paint consists of three components: a binder, pigments, and additives. The *binder* comprises the matrix that holds the paint together, within which are suspended very small particles of pigments and additives. The binder consists of an organic polymer that not only forms a strong film but also adheres to the substrate to which it is applied. A *pigment* is a finely powdered compound or mixture that imparts color, opacity, or other essential properties (such as luster or texture) to the paint. An *additive* is a substance added to a paint (usually in a small quantity) to improve its properties; it includes components such as corrosion inhibitors, catalysts, ultraviolet absorbers, and *plasticizers*. Plasticizers are moderate-weight organic compounds (often esters) or polymers that serve to fill the spaces between adjacent polymer chains, making the product more flexible and less brittle.

5.2.1 Binders

The polymers that are used for paint binders include acrylics, alkyds, polyesters, urethanes, epoxies, vinyls, cellulosics, and silicones (Morgans, 1990; Ryland, 1995; Thornton, 2002). Binder polymers may consist of single repeating units (homopolymers); two or more units, usually in a random order (copolymers); different homopolymer or copolymer units joined together (block copolymers); or homopolymers, copolymers, or block copolymers associated in the same binder matrix but without covalent bonds between them. Depending on the binder, the polymer chains may be linear, branched, cross-linked, or a combination of these, and many of the polymers cited may constitute the main backbone chain, a side chain, or a cross-linking chain (Oil and Colour Chemists' Association, 1983). Binders are typically manufactured by a chemical company, such as Rohm & Haas or Union Carbide, and then sold to a paint manufacturer, such as Sherwin-Williams or Inmont, for incorporation into the paint batch. There are exceptions to this trend with large companies such as DuPont, which has facilities to manufacture the binders as well as divisions devoted solely to paint manufacture.

Several different processes are used to create binders from polymers or monomers. One straightforward method involves a preformed polymer, usually linear, which is dissolved in an appropriate solvent. A film of the paint is formed by simple solvent evaporation and this type of paint is referred to as a *lacquer*. The adjacent polymer chains of the binder are held together by relatively weak forces (van der Waals and polar forces), so such paints can be redissolved in appropriate solvents.

One lacquer that many readers may be familiar with is nitrocellulose, since it is the main binder of most fingernail polishes. Nitrocellulose is a linear polymer that is produced by forming nitric acid esters with roughly two-thirds of the free hydroxyl groups of cotton (cellulose). It is soluble in acetone, ethyl acetate, and other organic solvents but forms a rather brittle film by itself, so a plasticizer is also used in the formulation. An additional polymer, such as tosylamide–formaldehyde, may be included to improve adhesion to the fingernail surface (Drahl, 2008). As a lacquer, fingernail polish is readily removed using acetone or other solvents. A nitrocellulose lacquer was used for the first production-line automobiles, manufactured in the 1920s (Fettis, 1995).

A second method of forming binders also involves a mostly preformed polymer, but in this case, small *micelles* of the polymer are suspended in the paint solvent (similar to fat globules dispersed in milk) rather than forming a true solution. These colloidal systems are known as *emulsion* or *dispersion finishes*, although when water is used as the medium and architectural coatings are involved, they are more commonly referred to as *latex paints*. Upon evaporation of the solvent, the micelles coalesce to form a film and further polymerization occurs. Depending on the specific polymer used, the film formed from this process may be an *enamel* (a finish that is not soluble in most solvents) or a lacquer. Solubility usually reflects polymer structure, with lacquers normally comprised of linear chains whereas enamels exhibit considerable cross-linking. Other methods of forming binder polymers are also used.

One of the most common automotive paint binders is the acrylic melamine enamel, which has served as a staple for North American automobile finish layers during the past 40 years or so (Rodgers et al., 1976a; Ryland, 1995; Suzuki, 1996a; Ryland et al., 2001). The polymer chain of this binder is comprised of a sequence of acrylic copolymers, and methylated, butylated, or isobutylated melamine–formaldehyde is used to cross-link the acrylic backbone (Nylen and

Sunderland, 1965; McBane, 1987; Fettis, 1995). This composition is shown schematically in Figure 5.1; eight different acrylic copolymers are depicted to illustrate the variety that may be used, although in practice, no more than five are normally found in a single formulation (Nylen and Sunderland, 1965). The various copolymers are chosen to impart specific properties, such as hardness, flexibility, durability, adhesion, gloss, and water and chemical resistance. Although not an acrylic, styrene is also frequently incorporated into the backbone. In addition to specific copolymer composition, the properties of this binder are determined by the extent of cross-linking. Original automotive paints are baked to cure the finish, that is, to promote the cross-linking process, and also to increase gloss.

5.2.2 Pigments

Pigments are distinguished from dyes in that they are insoluble in most solvents, and in particular they are insoluble in the solvent used for the paint itself. Dyes, in contrast, are soluble, and they are not used in most paints, as they are less durable than pigments and also because they are usually too transparent; it would be difficult and probably cost prohibitive to try to incorporate enough dye into a paint to make it opaque. Dyes also suffer from one other serious drawback regarding tint control. The optical properties of a color-imparting agent (such as hue and *tinctorial strength*) depend on particle size, particle-size distributions, and crystal structure (Braun, 1993). Many common paint pigments exhibit *polymorphism*; that is, they can occur in more than one crystal structure, and in some cases, two different crystal structures of the same pigment will exhibit quite different hues. Particle size, particle-size distributions, and specific crystal structures cannot be controlled when a dye recrystallizes from solution; hence, a color-imparting agent that is insoluble in the paint must be used.

Even with the use of pigments, paint formulators must be aware of the effects of polymorphism. Copper Phthalocyanine Blue (Figure 5.2) is a very common organic pigment used in automotive and other paints and it can occur in any of four different crystal forms (Lewis, 1995). The polymorph of this pigment that is used in most paints is treated to prevent its slow conversion to another crystal form that has a green rather than a blue hue, as most consumers do not appreciate their blue vehicle turning green with age.

The most common pigment used in paint is titanium dioxide. This compound occurs in three polymorphic forms, rutile, anatase, and brookite, but only the first two are used in paints. Rutile is, by far, the more common of the two, both because of its slightly greater index of refraction and, especially, its greater durability to light exposure and heat. Rutile serves to make a paint opaque, that is, to make it white (or a lighter color—adding a large amount of rutile to a red paint, for example, makes it pink). A binder that is cured to form a dried film is normally transparent, but the presence of numerous small particles of rutile causes light incident on this medium to scatter in all directions. Rutile is effective at scattering light because its index of refraction (~ 2.7) is quite high compared to that of paint binders (~ 1.5).

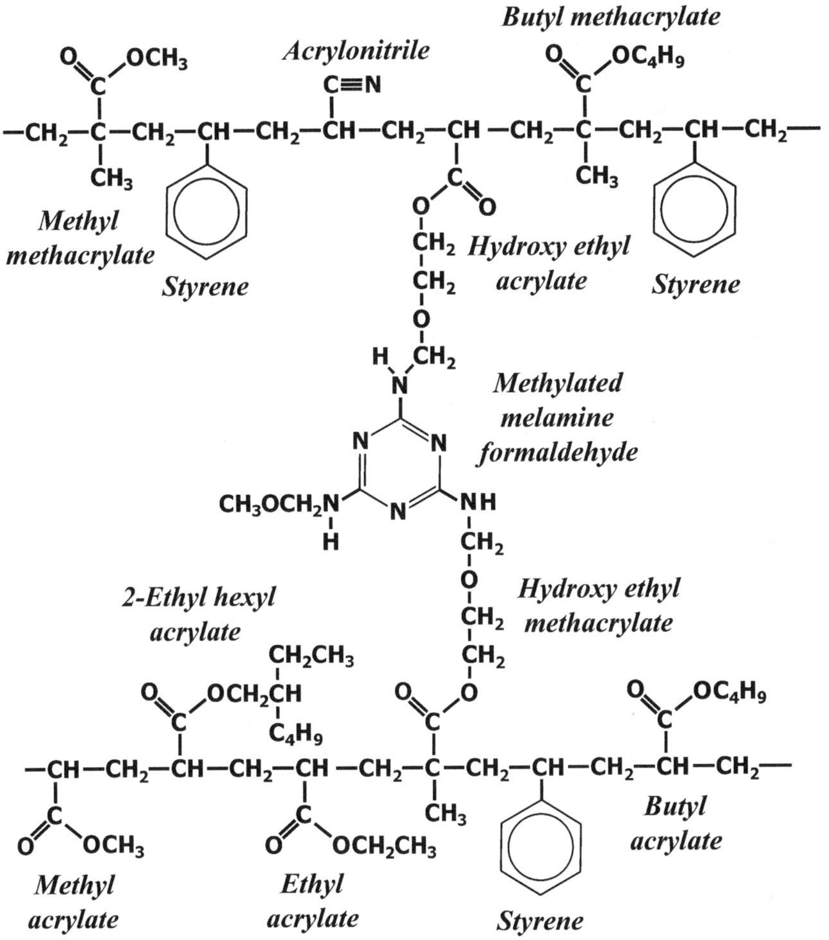

Figure 5.1 Composition of an acrylic melamine enamel binder commonly used in automotive finish layers.

Particles of a substance that have the same index of refraction as the paint binder will not produce any optical effects, but they do serve to add bulk to a paint. Pigments that have indices of refraction that are close to that of the binder are thus referred to as *extender pigments*. They are sometimes used to modify the optical, physical, or chemical characteristics of paint, such as gloss, surface texture, viscosity, strength, chemical resistance (weatherability), or abrasion resistance. Some of the more common extender pigments used in paints include talc, kaolin (clay), calcite (calcium carbonate), quartz (silicon dioxide), mica, and barytes (barium sulfate), all of which have indices of refraction near 1.6.

The vast majority of pigments used in paint produce color by selective absorption of certain wavelengths of visible light. Most inorganic pigments are comprised of transition metal complexes or salts, and their visible absorption properties arise from electronic transitions of valence electrons in d orbitals (Judd and Wyszecki, 1975). Electronic transitions between d orbitals of isolated atoms are not allowed by *selection rules*, but they may become "weakly allowed" when these orbitals form ligand bonds.

Organic pigments are flat planar systems with extensive conjugation (Figure 5.2) and light absorption involves π-electron transitions. These are allowed transitions, and organic pigments are strong absorbers of visible light. Organic pigments therefore have higher tinctorial strengths than inorganic pigments. In regard to other pigment properties, organic pigments tend to produce brighter and more vivid colors, have smaller particle sizes, are less opaque, are less durable to long-term light exposure, and are more expensive than inorganic pigments (Oil and Colour Chemists' Association, 1983).

In attempting to mimic the iridescence of pearls, butterfly wings, and some other visually striking objects in nature, chemists have devised several specialized pigments for use in cosmetics, paints, and other products. Most such *effect pigments* produce colors by means of light interference (Droll, 1999). A common example of this is the change in color of an oil slick when viewed from different angles. The oil slick consists of a very thin layer of hydrocarbons on water, and interference between light waves reflected from the top of this film and from the oil–water interface produces different colors, depending on the angle at which the film is viewed, the thickness of the film, and its index of refraction. This results from constructive and destructive interference between the two reflected rays. Pearlescent and certain colored and color-shifting pigments are based on this principle, and they consist of refractive laminated flakes that are oriented mostly horizontally in the paint. Color-shifting pigments based on very small

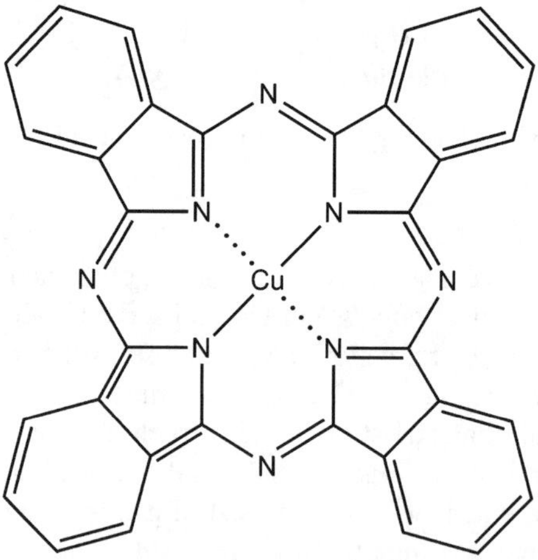

Figure 5.2 Structure of Copper Phthalocyanine Blue (Pigment Blue 15), a very common organic pigment used in automotive and other paints. Note that this structure is only one of several canonical structures and that the four isoindole ring systems are equivalent.

striated plates acting as miniature diffraction gratings have also been introduced recently.

The types of pigments and pigment combinations that are used in a particular paint depend very heavily on the applications of the paint and the visual effects that are desired. All white and light-colored exterior paints normally contain large amounts of rutile, while varnishes, automotive clearcoats, and other transparent finishes contain little or no pigmentation. Metallic paints must have a semitransparent finish so that the metal flakes, which are usually composed of aluminum metal, can be observed. Organic pigments are usually more prevalent in such paints, as they are more transparent than inorganic pigments, owing to their smaller particle sizes and lower indices of refraction. Extender pigments are normally used to control luster in paints, and architectural finishes with low lusters contain large amounts of such pigments. More details regarding paint composition, manufacture, and use are described by Nylen and Sunderland (1965), Oil and Colour Chemists' Association (1983, 1984), Morgans (1990), Lambourne and Strivens (1999), and Bentley (2001).

5.3 TYPES OF PAINT

There is no such thing as a universal finish, as even when discounting consumer tastes regarding color and type of finish, there is no single binder or binder and pigment combination that will meet all of the varied demands of every paint application. Certainly, this is a beneficial situation for the forensic scientist since, as we saw, the composition of a paint can tell us a lot about its uses, and it is this very wide diversity that makes paint so valuable as evidence.

Paint composition is ultimately a compromise between various conflicting performance characteristics, economic issues, and environmental and safety concerns. It is not possible, for example, for a paint to have an extremely hard finish and to be quite flexible at the same time. An epoxy with a bisphenol A component is often used for automotive undercoats and many indoor applications because of its excellent adhesion and other desirable properties, but it is not durable enough with respect to light exposure to be used in automotive finish layers. Acrylics and alkyds have been used extensively for automotive finish layers, providing a hard durable finish on rigid metal and plastic substrates, but they are not used (by themselves) on flexible plastic automobile bumper covers because they are too brittle and would crack readily when subjected to the minor collisions that the bumper itself is designed to withstand.

5.3.1 Automotive Finish Systems

The demands of an automotive finish are among the most stringent of any type produced in the coatings industry. Appearance is paramount for a consumer product, where utility often takes a back seat to status, and visually, automotive finishes are the most diverse in the industry (Panush, 1975; Ehlich, 1988). Such paints span the entire spectrum of colors from black to white (both of which are also popular choices), with virtually everything in between, and more than half of the original finishes contain some sort of "effect" pigment flake (metallic, pearlescent, interference, or diffraction). About the only color that is not used in an original automotive finish is nonmetallic pink ("shocking pink"), but even that hue may be found in some fleet or special-order original finishes. Flat and satin finishes are rare and the overwhelming majority of automotive finishes have high gloss. Durability (color and gloss retention) is a very important consideration, and vehicle paints also have very demanding mechanical requirements, such as chip, corrosion, and chemical resistance. They are carefully applied as highly engineered layer systems, with each layer serving a specific purpose. Binder formulations are highly complex and a broad variety of polymer types are employed.

5.3.2 Architectural Coatings (Structural Paints or House Paints)

Architectural paints are the second most common type of paint encountered in the forensic science laboratory. Unlike automotive paints, they are seldom designed to be applied as layer systems; consequently, layer colors and sequences will vary with the whim of the customer. They are designed either for interior or exterior application, and a good portion of their design is targeted at ease of use. This affects choices as to the binders and pigments to be used, the additives to control viscosity and, consequently, film buildup; and the additives to control stability during storage after use. There are two general chemical classifications for the binders; water-based (latexes) and oil-based. The latexes permit easy cleanup with soap and water, a big selling point for the home user. The oil-based systems require mineral spirits or turpentine for cleanup. Only three major binders are used, although there are many variations within each major class. The latex binders include the poly(vinyl acetate)–acrylic (PVA–acrylic) and acrylic resins, while the oil-based binders are almost solely alkyd resins. There are some poly(vinyl acetate)–polyethylene binders on the latex market, but they command a very small share of the sales (less than 2%), as do some newer latex paints with low concentrations of alkyd resin incorporated. Unlike the case of automotive finishes, appearance other than color is not paramount. This results in a highly competitive market with many more manufacturers than exist for automotive paints. Almost an infinite number of colors are available, but many fall into the off-white or pastel classes, employing very low concentrations of coloring pigments. The paints are pigmented at the retail store and offered in a series of gloss levels, including flat, satin, eggshell, semigloss, and high gloss. The pigment is added to a tint base, which is a 1- or 5-gallon can of white paint of the chosen binder type and gloss type. Even the high-gloss paints do not approach the gloss levels of automotive finish coats. It is extremely unusual for these paints to include effect pigments.

It is difficult to specifically place certain finishes in this general class of coatings; however, some applications deserve mention. Wood coatings, such as colorless varnishes, are of note, with most binders being alkyds or urethanes and some having acrylic copolymers incorporated as well.

5.3.3 Other Coatings

The third most common type of paint encountered in the forensic science laboratory falls into the class of maintenance paints. These finishes are designed primarily for their protective properties and functional characteristics, with color and gloss quality being of secondary importance. They are quite diverse and may be found as coatings on tools, appliances, lawn and garden tools, circuit boxes, safes, ATM devices, valves, bridges, traffic signs, and so on. Considering the applications, it does not take too much imagination to deduce why they are encountered as evidence. The binders and pigments used cover a broad range, with original finishes demanding durability similar to that of automotive coatings. They may be high gloss and may also have effect pigment in them, but rarely interference decorative flake.

Trade-sale spray paints, the type found on the shelves of your local hardware or home improvement store, are yet another general class of end-use product encountered in the forensic science laboratory. Binders typically include alkyds, acrylics, nitrocelluloses, urethanes, and epoxies. Some silicone binders may be found as well. The finishes may have effect pigments incorporated in them; however, application methods are obviously not nearly as controlled as in the automotive industry.

5.4 PAINT EVIDENCE INTERPRETATION CONSIDERATIONS

Although one often thinks of paint as a "wet" medium that is applied and allowed to dry, the forensic paint examiner deals predominately with cured in-service paint films. There may, however, be occasions where a can of paint or spray paint is submitted as evidence and the paint examiner is asked to determine if the contents of the can could have been the source of a paint film. In these instances, the paint examiner will prepare a film of dried paint for comparison.

In a comparison case, two samples are typically received for analysis, referred to as *known* and *questioned* specimens. The former denotes the reference sample, which usually has a known origin (e.g., a vehicle suspected of being involved in a hit-and-run incident), and this is to be compared to sample recovered from a scene (a paint chip from a hit-and-run location) or a suspect, or some other source that may provide a link between events and individuals.

Although it may seem like a straightforward process to compare two paint samples, the forensic paint examiner deals with real-world specimens that are often far from ideal for analysis. Because of the nature of the events that usually produce paint evidence, the questioned specimens are often quite small, such as those recovered in the Ridgway case. The paint may have been in service for many years, subjected to weathering, marring, contamination, repaints, and so on. These processes result in what is decidedly a double-edged sword as far as interpretation of evidence is concerned. They can make obtaining and interpreting data a very demanding task; at the same time, however, some of the unique characteristics that they might confer to a paint can be quite significant if similar features are found on both known and questioned samples.

There are many factors that the forensic paint examiner must consider when interpreting data for paint comparisons. Many of these arise from the inhomogeneities in paint, not only at the microscopic level but at the macroscopic level as well. Many readers have probably seen the effects of weathering on an interior house paint arising from differences in exposure to light: the paint behind a picture on the wall may have a different hue compared to the same paint that has been exposed to sunlight. Bleaching and other effects may be even more pronounced for automotive paints, where, for example, the roof has received considerably more exposure to sunlight and other environmental conditions than has the vertical side of a fender. The original paint systems on many automobiles are likely to be varied to begin with, considering that some have two tone colors, and as noted, paint systems used on metal substrates may differ from those used on plastic bumpers and other plastic parts. Another possible source of inhomogeneity of an original automotive finish commonly arises from factory touch-ups. These occur when a portion of the vehicle is damaged during the assembly process and must be repaired and refinished before leaving the plant. A nonfactory repaint will usually result in even more heterogeneity when only a portion of the vehicle is refinished, as typically occurs. Refinished areas may also differ due to removal of one or more previous layers as a result of sanding. Paint systems therefore cannot automatically be excluded as having a common origin, even if the number of layers does not match. For cases involving paint from a spray can, differences in paint composition can also occur, depending on whether the can has been shaken or not prior to use (Zeichner et al., 1992). Assessing heterogeneity in samples—and its significance—is a major portion of the forensic paint examiner's task.

These considerations emphasize the importance of having appropriate reference samples when comparisons are made, and the various circumstances of which the forensic paint analyst must be cognizant when interpreting paint evidence. Unfortunately, appropriate reference samples are not always available, and the condition of the questioned samples can be an analytical chemist's nightmare. Sample size, sample heterogeneity, layer smearing, or contamination by the substrate onto which the specimen was transferred epitomize some of the difficulties that may confront the paint examiner. Such realities, as well as the lack of comprehensive databases for all types of paints, underscore the reasons why the conclusions that the forensic paint examiner can draw run the gamut from unequivocal to simply unattainable.

5.5 ANALYTICAL METHODS

As discussed, dried paint is a complex matrix comprised of a variety of mostly nonvolatile and inert inorganic and organic ingredients spanning a very wide range of concentrations, and most paints that forensic analysts examine are insoluble enamels. It is thus not surprising that paint examiners typically use a battery of different methods to obtain as much information as practical to characterize, differentiate, identify, and compare such evidence. Methods that provide maximum discrimination between paints are sought and it is desirable that the suite of methods chosen provides complementary information. At the same time, the forensic paint examiner finds it helpful to have some overlap in this information, because by themselves, the methods may provide only limited data for identifying certain components or else provide general class characteristics rather than unequivocal data to identify each component conclusively. Methods that are nondestructive are preferred, both to preserve the evidence and to allow further testing using other techniques. If destructive methods are used, they are normally performed after the nondestructive methods. The choice of methods that are used may also be dictated by the type of paint that is encountered, as well as its size and nature (e.g., the analysis of a paint smear versus the analysis of paint from the entire door of a vehicle).

5.5.1 Microscopic Examinations

As mentioned previously, paint evidence can be found as smears on clothing, smears on objects, minute fragments fused to the surface of an object, or fragments contained in the surface debris recovered from an object. As with any trace evidence, microscopy is necessary in finding and examining the specimen. If you cannot find the material, you obviously cannot analyze it. Furthermore, if you cannot accurately characterize its form and structure, no matter what sophisticated analytical techniques you may have at your disposal, the analysis will ultimately not generate complete and accurate results. The initial instrument of choice for searches and preliminary characterization of recovered specimens is the stereomicroscope. This technique provides a three-dimensional view, with magnifications on the order of 5 to 60 times the object's actual size. It has a broad field of view, permitting relatively large areas to be searched microscopically. Furthermore, it offers long working distances providing ample room to manipulate the specimen for observation and sample preparation for subsequent examination techniques.

Unlike other common types of materials encountered as trace evidence, such as fibers, glass, plastics, adhesives, rubbers, foams, metals, and most composites, paint evidence often has a laminate structure comprised of multiple complex layers. Stereomicroscopy provides the initial view of this layer structure. Layer colors, order, textures, and relative thicknesses are often the initial clue as to the general source of origin of the specimen. Did it originate from a motor vehicle or from an architectural application? Is it more likely a maintenance coating, such as might be found on tools, commercial door frames, or electrical junction boxes? This initial information is important for several reasons. First, it lets the examiner know what to expect from the sample. Thorough background knowledge of all types of coatings is essential. If you are going to perform a forensic comparison of any material, first know your product. In the case of paint, it is important to know how it is manufactured, how it is applied to the substrate, of what is it composed, and what types of common manufacturing irregularities may be encountered. All become important in knowing what to look for and in recognizing what is unusual.

Second, it permits the examiner to deduce from where the sample may have originated. This is commonly an integral question in the investigation, whether it be a hit-and-run investigation or a homicide investigation involving forced entry to a structure. For example, there is usually little benefit in analyzing an architectural paint in detail when it comes from the debris recovered from the surface of a hit-and-run victim. Finally, it permits the examiner to know which pool of background knowledge to access when evaluating the evidential significance of a sample. Just as one does not want to testify that the layer structure of an architectural paint discovered in a homicide investigation is unusual based on their background knowledge of automotive paint layer structures, one equally does not want to attempt to determine the year/make/model vehicle origin of an architectural paint sample discovered in debris recovered from the garments of a hit-and-run victim. All these points may seem unwittingly basic, but striving to accomplish them is anything but.

To appreciate the microscopical methods used in forensic paint examinations, it is probably necessary first to discuss some of the attributes of the various end-use categories described previously (e.g., automotive, architectural, maintenance, and trade-sale spray paints). Due to the specialized nature of the forensic examination of cultural artifacts, this topic is not addressed, although many of the techniques to be discussed are employed in that endeavor as well.

Automotive Paints Automotive paints consist of finish coats applied over primer coats and generally fall into two distinct categories, original finishes and refinishes. Original finishes consist of the multiple paint layers applied at the factory at the time of manufacture. Refinishes commonly refer to either single or multiple layers applied after the vehicle leaves the factory (i.e., aftermarket applications). Refinishes applied at the factory, however, present a third category. Although these systems employ a reapplication of paint layers, the material used commonly has the same composition as that used in the original application. As might be expected, it is valuable to be able to recognize these three categories when assessing the evidential significance of a transfer, and various types of microscopy play an integral part in making those decisions.

Finish coats impart the visual appearance to the paint system and are the uppermost layers of the coating. They are found as either monocoats, clearcoat/basecoats, or tricoats. Monocoats consist of opaque layers designed to have high gloss and color. Both the clearcoat/basecoat and tricoat systems have a transparent layer on top that is typically colorless but may be lightly tinted. This is followed by a single colored and opaque basecoat layer in the former system or a transparent or translucent colored layer followed by the opaque basecoat layer in the latter system. Any one of these finish coat layers other than the clearcoat may have decorative flake dispersed throughout. The decorative flake is commonly either some form of aluminum flake, pearlescent flake, or interference flake. Stoecklein (2002) describes the latter in detail.

Below the finish coats are primers, designed to be an interface between the finish coats and the substrate material. They serve several functions, including adhesion, ease of sanding for leveling out imperfections, and corrosion resistance. There may be anywhere from one to four or five of these layers present; however, the most common layer structures employ only one or two. The primers may usually be distinguished from the finish coats, as they are opaque and include much higher concentrations of extender pigment. This in turn gives them a lower gloss and more particulate appearance. Color-coordinated primers are designed to have a similar color to the basecoat above them and often have slightly lower concentrations of extender pigment present. Figure 5.3 depicts the most common original finish layer systems. The primer-surfacer is a primer that is often present between the base primer and the finish coats. The term *surfacer* is added to the name in order to describe its primary function, a layer that is easily sanded to produce an acceptably smooth surface on which to apply the finish coats. Traditionally, it appeared in a rather limited color palette, consisting of various shades and tints of gray, black, or red-browns. Since the mid-1990s, some vehicle manufacturers began using colors coordinated to the general color of the monocoat or basecoat. Further descriptions of these systems may be found in Fettis (1995), Ryland (1995), Thornton (2002), and Streitberger and Dossel (2008). Anticorrosion treatment of the steel used in automotive panels leaves a characteristic pattern on the bottom of original primers commonly referred to as "orange peel," similar in appearance to that shown in Figure 5.4. Weathering of the surface of the top layer can result in oxidation of the resin and pigment and sometimes cracking of the surface. Improper refinish techniques may result in dirt or foreign material trapped between layers and visible at the interface between the layers. Occasionally, poor painting procedures result in foreign overspray paint droplets deposited on the surface of a topcoat.

Architectural (Structural) Paints Forensic laboratories may also have a substantial caseload, dealing with architectural paints ranging from cases involving

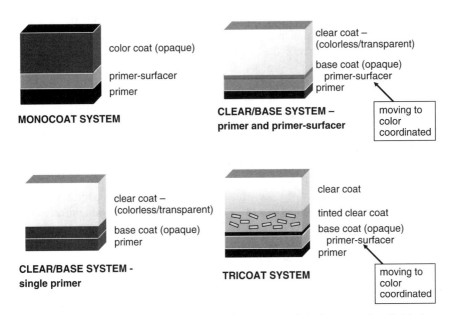

Figure 5.3 Diagrammatic representation of common original automotive finish layer structures. *(See insert for color representation.)*

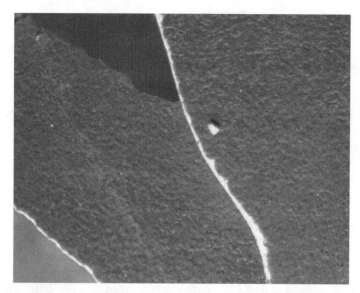

Figure 5.4 Stereomicroscopic view of the underside of the original base primer from an automotive application revealing the characteristic "orange peel" appearance. Magnification of 20×. *(See insert for color representation.)*

forced entry to those dealing with debris recovered from either a suspect's or a victim's garments. There is seldom a predesigned system of layers in these samples—the layer colors, textures, and chemistry changing with the whim of the consumer. The variety of finishes encountered can range from flat wall paints to colorless glossy varnishes on wood. The flat wall paints contain very high concentrations of extender pigments. Even semigloss to high-gloss wall paints are characterized by the presence of some extender pigment, with the grain size being much larger than that encountered in vehicle primers. The surfaces are typically somewhat rough, even though they may have a relatively high gloss, and the pigment dispersion is heterogeneous compared to automotive finishes. One can often see clumps of undispersed coloring pigment throughout the paint layer. Latex paints have a softer texture than automotive paints and when treated with a bit of chloroform form a gummy residue during solvent evaporation, quite unlike automotive paints treated in the same manner. It is unusual to encounter decorative flake, except for the occasional use of retail spray paints for finishing architectural accessories. In those cases, the decorative flake is usually quite dense in distribution compared to that found in most automotive applications. On occasion, a portion of the substrate material will cling to the bottom layer of the paint fragment. Discovery of wood or wallboard obviously indicates that the application was not automotive. Consecutive applications of light pastel or off-white colors that are similar to one another in color and texture are sometimes encountered and present difficulty discerning between layer interfaces. Microscopy is the basic tool for recognizing these characteristics.

Maintenance and Trade-Sale Spray Paints Maintenance and trade-sale spray paints have microscopic characteristics that lie somewhere between architectural finishes and automotive finishes. In the case of tools or commercial metal door frames, they are often coated with both a primer and a finish coat. The finish coats may be high gloss, resembling automotive monocoats. Their surfaces typically will have rough pitted spots over larger areas, but these may not be noticeable in small samples. The chemistry used in the paints is similar to that employed in the automotive industry and hence the textures are quite similar. Often, the layers are quite a bit thinner than automotive paint layers and may be brittle, giving rise to a suspicion that the origin is both nonautomotive and nonarchitectural. In the case of trade-sale spray paints, they are usually encountered as thin layers applied to the surface of existing finish systems and frequently display significant heterogeneity in pigment dispersion. This layer structure suggests nonautomotive applications.

5.5.2 Physical Nature of the Transfer

In addition to the various morphological characteristics attributable to different end-use types of paint, the physical nature of paint transfers encountered in forensic investigations plays a significant role in the choice of examination techniques. In a majority of automotive paint cases, the samples are quite small and multilayered, as depicted in Figures 5.5 and 5.6. Both the microscopical and analytical techniques chosen for comparisons must be compatible with this sample size and morphology, keeping in mind that each layer contains a wealth of morphological and both organic and inorganic chemical information. Some evidence will have numerous layers, suggesting the possibility of encountering another source with a corresponding

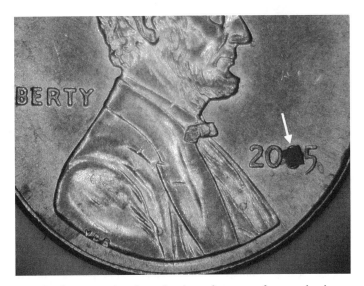

Figure 5.5 Paint fragment placed on the date of a penny for sample size perspective. *(See insert for color representation.)*

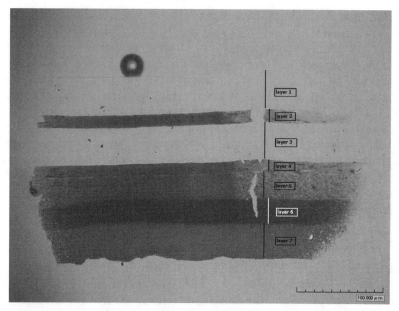

Figure 5.6 Cross section of a seven-layer automotive paint sample mounted on a microscope slide with a 1.56 refractive index mounting media and coverslip and viewed with transmitted brightfield light on a polarized light microscope. Layers 1 and 3 are clearcoats, layers 2 and 4 are basecoats with a decorative flake in them, and layer 5 is a color-coordinated primer followed by two additional primers (layers 6 and 7). *(See insert for color representation.)*

layer structure is extremely remote (Tippett et al., 1968; Gothard, 1976; Ryland and Kopec, 1979; Gothard and Maynard, 1996; Willis et al., 2001; Edmondstone et al., 2004; Wright et al., 2011). In such instances, further chemical analyses may not be deemed necessary (ASTM, 2007). In the case of larger paint fragments, be they automotive or not, the examiner must guard against damaging the edges and surfaces of the evidence since the potential for comparison of the fractured edge contours and the paint chip surface configurations with an appropriate known source hold the possibility of a physical correspondence that is definitive, unlike any compositional comparison. If such a correspondence is found, the analyst will conclude that the paint fragment whose origin is in question originated from and was at one time a part of the paint on the known exhibit.

Paint transfers also occur in the form of smears onto substrates. One such example is a multiple-layer architectural paint smeared onto a tool used to gain entry to premises. Mixing of the smeared layers and contamination of the transfer by the material on the tool's surface become serious analytical issues. Microscopy offers the initial means to evaluate and mitigate the impact of these problems on subsequent analytical techniques. Another such example is an automotive paint smeared onto the clothing of a hit-and-run victim. In such instances, it is often difficult to discern by stereomicroscopic examination alone whether the thin transfer is paint or plastic from a vehicle part. Other microscopic techniques and perhaps analytical techniques are usually necessary to draw a conclusion. Application of those techniques is often thwarted by contamination of the paint smear by the underlying synthetic fabric. Polyester and acrylic fibers have chemical characteristics similar to those of many paint binders, and the abrasion transfer process often generates heat that melts, mixes, and fuses the substrate fiber polymers into the paint transfer.

As described earlier, paint transfers may also occur with uncured paint. If the transferred paint dries following the transfer, microscopy is the only way to ascertain that it was wet when deposited. That little bit of additional information can be extremely valuable in an investigation, as evidenced in the Green River cases. Occasionally, tiny droplets of spray paint will float through the air at an automotive refinish facility and land on the painted surface of a nearby vehicle. This is obviously not good quality control and represents an unusual situation, not to mention that the color and chemistry of the contaminating paint is somewhat random. Recognizing these tiny cured droplets on the surface of both questioned and known paint chips represents a very unusual characteristic that adds immensely to the associative evidence afforded by the corresponding paint fragments alone. The transfer may also occur from a person rubbing against a freshly painted object. A transfer such as this seldom finds its way into the forensic laboratory, but it may. Recognizing this type of transfer is often of value when searching for automotive paint transfers on the clothing of a hit-and-run victim. If a paint smear is found that was transferred in a wet condition, it is probably not from the striking vehicle and routinely turns out to be an architectural paint. As another example, one of the authors (SGR) examined a case where the suspect of a sexual assault rolled over the victim's palette of paints she was using while painting at a secluded area on the beach. The significance of the numerous corresponding colors and chemistries of paints discovered on the suspect's clothing was reinforced further by the fact that the paints were transferred in an uncured condition. All of these transfers demonstrate very smooth surfaces and rounded edges at the perimeters of the deposits upon stereomicroscopic examination. The rounded edges are a result of flow during the curing process and are not present in abraded transfers of cured paint. Even when glossy automotive paint partially melts onto the fabric of a hit-and-run victim's garments as a result of heat generated from friction, the surface of the paint is not smooth nor are the corners at the perimeters rounded from flow.

5.5.3 Microscopy

With an understanding of the sample attributes present in a forensic paint examination, let us now focus in a bit more detail on the types of microscopy and sample preparation available to permit evaluation of those attributes. Stereomicroscopy affords views of the bulk sample on all sides, including the intact bulk layer structure. The small surface area of thin layers, the rough surface of neighboring layers, and the interplay of neighboring layer colors make it difficult to view all layers definitively. Smoothing the surface of the cross-sectional view will help

Figure 5.7 Peeling of successive layers to expose larger surface areas permits visualization of unusual characteristics, such as this thin mottled primer–surfacer between the white finish coat and the light gray primer. Stereomicroscopic view at 10× magnification. *(See insert for color representation.)*

somewhat, whether done manually with a sharp blade or polished using abrasives. This does not solve the problem of color rendition of small surface areas or the interplay of neighboring layer colors. Typically, the sample is carved manually from either the top or bottom layer to expose larger areas of each layer, including those on the interior of the sample. Portions of these carvings may be saved for subsequent testing. The carving and subsequent increased layer-area exposure permit stereomicroscopic evaluation of colors, textures, pigment heterogeneity, decorative flake size, shape and color in reflected light, and layer interface materials and their distribution (such as the mottled primer-surfacer between the topcoat and base primer shown in Figure 5.7). However, very thin layers may be missed by this method, especially if they are colorless and transparent. It is also difficult to see the transition from a basecoat to a color-coordinated primer (especially black on black or white on white) or the transition from one layer to another in a multilayered white architectural paint fragment.

To solve this problem, it is advisable to prepare thin cross sections like that in Figure 5.6. This can sometimes be done manually or by embedding a specimen in a plastic medium and then cutting very thin cross sections with a microtome. The sample can be observed on the stereomicroscope, but it is often advisable to view it at greater magnification on a transmitted light compound microscope. This can be accomplished on a simple compound microscope, on a polarized light microscope, or in the case of comparisons on a properly color-balanced compound comparison microscope. The cross sections are placed on a microscope slide and covered with a drop of mounting medium having a refractive index substantially different from the 1.50 to 1.52 of most paint resins. This permits the boundaries of colorless clearcoat layers to be observed distinctly. A coverslip is then placed on top, and the preparation is ready for viewing. It permits critical evaluation of layer structure, layer inclusions, and pigment color and distribution throughout the thickness of the layers. It can often allow classification of an automotive paint layer as a finish coat or a primer, primers having a much more particulate appearance, and is invaluable in discovering tricoat layer structures. In the case of some multiple-layer low-gloss white architectural paints, even this method does not reveal layer interfaces clearly. In that case, it is often of value to observe the cross section using a reflected light fluorescence microscope with a variety of excitation wavelengths, as reported by Allen (1994). This technique may also be of value in distinguishing similarly colored automotive paint primers, as described by Stoecklein (1994).

Aluminum decorative flakes and some interference decorative flakes cannot be distinguished when paints containing them are viewed with reflected light using a stereomicroscope. Transmitted light brightfield microscopy, however, can be used to differentiate the two types by observing thin peels cut from the finish coats that have been placed on a microscope slide with mounting medium and a coverslip. As reported by Stoecklein (2002), many types of interference flakes are based on mica substrates, which are transparent in transmitted light, unlike the opaque aluminum metallic flake (see Figure 5.8). Furthermore, the interference flakes that are opaque typically have sharp-lined geometric shapes, unlike the opaque aluminum flake profiles, which resemble a flake of cornflakes cereal or a sand dollar.

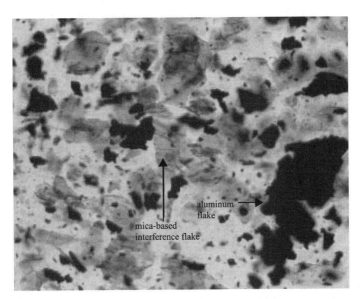

Figure 5.8 Transmitted light microscopic view of a thin peel of an automotive finish coat having two types of decorative flake: mica-based interference flake and aluminum flake. The hand-cut peel is mounted in Norland Type 64 optical adhesive on a microscope slide with coverslip and viewed at 250× magnification on a polarized light microscope with the polars uncrossed. *(See insert for color representation.)*

Observation of thin sections of paint on a polarized light microscope employing crossed polars can also aid in recognizing or even identifying anisotropic extender pigments. If noted in an automotive paint layer, it is a strong indication that the layer is a primer, not a finish coat. If observed in a thin peel of an architectural or spray paint, the grain size may be large enough to permit identifications using basic optical characteristics, as demonstrated by Kilbourn and Marx (1994).

The value of using various microscopic techniques in analyzing paint samples has been demonstrated over the years by a number of authors. McCrone (1979) describes the use of polarized light microscopy in characterizing particulate matter encountered in the forensic science laboratory, while Hammer (1982) describes the use of transmitted light microscopy for automotive paints. Stoecklein (1994) discusses the value of various microscopic techniques for examining automotive paints, and the ASTM *Standard Guide for Forensic Paint Analysis and Comparison* (ASTM, 2007) further details the basic use of microscopy in forensic paint examinations in general.

5.5.4 Microspectrophotometry

Although not all coatings are colored, paint is generally considered to be a pigmented liquid or liquefiable composition which is converted to a solid film after application as a thin layer. Coloring pigments may be either inorganic or organic, and both impart color to the film. Color by its basic definition involves the absorption and reflection of visible light. Beyond observation with the human eye, what more intuitive way is there for the chemist to measure and compare this characteristic than with visible light absorption spectrophotometry?

The comparison of color is one of the first steps taken in a forensic paint comparison. The human eye is a very discriminating sensor, being able to distinguish anywhere from over 100,000 to several million colors (Hunter, 1975; Boynton, 1979); hence, its value as a discriminating tool for paint evidence cannot be overemphasized. However, it is possible to formulate paints that appear the same color to the human eye, yet are comprised of different pigment combinations. Analytical instruments that permit the chemist and industrial technologist to discriminate such colors have been available for six decades. The ultraviolet–visible (UV–Vis) absorption spectrophotometer is known to every undergraduate chemistry and biology major. It is sensitive to the different degrees to which various wavelengths of visible and UV light are absorbed by a material, most typically a solution of the material contained in a UV-transparent cuvette. If two solutions are the same color but contain different chromophores, they will typically produce differing visible absorption spectra. And, of course, the solutions may have no color at all, yet contain organic molecules that have differing aromatic structures and thereby produce differing UV spectra. The chromophores in the organic pigments used in paints are typically rich in aromatic and conjugated ring systems and as such often produce characteristic UV absorption spectra.

UV–Vis spectrophotometry therefore seems to be a natural choice to include in an analytical scheme to compare paint samples. However, the task is not as simple as it may seem. As mentioned previously, the sample is typically dissolved in a solvent to decrease its absorption to a level within the dynamic range of the spectrometer's detector when performing UV–Vis absorption spectrophotometry. Many paint binders, the resins that bind the pigments together and adhere to the substrate, are insoluble in most common organic and aqueous solvents that do not affect the chromophore chemically. In those instances the only way to achieve the aforementioned goal is to make the solid sample very thin, which is often done by slicing or crushing the specimen. However, we also know that it is important to control the pathlength of the samples being compared, because Beer's law states that the absorbance is proportional to the pathlength. This is no easy task when slicing or crushing two samples. Hence, the *Standard Guide for Microspectrophotometry and Color Measurement in Forensic Paint Analysis* (Scientific Working Group for Materials Analysis, 2007) recommends that for detailed comparisons of thin sections of paint the specimens be prepared by cutting the two simultaneously using microtomy. This procedure involves embedding the two samples in a liquid resin, curing the resin, and then mounting the block in a microtome and cutting at controlled thicknesses to produce the same pathlength for both sections. This solves the first problem of dealing with solid specimens; however, most forensic samples present the examiner with a second issue. The samples are quite small and often multilayered. How is one to introduce this sample to a UV–Vis spectrophotometer?

Microspectrophotometers (MSPs) allow for objective measurement of the color of small or microscopic samples and are more precise and quantitative than the more subjective results of visual microscopical color comparisons. As the name suggests, they are UV–Vis spectrophotometers incorporated into microscopes (see Figure 5.9). The design of such instruments is difficult, for typical microscope

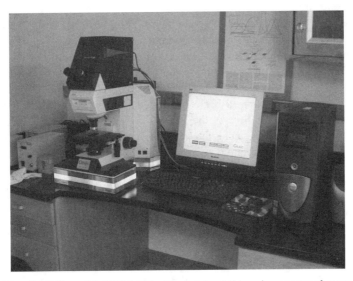

Figure 5.9 CRAIC model QDI 1000 ultraviolet–visible microspectrophotometer with Cassegrainian optics and a semiconductor array detector.

objectives are made of glass, which does not transmit UV light. Additionally, the amount of light transmitted through the small apertures required to define the area of interest in microscopic samples is quite limited and the sensitivity of the photometric detectors must be high while at the same time controlling deterioration of the signal-to-noise ratio. The first problem has been solved either by using non-UV-absorbing glass to make the microscope objectives (which is very expensive) or by using Cassegrainian objectives, which are constructed of mirrors and never pass the beam of light through any glass. The latter suffer from inferior image quality and control of the light path. The second problem has been solved by using very sophisticated focusing optics and highly stabilized photomultiplier tubes or by using fast-response semiconductor array detectors, which permit co-addition and averaging of the signals collected. In either design, the specimen is mounted in a non-UV-light-absorbing mounting medium (typically, glycerin), placed on a quartz microscope slide with a quartz coverslip, and then placed on the MSP's stage. The area of interest is physically defined by adjustable or fixed apertures viewed in the same field of view as the specimen and then background and sample measurements are made. The instrument software ratios the two, and plots of the resulting absorption or percent transmittance spectrum are obtained.

As might be expected, analysis of solid samples is not as straightforward as solution absorption spectrophotometry. There are microscopical concerns, such as light path focusing, sample focusing, and diffraction at specimen edges; spectrometer concerns, such as appropriate calibration standards for wavelength and photometric accuracy; and specimen concerns, such as microheterogeneity and sampling area. Eyring (2002), Stoecklein (2001), and the *Standard Guide for Microspectrophotometry and Color Measurement in Forensic Paint Analysis* (Scientific Working Group for Materials Analysis, 2007) discuss all these concerns in detail. Given the concern for controlling path length and the effect that it has on the interpretation of results, perhaps a brief discussion of sample microtomy might be in order.

Microtomy is a technique employed to prepare thin sections of controlled thickness using a microtome, a mechanical instrument designed especially for this purpose. The discipline finds its primary roots in the histology laboratory; however, its application also finds its way into the field of materials science. For the purpose of paint fragment sectioning, a basic instrument will suffice. Most laboratory or medical supply vendors will carry the array of available equipment, ranging from basic manual models using steel knives to motorized automatic models using cryostatic freezing to attain submicrometer sections. A used rotary microtome with a sharp steel blade will perform adequately, although the use of glass knives will produce sections that do not have striae (from knife blade sharpening) and are much less prone to curling while being cut. The use of glass knives will require a knife maker, special microtomy glass, and a glass knife microtomy stage. A typical rotary microtome with a glass knife can be seen in Figure 5.10 and with a steel knife in Figure 5.11.

The choice of embedding medium is just as important as the choice of microtome. The sample(s) will be placed in this liquid resin and then the resin will be cured to form a rigid block that is placed into the microtome sample jig for

ANALYTICAL METHODS **155**

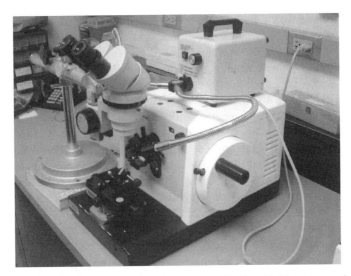

Figure 5.10 Microm manual rotary microtome with glass knife stage and universal sample holder.

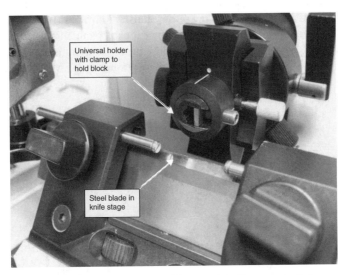

Figure 5.11 Microm rotary microtome with a steel knife stage and the sample block held in a universal specimen holder.

sectioning. It is recommended that the embedding medium be approximately the same hardness as the specimen to be sectioned for optimum results. If it is too soft, the specimen will be dislodged from the embedding medium when encountering resistance with the blade. If it is too hard the blade will quickly be dulled and produce poor-quality sections. Both Stoecklein (2001) and Derrick (1995) provide

156 ANALYSIS OF PAINT EVIDENCE

guidance on these topics. For those familiar with this technique, one of the authors of this chapter SGR finds that a UV-curing optical adhesive (UV-74 Lens Bond, Summers Optical Co., Fort Washington, PA) works quite nicely, permitting the examiner to place some embedding medium resin in the well of the mold and to cure it for approximately 6 s with a high-intensity UV light to form a thin skin on which to place the first paint fragment. This permits critical positioning of the paint chip so that it will be parallel to the remaining chips placed in the block. After positioning the first chip, some additional embedding resin is added and then flash-cured for about 6 s to form another partially cured surface. The next paint chip is then positioned and more resin added. This process is continued until the block has been completed. After adding the final aliquot of resin, the mold is cured under high-intensity UV light for approximately 6 to 10 min. It is then removed from the mold and all sides are exposed to the UV light to remove the tack from the optical adhesive. The block is then ready to be cut at this time or within a day or two, as depicted in Figure 5.12. Typically, excess resin is trimmed away manually to expose the specimens near the surface. The block is then secured in the microtome jig and trimmed even further using the steel blade until a position very close to the specimens is reached. The microtome knife stage is then switched to the glass knife stage, where the final sections are produced. Three-micrometer sections are cut for brightfield microscopy and UV–Vis microspectrophotometry, while 5-μm sections may be cut for Fourier transfer–infrared (FT–IR) microspectroscopy. Either the remaining block with the exposed paint chip surfaces or a very thick section (20 μm) may be used for subsequent scanning electron microscopy in conjunction with energy-dispersive x-ray spectrometry (SEM–EDS) analyses.

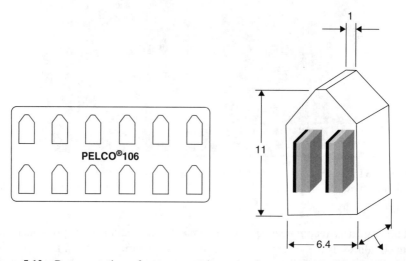

Figure 5.12 Representation of two automotive paint fragments embedded in the same microtome sectioning block (right) and of the top view of the silicone mold used to form the block (left). Dimensions are in mm. *(See insert for color representation.)*

Although this sample preparation approach may seem very appealing given the minimal sample manipulation of the native paint fragment, the production of controlled-thickness paint sections, and the control of surface topography and orientation (critical factors in SEM–EDS analyses), there are drawbacks to the preparation of samples by microtomy alone. First, very small areas of the sample are exposed in cross section. This forces the analyst to deal with microheterogeneity problems, since large analytical areas cannot be used to average heterogeneity. Cured paint is simply not homogeneous on the microscopic scale. Second, difficulties in control of the analytical area may be encountered when using an infrared microscope, due to diffraction produced with the small imaging apertures required, and in SEM–EDS analyses due to excited volumes crossing over into neighboring layers. These concerns are discussed by Ryland (1995), Henson and Jergovich (2001), Bartick (2010), and in the *Standard Guide for Using Scanning Electron Microscopy/X-Ray Spectrometry in Forensic Paint Examination* (Scientific Working Group for Materials Analysis, 2002).

As with most spectrophotometric methods of analysis of solid specimens, visible-light microspectrophotometry can be performed in the reflectance mode as well as in the transmission mode. This approach offers the lure of apparent minimal sample preparation but has notable disadvantages compared to transmission analyses. The signal-to-noise ratio is typically lower in reflectance measurements, and there are sample geometry, spectral distortion, and reproducibility concerns. The sample surface must be sufficiently smooth so as to have ridges and valleys measuring less than the size of the particles contained therein. If not, light is scattered randomly and signal targeting and signal-to-noise quality are lost. To achieve this goal, the sample must be polished on a microscopic level using a series of abrasives typically performed on embedded samples, much as would be encountered in the preparation of samples for microtomy. In addition, it is critical that the specimen(s) surfaces be oriented parallel to the reflectance objective of the microspectrophotometer, as failure to do so results in variation in spectral features. For certain monocoats containing high concentrations of strongly absorbing pigments (such as Copper Phthalocyanine Blue), specular reflection distortion of the measurement can occur (Suzuki, 2010). This process produces *more* reflection for strongly absorbing wavelengths and results in a *bronzing effect* (this phenomenon can also occur for some dark blue inks, which appear to have a metallic bronze sheen when viewed at certain angles). For metallic paints, the incident beam can undergo either direct reflection from a metal flake or diffuse reflectance from the paint matrix, and differences in the ratios of these two processes also affect spectral reproducibility. Hence, it is much more difficult to achieve precision with reflectance techniques than with transmission techniques. Discussion of this shortfall can be found in Eyring (2002), Stoecklein (2001), and in the *Standard Guide for Microspectrophotometry and Color Measurement in Forensic Paint Analysis* (Scientific Working Group for Materials Analysis, 2007).

Despite the intuitive application of microspectrophotometry to forensic paint comparisons, the technique is not widely used. Reports from paint proficiency tests offered by Collaborative Testing Services (2007) over a three-year span indicate only 7 to 15% of the participating laboratories used microspectrophotometry in the examinations. These rates are similar for both architectural and automotive paint samples. The limited application is understandable for either white or pastel flat architectural paints, for they have little color present and are loaded with extender pigments. However, this does not explain the limited use for automotive exemplars. Perhaps this can be traced to the fact that little work has been published to demonstrate the improved discrimination of this method over the classical microscopical and instrumental techniques. This is not surprising given the complexity of automotive paint formulations and layer systems. If different paint manufacturers used different pigments to achieve the same color, it is quite likely that other chemical differences would be present in the formulations as well (Contos and Ryland, 2001). That is not to say that very careful analyses may not have the ability to distinguish between batches, where slight color adjustments have been made to achieve an acceptable final color and appearance, as presented by Stoecklein and Palenik (1998). Similar analyses may even be able to discriminate between the quantity of UV absorbers remaining in the clearcoats of two cars painted with the same paint-layer system but subjected to varying degrees of exposure to the sun (Stoecklein and Fujiwara, 1999). In the case of automotive paints, Kopchick and Bommarito (2006) demonstrated that apparently achromatic finish coats can be discriminated by visible microspectrophotometry, although the study did not evaluate which of those finishes were also distinguishable by other routine analytical examinations. This rationale for automotive finish coats should not be carried over into the realm of other end-use paints, such as household spray paints, where the variation in binders and layer structures is much more limited.

As the number of available organic pigments expands and the diversity in binder formulations appears to be diminishing due to company mergers, international markets, and environmental concerns, it is argued that pigment identification may take on a new role in forensic paint examinations. This may be in the arena of comparisons or even in the task of provenance determinations. Certainly, with some care in sample preparation and access to high-magnification objectives, microspectrophotometry has the ability to focus on individual pigment grains in situ and effect identifications based on collections of known pigment spectra. The future will tell what benefits and acceptance this approach holds.

5.5.5 Infrared Spectroscopy

This technique involves measuring the absorptions of a sample in the infrared region, which consists of electromagnetic radiation with frequencies lower than that detected by the human eye. The frequencies in this region are denoted by the wavenumber, which is the reciprocal of the wavelength given in centimeters (cm^{-1}); the region of primary interest is between 4000 and 200 cm^{-1}.

The infrared absorption process results in an increase in the quantized vibrational energies of the molecules of the sample. Some of the energies are localized in vibrational modes involving specific chemical bonds of the compound, and these absorptions often occur over a narrow and characteristic range in the infrared region. Other vibrational modes may involve the entire molecule or large portions of the molecule, and their absorptions occur over a wide range of frequencies and tend to be characteristic of the molecule as a whole. For organic compounds, the former class of absorptions allows an analyst to determine the presence of functional groups, whereas the latter allows the differentiation—and hence the identification—of even closely related compounds. Organic chemists thus use this technique for both diagnostic and identification purposes.

For inorganic compounds containing predominantly covalent bonds (such as silica and other framework silicates), a somewhat similar situation occurs, although these infrared absorptions tend to occur at lower frequencies than for organic compounds with absorption breadths that are often broader. For salts involving cations or anions with covalent bonds (such as NH_4^+ or CO_3^{-2}), characteristic absorptions of these ions also occur, although they tend to be less distinct than those of organic compounds since they are often broader and far fewer in number. The only compounds that do not absorb in the region examined by infrared spectroscopy are homonuclear diatomic molecules (most of which are gases such as N_2 and O_2), and simple inorganic salts that do not contain covalent bonds (such as KBr and CsI, which are used as windows for infrared spectroscopy).

All of the components of paint thus produce infrared absorptions, and infrared spectroscopy is one of the main tools used by both the coatings chemist and the forensic analyst to determine the overall composition of paint. For most paints, the binders that are used can be identified by these means, as well as the pigments that are present in high concentrations. The notable exception to this occurs for some architectural coatings with low lusters, as these typically contain large quantities of inorganic extender pigments, the strong broad absorptions of which typically obscure most binder absorptions.

Paint Analysis Methods The two main sampling accessories that are used with FT–IR spectrometers to obtain paint spectra are the diamond anvil cell (DAC) and the infrared microscope. Less often, attenuated total reflectance (ATR) may also be used. Both high- and low-pressure DACs are used for the analysis of paint evidence, although the latter has become the main device used in the United States, due to its thinner windows, ability to be used on an infrared microscope, and lower cost. Paint samples are pressed between the two 1-mm-thick diamond windows (known as *anvils*) of the DAC, or they may be pressed onto a single window. Although the DAC requires much larger samples than can be analyzed with the infrared microscope, one of the main advantages of this accessory is that when used on an extended-range FT–IR spectrometer, the lower frequency (far-infrared) absorptions of the sample can be observed.

An infrared microscope accessory allows the paint examiner to choose a small rectangular area of the sample to be analyzed by the spectrometer while viewing the paint specimen optically. Areas as small as 20 by 20 μm can be analyzed, and multilayered paint specimens can be cross-sectioned, then each layer analyzed

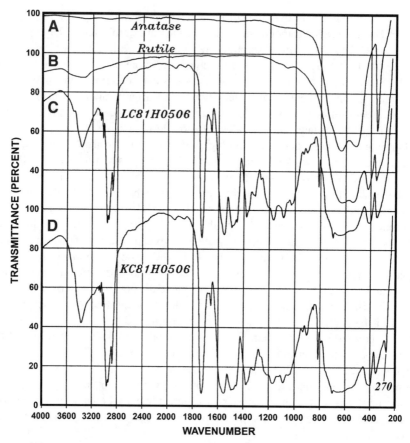

Figure 5.13 Infrared spectra of rutile, anatase, and two yellow nonmetallic automotive monocoats from the *Reference Collection of Automotive Paints*. The monocoats are identified by their *Reference Collection* codes: (A) rutile; (B) anatase; (C) LC81H0506; (D) KC81H0506. Both monocoats have acrylic melamine enamel binders with styrene and contain rutile, Nickel Titanate, and an organic pigment. KC81H0506 also contains hydrous ferric oxide. [From Suzuki (1996b); copyright © ASTM International; reprinted with permission.]

sequentially. The main advantages of the infrared microscope are thus its size and sequential analysis capabilities. Care must be exercised when performing a sequential analysis to set the edge of the aperture as far from the adjacent layer or layers as possible. This minimizes the amount of stray light from an adjacent layer or layers (caused by diffraction) that might contribute to the spectrum of the layer of interest. This is not an easy task in the case of thin layers.

One drawback of the infrared microscope arises from the sensitive detectors that must be used with this accessory, as they have limited spectral ranges. Their low-frequency cutoff points vary between 700 and 450 cm^{-1} and there is an inverse relationship between sensitivity and range; the wider the range, the less sensitive

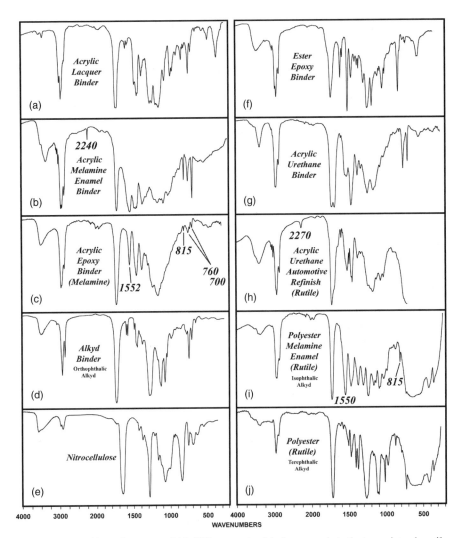

Figure 5.14 Infrared spectra of 10 different paint binders or paints that consist primarily of binders; three of the paints also contain significant amounts of rutile. (A) Acrylic lacquer. (B) Acrylic melamine enamel that contains styrene and an acrylonitrile copolymer. (C) Acrylic epoxy clearcoat (an acrylic melamine enamel with a small amount of styrene; the minor aliphatic epoxy component does not produce a characteristic infrared absorption). (D) Alkyd (orthophthalic alkyd). (E) Nitrocellulose. (F) Epoxy (aromatic) with an ester modifier. (G) Acrylic urethane with a small amount of melamine. (H) An acrylic urethane used in automotive refinishes that contains rutile; the weak absorption at 2270 cm^{-1} is due to unreacted isocyanate starting material. (I) Polyester melamine enamel (isophthalic alkyd) that contains rutile. (J) Polyester (terephthalic alkyd) that contains rutile.

the detector. Many infrared microscopes are thus equipped with the most sensitive detector, even with its narrower range. This does not adversely affect the identification of most paint binders, but it does limit the identification of those inorganic pigments that have most or all of their absorptions below 700 cm^{-1}. Differentiation of the two polymorphs of titanium dioxide used in paints (rutile and anatase), for example, requires the observation of low-frequency absorptions between 500 and 300 cm^{-1} (see Figure 5.13A and 5.13B). The infrared microscope spectrum of a white automotive monocoat that contains rutile is depicted in Figure 5.14H, and only the shoulder of its absorption is observed (the detector used for this analysis has a cutoff point at 680 cm^{-1}). Spectra of two other paints that were collected using a DAC on an extended-range instrument are shown in Figure 5.14I,J, and as can be seen, both contain rutile and not anatase. Another very common inorganic pigment used in paints is ferric oxide (Fe_2O_3, rust-colored). The three absorptions of this pigment occur below 700 cm^{-1} (Figure 5.15A) and they may be observed in the spectrum of a maroon nonmetallic automotive monocoat that contains this pigment (Figure 5.15B). The number of Fe_2O_3 absorptions that are observed using an infrared microscope will depend on the specific range of the detector, but none will be observed using the most sensitive detector.

ATR is a reflectance method that is utilized in either a single-pass accessory or as an objective of an infrared microscope. Paint samples are simply pressed against an ATR element, which is a refractive infrared-transparent material such as zinc selenide or diamond. ATR has a very shallow penetration depth that is proportional to wavelength, so ATR spectra have stronger relative intensities for the lower-frequency absorptions than those of transmission spectra. ATR spectra of paints containing inorganic constituents may also exhibit considerable distortions of the absorptions of these constituents. This occurs because ATR penetration depths also depend on indices of refraction of the sample, and inorganic compounds typically exhibit large changes in their indices of refraction across an absorption band.

Because of the very shallow ATR penetration depths, thin slices of paint do not have to be prepared and paint specimens can be analyzed intact, allowing spectra of the outer layers of a multilayered paint chip to be readily obtained. An ATR objective of an infrared microscope also provides the best means to analyze paint smears that are too thin to be physically removed from the substrate. Such smears can be analyzed in situ, although the spectra obtained by this method should always be compared to those of the substrate since, even with a shallow penetration depth, absorptions of the substrate may be observed in the resulting spectra. If this occurs, it is possible that only the lower-frequency absorptions of the substrate will be observed since these frequencies are more likely to reach the substrate. An ATR spectrum of a thin smear of an automotive clearcoat layer onto a light-colored paint, for example, may well exhibit a "tail" of a rutile absorption (as seen in Figure 5.14h). Although the clearcoat does not contain rutile, the substrate paint does, so the differential nature of the analysis process must always be considered when interpreting such spectra. This point was emphasized in a case involving several smears of automotive paint, which were analyzed using ATR (Giang et al., 2002).

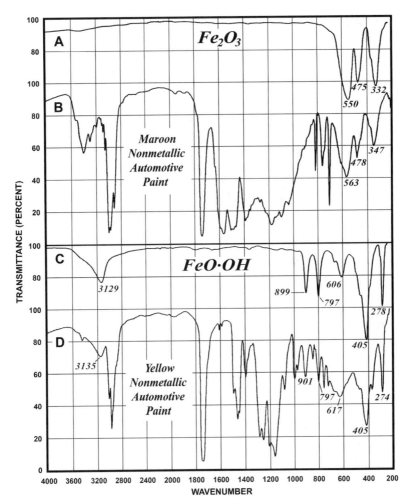

Figure 5.15 Infrared spectra of two iron oxide pigments and two automotive monocoats from the *Reference Collection of Automotive Paints* that contain these pigments. (A) Ferric oxide, Fe_2O_3. (B) Maroon nonmetallic acrylic melamine enamel with styrene that contains ferric oxide. (C) Hydrous ferric oxide, $FeO \cdot OH$. (D) Yellow nonmetallic acrylic lacquer that contains hydrous ferric oxide. [From Suzuki (1996b); copyright © ASTM International; reprinted with permission.]

More details regarding the ATR analysis process are described by Mirabella (1993), while the differences between ATR and transmission paint data are noted by Ryland et al. (2001) and discussed in general by Suzuki (2010). Bartick et al. (1994) describe some forensic applications using an ATR objective of an infrared microscope. Paint analyses using the DAC (Schiering, 1988; Suzuki, 2010) and the infrared microscope (Allen, 1992; Ryland, 1995; Bartick, 2010) are also described in more detail elsewhere.

Binder Analyses

One of the most common binders that the forensic paint examiner is likely to encounter is the acrylic melamine enamel (Figure 5.1) since, as noted, it is often used in automotive finish layers (basecoats and clearcoats). The absorptions of this binder are thus discussed in some detail to illustrate the types of information that forensic paint examiners can deduce from infrared spectra, as well as the limitations of this technique.

The spectrum of a typical acrylic melamine enamel binder is shown in Figure 5.16, and the characteristic absorptions that paint analysts use to identify components of this binder are indicated. The main absorptions of the acrylic component are the ester carbonyl stretching absorption at 1730 cm^{-1} and the series of C—O stretching absorptions between 1300 and 900 cm^{-1}. The latter are somewhat more characteristic of acrylics than the ester carbonyl absorption, since spectra of almost all paint binders contain an ester absorption (produced either by the binder or binder-related components). Melamine produces a broad absorption near 1550 cm^{-1} and a sharper, weaker absorption at 815 cm^{-1} due to vibrations of the triazine ring (Figure 5.1), and these two absorptions are observed in spectra of other binders containing melamine (such as the polyester melamine enamel of Figure 5.14I). Styrene, present in most acrylic melamine enamels, may produce three sets of absorptions (Figure 5.16): (1) weak or very weak =C—H stretching absorptions above 3000 cm^{-1}; (2) two very weak overtone-combination bands between 2000 and 1800 cm^{-1} which appear as "blips"; and (3) weak to medium =C—H out-of-plane bending vibrations at 760 and 700 cm^{-1} (Figure 5.16).

Some of the variations that are observed between spectra of acrylic melamine enamels arise from differences in the relative amounts of melamine or styrene present, as well as differences in the acrylic copolymers that are used. One means

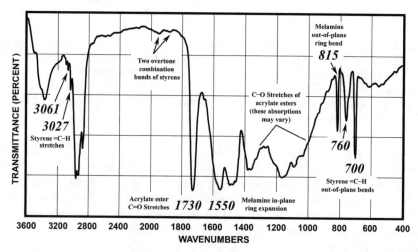

Figure 5.16 Infrared spectrum of a typical acrylic melamine enamel binder with styrene.

to gauge the relative amount of melamine present is to compare the intensity of the carbonyl stretching absorption at 1730 cm^{-1} (which reflects the presence of all of the acrylic copolymers that comprise the backbone) to that of the adjacent melamine 1550 cm^{-1} absorption. Care must be exercised when judging relative infrared intensities, as a nonlinear logarithmic scale is involved in the percent transmittance presentation. Variations may be introduced when working with samples of differing thicknesses or nonpristine samples of limited sizes. Nonetheless, noticeable differences in this ratio may occur, as seen by comparing the spectra of the two automotive monocoats depicted in Figures 5.17A, and 5.17B; a more dramatic difference is seen by comparing Figure 5.14B and 5.14C. Although the binder of Figure 5.14C is referred to as an acrylic epoxy by PPG, the manufacturer of this paint, the aliphatic epoxy component does not produce significant infrared absorptions for characterization, and in terms of its composition, the spectrum is indicative of an acrylic melamine enamel with a relatively small amount of melamine (and a small amount of styrene).

Styrene is found in the vast majority of acrylic melamine enamels used for automotive finishes, and its levels may also vary. The 760/700 cm^{-1} styrene pair may appear to be quite weak when there is a large amount of rutile present, as seen in Figure 5.13C,D. This is a consequence of the logarithmic nature of the percent transmittance scale, which produces a compression of absorption intensities when baselines are low. The other absorptions of styrene (=C—H stretches above 3000 cm^{-1} and the two overtone-combination band "blips" between 2000 and 1800 cm^{-1}), however, are easily seen in Figure 5.13C,D. The three sets of styrene absorptions can be seen in spectra of the acrylic melamine enamels depicted in Figures 5.14B, 5.15B, 5.17B and 5.18C. They may not all be observed, however, when small amounts of styrene are present (Figures 5.14C, 5.17A, and 5.18B) or when large amounts of rutile or other pigments are present (Figure 5.19A, 5.19F). Styrene is used in other binders, and this same set of absorptions may be observed in their spectra; styrene =C—H stretches and the two "blips" are seen, for example, in the spectrum of the acrylic urethane automotive refinish of Figure 5.14H.

Several different acrylic copolymers are used to create the backbone of an acrylic melamine enamel (Figure 5.1), and in some cases this produces differences in their infrared spectra (Rodgers et al., 1976a). The differences occur mainly in the C—O stretching region between 1300 and 900 cm^{-1} (Figure 5.16), as can be seen by comparing these absorptions in Figure 5.17A and 5.17B and Figure 5.18B and 5.18C. These differences are empirical, that is, they do not permit one to identify which set of copolymers is present. However, acrylonitrile, which contains the —C≡N group (see Figure 5.1), produces a sharp nitrile stretching peak at 2240 cm^{-1}, which does allow one to identify the presence of this copolymer (Suzuki, 1996a). Although weak, this absorption is readily identified (see Figure 5.14B) because it occurs in a region usually devoid of other significant features.

Another common acrylic binder that was used for many General Motors vehicles from the 1960s to the early 1990s is acrylic lacquer (Ryland, 1995). Although it, too, contains a mixture of acrylic copolymers, the main copolymer is poly(methyl methacrylate) (Plexiglas) and the infrared spectrum of this binder

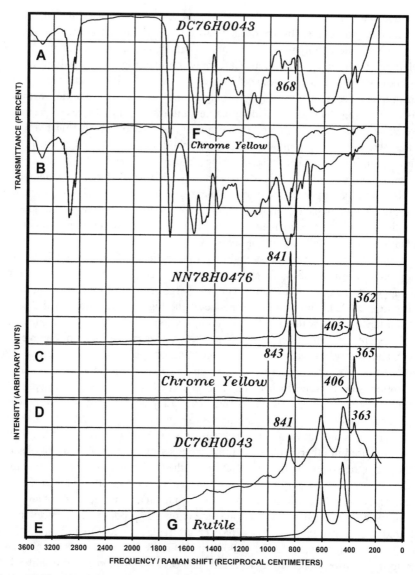

Figure 5.17 (A) Infrared spectrum of a yellow nonmetallic acrylic melamine enamel automotive monocoat with styrene, DC76H0043, which contains a large amount of rutile and a small amount of Chrome Yellow (the Chrome Yellow absorption is marked with its frequency). (B) Infrared spectrum of a yellow nonmetallic acrylic melamine enamel automotive monocoat with styrene, NN78H0476, which contains a large amount of Chrome Yellow. (C) Raman spectrum of NN78H0476. (D) Raman spectrum of Chrome Yellow. (E) Raman spectrum of DC76H0043. (F) Infrared spectrum of Chrome Yellow. G. Raman spectrum of rutile. [From Suzuki and Carrabba (2001); copyright © ASTM International; reprinted with permission.]

ANALYTICAL METHODS 167

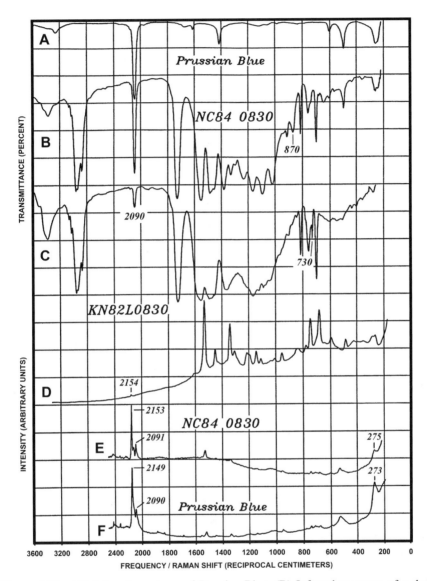

Figure 5.18 (A) Infrared spectrum of Prussian Blue. (B) Infrared spectrum of a dark blue nonmetallic acrylic melamine enamel automotive monocoat with styrene, NC84 0830, which contains a large amount of Prussian Blue. (C) Infrared spectrum of a dark blue nonmetallic acrylic melamine enamel automotive monocoat, KN82L0830, which contains a small amount of Prussian Blue; this paint has the same color as NC84 0830. (D) Raman spectrum of KN82L0830. (E) Raman spectrum of NC84 0830. (F) Raman spectrum of Prussian Blue. [From Suzuki and Carrabba (2001); copyright © ASTM International; reprinted with permission.]

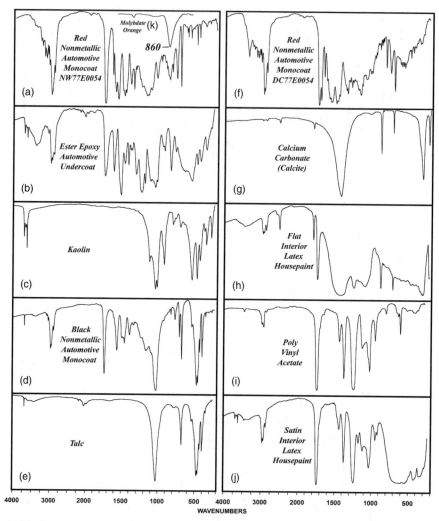

Figure 5.19 Infrared spectra of paints and pigments. (A) Red nonmetallic acrylic melamine enamel monocoat with styrene that contains Molybdate Orange and Quinacridone Red Y. (B) Automotive undercoat with an ester epoxy binder that contains rutile and kaolin. (C) Kaolin. (D) Nonmetallic black automotive monocoat with styrene that contains talc. (E) Talc. (F) Red nonmetallic acrylic melamine enamel monocoat with styrene that contains rutile, ferric oxide, Quinacridone Red Y, and Benzimidazolone Orange. (G) Calcium carbonate (Calcite). (H) Flat interior latex house paint that contains calcite, a silicate, and titanium dioxide or some other oxide. (I) Poly(vinyl acetate). (J) Satin interior latex paint that contains rutile and a small amount of kaolin. (K) Molybdate Orange.

(Figure 5.14A) bears a gross similarity to that of this plastic. Binders referred to simply as "acrylics" are also used in automotive and other paints and have spectra similar to that of Figure 5.14C (the "acrylic epoxy"), except that melamine may not be present, or there may be more styrene (Rodgers et al., 1976a).

The spectra of some other paint binders are depicted in Figure 5.14D to J and Figure 5.19I to illustrate the diversity of their absorption patterns. These are spectra of the binders themselves or of paints that contain little or no observable pigment absorptions (except for the paints of Figure 5.14H to J, which contain large amounts of rutile). Figure 5.14D,I,J are spectra of an alkyd (orthophthalic alkyd) binder, an automotive undercoat with a polyester (isophthalic alkyd) melamine binder and rutile, and a beige paint from a three-ring binder hole punch consisting of a polyester (terephthalic alkyd) binder and rutile. The compositions of the three binders differ primarily in the substitution pattern of two ester groups on an aromatic ring, which are *ortho, meta*, and *para*, respectively, and they produce distinct absorption patterns. All three binders are polyesters, but for historical and chemical reasons, different nomenclatures have been used to describe them, and the term *polyester* has been used for two distinct compositions.

The nominal polymers of three other paint binders—nitrocellulose, epoxy, and urethane—do not by themselves contain an ester carbonyl group, but when used in paints, they almost always have a modifier (copolymer, block copolymer, cross-linking chain, side chain, or plasticizer) that does. The spectrum of nitrocellulose is depicted in Figure 5.14E. As noted, a film of this polymer by itself is quite brittle and it is always used with a block copolymer such as an alkyd or a plasticizer; the absorptions of nitrocellulose, however, are often the strongest in spectra of such binders. The epoxy binders used in paints are aromatic epoxies based on bisphenol A and are actually polyethers; they always contain a modifier such as an acrylic, alkyd, or polyester, so an ester carbonyl absorption is observed in their spectra (Figure 5.14F).

Urethanes are carbamates with the R_1O—(C=O)—NHR_2 functional group, which has a C=O stretching absorption that occurs below 1700 cm^{-1}. This characteristic absorption of urethanes is usually manifested as either a doublet with the ester carbonyl absorption (Figure 5.14G) or as a shoulder to the ester peak (Figure 5.14H), and is accompanied by the urethane C—N stretch absorption at 1520 to 1530 cm^{-1}. The spectra of a few urethanes may also have a weak to medium absorption in the region 2270 to 2260 cm^{-1} (Figure 5.14H) produced by an isocyanate (R—N=C=O) functional group (Suzuki, 1976a). An isocyanate compound is used as the starting material for most urethanes, and its presence in the paint represents unreacted precursor. Like the nitrile group of acrylonitrile (Figure 5.14B), this absorption is usually readily identified because so few compounds absorb in this region. (The very weak triplet of absorptions near 2200 cm^{-1} in Figure 5.14C,G,I, are artifacts; the paints were sandwiched between both DAC anvils, and this may result in what appear to be weak absorptions caused by differences in reflection from the diamond faces.) The acrylic urethane of Figure 5.14G also appears to contain a small amount of melamine, evidenced by the shoulder band near 1550 cm^{-1} and the very weak sharp peak at 815 cm^{-1}. Figure 5.19I is the spectrum of poly(vinyl acetate) (PVA); a PVA–acrylic binder is the most common used in latex house paints, and the spectrum of this binder is primarily that of PVA since the acrylic component is minor and its absorptions generally overlap those of PVA.

The identification of paint binders based on their infrared absorptions is treated in more detail by Ryland (1995), who also presents an automotive paint binder classification flowchart for their identification based on certain key absorptions. Flowcharts are intended primarily for use with paints lacking significant pigment absorptions or for use by analysts who already possess some pattern recognition skills for absorptions of common binders and pigments. Some caution should thus be exercised in their use and it should be remembered that as new products evolve, their spectral characteristics may not be reflected in existing flowcharts. Nonetheless, they may be a useful aid when used with the proper precautions, and an updated chart used for the identification of automotive paint binders is presented in Figure 5.20.

Several new binders were introduced for automotive paints during the period from roughly the mid-1980s to the mid-1990s. A collaborative study using infrared spectroscopy was therefore conducted to determine the discriminating characteristics of this technique for black nonmetallic basecoat/clearcoat finishes used in the 1990s (Ryland et al., 2001). Such finishes lack most of the microscopic discriminating features described previously, and carbon black, which does not produce

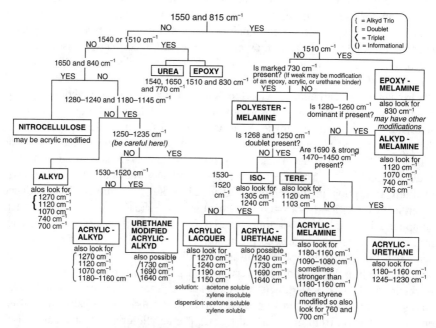

Figure 5.20 Automotive paint binder classification flowchart based on infrared absorptions. Updated from an earlier version published in Ryland (1995).

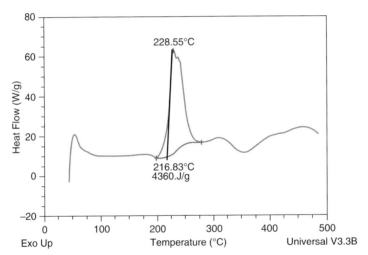

Figure 2.2 Differential scanning calorimetric scan of unknown black solid (suspected of being an explosive).

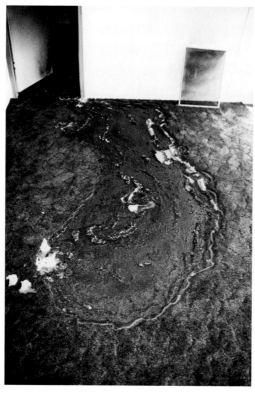

Figure 3.1 "Obvious pour pattern." The only surface that was burned in this mobile home was the floor. There were no furnishings in the house. The carpet was tested and found to be positive for the presence of a medium petroleum distillate such as mineral spirits or charcoal lighter fluid. This is a unique case where visual observation alone can lead to valid conclusions about what caused the pattern.

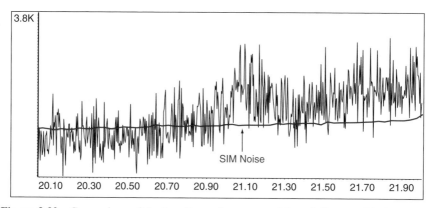

Figure 3.32 Comparison of the baseline noise generated by a full scan with that generated in the SIM mode.

Figure 4.1 Pond lined with bentonite clay where Janice Dodson got clay on her shoes and pants when she went to steal the rifle that she used to kill her husband, John.

Figure 4.2 Two soil samples available for examination for color comparison.

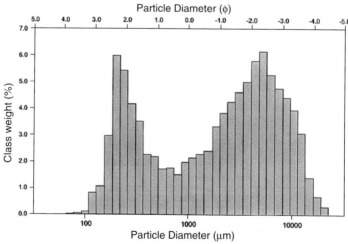

Figure 4.3 Nest of sieves used to separate particles into different size grades and a diagram showing the size distribution of particles.

Figure 4.6 Photomicrograph of a rock thin section as viewed through a petrographic microscope. Note the different minerals with different sizes and shapes.

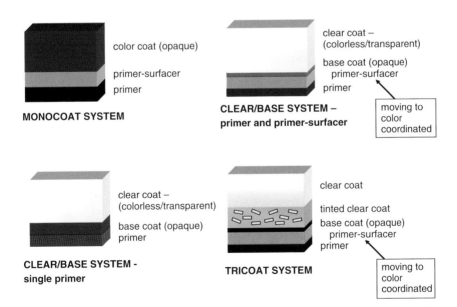

Figure 5.3 Diagrammatic representation of common original automotive finish layer structures.

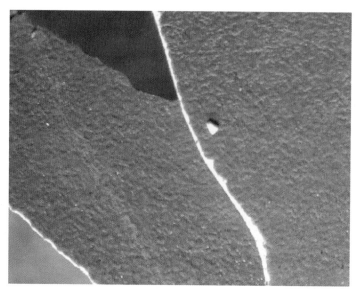

Figure 5.4 Stereomicroscopic view of the underside of the original base primer from an automotive application revealing the characteristic "orange peel" appearance. Magnification of 20×.

Figure 5.5 Paint fragment placed on the date of a penny for sample size perspective.

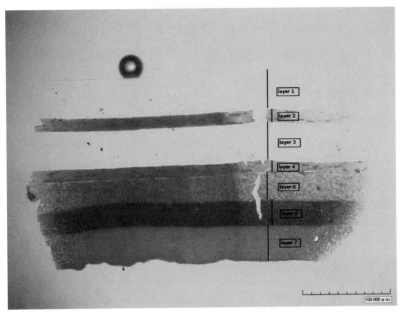

Figure 5.6 Cross section of a seven-layer automotive paint sample mounted on a microscope slide with a 1.56 refractive index mounting media and coverslip and viewed with transmitted brightfield light on a polarized light microscope. Layers 1 and 3 are clearcoats, layers 2 and 4 are basecoats with a decorative flake in them, and layer 5 is a color-coordinated primer followed by two additional primers (layers 6 and 7).

Figure 5.7 Peeling of successive layers to expose larger surface areas permits visualization of unusual characteristics, such as this thin mottled primer–surfacer between the white finish coat and the light gray primer. Stereomicroscopic view at 10× magnification.

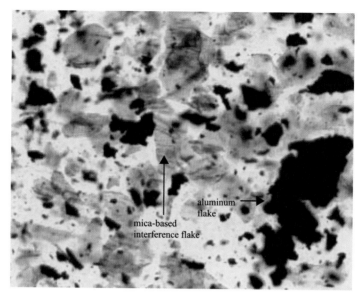

Figure 5.8 Transmitted light microscopic view of a thin peel of an automotive finish coat having two types of decorative flake: mica-based interference flake and aluminum flake. The hand-cut peel is mounted in Norland Type 64 optical adhesive on a microscope slide with coverslip and viewed at 250× magnification on a polarized light microscope with the polars uncrossed.

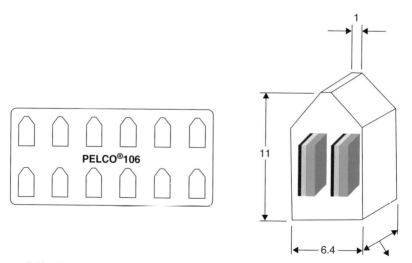

Figure 5.12 Representation of two automotive paint fragments embedded in the same microtome sectioning block (upper) and of the top view of the silicone mold used to form the block.

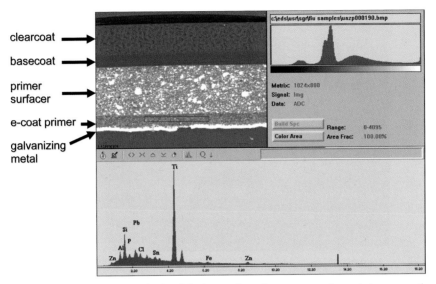

Figure 5.33 SEM–EDS analysis of the base primer in an automotive paint cross section block cut by microtomy. The red box superimposed on the e-coat primer defines the visual boundaries of the analytical area.

Figure 5.34 SEM–EDS spot analysis of interference pigment flake in situ. The flake is visualized by viewing the exposed layer in the backscatter mode. The red + indicates the beam target spot.

Figure 5.36 The suspect 1977 GMC van recovered from a salvage yard.

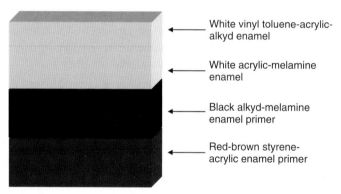

Figure 5.37 Layer structure of the paint from the passenger side door and quarter panel of the 1977 GMC van.

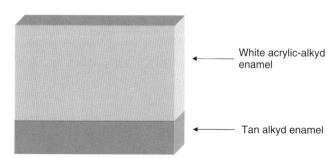

Figure 5.38 Layer structure of the paint from the right side of the hood of the 1977 GMC van.

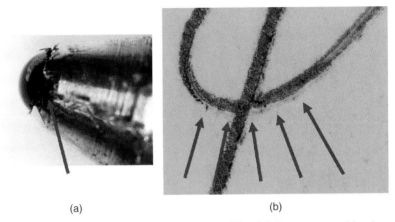

Figure 6.2 (a) Damaged ballpoint pen casing. (b) Visual defects generated by damaged writing instrument.

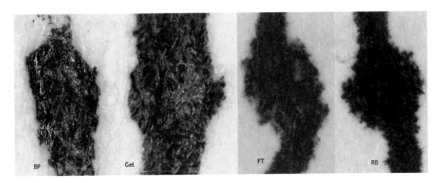

Figure 6.3 Characteristics of the written line from various writing instruments.

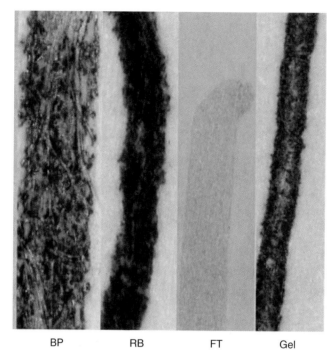

Figure 6.4 Written lines from various writing instruments.

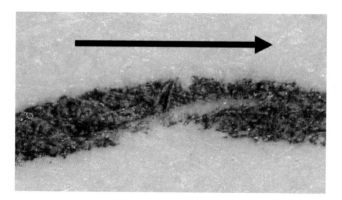

Figure 6.5 Ballpoint pen striations.

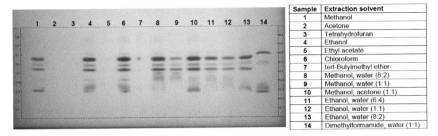

Figure 6.7 Different banding patterns based on changes to extraction solvent.

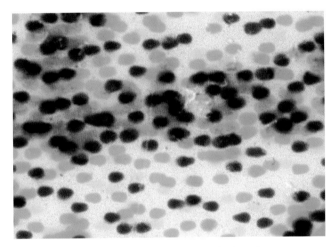

Figure 6.8 Inkjet droplets on a plastic substrate.

(a) (b)

Figure 6.9 (a) Conventional toner. (b) Chemically prepared toner.

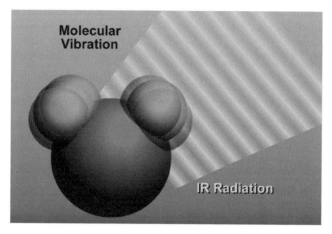

Figure 7.1 Basic principles of IR spectroscopy.

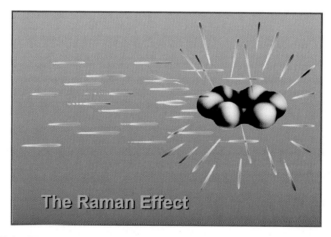

Figure 7.6 The Raman effect: Illuminating light is not absorbed but is scatterred.

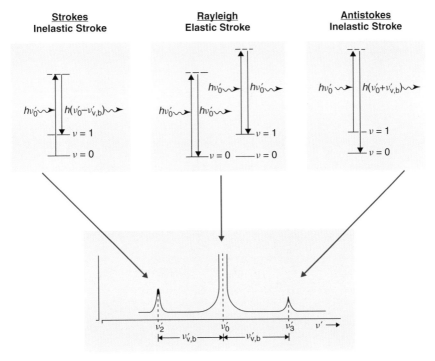

Figure 7.7 The Raman principle.

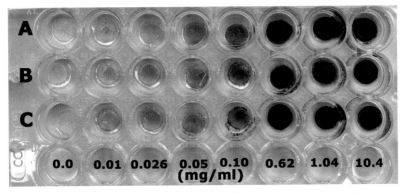

Figure 11.6 Prototype of a dipstick assay for GHB. Ten-microliter samples containing the concentrations of GHB indicated were spotted onto white-paper circles lying in wells of a microtiter plate. Ten microliters of assay reagent containing buffer, GHB dehydrogenase, NAD$^+$, diaphorase, and MTT prodye was spotted, and color was allowed to develop for 2 min at 23°C. Row A, GHB in water; row B, GHB in normal human urine; row C, GHB in normal human urine containing 0.63% (w/v) ethanol. [From Bravo et al. 2004; used with permission.]

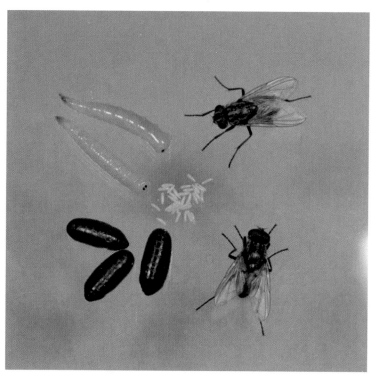

Figure 14.1 Life cycle of *Musca domestica*, showing complete metamorphosis. These life stages are typical of all Muscidae and Calliphoridae. (Photo courtesy of Clemson University; U.S. Department of Agriculture Cooperative Extension Slide Series.)

discrete infrared absorptions, is typically the only pigment used. Binder composition thus assumes a much greater role as a means to distinguish such finishes, and this collaborative study confirmed the suitability of using infrared spectroscopy for this purpose and demonstrated the reproducibility of infrared spectra acquired by numerous laboratories using different instrumentation.

Inorganic Pigment Analyses Because colored inorganic pigments are generally weak absorbers of visible light and strong absorbers of infrared radiation, high concentrations are used in some paints, and their absorptions may be prominent in the spectra of these paints. In addition to ferric oxide (Figure 5.15A,B), there are two other common iron-containing inorganic pigments, hydrous ferric oxide and Prussian Blue. The spectrum of hydrous ferric oxide (FeO•OH), a yellow pigment, is shown in Figure 5.15C. Absorptions of this pigment may be seen in the spectrum (Figure 5.15D) of a yellow nonmetallic automotive paint with an acrylic lacquer binder (compare to Figure 5.14A).

Prussian Blue, which has a dark blue hue (Navy Blue), is comprised of iron and other salts of the $[Fe(CN)_6]^{4-}$ anion; its spectrum is depicted in Figure 5.18A (Suzuki, 1996a). The strongest absorption of this pigment is the cyano (C≡N$^-$) stretching peak at 2090 cm^{-1}, which like the acrylonitrile (Figure 5.14B) and isocyanate (Figure 5.14H) peaks, occurs in a region where it can readily be observed. It is easily seen in the spectra of two blue nonmetallic automotive monocoats (Figure 5.18B,C) that contain this pigment; both paints have acrylic melamine enamel binders with styrene.

The lead chromates are another important family of colored inorganic pigments, comprised of Chrome Yellow, Molybdate Orange, and silica-encapsulated versions of the two (Suzuki, 1996b). The spectrum of Chrome Yellow, a bright yellow–orange inorganic pigment that consists of lead chromate with a lesser amount of lead sulfate ($PbCrO_4 \cdot xPbSO_4$), is shown in Figure 5.17F. The absorptions of this pigment can be seen in the spectrum (Figure 5.17B) of a bright yellow–orange nonmetallic automotive monocoat, which has an acrylic melamine enamel binder with styrene. Molybdate Orange is comprised of lead chromate, lead molybdate, and lead sulfate ($PbCrO_4 \cdot xPbMoO_4 \cdot yPbSO_4$) and its spectrum consists of one main absorption near 860 cm^{-1} (Figure 5.19K), in the same region as the Chrome Yellow absorption (Figure 5.17F). When present in high concentrations, the minor differences in the absorptions of Chrome Yellow and Molybdate Orange may be observed in some paint spectra, although this is usually not the case when low concentrations are used. Figure 5.19A is the spectrum of a red nonmetallic automotive topcoat with an acrylic melamine enamel binder and styrene that contains Molybdate Orange together with an organic pigment.

Three of the most common inorganic extender pigments used in paints are kaolin (Figure 5.19C), talc (Figure 5.19E), and calcium carbonate (Figure 5.19G). Kaolin and talc are silicates that contain hydroxyl groups involving little or no hydrogen bonding, hence the frequencies of the O—H stretches are relatively high (above 3650 cm^{-1}) and the bands are quite sharp. Although weak, these peaks are thus conspicuous and are often easy to detect in paint spectra. The spectrum of an automotive undercoat with an epoxy binder is depicted in Figure 5.19B, and absorptions of both rutile (Figure 5.13B) and kaolin can be seen. The spectrum of the satin latex paint depicted in Figure 5.19J also contains weak kaolin absorptions, along with strong absorptions of rutile. All of the absorptions of talc (Figure 5.19E) are observed in the spectrum of the black nonmetallic acrylic melamine enamel automotive monocoat with styrene shown in Figure 5.19D. This paint is quite unusual in that it has a flat finish, whereas almost all automotive finishes have a high gloss. The presence of such a large amount of talc is, in fact, the reason for this lack of luster, as this extender pigment is serving as a flatting agent.

Calcium carbonate has two common polymorphic forms, calcite and aragonite, both of which may be used in paints. The spectra of the two can be distinguished (Infrared Spectroscopy Committee of the Chicago Society for Coatings Technology, 1980), and the spectrum of calcite, the more common of the two, is shown in Figure 5.19G. This spectrum lacks the sharp O—H "handles" seen in the spectra of kaolin or talc, but it does have two very weak peaks at 2515 and 1797 cm^{-1}. When calcite is used in high concentrations in paint, the two are often conspicuous in spectra because the former is too low in frequency for a C—H stretch and the latter is quite high for a carbonyl stretching absorption. The two can be seen in the spectrum of a flat interior latex house paint (Figure 5.19H), which contains a very large amount of calcite, a silicate (probably silica), and titanium dioxide or some other oxide. As noted, paints having low lusters typically contain large amounts of inorganic pigments, and their broad and strong absorptions often obscure binder features in paint spectra. Only two of the PVA absorptions below 2000 cm^{-1} (Figure 5.19I) are observed in Figure 5.19H. In contrast, most of the PVA absorptions are observed in the spectrum of an interior latex house paint that has a satin finish (Figure 5.19J). Although characterization of the binder is hampered for such low-luster paints, the prominent absorptions of the extender pigments are an important means to differentiate such paints, since the concentrations and combinations of these pigments vary considerably.

Organic Pigment Analyses Organic pigments can be identified by their absorptions in certain paint infrared spectra, although these features are typically weaker than those of inorganic pigments. Organic pigment peaks are narrower than most binder or inorganic pigment absorptions, and they occur over a wider spectral range than inorganic pigments. They are mostly observed in spectra of some paints that have bright vivid colors (characteristic of many automotive finishes), however, and are rarely seen in spectra of paints with pastel or pale shades and blue hues (Suzuki, 2010).

Figure 5.19A,F show spectra of two bright red automotive monocoats that contain absorptions of both organic and inorganic pigments (Suzuki and Marshall, 1998); both paints have acrylic melamine enamel binders with styrene, and the organic pigment absorptions are the sharp features not observed in the spectrum of this binder (Figure 5.16). Analysts who are not familiar with binder absorptions, however, would probably find it very difficult to distinguish binder features from those of either inorganic pigments or organic pigments. The spectrum of Figure 5.19F, in fact, contains significant absorptions of two inorganic pigments and two organic pigments. Familiarity with binder absorption patterns is thus the most important skill for paint analysts to acquire, followed by familiarity with common inorganic pigment absorptions. Organic pigments, in contrast, are best identified by comparisons to literature reference spectra rather than by trying to memorize their absorptions.

Some of the organic pigments or classes of organic pigments that have been identified in automotive finishes using infrared spectroscopy include benzimidazolones (Suzuki and Marshall, 1997), quinacridones (Suzuki and Marshall, 1998), DPP Red BO and Thioindigo Bordeaux (Suzuki, 1999a), Isoindolinone 3R, Isoindoline Yellow, and Anthrapyrimidine Yellow (Suzuki, 1999b), and perylenes (Massonnet and Stoecklein, 1999a). Other spectra of organic pigments may be found in the various editions of *An Infrared Spectroscopy Atlas for the Coatings Industry* (Infrared Spectroscopy Committee of the Chicago Society for Coatings Technology, 1980; Infrared Spectroscopy Atlas Working Committee, 1991). These volumes are also the most comprehensive compilations of spectra of other paint ingredients, including binders, inorganic pigments, and additives.

Color is the most distinguishing feature of certain types of paints, particularly automotive finishes (Ryland and Kopec, 1979). It is thus desirable that the analytical methods that are used in addition to microscopy and microspectrophotometry provide as much discrimination as possible between paints having very similar colors. For paints of the same color where different binders are used, there will clearly be differences in their infrared spectra. Similar pigment formulations are often used to produce the same color; however, this is not always the case, even when similar binders are used. Different pigments or pigment combinations may be used to produce the same color, as illustrated by the spectra of Figures 5.13C,D,B,C, and Figures 5.19A,F. These are spectra of pairs of original automotive monocoats from two different suppliers used for a specific color on a particular vehicle model. As discussed previously, the monocoats of Figure 5.18B,C both contain Prussian Blue, but with significantly different concentrations. Figure 5.19A, 5.19F show spectra of two automotive monocoats with bright orange–red nonmetallic hues (referred to in the industry as "Fire Engine Red"). The first paint (Figure 5.19A) contains Molybdate Orange and Quinacridone Red Y; the second (Figure 5.19F) contains rutile, ferric oxide, Quinacridone Red Y, and Benzimidazolone Orange. The main difference between the spectra of Figure 5.13C and 5.13D is a peak of hydrous ferric oxide (Figure 5.15C) in the latter spectrum (the other peaks of this pigment are mostly obscured).

Rodgers et al. performed most of the early work on the forensic identification of automotive paint constituents using infrared spectroscopy, and their series of papers serves as an excellent introduction to this subject (Rodgers et al., 1976a–c). A statistical method has been applied to evaluate the significance of minor differences in the infrared spectra of some automotive topcoats to determine their value as evidence (Zieba and Pomianowski, 1981). The infrared spectra of various automotive refinish paints were examined and considerable variations in their compositions were noted, as well as differences in the types of refinish binders compared to those of original finishes (Percy and Audette, 1980). A collaborative study involving the infrared analysis of automotive undercoats has demonstrated that even when two or more pigments are present, it is usually possible to identify the pigments and the major binder components based on their absorptions (Norman et al., 1983). Buzzini et al. (2005) used infrared spectroscopy to analyze paint on 207 crowbars that had been seized as evidence. The purpose of this study was to determine the types and frequencies of such paints. Simulations involving crowbars and painted wood surfaces were also conducted, and it was found that paint transfers, including surface to tool and tool to surface, were common.

The infrared analysis of paint evidence has been described in review treatments by Ryland (1995) and Beveridge et al. (2001). The analysis of paint using FT–IR spectroscopy has also been reviewed from the perspective of the coatings chemist (Hartshorn, 1992). General guidelines for forensic paint analyses are outlined in the *Standard Guide for Using Infrared Spectroscopy in Forensic Paint Examinations* (Scientific Working Group for Materials Analysis, 2009).

Vehicle Identification The identification of an unknown paint recovered from the scene of a hit-and-run incident requires a suitable reference collection or database. Until 1990, paint examiners in the United States relied on the *Reference Collection of Automotive Paints* (Collaborative Testing Services, 1989), a comprehensive collection of original finishes used on American automobiles manufactured between 1974 and 1989. Many of the automotive paints discussed in this chapter are from this collection, and they are referred to by their *Reference Collection of Automotive Paints* identification codes. The codes provide information about the paint manufacturer, binder composition, and the year that formulation was first used. The *Reference Collection of Automotive Paint Technical Data* booklet (Collaborative Testing Services, 1989) provides a list of the vehicle models for which each such paint may have been used.

As an example, the two dark blue nonmetallic monocoats with spectra shown in Figure 5.18B (NC84 0830) and 5.18C (KN82L0830) have the same color, which is indicated by the "0830" numerals. Older paints have a letter color code and the "L" of KN82L0830 signifies that this paint was classified as blue. The "N" of NC84 0830 indicates that the paint was manufactured by BASF Inmont, while the "C" denotes an acrylic enamel binder; this paint was first manufactured in 1984 based on the "84." KN82L0830 was manufactured by Glasurit America, has a nonaqueous dispersion enamel binder, and was first produced in 1982. This paint was used on some 1982 Jeeps and on American Concords, Eagles, and Spirits. NC84 0830 was used on some 1984 Jeep Cherokees and on American Alliances, Eagle 30s, Eagle SX4s, and Encores.

Although a useful aid for identifying domestic vehicles, the *Reference Collection of Automotive Paints* had two major limitations. It did not include the complete finish system, including primers and other undercoats, nor did it include paint from imported vehicles. In addition, after 1989 the only comprehensive original automotive paint finish coat collection in the United States was maintained by the FBI Laboratory. In 1994, a committee of international forensic paint examiners, then known as TWGPAINT (now part of SWGMAT, the Scientific Working Group for Materials Analysis), addressed this issue. TWGPAINT obtained funding to expand an existing database maintained by the Royal Canadian Mounted Police (RCMP) Laboratories, known as the Paint Data Query (PDQ) system.

The goal of the PDQ system is to compile data for the complete original finish systems of all vehicles in North America, domestic and import, with infrared data collected for each individual layer of the finish system. The system includes both text and spectral search capabilities, and users input data about binders and pigments, as well as color information, from an unknown original automotive finish. When all of these factors are included in the search, it has generally been found that the list of possible vehicles is quite short. Often, the list consists only of vehicles manufactured during a short time period at one specific assembly plant. More information regarding the PDQ system is discussed elsewhere (Buckle et al., 1997; Beveridge et al., 2001; Ryland et al., 2006).

5.5.6 Raman Spectroscopy

Like infrared spectroscopy, Raman spectroscopy involves vibrational transitions of the molecules of a sample. Raman spectra are generated by an entirely different mechanism, however, involving *inelastic scattering* of a monochromatic laser source. The laser may have any frequency in the near-ultraviolet, visible, or near-infrared region, and the interaction of this radiation with the molecules of the sample can be viewed as an oscillating electric field driving a dipole (the molecules). This generally results in *elastic scattering* (*Rayleigh scattering*), where the scattered radiation has the same frequency. Inelastic scattering (Raman scattering) occurs when some of the energy is transferred to vibrational modes of the molecules of the sample. As a result, the scattered radiation has less energy, and this difference is equal to the energy of the vibrational transition that occurs. A Raman spectrum is a plot of the intensity of the scattered radiation versus its frequency, with the abscissa depicting the *Raman shift*, the difference in frequency (in wavenumbers) between the laser line and the scattered radiation.

The Raman effect is quite weak, and the vast majority of the scattered light (by a factor or 10^5 or so) occurs as Rayleigh scattering. Florescence is a much more efficient process than Raman scattering, and if it occurs (caused either by the analyte itself or by impurities), it may overwhelm the much weaker Raman peaks. In addition to fluorescence, the other main difficulty that may occur when attempting to obtain a Raman spectrum of paint arises from the absorption of the laser light by the sample. This can cause localized heating and possibly destruction of the sample, so that low laser power levels and defocused beams are normally used.

Dispersive charge-coupled device (CCD) array and Fourier transform (FT) Raman spectrometers are the two main types of Raman instruments commercially available. The former uses gratings to separate different wavelengths of Raman-scattered light, which is then dispersed onto a CCD detector array. Dispersive CCD spectrometers use near-ultraviolet or visible lasers. FT–Raman spectrometers employ an FT–IR interferometer in lieu of a dispersing device, a near-infrared laser, and a separate near-infrared detector. Because Raman scattering is proportional to the fourth power of the frequency of the scattered radiation, Raman spectral peaks are stronger when they occur in the visible region compared to the near-infrared region. CCD detector arrays are also more sensitive than near-infrared detectors, so dispersive CCD instruments are inherently more sensitive devices than are FT–Raman spectrometers. Since fluorescence is the main problem encountered with Raman spectroscopy, however, FT–Raman systems may be more applicable for some samples. Both types of instruments may be fitted with a microscope attachment that allows selection of the area to be examined while viewing the sample optically, analogous to an infrared microscope.

Essentially no sample preparation is required with Raman spectroscopy when examining many paints, other than ensuring that the surface is clean. The specimen is simply positioned at the focal point of the instrument collection optics. Using a Raman microscope, sequential analyses of individual layers of cross-sectioned multilayered paint can be conducted. Unlike infrared microscopy, thin sections do not have to be prepared, and spectral contributions from adjacent layers are much less of a concern. For the most part, the technique is nondestructive, although it is possible to create small craters in the paint when laser power levels are too high. Because clearcoat layers are mostly transparent to the incident laser, Raman spectra of basecoats can often be obtained without removing the clearcoat.

A Raman spectrum is generally quite distinct from an infrared spectrum of the same compound. For a few very symmetric compounds, in fact, the transitions that are observed in a Raman spectrum cannot be observed using infrared spectroscopy, and vice versa. Although this is usually not the case, the two techniques are still very complementary, particularly when they are applied to a complex matrix such as a paint that contains inorganic and organic components spanning a wide range of concentrations. Symmetric vibrational modes may produce no infrared absorptions or only weak ones, whereas they often give strong Raman scattering peaks. As noted for pigments, inorganic compounds may produce broad infrared absorptions, but their Raman scattering peaks are usually narrow. Many inorganic pigments are much stronger Raman scatterers than binders. In certain instances, pigments (usually organic) that can absorb the laser light produce significantly enhanced Raman scattering, known as the *resonance Raman effect* (Johnson and Peticolas, 1976), and even when present in very low concentrations, Raman peaks of such pigments may be observed. Raman spectra of paints are thus typically much simpler than corresponding infrared spectra, as they are usually dominated by pigment features.

Some of the differences that are observed between infrared and Raman spectra of paints can be seen by comparing Figures 5.17A, and 5.17E, 5.17B and 5.17C, 5.18B and 5.18E, and 5.18C and 5.18D. These are infrared and Raman spectra, respectively, of four automotive monocoats, and the Raman spectra were acquired using a dispersive CCD system (Suzuki and Carrabba, 2001). As discussed previously, Figure 5.17B is the infrared spectrum of a yellow acrylic melamine enamel with styrene that contains Chrome Yellow. The Raman spectrum of this paint (Figure 5.17C), however, is predominantly that of Chrome Yellow (Figure 5.17D), and binder peaks are almost totally absent. Figure 5.17A is the infrared spectrum of another yellow acrylic melamine enamel with styrene that contains a large amount of rutile. A small amount of Chrome Yellow is also present, but the weak infrared absorption at 868 cm^{-1} cannot be definitely attributed to this pigment since it may also be due to either Molybdate Orange (Figure 5.19K) or a binder feature (Figure 5.18B). The Raman spectrum of this paint, however, indicates clearly that Chrome Yellow is present along with rutile (Figure 5.17G), as the Raman spectrum of Molybdate Orange has a doublet with equal intensities near 362 cm^{-1} (Suzuki and Carrabba, 2001) rather than a singlet. The sloping baseline of Figure 5.17E is caused by fluorescence.

Copper Phthalocyanine Blue is by far the most common blue pigment used in automotive paint as well as many other paints. Copper Phthalocyanine Blue has a very high tinctorial strength (unlike Prussian Blue), so the concentrations of this pigment used in paint are quite low. Consequently, infrared absorptions of Copper Phthalocyanine Blue are rarely observed in paint spectra. As discussed previously, Figure 5.18B,C show infrared spectra of two dark blue nonmetallic automotive monocoats with the same color, but they have noticeably different intensities for the Prussian Blue cyano absorption. The difference in concentration of Prussian Blue is actually greater than might be inferred from the two spectra, as a logarithmic scale is involved and weak peaks appear stronger than in a linear scale. How can two paints have the same color if there is much more pigment present in one of them? A good guess would be that there is another blue pigment in the paint of Figure 5.18C. In fact, the pigment is Copper Phthalocyanine Blue, but the only manifestation of its presence in the infrared spectrum of Figure 5.18C is a weak sharp absorption at 730 cm^{-1} (sandwiched between the two styrene absorptions at 760 and 700 cm^{-1}). The Raman spectrum of this paint (Figure 5.18D), however, is predominantly that of Copper Phthalocyanine Blue, which produces resonance-enhanced peaks (Palys et al., 1995). Prussian Blue, in contrast, is not a particularly strong Raman scatterer. Its peaks (Figure 5.18F) are seen in the Raman spectrum of the paint with the larger quantity of Prussian Blue (Figure 5.18E), but are barely perceptible in the Raman spectrum with the lesser amount (Figure 5.18D).

Massonnet and Stoecklein (1999b) collected FT–Raman spectra of some yellow, orange, and red organic pigments used in automotive finishes along with spectra of some red paints that contain some of these pigments. De Gelder et al. (2005) also examined automotive paints using FT–Raman spectroscopy and concluded that the Raman peaks of organic pigments in basecoats provide the best discrimination. Raman peaks of rutile, calcite, and barium sulfate in spectra of undercoats were also found useful for characterization. Unlike infrared spectroscopy, however, Raman spectroscopy is not useful for detecting a wide variety of inorganic pigments. In their study of pigments in automotive paint, Suzuki and Carrabba (2001) found that silicates (talc, kaolin, quartz, and others) do not produce significant Raman peaks, even when present in high concentrations.

In three studies, Bell et al. examined the discriminating capabilities of Raman spectroscopy for house paints using a dispersive CCD spectrometer. In the first study (Bell et al., 2005a), the spectra of 51 lilac-colored house paints were collected and compared. The three components producing the strongest Raman peaks of these paints were rutile, Copper Phthalocyanine Blue, and a bluish-violet dioxazine pigment, Pigment Violet 23. In their two other studies, a combination of infrared and Raman spectroscopy was used to study the discrimination between binders used in house paints (Bell et al., 2005b) and between white house paints (Bell et al., 2005c). These studies reiterated the complementary nature of the two techniques.

Kendix et al. (2004) examined some historical (nineteenth century) house paints using a Raman microscope. Some of the inorganic pigments that they identified include calcium carbonate, barium sulfate, lead carbonate, rutile, and anatase. The rutile-to-anatase ratio varied considerably in these older paints, and differences in this ratio were readily detected because the two pigments produce discrete narrow Raman peaks; in contrast, such differences are difficult to observe using infrared spectroscopy—compare Figure 5.13A and 5.13B. Buzzini and Massonnet (2004) analyzed 40 green spray paints using a combination of infrared and dispersive CCD Raman spectroscopy and found that this combination allowed discrimination of most of the 40, although when used alone, infrared spectroscopy provided a higher discrimination power than did Raman spectroscopy. Resonance Raman peaks of two green phthalocyanine pigments, Pigment Green 7 and Pigment Green 36, were identified in some spectra. Buzzini et al. (2006) discussed six case examples involving automotive, household, and spray paints in which the combination of infrared and dispersive CCD Raman spectroscopy also proved useful.

5.5.7 Pyrolysis Gas Chromatography and Pyrolysis Gas Chromatography–Mass Spectrometry

Pyrolysis gas chromatography (PyGC) and pyrolysis gas chromatography in conjunction with mass spectrometry (PyGC–MS) are additional techniques used to examine and compare the organic portion of a cured coating: the binder and additives. In *analytical pyrolysis*, the sample is subjected to a sufficiently high temperature under controlled conditions in an oxygen-free environment so that its constituent molecules break down into smaller fragments. Those molecular fragments (pyrolyzates) are then swept into an analytical instrument, most commonly a gas chromatograph, where they are separated chromatographically and detected by either a flame ionization detector (FID) or a mass selective detector (MSD) upon eluting from the chromatography column. The FID offers a relatively inexpensive, sensitive, and broadly applicable means for the detection of analytes eluting from the column. The data obtained from such a detector can provide a pattern of pyrolyzates that can be used either to identify the class of polymer that is present or to perform a detailed comparison of resin compositions. If a MSD is used in

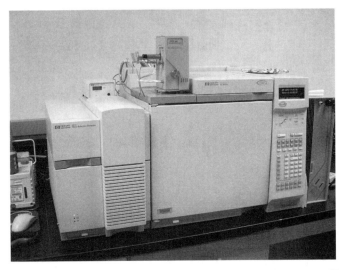

Figure 5.21 Typical pyrolysis gas chromatograph–mass spectrometer (PyGC–MS) equipped with an inductively heated pyrolysis probe, pyrolysis interface, and single capillary column having a mass selective detector (MSD).

place of a FID, much more information can be obtained from the sample. In addition to the pattern information obtained with the FID, the data acquired from a MSD can be used for the potential identification of select compounds of interest. This often aids in identifying the class of polymer that is present. Of course, this additional information comes at a price, as the MSD is considerably more expensive than the FID. A typical PyGC–MS system is shown in Figure 5.21, and an inductively heated pyrolysis probe that inserts into the pyrolysis interface on the gas chromatograph in Figure 5.22.

Why pyrolyze the sample in the first place? Paint binders consist of high-molecular-weight polymers that are simply too large to volatize and introduce into a gas chromatograph. Even if the resin is soluble in a volatile organic solvent such as acetone or chloroform, the polymer chains are still of extremely high molecular weight and cannot be volatized when injected into the standard injection port of a gas chromatograph. By thermally fragmenting the polymers, pyrolyzates are produced that are amenable to volatization and chromatographic separation in a vaporous state. The technique requires little sample (on the order of 10 to 20 µg) and can be performed on most solid polymeric specimens. One may ask why gel permeation chromatography or high-pressure liquid chromatography could not be used, to avoid fragmentation of the polymers. These techniques may be suitable in polymer laboratories but are less so when dealing with real-world forensic samples. First, they require the sample to be in solution, a feat unachievable for a majority of cured, cross-linked paint resins. Second, these techniques typically require a substantial quantity of polymer, which is not available from each layer of most multilayered forensic paint specimens. Finally, these techniques do not

180 ANALYSIS OF PAINT EVIDENCE

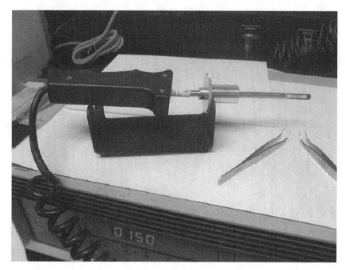

Figure 5.22 Chemical Data Systems inductively heated pyrolysis probe which inserts into the pyrolysis interface mounted on the gas chromatograph injection port.

provide chromatographic characteristics that are specific to the component polymeric species and hence do not maximize discrimination.

Perhaps the more pressing question is why use PyGC or PyGC–MS at all? As we have seen earlier, infrared spectroscopy is a versatile nondestructive technique available to most laboratories which provides spectral data that permit both discrimination and classification of most binders. Furthermore, often it simultaneously provides spectral data characterizing the major inorganic constituents in a paint. Demonstrating its discriminating capabilities, Contos and Ryland (2001) reported 97.5% of the 80 possible pairs of similarly colored *Reference Collection of Automotive Paints* (Collaborative Testing Services, 1989) finish coats having no decorative flake were discriminated by infrared spectroscopy. Edmondstone et al. (2004) reported on a discrimination study of 260 randomly acquired automotive paints, yielding 32,670 possible pairs for comparison. Samples were initially discriminated by microscopical comparison of their finish coat colors, which differentiated all but 28 of the 32,670 possible pairs. Following attenuated total reflectance (ATR) FT–IR microspectroscopic analyses of these 28 indistinguishable pairs' clearcoats, only two pair could not be distinguished. Subsequent transmission FT–IR microspectroscopy of the two pairs' primers served to discriminate another one of the pairs. Obviously, the technique is capable of substantial discrimination.

Infrared spectroscopy does have limitations in the analysis of coating samples. The first is in the discrimination of binders consisting of a mixture of similarly structured polymers. This occurs in the acrylics and alkyds used in automotive paints, in the acrylics used in architectural latex paints, and in the alkyds used in architectural oil-based paints and household spray paints. When performing infrared spectroscopy on a paint resin, the spectrum is comprised of the resulting absorptions of each component in the polymeric mixture, overlaid upon one another. Detection of characteristics arising from minor components in the mixture is difficult to achieve when the spectra of the individual polymers are all similar to one another. PyGC, on the other hand, employs a chromatographic separation of the pyrolyzates generated from the mixture, permitting detection of characteristics produced from even small amounts of organic components. One such example can be seen in Figures 5.23 and 5.24. The transmission infrared spectra of two different acrylic–melamine enamel automotive finish coats are displayed in Figure 5.23.

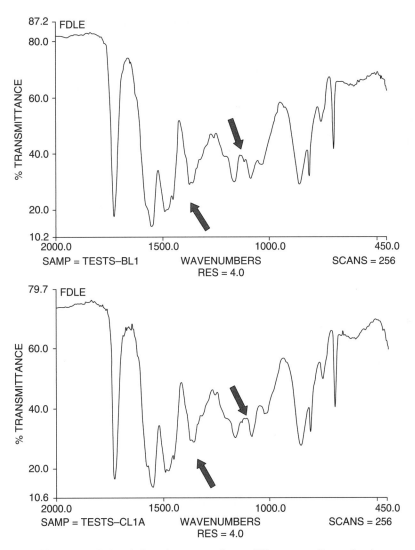

Figure 5.23 Transmission infrared spectra of two different acrylic–melamine enamel automotive finish coats that have the same color. The arrows point out reproducible spectral differences.

182 ANALYSIS OF PAINT EVIDENCE

They are quite similar, with small differences noted by the arrows. The pyrograms of the same two are displayed in Figure 5.24, presenting no difficulty in differentiating the two. Another example is given in Figures 5.25 and 5.26, these also being two automotive acrylic–melamine enamel finish coats produced by the same paint manufacturer in the same year in the same color. Again, the pyrograms demonstrate little difficulty in differentiating the two. As mentioned previously, the same type of situation can occur in the alkyd enamels, where the same diacid and polyol are used as precursors, but the drying oils are different. Burke et al. (1985), Fukuda (1985), Cassista and Sandercock (1994), and Ryland (1995) all have noted the capacity of PyGC to detect binder variations that go unnoticed, or are at least questionable, when samples are analyzed by infrared spectroscopy alone.

The second major limitation of infrared spectroscopy is a consequence of one of the aforementioned benefits of the technique. In paints that are heavily loaded with extender or coloring pigment, the absorptions resulting from the binder are masked by those that are detected simultaneously from the pigment (Figure 5.19H). Thus, little is present to discriminate between differing binders when performing comparisons. These include some automotive primers, flat architectural paints, flat household spray paints, and some low-gloss maintenance finishes. On the other hand, PyGC in effect performs an automatic extraction prior to analysis. Upon pyrolysis

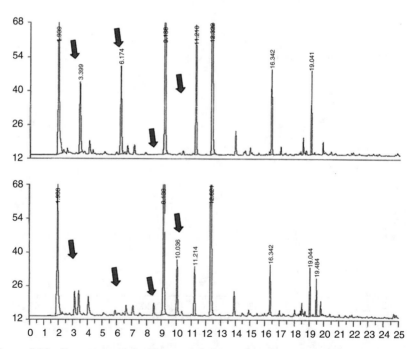

Figure 5.24 Pyrograms of the same two finishes whose infrared spectra are depicted in Figure 5.23. Arrows indicate obvious reproducible differences between the two. The X axis is labeled in minutes and the Y axis is relative intensity.

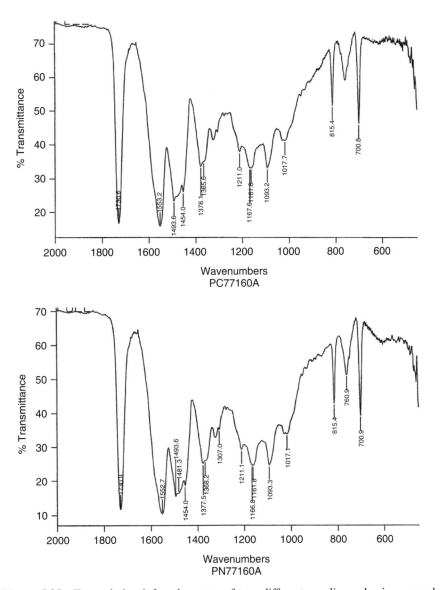

Figure 5.25 Transmission infrared spectra of two different acrylic–melamine enamel automotive finishes having the same color. Both were manufactured by PPG for use on the same year vehicle line; however, one is a low-solvent-emission formulation while the other is not.

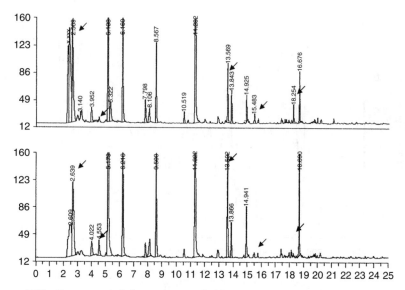

Figure 5.26 Pyrograms of the same two finishes whose infrared spectra are depicted in Figure 5.25. Arrows indicate some of the reproducible differences between the two finishes. The X axis is labeled in minutes and the Y axis is relative intensity.

at an appropriate temperature, the organic portion of the paint is volatilized, fragmented, and passed onto the gas chromatograph while the inorganic portion of the paint remains behind, effectively eliminating any inorganic interference.

It becomes apparent that unlike many other types of trace evidence, paint presents some complex analytical chemistry challenges that demand the use of several analytical techniques to properly characterize its composition. PyGC offers distinct advantages over infrared spectroscopy, but is not without its limitations as well. It requires approximately 10 to 20 μg of sample, on the order of 10 times more than is required for an infrared microscope. In the case of smeared architectural exemplars or very tiny multilayered automotive paint fragments, which are increasingly becoming the norm, PyGC is simply not an option. Furthermore, it is destructive and the organic portion of the sample cannot be reclaimed for further analytical examinations. The inorganic portion of the sample can be reclaimed; however, it has suffered potential deterioration due to exposure to high temperatures. It is much more time consuming than infrared spectroscopy, as the typical run takes approximately 30 to 45 min plus an additional 15 min or more to clear out the column between analyses. It has rather poor long-term reproducibility, making it a poor choice for databasing. However, the discrimination benefits are compelling given certain types of specimens, particularly when used in a comparative mode.

A variety of pyrolysis accessories are available on the market, including inductively heated filament, Curie point, and furnace. Their design, benefits, and shortfalls are discussed by Wampler (2006). In the inductively heated filament type, the sample is placed either on a platinum ribbon probe or in a quartz tube contained within a platinum filament coil probe and inserted into a heated interface attached to the injection port of the gas chromatograph. Heat is supplied by electrical resistance in the platinum element with both the rate of heating and the final temperature being controlled by the electrical power supplied by the control module. Sophisticated electrical circuitry is used to control the rate of heating reproducibly, and the final temperature is calibrated for each pyrolysis probe. Upon completion of analysis, the sample is heated to a very high temperature to clean the platinum ribbon or the quartz tube. An overwhelming majority of instruments used in forensic laboratories in the United States are of the inductively heated filament type, such as that depicted in Figure 5.22. In the Curie point type, the sample is either wedged into a crimped metal alloy wire or placed inside a neatly folded metallic foil pack. The metal carrier along with the sample is then placed into the pyrolysis probe, which houses a radio-frequency (RF) coil. When power is supplied to the RF coil, the metallic carrier quickly heats to its Curie point temperature, that temperature being determined by the metal alloy being used in the carrier. A number of alloys are available, providing an array of fixed final temperatures. Upon completion of analysis, the Curie point wire is either cleaned of any residual inorganic material or the Curie point foil is simple removed and discarded. In the furnace type, the sample is placed in a carrier which drops into an inductively heated furnace held at an adjustable final pyrolysis temperature. Following analysis, the residual sample containing any inorganic residue is removed with the carrier.

No matter what the design, the basic principles of effective, reproducible pyrolysis are the same. The specimen must be small enough to be pyrolyzed completely, yet large enough to provide sufficient pyrolyzates to be detected by the analytical instrument following pyrolysis. Precision will suffer if the samples are not pyrolyzed completely, since differing quantities of pyrolyzates will be produced from run to run. The specimen must be heated to the final pyrolysis temperature both quickly and at a reproducible rate. An overwhelming amount of the fragmentation occurs in the first few milliseconds of the run, and the temperatures experienced by the samples must be reproducible at this point as well as at the final temperature, or the pyrolyzates will differ. The final temperature will determine the amount of energy available for cleaving the bonds and obviously must be reproducible as well. It is typically minimized so that higher-molecular-weight pyrolyzates are produced, being more characteristic of their polymeric source, while being kept high enough to assure complete pyrolysis. The latter can be checked quickly by initiating another run following the sample run and its "bakeout" to see if any additional material is eluted. If pyrolysis was complete, no material should be eluted. Most paint resins are pyrolyzed in the range 650 to 750°C. In addition, small thin specimens are more desirable than large thick specimens. The outer surface of the specimen is exposed to the pyrolysis heat first and this is where the initial pyrolyzates are formed. If the specimen is thick, the inner portion will remain cooler and the newly formed species will have a tendency to condense on these surfaces and form secondary pyrolysis species, which tend to be much less reproducible than the initial pyrolyzate species. It is best that the energy sweep rapidly through sample to prevent the formation of temperature gradients. Finally, the placement of the samples must be reproducible such that they are exposed to the same thermal environment every time. This is quite critical in the inductively heated filament pyrolysis probes.

These concepts should be kept in mind when preparing paint samples for PyGC analysis. Samples should be single-layer peels if possible, not thick, intact layered fragments. Replicate analyses are essential if one is to establish precision and then use that range of variation to judge if one material produces a pyrolysis pattern similar to that of another. Precision will change from one type of resin to another. For example, pyrograms of acrylic resins are much more reproducible than those of alkyds. Samples should also be of approximately the same mass. This can usually be deduced from size as long as the morphology of the specimens is similar.

The data produced are in the form of the familiar chromatogram, with eluted material producing peaks displayed on an x-axis of time. The y-axis is proportional to the quantity of material eluted, although this is reflected more accurately in individual peak areas. The chromatograms are typically quite complex, as can be seen in the Figure 5.27 flame ionization–detected pyrogram of an automotive acrylic–melamine binder. The *thermal fragmentation* of one polymeric species will produce numerous characteristic pyrolyzate fragments, not just one peak. The concept is akin to electron impact mass spectrometry, where a beam of electrons breaks the molecule apart into fragments which are then recorded by the mass spectrometer's analyzer. This fragmentation pattern is quite characteristic of the parent molecule and is often thought of as being a definitive identification of the material. In PyGC the polymer is fragmented by thermal energy and the resulting fragments (*pyrolyzates*) are quite characteristic of the parent polymer. It is not simply gas chromatography with a liquid injection of a solution of intact molecules; it is a fragmentation process that imparts uniqueness to the pattern. This pattern (the pyrogram) can be used for comparison (i.e., to ascertain if one paint is like the other); or it can be used for classification (i.e., to ascertain what kind of material it is). To achieve the latter, the analyst must have pyrograms collected from the full array of materials of interest since the classification is based on pattern similarity to a known material, not primary molecular information. This shortfall of PyGC is often overcome by incorporating a mass spectrometer as the detector. Primary molecular information can then be acquired for the pyrolyzates, which leads to pyrolyzate identification and subsequently, to polymer classification. This is not to say that PyGC–MS is more discriminating than PyGC when comparing two resins, merely that it will produce additional information that often permits classification (identification) of the polymer. Remember, pyrolysis of one polymeric species does not just produce one fragment. It produces many fragments, as demonstrated in the pyrogram of pure polystyrene in Figure 5.28. One would be hard pressed to find another material that produces the same thermal fragmentation pattern.

In using PyGC–MS to classify polymers, it is important to keep in mind that the monomer used to construct the polymer is not necessarily what is detected following pyrolysis. Some copolymers, such as the acrylates, fragment primarily back to their monomeric form. An example can be seen in the Figure 5.29 pyrogram of an acrylic–melamine enamel automotive finish (Figure 5.1). The various acrylates and styrene used to construct the binder are readily apparent. On the other hand,

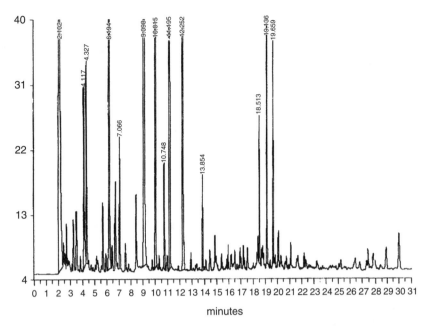

Figure 5.27 Pyrogram of an automotive acrylic–melamine enamel analyzed on a pyrolysis gas chromatograph equipped with an inductively heated pyrolysis probe, a 25-m medium-bore free fatty acid phase (FFAP) fused-silica capillary column and a flame ionization detector.

there is no indication of the presence of melamine other than the presence of *n*-butanol, a cleavage fragment from the butylated melamine cross-linker. Another example can be found in the pyrolysis of poly(vinylacetate). Vinyl acetate is not the major pyrolyzate detected, but instead it is acetic acid. One can deduce why that pyrolysis fragment is produced in the degradation of poly(vinylacetate), but the classification is not straightforward. In other instances, the deduction is not nearly as easily understood. Pyrolysis of an isophthalate-based polyester resin yields benzene, toluene, and benzoic acid as primary pyrolyzates, with no intact indication of the *meta*-substituted aromatic monomer. This point has been demonstrated by numerous authors, such as Challinor (2001, 2006) and Learner (2004).

In an effort to solve this problem to some degree, Challinor (1995) reviewed previous work he published using *tetra*-methylammonium hydroxide (TMAH) in a methanol solution to stabilize pyrolysis frgments as soon as they form (Challinor, 1989, 1991a,b). A small quantity of the solution is placed on the sample prior to introducing it into the pyrolysis chamber, and upon pyrolysis the reagent forms the stabilized methyl esters of the fragile pyrolysis fragments most characteristic of the mother polymer's precursors. For example, this approach permits recognition of the intact methyl esters of the diacids and polyols used to form polyester and alkyd resins as well as the fatty acids comprising the various drying oils used to cross-link alkyd resins. This information is difficult to come by using infrared

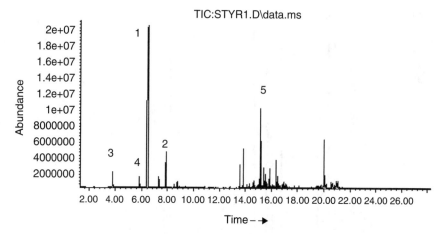

Figure 5.28 Pyrogram of pure polystyrene pyrolyzed at 750°C in an inductively heated pyrolysis probe and chromatographed on a PyGC–MS system using a 30-m narrow-bore 5% phenyl methyl siloxane fused-silica capillary column. The pyrogram consists of many other pyrolyzates in addition to the styrene peak (1), including methyl styrene (2), toluene (3) and ethyl benzene (4). Some of the longer retention time pyrolyzates are dimmers and trimers of these pyrolyzates (5).

spectroscopy alone, especially in the case of polyester resins employing a mixture of ortho-, iso-, and tere- substituted phthalates or alkyd resins formed from similar diacids and polyols but different drying oils.

5.5.8 Elemental Analysis Methods

The two yellow nonmetallic automotive monocoats with spectra depicted in Figure 5.13C and D appear to contain rutile, but both actually contain a mixture of rutile and Nickel Titanate, a yellow inorganic pigment closely related to rutile (Suzuki and McDermot, 2006). Nickel Titanate and other titanate pigments are formed by partial replacement of titanium in the rutile crystal lattice by two other transition metals (Hackman, 1988). For Nickel Titanate (which has the formula $20TiO_2 \cdot Sb_2O_5 \cdot NiO$), antimony and nickel are used. The elements present in some other titanate pigments are apparent from their names: Chromium Antimony Titanate, Nickel Niobium Titanate, Chromium Niobium Titanate, and Manganese Antimony Titanate. The titanates and rutile all have similar infrared spectra, so titanate absorptions in paint spectra can easily be mistaken for those of rutile. Elemental analysis, used in conjunction with infrared spectroscopy, is therefore required to detect and differentiate members of this pigment family.

Based on the infrared data of Figure 5.19H, calcium from the calcium carbonate and silicon from the silicate should be present. If silicates other than silica are present, other silicate elements might be observed, such as aluminum from kaolin, magnesium from talc, or aluminum and potassium from mica. Titanium

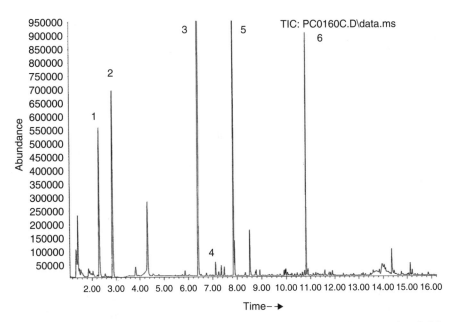

Figure 5.29 PyGC–MS program of an acrylic–melamine enamel automotive finish. The major pyrolyzate peaks are (1) n-butanol, (2) methyl methacrylate, (3) styrene, (4) 2-hydroxyethyl acrylate, (5) n-butyl methacrylate, and (6) 2-ethylhexyl acrylate.

dioxide (Figure 5.13A,B) appears to be present in this paint, but the additional low-frequency absorptions of calcium carbonate and the silicate render interpretation of this region ambiguous at best. Other oxides, such as zinc oxide and antimony oxide, might be used in house paint, and they also produce broad absorptions in this region.

As noted previously, the detectors used on infrared microscopes have cutoff energies that deny the analyst important information about inorganic pigments (both coloring and extender) in the region below 700 cm^{-1}. Elemental analysis aids in counteracting this limitation by providing comparative data related to the elements comprising these extender pigments. For example, if a paint contains a mixture of kaolin, hydrous silica, and mica, detection of the silica and mica will be difficult, if not impossible, using an infrared microscope. However, elemental analysis will not only reveal differences in the quantities of kaolin and silica present based on the relative intensities of the silicon and aluminum signals, but will also reveal characteristics of the mica based on the potassium signal.

Elemental analyses therefore serve to confirm, augment, and clarify the data obtained by infrared spectroscopy and other methods. Further information about the paint may be provided by the detection of elements not revealed by other methods. When such elements are found, however, the lack of definitive corresponding molecular structural information necessarily limits the degree to which they can be attributed to specific compounds. Since paint data are typically used in a comparative mode, this limitation is not usually a serious issue. At the same time, the presence of some elements can be revealing when trying to determine the type of coating involved, as some elements are found only in certain types. Information about the possible sources of various elements can also aid in interpretation when differences between known and questioned samples occur, as may arise from contaminants or from actual differences. Some knowledge of the possible origins of paint elements is therefore essential, as this can have a bearing on the nature and strengths of the conclusions that can be drawn. In the following discussion, only elements with atomic numbers of 9 (fluorine) or higher are considered since lower atomic number elements are of more limited value for forensic paint examinations.

Detectable elements in paint may arise from binders, although they are much more prevalent in pigments and additives. Except for acrylic silanes and silicones, which contain silicon, other common binders do not contain detectable elements that are an integral component of the polymer structure (although such elements might be present in a binder formulation as a catalyst or other additive). In contrast, virtually all inorganic pigments (both coloring and extender) and many organic pigments contain elements that can be detected. Tabulations of these materials can be found in sources such as Morgans (1990) and Thornton (2002). Some of the elements that are observed for a particular pigment, however, are actually from compounds closely associated with their formulation and not from the pigment itself. Two of the most common paint pigments, rutile and Copper Phthalocyanine Blue, illustrate this point.

The surface of a rutile (TiO_2) particle is capped with hydroxyl groups. In the presence of water and oxygen and with exposure to ultraviolet light, the hydroxyl groups catalyze the production of free radicals, which can lead to degradation of the binder (Morgans, 1990). Individual particles of rutile intended for outdoor use (including all automotive paints) are therefore coated (encapsulated). The most common encapsulating agent is silica (SiO_2), but alumina (Al_2O_3) or zirconia (ZrO_2) may also be used. In addition, rutile is usually treated with alumina to prevent flocculation (the formation of pigment agglomerates or clusters). Therefore, silicon and aluminum will usually be observed along with titanium when analyzing outdoor paints containing rutile. Rutile that is encapsulated with zirconia may also contain tin oxide (Tyler, 2000), and titanium, zirconium, and tin were found in a zirconia-encapsulated rutile product that was analyzed by XRF (Suzuki and McDermot, 2006). Some formulations of lead chromate pigments are also silica-encapsulated to minimize reactions of the chromate anion, an oxidizing agent.

As noted, Copper Phthalocyanine Blue can exist in four different crystal forms, and the most common formulation (known as Pigment Blue 15:2) is stabilized to prevent both polymorphic conversion and flocculation. The exact nature of the stabilization process is proprietary, but it usually includes incorporation of chlorinated and sulfonated forms of this pigment (Lewis, 1995). Consequently, in addition to copper, chlorine and sulfur are usually found in paints containing Copper Phthalocyanine Blue. The green phthalocyanine pigments, Pigment Green 7 and Pigment Green 36, have the same structures as Copper Phthalocyanine Blue (Figure 5.2) except that the aromatic rings are highly chlorinated, or highly chlorinated and brominated. These pigments are normally not treated, so paints containing Pigment Green 7 or Pigment Green 36 will contain copper and chlorine, or copper, chlorine, and bromine, respectively.

Encapsulation coatings, antiflocculating agents, and stabilizers are often bound intimately to the pigment particles. Some pigments, however, might also contain small quantities of a diluting agent. These may be added to adjust the tinctorial strength, which is done to ensure that the hue of a particular pigment lot falls within a specified tolerance range. Barium sulfate is commonly used for this, and since it is also a common extender pigment and antisettling additive, its levels in paint can span a very wide concentration range.

Metallic and pearlescent pigments, which are common in automotive finishes, always contain one or more detectable elements. The most common metallic pigment consists of aluminum flakes. Aluminum flakes used in aqueous dispersion basecoats are treated with phosphate surfactants (Bentley, 2001), so phosphorus may be detected in addition to aluminum. Gold-bronze metallic pigments have a composition ranging from pure copper to pure zinc or various alloys (brass) of the two, and these produce various hues (Thornton, 2002). Stainless steel metallic pigments contain chromium, nickel, molybdenum, and manganese, in addition to iron. Zinc metallic pigments may contain traces of iron, lead, and cadmium, and nickel metal pigments may contain traces of iron and sulfur.

Mica-based pearlescent pigments may include laminates of rutile, ferric oxide, chromium oxide, or zirconia, so in addition to silicon and other elements of mica, titanium, iron, zirconium, or a combination may be observed. Some pearlescent pigments are based on bismuth oxychloride, so bismuth and chlorine may be observed in addition to elements of the laminates. The elements of mica (muscovite, the most common form) include potassium, aluminum, and silicon, but micas often include several other minerals, and elements of these might also be present. Chromatic shift interference pigments consist of a core low-refracting dielectric material with a thin coating of a semitransparent highly reflecting layer. The core may transmit light or it may be reflective. Core materials include silicon dioxide, aluminum oxide, magnesium fluoride, and aluminum, while outer coatings include aluminum, chromium, or even molybdenum sulfide, thus expanding the list of detectable elements even further.

Driers are used to catalyze reactions involved in cross-linking between double bonds of a substance, such as unsaturated fatty acids. Elements of driers may thus be found in paints with binders such as alkyds, which use the fatty acids found in drying oils. Driers are probably the source of the greatest variety of elements in paint, and some of the elements that have been used or are currently used include (in order of increasing atomic number) potassium, calcium, vanadium, chromium, manganese, iron, cobalt, zinc, zirconium, cadmium, tin, barium, lanthanum, cerium, and lead. Other transition metals probably have also been used. Unfortunately, their concentrations are quite low, typically 0.1% by weight or less. Some of the elements cited do not function as driers by themselves but do so only when used with others, as discussed by Morgans (1990) and Thornton (2002).

Contaminants that might be introduced during the paint manufacturing process are another possible source of elements. Pigment dispersion techniques rely on various grinding media, which have included steel or porcelain balls, glass or zirconia beads, sand, and other objects (Bentley, 2001). Such media require periodic replacement due to wear, and their fragments are introduced into the paint. Although this might be expected to result in a very minor component, highly variable levels of zirconium were found in a number of automotive paints of various colors using XRF analysis (Suzuki and Marshall, 1998). Various sources of this element were thus considered. Most of the paints with zirconium contained little or no rutile, and tin was not found, so a zirconia-encapsulated rutile product was ruled out. Driers were another possible source of zirconium, but for an automotive paint, they would probably be found in an alkyd; all of the paints involved had either acrylic melamine enamel or acrylic lacquer binders. In addition, if zirconium were used as a drier, it would usually occur with cobalt (Morgans, 1990), but cobalt was not detected. Wear from zirconia dispersion beads thus became the prime suspect, and the paint manufacturers who were contacted concurred that this was the most likely source.

Preservatives, particularly fungicides used in exterior architectural coatings, are another source of elements. Some of the elements that have been used, or are currently used, in fungicides and other preservatives include chlorine, copper, chromium, zinc, iodine, mercury, and other heavy metals, although the recent trend has been to avoid heavy metals due to health concerns.

Silicates (talc, kaolin, quartz, diatomaceous silica, synthetic silica, mica, bentonite, asbestine, wollastonite, montmorillonite, etc.) are very common extender pigments. As noted, extender pigments are frequently used in low-luster finishes and automotive undercoats since in addition to lowering the cost of the paint by adding bulk, they serve as flatting or delustering agents. Synthetic silica and treated montmorillonite are used as thickening agents for paints with low viscosities, such as automotive clearcoats, where they help to keep the wet paint of a thick layer on a vertical surface from dripping; talc is used as an antisettling agent to help keep other pigments suspended in the liquid medium; mica is used in pearlescent paints; silica is used to encapsulate certain pigments; silicates are minor components of some inorganic pigments, such as ochres (which are mainly ferric oxide or hydrous ferric oxide); and so on.

Not surprisingly, silicon is a common element in paint. In some cases, the presence of other elements associated with the source of the silicon might serve to clarify its origin. Magnesium, for example, is present in talc [$Mg_3Si_4O_{10}(OH)_2$], and aluminum in kaolin [$Al_2Si_2O_5(OH)_4$]. For some minerals, including talc, cation substitution is a common occurrence. Magnesium (Mg^{2+}) and iron (Fe^{2+}) can occur interchangeably in these minerals since they have the same charge and the same size. Talc may thus contain iron even though its nominal formula may not indicate this. Since the function of extender pigments is not usually affected adversely by minor impurities, there are probably many other impurities in them as well, although they may be present in levels below the detection limits of the method used for analysis.

Silicon in paint can thus originate from any number of components, including binders (silanes), extender pigments, mica-containing pearlescent pigments, impurities in other pigments, encapsulating agents, products of dispersion bead wear, and additives serving a wide variety of functions. This element can therefore occur in virtually any type of coating, including every layer of an automotive finish. It is also a very common element in soil, so the condition of the sample(s) must be evaluated carefully to decide if it may be present as a result of contamination from material that is *external* to the paint.

Although detectable elements are less common for organic pigments, they do occur. Twenty-nine pigments were examined in four studies of organic pigments used in automotive paint (Suzuki and Marshall, 1997, 1998; Suzuki, 1996a,b). Ten contain chlorine, one chlorine and sulfur, one chlorine and nickel, one sulfur, one nickel, one copper, and one fluorine. Chlorine is therefore relatively common (12 of 29) in organic pigments. This element, which is usually present as a substituent on an aromatic ring, is serving at least two functions (Lewis, 1995). It imparts insolubility to a pigment and is used to modify the color of a base structure by changing its chromophore (as evidenced by the effects of chlorine and bromine substitution on the phthalocyanine ring).

Extensive lists of pigments are presented in compilations by Crown (1968) and Eastaugh et al. (2004), and the detectable elements of pigments can be obtained from their chemical formulas. It should be noted, however, that components that might be associated with some of their formulations are not included, and these may contain additional elements. Nolan and Keeley (1979) also compiled classes of paint elemental constituents based on their frequency encountered in SEM–EDS analysis of random samples.

Scanning Electron Microscopy/Energy-Dispersive X-Ray Spectrometry

SEM–EDS is the main method used in U.S. forensic science laboratories for elemental analysis of paint evidence (Henson and Jergovich, 2001; Ryland et al., 2006). This instrument can be used both as an imaging device with very high magnifications (greater than 100,000× in some cases) and as a means to obtain elemental profiles for selected areas of a sample. Elemental data for individual layers of multilayered paint specimens can be collected sequentially and the high magnifications of the microscope permit the analysis of very small samples. The range of elements that can be detected simultaneously with an EDS analysis is extremely wide and, with an appropriate detector window, all elements of atomic number six (carbon) or higher can be identified. The detector has a broad dynamic range and the technique may be nondestructive to the specimen. It is, however, limited to a minimum detection level of approximately 0.1% by weight for median atomic weight elements.

The terminology *energy-dispersive* x-ray spectrometry is intended to distinguish it from a related technique, wavelength-dispersive x-ray spectrometry (WDS). These two methods employ different mechanisms to measure x-ray energies. Both may be mounted on a scanning electron microscope. In a WDS instrument, x-rays are separated by wavelength using a crystal, which serves as a diffraction grating. Multielemental analysis is much slower than with EDS and higher beam currents are required to produce enough x-rays to be detected; however, sensitivity is increased tenfold and the improved resolution provides a solution to the peak overlap problems experienced in the EDS analysis of some elements. In an EDS system, a solid-state silicon–lithium (SiLi) detector not only counts the number of x-ray photons that arrive at the detector, but also measures their energies. An x-ray photon in the SiLi detector loses its energy in a series of steps, after which a number of electrons are promoted from the valence band to the conduction band. This number is proportional to the original energy of the photon and is measured electronically. The total number of photon counts for a given characteristic energy is proportional to the quantity of the element present in the sample.

For an SEM–EDS analysis, samples that are nonconductive, such as paint, must be coated with a film of a conducting material, usually carbon. This prevents the material from developing a negative charge during analysis, which would cause it to repel further electrons. Most instruments are operated in a high vacuum, as atmospheric gases not only impede the electron beam, but also absorb the lower-energy x-rays. Variable-pressure SEMs are available at a nominal increase in cost and make it unnecessary to coat a specimen. By leaking just a small amount of air or selected gas into the specimen chamber, the charge on the uncoated sample is dissipated. This does, however, result in a slightly diminished signal from the sample and a slight loss in control of the analytical area due to spread of the electron beam diameter.

The operation of an SEM–EDS instrument is not unlike that of a cathode ray tube (CRT) monitor. An electron beam, directed and focused by electric and magnetic fields, traces a raster pattern on the selected rectangular area of interest of the sample. An image is formed by measuring the intensity of electrons ejected by the sample as the beam traverses the selected area. Analogous to an optical image, lighter areas of the image correspond to a higher flux of electrons. Two detectors located at different positions relative to the sample are used to collect *secondary electrons* and *backscattered electrons*. Secondary electrons result from ionization of atoms of the sample near the surface and have relatively low energies. Backscattered electrons, in contrast, have nearly the same energy as that of the incident beam. Backscattering results when an incident electron passes very close to an atomic nucleus, and the strong electric field of the nucleus swings the much lighter electron around, sending it in the opposite direction. Backscattering increases with atomic number, so an SEM image based on backscattered electrons provides a compositional map, with lighter areas representing domains comprised of heavier atoms. Imaging of paint specimens in this mode provides a considerable advantage over other elemental analysis techniques since it provides a rapid visual assessment of heterogeneity (Figure 5.30) and aids in selecting appropriate analytical sizes and locations.

The energy of the electrons striking the sample is set by the analyst, and this is an important parameter for x-ray analysis because it determines the energy range of the x-rays that can be produced. An electron that is accelerated through a voltage

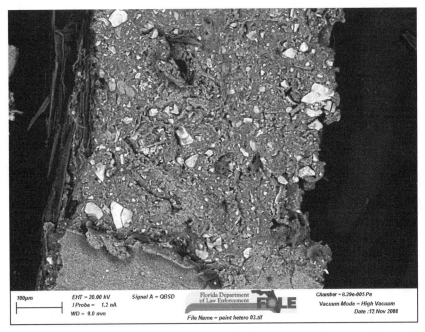

Figure 5.30 SEM backscatter image of an architectural paint layer exposed by peeling back the layer above. Note the visual rendering of the large extender pigment grain heterogeneity.

potential of 20,000 V (20 kV) acquires an energy of 20,000 electron volts (20 keV). The range of possible voltages for an SEM–EDS is typically between 1 and 40 kV, although voltages higher than 25 kV are generally not used with paint because of the excessive heat that may be generated and the increased depth of the beam's penetration. X-ray energies are also measured in electron volts, and the range of an SEM–EDS x-ray detector is typically from 0.1 to 40 keV. The flux of x-rays reaching the detector is controlled by the electron beam current.

When the electron beam strikes the sample, x-rays are generated by two different mechanisms. The first involves a process known as *bremsstrahlung*, a German word meaning "braking radiation." When a charged particle accelerates (meaning that it increases or decreases its velocity or is deflected from a straight path), it can produce radiation. In the case of the electron beam of an SEM–EDS instrument, electrons are decelerated and their paths deflected as they interact with the sample. The bremsstrahlung x-rays that result from this have a continuum of energies, ranging from zero to that of the incident electron, and this is manifested as a broad background (having a roughly skewed semicircular shape) in all EDS spectra. Bremsstrahlung backgrounds can be seen in the EDS spectra of Figures 5.31A,B, 5.32C, and 5.33. In general, a matrix having a low average atomic number composition, such as an automotive finish coat, will have a much higher bremsstrahlung

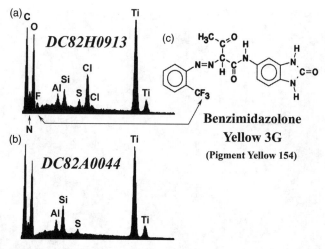

Figure 5.31 (A) SEM–EDS spectrum of a yellow nonmetallic automotive monocoat, DC82H0913, which contains rutile and Benzimidazolone Yellow 3G. (B) SEM–EDS spectrum of a white automotive monocoat, DC82A0044, which contains rutile. An electron beam voltage of 20 kV was used for both SEM–EDS analyses. (C) Structure of Benzimidazolone Yellow 3G. [From Suzuki and Marshall (1997); copyright © ASTM International; reprinted with permission.]

background than that having a relatively higher average atomic number composition, such as an automotive primer or architectural paint. This decreases the elemental minimum detection levels in the former.

The second type of x-rays produced is referred to as *characteristic x-rays*. This process begins with ionization of the atoms of the sample, and the electrons of the incident beam must have sufficient energies to remove electrons from one of the first three shells of an atom. Atomic shells are designated as K, L, or M, corresponding to electrons in orbitals of the first three principal quantum numbers. The K shell is comprised of 1s electrons, the L shell of 2s and 2p electrons, and the M shell of 3s, 3p, and 3d electrons. Following removal of an electron from one of these shells, a "hole" is created in the atom and an electron from a higher-energy shell decays to fill the vacancy and bring the atom to a lower-energy state. In so doing, it emits an x-ray having an energy equal to the energy difference between the two shells. The 2s and 2p electrons have different energies, as do the 3s, 3p, and 3d electrons. Consequently, the L to K shell decays represented by 2s → 1s and 2p → 1s have different energies, as do the various transitions from the M to the L shell. There is therefore more than one x-ray energy produced when an electron is removed from a particular shell. The x-rays emitted by electrons decaying to fill the K shell are referred to as the K series, with analogous definitions for the L and M series.

X-rays produced by individual transitions of a series may have very similar energies that cannot be resolved by the SiLi detector, In the case of heavier elements, separate peaks occur for the K series and for the L series. The lower-energy peak

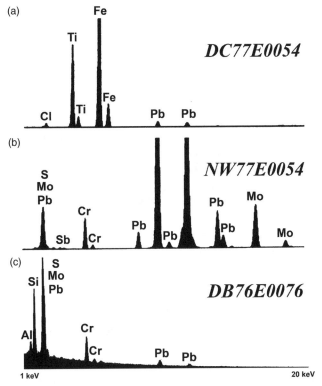

Figure 5.32 XRF and SEM–EDS spectra of three red nonmetallic automotive monocoats depicted between 1 and 20 keV. The two XRF spectra were obtained using a tin secondary target. (A) XRF spectrum of DC77E0054; the infrared spectrum of this paint is shown in Figure 19f. (B) XRF spectrum of NW77E0054; the infrared spectrum of this paint is shown in Figure 19A. (C) SEM–EDS spectrum of DB76E0076, acquired using an electron beam voltage of 22 kV. (From Suzuki and Marshall, 1998; copyright © ASTM International; reprinted with permission.)

of the K series (known as K_α) typically has an intensity roughly six times that of the second peak (known as K_β), whereas the two peaks of the L series have more nearly equal intensities. All of the peaks observed in Figure 5.31A are from the K series, and the two L-series peaks of lead can be seen in Figure 5.32C (compare to the same peaks of lead obtained by XRF in Figure 5.32A and 5.32B). M series transitions cannot be resolved and only one peak is observed. Experienced paint examiners can usually distinguish between peaks of the three series, but instrument software may also be used to aid in this task, as they include markers that indicate the positions and relative intensities of K, L, and M peaks for each element.

As the atomic number increases, the inner-shell electrons become more tightly bound to the nucleus, lowering their energies, and the energy differences between adjacent shells increase. There is thus a regular increase in the energies of each series as the atomic number increases (*Mosley's law*); the differences in energies between peaks of adjacent elements also increases with atomic number. Consequently, there is no element for which K, L, and M peaks can all be observed in the range 0.1 to 40 keV, although for some elements, two series are observed. As noted, electron energies much above 25 keV may damage the paint specimens or penetrate too deeply into its surface, so only about half of the detectable range is used in any case. Even with this more limited range, detection of a single element present in sufficient concentration is not a problem, and this is also often true when several elements are present. However, due to overlap of K, L, and M peaks in the low-energy region, certain combinations of elements may be difficult to determine.

This is illustrated by the EDS spectrum (Figure 5.32C) of a red nonmetallic automotive monocoat that contains Molybdate Orange ($PbCrO_4 \cdot xPbMoO_4 \cdot yPbSO_4$), for which peaks of lead, chromium, molybdenum, and sulfur are expected. L-series peaks of lead and K-series peaks of chromium are observed, but the K-series peaks of molybdenum are not (compare to Figure 5.32B, which is the XRF spectrum of another paint that contains Molybdate Orange). L-series peaks of molybdenum do occur in the range of the instrument, but in this case they are buried in the strongest peak of Figure 5.32C. This peak represents an overlap of the M peak of lead, the L peaks of molybdenum, and the K peaks of sulfur, so the presence of molybdenum and sulfur cannot be definitely established in this case. This confluence of K, L, and M peaks of three elements of a single pigment is *extremely* unusual (if not unique), but it does illustrate the interpretation difficulties that overlapping peaks can present.

Secondary electrons arise from atoms very near the surface of the sample, and backscattered electrons originate from somewhat greater depths. SEM images thus reveal predominately surface characteristics. The electron beam, however, penetrates beyond these relatively shallow depths. As it interacts with the sample, it begins to lose its directionality and individual electron paths become more random. Consequently, the morphology of the volume from which x-rays originate is not a well-defined narrow cylinder, but rather, has a hemispherical or a teardrop shape, depending on beam energy, sample composition, and x-ray energies.

The interactions of the electrons with the sample also affect their energies. An electron beam energy of 22 keV was used to acquire the EDS spectrum of Figure 5.32C, and this energy is sufficient to ionize a K-shell electron of molybdenum, but no K-series peaks of this element are observed. The sample contains much more lead than molybdenum, and lead requires even less energy to ionize its L-shell electrons, yet the lead L-series peaks are quite weak. The intensities of the lead L-series peaks and the molybdenum K-series peaks can be increased slightly by employing a higher beam excitation potential, ideally 2.5 times the minimum required to ionize the electrons in those shells. However, as mentioned earlier, there are drawbacks to doing this.

The preferential excitement of the lower-atomic-number elements of a sample is a characteristic feature of an SEM–EDS analysis. It results from the numerous interactions that the electron beam experiences as it travels through the sample, constantly losing energy. Consequently, most of these electrons can only excite elements of lower atomic numbers. Low-atomic-number elements are thus readily observed in SEM–EDS spectra, as may be seen for the K-series peaks of carbon, nitrogen, oxygen, and fluorine of Figure 5.31A (which is the spectrum of a yellow nonmetallic automotive monocoat). Some of the carbon is from the thin coating used to make the sample conductive, while the fluorine is from a yellow organic pigment, Benzimidazolone Yellow 3G (Figure 5.31C).

Figure 5.31B is the SEM–EDS spectrum of a white nonmetallic automotive monocoat that contains a large quantity of rutile, and in addition to the titanium peak of rutile, the silicon and aluminum from the encapsulation coating and the antiflocculating agent, respectively, are also observed. Similar ratios of these three elements are observed for the spectrum of the yellow monocoat (Figure 5.31A), which also contains a large amount of rutile. It is also instructive to note that the aluminum and silicon are not originating from a clay extender pigment that also might be suspected as a source from the SEM–EDS spectrum alone.

Because of the differences in the sample domains that are responsible for image formation and generation of x-rays, SEM–EDS analysts should always remember the caveat, "What you see is not always what you get." It is especially important to remember when analyzing individual layers of an intact multilayered paint.

Four different methods have been used to obtain SEM–EDS data for individual layers of multilayered paint samples, each with its merits, limitations, and degree of skill (and patience) required to obtain optimal results. In the first, each paint layer is excised and analyzed separately. This requires a fair amount of sample preparation, but it is the only method for which analysts can be assured that elements of adjacent layers are not being observed. A second procedure involves what is referred to as the stair-step method since after preparation of a specimen, it has a steplike appearance when viewed from the side. This method also requires a considerable amount of sample preparation, as beginning with what will be the bottom layer, a rectangular portion consisting of all of the overlaying layers is removed, exposing a portion of the bottom layer. An adjacent rectangle is then removed, exposing a portion of the next-to-bottom layer, and so on. The sample is positioned in the SEM–EDS instrument, so that all of the exposed layers are perpendicular to the beam. Analysis of very thin layers may yield x-rays from a underlying layers, and it may help to reduce electron beam energies to determine if this is the case.

The last two methods involve cutting a cross section through the layers and situating this cut surface perpendicular to the SEM–EDS beam. This can be done without any further sample preparation, or the sample can be embedded in a medium that hardens and the surface of the cross section polished. Care must be taken in either case to ensure that the surface is situated horizontally and that the cut was made perpendicular to the layers, so that electrons traveling straight down do not encounter another layer. Embedding and polishing require considerable sample preparation, but this process also yields the most reproducible results regarding relative ratios of EDS x-ray peaks. Embedding, however, makes removal of the sample difficult, if not impossible, should there be a need for further analysis. For either type of cross-sectional analysis, a major concern is the possibility that the EDS analysis volume may extend into an adjacent layer or into two adjacent layers. To minimize this possibility, long narrow rastering areas with edges as far from

200 ANALYSIS OF PAINT EVIDENCE

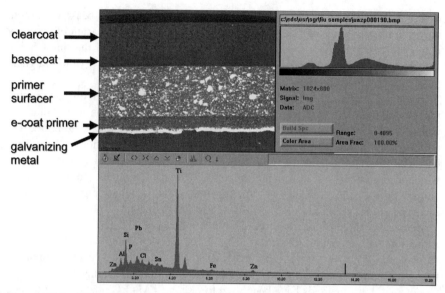

Figure 5.33 SEM–EDS analysis of the base primer in an automotive paint cross section block cut by microtomy. The red box superimposed on the e-coat primer defines the visual boundaries of the analytical area. *(See insert for color representation.)*

the adjacent layers as practical are used, as depicted in Figure 5.33. Regardless of which of the four methods is used, it is important that both questioned and known samples be subjected to the same procedures.

As mentioned, the analytical area is typically chosen by scanning the electron beam over a large defined area to average out the effects of heterogeneity, a constant concern in paint samples. The relative success of this endeavor is assessed by comparing spectra collected from one area of a layer to another of the same layer. If the analytical area is large enough to average out the lack of precision introduced by heterogeneity, the spectra will appear like one another. The SEM offers another mode of sample collection called the *spot mode*. This approach is quite valuable in analyzing particulate material in the matrix, such as metallic flake and interference pigment flake (the rich elemental and morphological information contained in these pigments was described previously). The beam is targeted on the particle of interest and is held stationary, as seen in Figure 5.34. The excitation voltage is typically reduced in an attempt to keep the excited volume within the boundaries of the target particle, keeping in mind the necessity to maintain enough energy to effectively excite the elements of interest. In addition to analyzing effect pigment flake in paints, the same method may be used to enhance the inferred identification of other pigments in the paint. If a large area is scanned and barium, sulfur, magnesium, aluminum, and silicon are detected, this merely informs the analyst that these elements are present in the layer in their respective relative ratios. Should spot mode be used in addition to this technique, it can be demonstrated that the barium and sulfur are associated with one particulate population, magnesium and silicon

Figure 5.34 SEM–EDS spot analysis of interference pigment flake in situ. The flake is visualized by viewing the exposed layer in the backscatter mode. The red + indicates the beam target spot. *(See insert for color representation.)*

are associated with another, and aluminum and silicon are associated with a third. This significantly strengthens the deduction that the paint layer contains barium sulfate, talc, and clay extender pigments. Although these are typically recognized in infrared spectra collected down to at least 450 cm^{-1}, it may not be apparent in situations where there is a significant quantity of titanium dioxide present, and one of the extenders is present in a low concentration relative to the others.

More information about SEM–EDS instrumentation is given by Goldstein et al. (1992), while the preparation of paint samples for SEM–EDS analysis is discussed in more detail by Ryland et al. (2006). An introduction to paint analysis using an SEM–EDS analysis is given by Wilson and Judd (1973), and Henson and Jergovich (2001) provide an extensive review of this topic. General guidelines for such analyses are outlined in the *Standard Guide for Using Scanning Electron Microscopy/X-Ray Spectrometry in Forensic Paint Examination* (Scientific Working Group for Materials Analysis, 2002).

X-Ray Fluorescence Spectrometry XRF spectrometry is the second most frequently used method in the United States for the elemental analysis of paint evidence (Henson and Jergovich, 2001; Ryland et al., 2006). *Fluorescence* refers to a process whereby atoms or molecules absorb electromagnetic radiation, then reemit radiation with lower energies. In XRF spectrometry, as opposed to SEM–EDS, x-rays are used to ionize atoms instead of electrons. It should be noted, however, that the emission process is the same for both techniques. XRF spectra might therefore be expected to mirror EDS spectra, but this is generally far from the case. As we saw, the forte of an EDS analysis is the detection of lighter elements. XRF spectrometry, in contrast, is better suited for the identification of heavier elements, so the two are complementary. There is no need to coat samples for an XRF analysis, and if only the heavier elements are sought, sample chambers do not have to be evacuated. XRF instruments with the lowest detection limits (tens of parts per million or lower in some cases) require considerably larger samples than are required for an SEM–EDS analysis, while micro-XRF systems, which have higher detection limits, allow smaller samples. There are concerns regarding the increased beam penetration experienced when using x-rays as opposed to electrons; consequently, the current approach for individual layer analysis favors preparation of individual layer peels. When comparing spectra of layers excised from small multilayered samples, one must also be concerned with the inherent imprecision introduced by critical depth effects experienced in samples of different thicknesses (Howden et al., 1977).

As with an EDS analysis, the energies of the x-rays used for excitation must be sufficient to ionize K, L, or M electrons. With x-rays, however, the most efficient means to do this is with an x-ray energy that is slightly greater than the minimum required (known as an *absorption edge*), and as x-ray energies increase beyond this value, the excitation efficiencies decrease.

X-rays for XRF excitation are generated in an x-ray tube. The tube consists of a metal anode or target, typically rhodium, housed in a vacuum. A voltage between 1 and 50 or 60 kV is used to accelerate a beam of electrons toward the target. Analogous to what occurs in an SEM–EDS instrument, bremsstrahlung and characteristic x-rays of rhodium are produced. Depending on the voltage used, Rhodium L peaks, or K and L peaks, are generated. All XRF instruments may use x-rays from the tube for excitation. Since the distribution of x-ray energies from the tube includes the broad bremsstrahlung continuum, this serves to excite a wide range of elements.

For some instruments, an alternative mode of excitation is possible using *secondary targets*, which consist of several elements spanning a range of atomic numbers. X-rays from the tube are directed toward a selected secondary target, causing it to emit K-series peaks. These x-rays are then used to excite the sample, and since K_α is more intense than K_β, this amounts to using mostly monochromatic radiation for excitation. Since this is the most effective means to excite a particular element or small group of elements with similar atomic numbers, this mode provides the lowest detection limits, particularly for heavier elements. To achieve this for a wide range of elements, however, analyses using several different secondary targets are required. The intensities of x-rays produced by secondary targets are limited, so relatively large samples and long collection times are normally required for paint samples. Secondary targets are not available on micro-XRF spectrometers, which use collimators or focus optics to produce beam diameters between 300 and 10 μm. Micro-XRF instruments with selectable filters that improve the detection limits for selected elemental ranges are now available. Appropriate thin elemental (metal) films are positioned selectively in front of the x-ray tube to reduce the bremsstrahlung continuum and/or characteristic tube x-ray lines striking the sample. This in turn reduces the x-ray scatter background or eliminates the scattered x-ray tube lines emanating from the specimen and still leaves enough higher continuum x-rays to excite the elements of interest. Like secondary target instruments, however, much longer data collection times are required.

The complementary nature of an EDS and XRF analysis can be seen by comparing Figure 5.32B,C, which are spectra of two red automotive monocoats (NW77E0054 and DB76E0076) that contain Molybdate Orange. For the XRF analysis (Figure 5.32B) a tin secondary target ($K_\alpha = 25.2$ keV) was used, while an electron beam energy of 22 keV was used to obtain the EDS spectrum (Figure 5.32C). DB76E0076, which was analyzed by EDS, contains more Molybdate Orange than NW77E0054, judging from relative infrared absorption intensities. Note that even with an XRF analysis (Figure 5.32B), the presence of the sulfur of Molybdate Orange cannot be determined because of the overlap discussed earlier. In these situations additional techniques may be employed, such as wavelength-dispersive x-ray spectrometry (WDS).

The XRF spectrum of a second automotive paint (DC77E0054) that has the same color as NW77E0054 is shown in Figure 5.32A. The infrared spectra of these two paints (Figure 5.19A,F) were discussed previously and it is informative to see how the infrared and elemental analysis data correlate. NW77E0054 (Figures 5.19A and 5.32B) contains Molybdate Orange and Quinacridone Red Y. Quinacridone Red Y does not contain any detectable elements, but peaks of all of the elements of Molybdate Orange, except for sulfur, are observed in the XRF spectrum of NW77E0054. DC77E0054 contains rutile, ferric oxide, Quinacridone Red Y, and Benzimidazolone Orange, and its XRF spectrum (Figure 5.32A) has peaks of titanium (rutile), iron (ferric oxide), and chlorine (Benzimidazolone Orange). A small quantity of lead is also observed, and analyses using other secondary targets and longer collection times detected very small amounts of molybdenum and chromium. The presence of a small quantity of Molybdate Orange is thus indicated, although its absorptions (Figure 5.19K) are not observed in the infrared spectrum of this paint (Figure 5.19F).

The XRF spectra have reiterated the value of elemental analysis when used in conjunction with infrared spectroscopy and other methods, and the additional information about the paint that may be obtained. They were obtained, however, with secondary targets using large pristine samples, and the data were collected overnight. Micro-XRF systems are currently better suited for most case samples, since much smaller specimens can be analyzed with sensitivities intermediate between those of an EDS and a secondary target XRF analysis. These sensitivities are still sufficient to detect low concentrations of tinting pigments or even drier metals having concentrations on the order of 0.005% by weight. Other innovations have continued to allow smaller samples to be subjected to XRF analyses. A micro-XRF attachment for an SEM–EDS instrument, for example, has recently been introduced. It consists of an x-ray tube using poly capillary focus optics to form an excitation beam diameter of 40 μm. Fluorescent x-rays are detected using the SiLi detector of the SEM–EDS system. There are, however, concerns for background scatter levels on samples mounted on carbon substrates and for isolation of layers given the increased penetration of x-rays versus electrons from the SEM.

Fiber optic x-ray guide tubes are also used on a new micro-XRF spectrometer that produces spot sizes as small as 10 µm. Kanngiesser et al. (2005) demonstrated the feasibility of analyzing individual layers of a multilayered paint sample using a micro-XRF instrument with a 10-µm spot size, but this required use of synchrotron radiation. This new system focuses x-rays from a conventional x-ray tube without the need for a synchrotron.

Jenkins (1999) provides more information about XRF spectrometry theory and instrumentation. Howden et al. (1977) examined the discrimination of white, red, and green single-layer household paints using XRF spectrometry. All of the red and green paints could be distinguished by this means and the white paints were divided into nine groups. Haag (1977) used XRF spectrometry to study elemental profiles of automotive paint chips. Zieba-Palus and Borusiewicz (2006) examined several multilayered paint samples using a combination of infrared and Raman microscopy and micro-XRF spectrometry. The discrimination of black (Govaert et al., 2001) and red (Govaert and Bernard, 2004) spray paints was examined using a combination of optical microscopy, infrared spectroscopy, and XRF spectrometry.

Other Elemental Analysis Methods Several other elemental analysis techniques have been used for the analysis of paint. Emission spectrography was one of the first such methods used in forensic science, but it has been supplanted by inductively coupled plasma emission spectroscopy (ICP-ES) and its even more sensitive offspring, inductively coupled plasma mass spectrometry (ICP-MS). ICP methods, however, require samples that are soluble—not an easy task for most coatings, which are enamels that contain pigments designed to be insoluble. Neutron activation analysis (NAA), which is gamma-ray emission spectrometry of nuclei in excited states, is probably the most esoteric of the methods that have been applied to paint. NAA requires a neutron source such as a nuclear reactor, and although not destructive, it produces, at least temporarily, a radioactive sample. Like an SEM, an electron microprobe analysis uses an electron beam to produce characteristic x-rays, but this method has few, if any, advantages over SEM–EDS analysis. A proton beam is used in a related technique known as proton-induced x-ray emission or particle-induced x-ray emission (PIXE). This method has lower detection limits for many elements compared to an SEM–EDS analysis, but like an electron microprobe instrument, it lacks the imaging capabilities of an SEM–EDS system. Thornton (2002) provides a more detailed review of these methods as applied to paint.

Two relatively new techniques show some potential as future tools for the examination of paint evidence, as both allow trace levels of a wide variety of elements to be detected. Major issues regarding reproducibility occur with both techniques, however, and they are currently used primarily for qualitative analyses.

Total-reflection x-ray fluorescence spectrometry (TXRF) is a variation of a conventional XRF analysis where a grazing incident angle ($<0.1°$) of an x-ray beam is used for excitation (Klockenkamper et al., 1992). A standing wave is produced, and this results in significantly enhanced (parts per billion) sensitivities. TXRF has been used for several elemental analyses of artists' paints and other pigmented historical or archeological objects (Klockenkamper et al., 2000).

Laser ablation inductively coupled plasma mass spectrometry (LA-ICP-MS) is an alternative method for detecting trace elements in materials that cannot easily be subjected to a conventional ICP-MS analysis. A pulse from a laser is used to vaporize a sample and create an aerosol, which is swept into the plasma torch of the ICP-MS instrument. The high temperatures of the plasma ($>6000°C$) produce mostly monoatomic ions, so that the mass spectrometer yields elemental analysis data. Hobbs and Almirall (2003) examined the feasibility of using this method for the analysis of automotive paint. Typically, a suite of suspected elements must be selected prior to analysis. While a number of elements were identified at trace levels, their relative ratios varied considerably between results obtained from successive pulses. More stability was found in primer layers; however, the diameter of the ablation crater had to be increased to approximately 200 μm to improve precision. There is a "charred" discolored zone that extends approximately 300 μm beyond this crater, and precision of analysis within this zone has currently not been assessed. This has a practical impact on sampling, as replicate analyses would require a sample much larger than 1 mm^2. Assuming that the precision concerns can be overcome, there is some potential for batch discrimination of automotive primers resulting from the variation of trace elements contained as contaminants in different lots of the minerals used as extender pigments. The same holds true for architectural paint tint bases, although full-thickness 1-mm^2 samples are seldom encountered in casework.

Deconinck et al. (2006) also examined automotive paint using LA-ICP-MS, although their work focused on demonstrating inaccuracies suffered by quadrapole-based ICP-MS instruments, due to spectral interferences during trace element depth profiling of multilayered paint samples. They concluded that a high-resolution sector field–based mass spectrometer would be required to avoid such problems. Like the earlier study by Hobbs and Almirall, this technique was found useful for qualitative analyses, but it did not resolve issues of reproducibility and quantitative analyses on small samples. Although it was stated that four-layered car paints of the same color produced by different manufacturers could be discriminated from one another using the technique proposed, improved discrimination over current methodology was not documented.

5.5.9 Other Methods

X-Ray Diffraction Like emission spectrography, XRD was one of the first methods used for the analysis of inorganic components in paint evidence. The term *x-ray diffraction* actually refers to two related methods. In one, usually referred to as *x-ray crystallography*, a single crystal is analyzed using an incident monochromatic x-ray beam, and the three-dimensional pattern of spots that results from reflection from various crystal planes is analyzed. The planes are formed by the particular arrangement that individual atoms or ions assume in the crystal, and they are not only peculiar to each compound but also differ for each crystal structure of a polymorphic substance. X-ray crystallography is used predominantly for molecular structural determinations, although it has also been used for analytical chemistry.

The technique that is used for the analysis of paint is *powder XRD*, which is applicable to materials that consist of, or contain, crystalline substances where the orientations of individual crystals of the sample are random. To further ensure that the orientations are completely isotropic, the sample is rotated during analysis. Instead of a pattern of spots, x-rays are reflected from the sample in cones with apices at the sample. The cones are characterized by their angles relative to the direction of the incident beam and their intensities. A graph of an XRF powder diffraction pattern, known as a *diffractogram*, is produced with the abscissa representing the angle of reflection (actually 2θ, or twice this angle), and the ordinate depicting the intensities of the reflected x-rays. In principle, all of the major crystalline components of paint can be identified by powder XRD. Some paint binders may exhibit localized regions of crystallinity, but because they lack large-scale regularities in their structures, binders cannot be identified by these means. Reflections from crystal planes arise from x-ray scattering from individual atoms or ions; such scattering is caused by the electrons of the sample, so the intensity of x-ray scattering increases with atomic number. Organic pigments are therefore weak scatterers of x-rays, so powder XRD analyses are limited primarily to inorganic pigments.

Because of its limited scope and past requirements for relatively large samples and long analysis times, powder XRD has not been widely used for forensic paint examinations. Newer instruments, however, are capable of analyzing paint samples as small as 20 μm in reasonable periods of time with essentially no sample preparation. Unlike an elemental analysis, powder XRD provides molecular structural information, and for some paints, it provides more definitive data than either infrared or Raman spectroscopy. Silicates, for example, cannot be identified by Raman spectroscopy, and when large quantities of extender pigments are present in paint (as typically occurs for low-luster house paints; see Figure 5.19H), their strong broad infrared absorptions usually overlap, hampering identification of individual pigments. Powder XRD diffractogram peaks, in contrast, are quite narrow and because there are no contributions from binders or most organic pigments, relatively simple patterns occur for paint. This is illustrated by Figure 5.35A, which is the diffractogram of a paint that contains significant quantities of four common extender pigments (Snider, 1992). Diffractograms of two of these, kaolin and quartz, are shown in Figure 5.35B and 5.35D, respectively, along with that of hydrous ferric oxide (Figure 5.35C). Different polymorphs of a given compound are readily distinguished by powder XRD, so mixtures of such polymorphs can also be identified easily.

Laser Desorption Mass Spectrometry LDMS is a relatively new technique that, unlike its cousin LA-ICP-MS, provides molecular structural information. Both techniques use a pulse from a laser to volatilize and ionize samples, but in LA-ICP-MS, all molecular species are destroyed in the plasma. In LDMS, much less

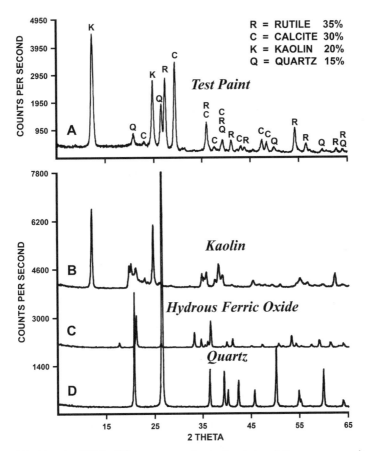

Figure 5.35 Powder XRD diffractograms of a test paint and three common inorganic paint pigments: (A) test paint that contains rutile (35%), calcite (30%), kaolin (20%), and quartz (15%); (B) kaolin; (C) hydrous ferric oxide; (D) quartz. [From Snider (1992); copyright © ASTM International; reprinted with permission.]

vigorous conditions are used to vaporize and ionize paint constituents, which are then introduced directly into a mass spectrometer. No sample preparation is required other than separating individual layers of a multilayered paint if they are to be subjected to analysis.

Stachura et al. (2007) demonstrated the feasibility of using LDMS to identify some organic and inorganic pigments in automotive paints. Positive and negative ions were examined, and both mass spectra were found to be useful since they provided different types of information. Molecular ions of Copper Phthocyanine Blue with expected isotopic ratios were identified in mass spectra obtained for a blue automotive paint. Lesser amounts of molecular ions of the mono- and dichloro-substituted compounds were also observed, from the stabilizers discussed previously. Of all the methods that have been described, this is the only one that has elucidated the nature of these stabilizers.

5.6 EXAMPLES

5.6.1 Example 1

The body of Veronica Hill was found several feet off the side of a little-traveled dirt road in Daytona Beach, Florida. Her clothed body was wrapped in a peach-colored sheet, a floral print white sheet, and an old blanket. The cause of death was determined to be blunt trauma to the head. There was no evidence of sexual assault. Subsequent investigation disclosed that she frequently had sex with men in exchange for money, and several of her recent acquaintances were identified by the clientele in establishments she frequented. Time lines pointed to one primary suspect, David Cartwright. He appeared quite cooperative upon questioning and agreed that the police could search his apartment and garage after obtaining the appropriate paperwork. He acknowledged being with Ms. Hill in the past, but claimed that he was not with her on the night of her death.

Mr. Cartwright's apartment appeared to be in order, with no signs of struggle or evidence of Ms. Hills' suspected presence. Even had such evidence been discovered, it would have been of little value, as Mr. Cartwright already said he had known her and she had previously been to his apartment. In searching the garage, a strong scent of chlorine bleach was noted along with signs of a recent acid wash of the concrete floor. A search for blood resulted in failure. Examination of his 1984 Toyota pickup truck revealed that it was customized both inside and out. It had a relatively new coat of deep purple paint with a distinctive sparkling effect. The interior had the headliner and seats covered with purple imitation fur and there was evidence that the bed area of the truck had at one time also had a similar treatment. The rear of the cab had previously been removed and a truck bed cover was welded to the cab, modifying the pickup truck into somewhat of a sport utility vehicle. The purple faux fur fabric lining was torn from the bed of the truck as well as from the sides of the compartment. The roof's interior still had the fabric present as a custom headliner. Mr. Cartwright explained that he was in the process of refurbishing the interior, and a search failed to reveal the presence of blood stains. Mr. Cartwright shared photos of the interior of the custom vehicle prior to his removal of the appointments and acquaintances related that the work must have been done recently, as the interior was intact just the prior week. He was an avid do-it-yourself customizer as evidenced by the paint and refinish supplies in his garage.

Witness statements and cell phone logs continued to point to Mr. Cartwright and Ms. Hill being together the night of her demise. He held to his claim of not being with her but had no alibi, since he was home alone that night. No murder weapon had been discovered and there was no DNA evidence. Examination of the sheets and blanket in which Ms. Hill's body was wrapped revealed an odd pattern of apparent oversprayed black paint on one of the sheets and the blanket. There were straight lines with no paint on one side and thin overspray on the other. One even had a curve at the top of it, as if something was placed on the fabric and spray-painted. There were two other round spray areas on the blanket, as if a can of spray paint was being tested or the nozzle cleared. These areas were cut out and the adhering surface debris from the remainder of the fabrics was collected and searched stereomicroscopically. This search disclosed the presence of numerous purple synthetic fibers, some appearing to be long and furlike in shape. Comparison of these fibers to various standards taken from the headliner, side compartment liners, and seat covers of Mr. Cartwright's customized truck revealed several corresponding types. There were three different types that were alike generically (acrylic, modacrylic, and polyester) in addition to their microscopical morphology and color. Morphology was compared by polarized light microscopy, fluorescence comparison microscopy, and brightfield comparison microscopy. Color was compared by comparison brightfield microscopy and microspectrophotometry. Of course, one could argue that Ms. Hill had probably been in Mr. Cartwright's truck and could have had some of the fibers on her clothing from previous contact. However, the number of fibers found was striking; and they were found not only on the victim's clothing but on the outer blanket and sheets as well. Also, there was the paint overspray.

Stereomicroscopical examination of the overspray areas on the outer blanket and the floral sheet used to wrap the body revealed tiny droplets of cured paint fused to the fibers comprising the fabrics. The surfaces of the droplets were smooth and tapered down to the shaft of the fibers, indicating that the paint had been applied wet and then cured on the fabric as opposed to being abraded onto the fabric's surface. Examination by polarized light microscopy (PLM) and FT–IR microspectroscopy revealed that the droplets consisted of a black urethane-modified alkyd enamel paint with interference decorative flake present, providing a purple tint. This paint was encased with a clear, colorless coating that was also a urethane-modified alkyd enamel, similar to the underlying black layer in binder chemical characteristics. The binder chemistry and compatible layer system of the two paints suggested a vehicle refinish paint. Refinish paints are particularly valuable evidence, due to the number of manufacturers and the variety of binder types marketed by each manufacturer. Considering the combination of these two factors, there are over 30 differentiable binder formulations for each color offered.

No potential sources of this paint were discovered in Mr. Cartwright's garage; however, there were numerous overspray areas on the concrete walls and cabinetry. One such area consisted of a black urethane-modified alkyd enamel paint containing interference decorative flake, providing a purple tint. It was the same color as the paint on the blanket around the victim's body. Examination of paint samples collected from his 1984 Toyota truck disclosed that the uppermost finish coat consisted of a clear colorless urethane-modified alkyd enamel with either metallic decorative flake or both metallic and pearlescent flake distributed sparsely throughout. This was followed by a black urethane-modified alkyd enamel containing interference decorative flake providing a purple tint. Numerous refinish layers were present below these topcoats, varying in layer structure from one panel of the vehicle to another. Comparison of the top two paints with the paint overspray found on the blanket and the garage wall revealed that they are like one another with respect to their colors, binder characteristics, and pigment characteristics, including the elemental characteristics of the interference decorative flake present in the black basecoat. Stereomicroscopy, comparison brightfield microscopy, comparison polarized light microscopy, infrared spectroscopy, and scanning electron microscopy in conjunction with energy-dispersive x-ray spectroscopy were all used in the comparisons. PyGC was not used, due to the limited quantity of paint present in the overspray transfers.

As with many cases involving trace circumstantial evidence, the two layers of paint overspray discovered on the blanket and sheet wrapped around Ms. Hill's body corresponding to the paint on Mr. Cartwright's vehicle and garage wall was certainly not definitive proof of the association in and of itself. Paint is a mass- and batch-produced material and typically escapes the realm of source individualization. But on the other hand, considering that the paint on the body's wrappings was applied as a wet spray of a compatible two-layer vehicle refinish paint system which coincided with the outermost finish on Mr. Cartwright's vehicle along with the presence of the numerous unusual fibers of varying types found on the body's wrappings that also coincided to the upholstery in Mr. Cartwright's vehicle makes for very compelling evidence of an association.

5.6.2 Example 2

By far the most common type of paint examination encountered in public forensic science laboratories involves vehicle paint transferred in hit-and-run cases. There are two types of investigations. In the first, a victim's clothing and/or possessions are examined for the presence of automotive paint in order to determine the year, make, and model of the striking vehicle. This investigative lead information is provided to authorities to aid in the search for a suspect vehicle. In the second, paint exemplars taken from a suspect vehicle (knowns) are compared to paint transfers found on the victim to determine whether or not the suspect vehicle could have been the source of the paint transferred. Occasionally, the investigations become very challenging, such as in a 2002 "cold case" from a city on the east coast of Florida, where the suspect vehicle of a hit-and-run fatality was a white 1977 GMC van that had been discovered in a salvage yard eight months after the incident occurred.

The suspect had sold the van for parts shortly after the accident, and many accessories were missing (see Figure 5.36). It had been stripped of all interior parts, front-end exterior lamps, lenses, and side rear-view mirrors. All windows had been broken and removed. No fabric marks were found on the vehicle. There was some body damage to the front passenger side of the vehicle, especially the right-front fender and hood; however, other panels on the vehicle were also damaged. The exterior passenger side of the vehicle had been spray painted white by hand, as evidenced by the thin coating where the paint had run down the surface of the door and quarter panel.

The accident scene had pieces of broken amber plastic side marker lens, pieces of broken side-view mirror glass, pieces of broken headlight glass, and a right exterior rear-view mirror assembly strewn about. The victim was injured primarily by blunt trauma to the rear of the head, apparently from being struck by the mirror assembly. Search of the victim's clothing revealed the presence of white paint

Figure 5.36 The suspect 1977 GMC van recovered from a salvage yard. *(See insert for color representation.)*

deposits fused to the back of the victim's jeans. Numerous borosilicate headlight glass fragments and nontempered window glass fragments were found in the surface debris recovered from the clothing, along with several fragments of multilayered paint fragments having a white topcoat.

Examination of the suspect vehicle revealed rust-encrusted mounting points for an exterior right rear-view mirror, which corresponded in general configuration to the mirror mount found at the scene. However, no detailed characteristics remained at the mounting sites due to the corrosion present. This was associative evidence, but far from definitive. The pieces of amber plastic side maker lens were similar to the red lenses remaining on the rear of the vehicle. Both they and the exterior side-view mirror assembly were consistent with the types used on mid- to late-1970 to early-1980 General Motors (GM) vans and panel trucks. No headlight, mirror, or window glass remained in the vehicle for use as standards to compare to the glass fragments recovered from the victim's garments. Following comparisons by stereomicroscopy, polarized light microscopy, FT–IR microspectroscopy of all layers, PyGC of the finish coats, and SEM–EDS of all layers, it was found that the paint chips fused to the rear of the victim's jeans had two-layer structures corresponding to the finishes on the right front fender and hood of the suspect vehicle. One panel has a three-layer original finish paint system underneath a poorly applied refinish paint (see Figure 5.37), while the other has a two-layer refinish paint system (see Figure 5.38). The two types of corresponding paint, along with the parts found at the scene, provide strong evidence of an association between the victim and the suspect 1977 GMC van.

The forensic examiner's task often does not end with the completion of examinations in the laboratory and authoring a report of the findings. He or she may be called to court to give testimony not only of what was found but also of its

212 ANALYSIS OF PAINT EVIDENCE

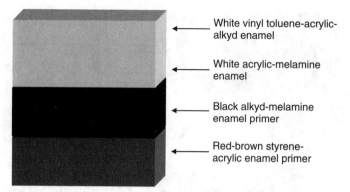

Figure 5.37 Layer structure of the paint from the passenger side door and quarter panel of the 1977 GMC van. *(See insert for color representation.)*

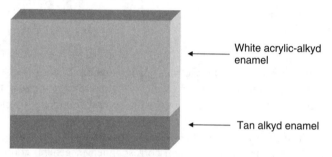

Figure 5.38 Layer structure of the paint from the right side of the hood of the 1977 GMC van. *(See insert for color representation.)*

significance. Although the work on this case could have stopped at this point, a search of the Paint Data Query (PDQ) International Original Automotive Paint database was undertaken to further evaluate how many other vehicles may have had the same original finish as that on the 1977 GMC van produced in GM's Lordstown, Ohio plant. The original finish paint layer structure consists of a white acrylic–melamine enamel topcoat over a black alkyd–melamine enamel primer-surfacer over a reddish-brown styrene–acrylic enamel primer. There is a thin white vinyl toluene–acrylic–alkyd enamel layer on the surface of the original topcoat. This surface paint binder is typical of a household spray paint. The characteristics of the remaining three original automotive finish layers were entered into the PDQ database and searched. Although there was no record of a finish system exactly like that on the Lordstown, Ohio–manufactured 1977 GMC van in this case, all queries did lead to General Motors products manufactured between 1977 and 1980. A further search of these "hits" removed the possibility of the finish coat being an acrylic lacquer, thus restricting the possibilities to GM trucks and vans, with the exception of several types of cars being manufactured in California. Insistence on the color and chemistry of the base electro-coat primer in the PDQ search, even without the black primer-surfacer, points to 1977 to 1980 GM trucks or vans.

To what end did this additional work bring the examiner? Testimony would surely stress that two types of paint were found on the victim that correspond to the coatings on the front passenger side of the suspect vehicle. One is a refinish and the other is an original finish with a repaint on its surface (determined by their layer structures and chemistry). Furthermore, the binder chemistry of the latter refinish is consistent with a household spray paint. When asked the question, "But how many other vehicles could have an original paint like this common white finish on them?" the examiner can now respond, "In the case of the three-layer original finishes found in the questioned and known samples in this case, a search of an automotive paint database containing over 14,000 original-finish layer systems used over the past 30 years resulted in similar paint layer systems being found only on vehicles produced by General Motors from 1977 to 1980, primarily vans and trucks."

5.6.3 Example 3

Human use of pigments for paintings, decorations, and other ritual purposes has a very long history. The earliest known cave paintings, for example, are well over 30,000 years old. The pigments used for these were predominantly inorganic compounds (especially ferric oxide and hydrous ferric oxide), although a few naturally occurring organic compounds, such as chlorophyll from plants, were also used. The vast majority of synthetic pigments were not available until the early twentieth century, and this is certainly the case for synthetic paint binders. When dealing with art and other painted or pigmented objects, the presence of synthetic pigments and binders is therefore one of the most definitive markers used by art conservation and provenance chemists to indicate either a modern restoration or a forgery. Introduction dates for earlier paint components, however, can also serve to limit the time periods when they could have been used. Although some inorganic pigments have been used for tens of thousands of years, their modern formulations have characteristic signatures that serve to distinguish them from their ground mineral counterparts. Modern pigments, for example, have definite particle size and particle-size distributions, morphologies, specific crystal structures in the case of polymorphic substances, additives, and so on.

The Shroud of Turin There has probably never been an object analyzed by a forensic science laboratory that has generated more controversy than the Shroud of Turin. The shroud is a linen cloth that appears to bear the image of a man. It is believed by some to be the cloth that was placed on Jesus Christ prior to his burial, with the image formed by blood, secretions, or other means. The shroud is kept at the Cathedral of Saint John the Baptist in Turin, Italy.

In 1977, a team of scientists was selected to study the shroud, with this investigation given the designation Shroud of Turin Research Project (STURP). One of the STURP team members was Walter McCrone (1916–2002), a leading microscopist and founder of the McCrone Research Institute and McCrone Associates in Chicago. McCrone's main tool for this investigation was polarized light microscopy, although he confirmed his findings using a variety of other analytical methods. In 1979, based on his analyses, McCrone concluded that there is no blood on the shroud and that the image is produced by a paint consisting of a collagen tempera binder pigmented with red ochre (mainly ferric oxide) and vermilion (mercuric sulfide) (McCrone and Skirius, 1980; McCrone, 1980, 1981). This was a common paint composition during the fourteenth century and McCrone stated that "the Shroud is a beautiful painting by an inspired medieval artist."

McCrone was the only STURP team member to reach this conclusion, and when other members learned that he disagreed with them, his samples were confiscated and he was removed from the team. When his conclusions became public, he received hate mail and death threats. In 1989, three independent radiocarbon studies of the shroud were conducted at the University of Arizona, Oxford University, and the Swiss Federal Institute of Technology. The data from all three laboratories were consistent and indicated that the shroud originated from the time period 1260 to 1390 (Damon et al., 1989), confirming McCrone's earlier claim. More details of this case are described in *Judgement Day for the Shroud of Turin* (McCrone, 1999b). In 2000, the American Chemical Society acknowledged McCrone for his contributions to chemical microscopy—specifically citing his studies on the shroud—by honoring him with the American Chemical Society Award in Analytical Chemistry.

The Vinland Map The Vinland map is purported to be a fifteenth-century document that was discovered in 1957. Its authenticity gained some credibility from the fact that at one time it was probably bound together with another ancient document of undisputed authenticity. The map depicts most of the world, including Europe and Asia, Africa, Greenland, and a large island west of Greenland labeled "Vinland" (North America). The map was donated to Yale University, but before accepting it, the university had it examined by two museum curators and a librarian, who pronounced it authentic. Unfortunately, this was done in secret and other specialists were not consulted.

In the scientific community, the Vinland map has probably generated more controversy than the Shroud of Turin, owing to both its content and its chemical composition. The map depicts Greenland as an island with a shape that is quite accurate, considering that in the fifteenth century it was thought to be a peninsula, and there were no recorded passages of ships around the island until the twentieth century. Some of the Latin names cited on the map were not believed to have been used until hundreds of years later.

One of the main factors that has continued to fuel the Vinland map controversy is the radiocarbon dating analysis, which indicates the map parchment to be from the 1423 to 1445 time period, consistent with its purported date of origin. The map, however, has a coating of an unidentified substance from the 1950s. The British Museum examined the map in 1967 and found that it appeared to have been drawn with two superimposed lines consisting of black and yellow inks, and was unlike any that had been examined previously.

In 1972, Yale University sent the map to McCrone Associates for chemical analysis of the ink. Walter McCrone found that the map was produced with a yellow line overlaid with a black line in the middle (McCrone, 1976). McCrone indicated that this was done with considerable skill, but he did find segments where the yellow and black lines were not parallel. The use of two lines was apparently done to simulate the natural weathering of medieval black iron-containing inks, which would develop such an appearance as the ink migrated slowly along the parchment fibers. Using polarized light microscopy, SEM–EDS, transmission electron microscopy (TEM), XRD, and other methods, McCrone identified anatase in the yellow ink, with the anatase having morphologies, particle sizes, and particle-size distributions consistent with those of modern formulations. Anatase was first used as an ink pigment in the 1920s and may have a yellow hue from iron impurities. McCrone thus concluded that the Vinland map could not have been produced when purported.

Proponents of the authenticity of the Vinland map have argued that since anatase is present in trace quantities in some of the components of inks used in the medieval period, McCrone's findings do not support a forgery. However, there have been no investigations into the microscopic characteristics of anatase found in such documents to show that these impurities cannot be distinguished from modern formulations, which was the key to McCrone's findings. In 1991, McCrone performed a second examination of the map ink. Using an infrared microscope, he identified the ink binder as gelatin, probably made from animal skin (McCrone, 1999a). Although this does not rule out a medieval origin, McCrone confirmed that significant quantities of anatase are present. This had become an issue since his first study, as an examination of the ink using PIXE suggested that the titanium levels were too low to be from anatase in the ink (Cahill et al., 1987). However, a 2002 study using a Raman microscope confirmed that a significant amount of anatase is present in the yellow ink, and found that most of the black ink consists of carbon (Brown and Clark, 2002). Such inks would not be expected to produce yellowing with age, which is characteristic of iron-containing inks.

Judging from the titles of two letters to *Analytical Chemistry* and a recent review article—*Evidence That the Vinland Map Is Medieval* (Olin, 2003), *The Vinland Map Ink Is NOT Medieval* (Towe, 2004), and *Analysing the Vinland Map: A Critical Review of a Critical Review* (Towe et al., 2008)—the heated controversy over the authenticity of this document will probably not end any time soon. McCrones's detailed examinations of the anatase particles of the yellow ink, however, provide convincing evidence for a modern pigment rather than a natural contaminant in a medieval pigment.

A Trove of Previously Undiscovered Paintings by Jackson Pollock?

In 2002, Alex Matter reported his discovery of more than 20 previously unknown paintings that he attributed to the American abstract expressionist artist Jackson Pollock (1912–1956). He found them in a storage locker belonging to his deceased father, who had been a friend of Pollock's. Based on the style of the paintings and an inscription on the paper wrapper in which the paintings were said to have been discovered, at least one leading expert on Pollock's work concluded that the collection was genuine. This expert felt that "there are too many things about them that are pure Jackson" (Kennedy, 2005). In preparation for an exhibition that featured the paintings, the owners sent them to the Harvard University Art Museum Laboratory, the Museum of Fine Arts (Boston) Laboratory, and to Orion Analytical LLC in Massachusetts for paint analysis. Scientists at these laboratories also examined previous known paintings by Jackson Pollock as a comparison (Martin, 2008).

The following are some of the paint components that were identified in most of the paintings that were analyzed: (1) an acrylic binder with styrene, (2) Benzimidazolone Yellow 4G (Pigment Yellow 151), (3) Benzimidazolone Orange (Pigment Orange 62), (4) rutile-coated mica pearlescent pigments, (5) iron oxide–coated mica pearlescent pigments, and (6) DPP Red BO (Pigment Red 254). These components were identified by infrared spectroscopy, with additional methods, including microscopy and elemental analysis, used to identify the pearlescent pigments. Benzimidazolone Yellow 4G, Benzimidazolone Orange, and DPP Red BO were first identified in automotive paints. Because they were fairly new, their introduction dates and their first known use for American vehicles were also reported to aid forensic paint examiners interpret data from unknown paint chips recovered from hit-and-run scenes. The art provenance chemists performing the Pollock studies relied on these automotive paint studies, as well as others (Martin, 2008).

None of the six paint components cited was found in previous known works by Pollock. More revealing, however, are the commercial introduction dates for the six: (1) 1961, (2) 1969, (3) 1971, (4) 1978, (5) 1979, and (6) 1983 (Martin, 2008). All six were therefore introduced *after* Pollock's death in 1956, and all three laboratories concluded that the paintings containing these compounds could not have been produced by Pollock.

Acknowledgments

We would like to thank JoAnn Buscaglia (FBI Laboratory Counterterrorism and Forensic Science Research Unit), Vincent Desiderio (New Jersey State Police Forensic Laboratory), Jim Parzych (Washington State Crime Laboratory), and John Reffner (John Jay College of Criminal Justice) for their helpful discussions; and Mr. Desiderio, Helen Griffins (Ventura County Forensic Laboratory), James Martin (Orion Analytical LLC), Erik Neilson (Washington State Crime Laboratory), Skip Palenik (Microtrace), Bill Schneck (Washington State Crime Laboratory), Donna Wilson (Washington State Crime Laboratory), and Daniel Van Wyk (Washington State Crime Laboratory) for taking the time to review this manuscript.

The co-author (EMS) is eternally grateful for the involvement of the principal author, who was extremely reluctant to get involved in yet another writing project. He only consented after considerable arm-twisting and cajoling, as—being a very tactful person—he was concerned about offending those who in the past have sought his considerable knowledge, skills, and writing ability, but whom he was unable to accommodate.

REFERENCES

Allen, T. J. (1992). Paint sample presentation for Fourier transform infrared microscopy. *Vibrat. Spectrosc.*, **3**:217–237.

Allen, T. J. (1994). *Effects of Environmental Factors on the Fluorescence of White Alkyd Paint*. Central Research Establishment Report #758. Aldermaston, UK: Home Office Central Research Establishment.

ASTM (2007). *ASTM E1610-02: Standard Guide for Forensic Paint Analysis and Comparison*. West Conshohocken, PA: ASTM International.

Bartick, E. G. (2010). Infrared microscopy and its forensic applications. In: Saferstein, R. (Ed.), *Forensic Science Handbook*, Vol. II1, 2nd ed. Upper Saddle River, NJ: Prentice Hall, pp. 252–306.

Bartick, E. G., Tungol, M. W., and Reffner, J. A. (1994). A new approach to forensic analysis with infrared microscopy: internal reflection spectroscopy. *Anal. Chim. Acta*, **288**:35–42.

Bell, S. E. J., Fido, L. A., Speers, S. J., and Armstrong, W. J. (2005a). Rapid forensic analysis and identification of "lilac" architectural finishes using Raman spectroscopy. *Appl. Spectrosc.*, **59**:100–108.

Bell, S. E. J., Fido, L. A., Speers, S. J., Armstrong, W. J., and Spratt, S. (2005b). Forensic analysis of architectural finishes using Fourier transform infrared and Raman spectroscopy: I. The resin bases. *Appl. Spectrosc.*, **59**:1333–1339.

Bell, S. E. J., Fido, L. A., Speers, S. J., Armstrong, W. J., and Spratt, S. (2005c). Forensic analysis of architectural finishes using Fourier transform infrared and Raman spectroscopy: II. White paint. *Appl. Spectrosc.*, **59**:1340–1346.

Bentley, J. (2001). Composition, manufacture and use of paint. In: Caddy, B. (Ed.), *Forensic Examination of Glass and Paint: Analysis and Interpretation*. New York: Taylor & Francis, pp. 123–141.

Beveridge, A., Fung, T., and MacDougall, D. (2001). Use of infrared spectroscopy for the characterization of paint fragments. In: Caddy, B. (Ed.), *Forensic Examination of Glass and Paint: Analysis and Interpretation*. New York: Taylor & Francis, pp. 221–232.

Boynton, R. M. (1979). *Human Color Vision*. New York: Holt, Rinehart and Winston.

Braun, J. H. (1993). *Federation Series on Coating Technology: Introduction to Pigments*. Blue Bell, PA: Federation of Societies for Coating Technology.

Brown, K. L., and Clark, R. J. H. (2002). Analysis of pigmentary materials on the Vinland map and Tartar relation by Raman microprobe spectroscopy. *Anal. Chem.*, **74**:3658–3661.

Buckle, J. L., MacDougall, D. A., and Grant, R. R. (1997). PDQ paint data queries: the history and technology behind the development of the Royal Mounted Canadian Police Forensic Laboratory Services automotive paint database. *Can. Soc. Forensic Sci. J.*, **30**:199–212.

Burke, P., Curry, C. J., Davies, L. M., and Cousins, D. R. (1985). A comparison of pyrolysis mass spectrometry, pyrolysis gas chromatography and infrared spectroscopy for the analysis of paint resins. *Forensic Sci. Int.*, **28**:201–219.

Buzzini, P., and Massonnet, G. (2004). A market study of green spray paints by Fourier transform infrared (FTIR) and Raman spectroscopy. *Sci. Justice*, **44**:123–131.

Buzzini, P., et al. (2005). Survey of crowbar and household paints in burglary cases: population studies, transfers and interpretation. *Forensic Sci. Int.*, **152**:221–234.

Buzzini, P., Massonnet, G., and Sermier, F. M. (2006). The micro Raman analysis of paint evidence in criminalistics: cases studies. *J. Raman Spectrosc.*, **37**:922–931.

Cahill, T. A., et al. (1987). The Vinland map, revisited: new compositional evidence on its inks and parchment. *Anal. Chem.*, **59**:829–833.

Cassista, A. R., and Sandercock, P. M. L. (1994). Comparison and identification of automotive topcoats: microchemical spot tests, microspectrophotometry, pyrolysis gas chromatography, and diamond anvil cell FTIR. *Can. Soc. Forensic Sci. J.*, **27**:209–223.

Challinor, J. M. (1989). Pyrolysis derivitization using tetraalkylammonium hydroxide. *J. Anal. Appl. Pyrol.*, **16**:323–333.

Challinor, J. M. (1991a). The scope of pyrolysis methylation reactions. *J. Anal. Appl. Pyrol.*, **18**:15–24.

Challinor, J. M. (1991b). Structure determination of alkyd resins by simultaneous pyrolysis methylation. *J. Anal. Appl. Pyrol.*, **18**:233–244.

Challinor, J. M. (1995). Examination of forensic evidence. In: Wampler, T. P. editor, *Applied Pyrolysis Handbook*. New York: Marcel Dekker, Inc., 207–241.

Challinor, J. M. (2001). Pyrolysis techniques for the characterization and discrimination of paint. In: Caddy, B. (Ed.), *Forensic Examination of Glass and Paint: Analysis and Interpretation*. New York: Taylor & Francis, pp. 165–182.

Challinor, J. M. (2006). Examination of forensic evidence. In: Wampler, T. P. (Ed.), *Applied Pyrolysis Handbook*, 2nd ed. Boca Raton, FL: CRC Press, pp. 175–200.

Collaborative Testing Services (1989). *Reference Collection of Automotive Paints (1974–1989) Technical Data*. Herndon, VA: Collaborative Testing Services.

Collaborative Testing Services (2007). Forensic program: paint analysis reports 2005 to 2007. http://www.collaborativetesting.com/forensics/report_list.html.

Contos, S., and Ryland, S. (2001). Microspectrophotometric discrimination study of CTS automotive paint collection non-decorative flake finish coats. Presented at the fall seminar of the Southern Association of Forensic Scientists, Sanibel, FL, Sept. 10–13.

Crown, D. A. (1968). *The Forensic Examination of Paints and Pigments*. Springfield, IL: Charles C Thomas.

Damon, P. E., et al. (1989). Radiocarbon dating of the Shroud of Turin. *Nature*, **337**:611–615.

Deconinck, I., Latkoczy, C., Günther, D., Govaert, F., and Vanhaecke, F. (2006). Capabilities of laser ablation-inductively coupled plasma mass spectrometry for (trace) element analysis of car paints for forensic purposes. *J. Anal. At. Spectrom.*, **21**:279–287.

De Gelder, J., Vandenabeele, P., Govaert, F., and Moens, L. (2005). Forensic analysis of paints by Raman spectroscopy. *J. Raman Spectrosc.*, **36**:1059–1067.

Derrick, M. R. (1995). Infrared microspectroscopy in the analysis of cultural artifacts. In: Humecki, H. J. (Ed.), *Practical Guide to Infrared Microspectroscopy*. New York: Marcel Dekker, pp. 287–322.

Drahl, C. (2008). Nail polish: classic formulas behind chip-free coatings slowly gets a makeover. What's that stuff? *Chem. Eng. News*, Aug. 11, p. 42.

Droll, F. J. (1999). New light-interference pigments expand the color palette. *Paint Coat. Ind.*, Apr., pp. 40–44.

Eastaugh, N., Walsh, V., Chaplin, T., and Siddall, R. (2004). *The Pigment Compendium*. New York: Elsevier Butterworth-Heinemann.

Edmondstone, G., Hellman, J., Legate, K., Vardy, G. L., and Lindsay, E. (2004). An assessment of the evidential value of automotive paint comparisons. *Can. Soc. Forensic Sci. J.*, **37**:147–153.

Ehrich, F. F. (1988). Pigmentation of automotive finishes. In: Lewis, P. (Ed.), *Pigment Handbook*, Vol. 1. New York: Wiley, pp. 5–24.

Eyring, M. B. (2002). Visible microscopical spectrophotometry in the forensic sciences. In: Saferstein, R. (Ed.), *Forensic Science Handbook*, Vol. I, 2nd ed. Upper Saddle River, NJ: Prentice Hall, pp. 322–387.

Fettis, G. (Ed.) (1995). *Automotive Paints and Coatings*. New York: VCH.

Fukuda, K. (1985). The pyrolysis gas chromatographic examination of Japanese car paint flakes. *Forensic Sci. Int.*, **29**:227–236.

Giang, Y. G., et al. (2002). Identification of tiny and thin smears of automotive paint following a traffic accident. *J. Forensic Sci.*, **47**:625–629.

Goldstein, J. I., et al. (1992). *Scanning Electron Microscopy and X-Ray Microanalysis*. New York: Plenum Press.

Gothard, J. A. (1976). Evaluation of automobile paint flakes as evidence. *J. Forensic Sci.*, **21(3)**: 636–641.

Gothard, J., and Maynard, P. (1996). Evidential value of automotive paint. *Proceedings of the 13th International Symposium of the ANZFSS* (Australian and New Zealand Forensic Science Society), Sydney, Australia, Sept. 8–13.

Govaert, F., and Bernard, M. (2004). Discriminating red spray paints by optical microscopy, Fourier transform infrared spectroscopy and x-ray fluorescence. *Forensic Sci. Int.*, **140**:61–70.

Govaert, F., De Roy, G., Decruyenaere, B., and Ziernicki, D. (2001). Analysis of black spray paints by Fourier transform infrared spectrometry, x-ray fluorescence and visible microscopy. *Probl. Forensic Sci.*, **47**:333–339.

Haag, L. C. (1977). Elemental profiles of automotive paint chips by x-ray fluorescence spectrometry. *J. Forensic Sci. Soc.*, **16**:255–263.

Hackman, J. R. (1988). Nickel antimony titanate yellow. In: Lewis, P. (Ed.), *Pigment Handbook*, Vol. 1. New York: Wiley, pp. 375–381.

Hammer, P. S. (1982). Pigment analysis in the forensic examination of paints: III. A guide to motor vehicle paint examination by transmitted light microscopy. *J. Forensic Sci. Soc.*, **22**:187–192.

Hartshorn, J. H. (1992). Applications of FTIR to paint analysis. In: Golton, W. C. (Ed.), *Analysis of Paints and Related Materials: Current Techniques for Solving Coatings Problems*. West Conshohocken, PA: ASTM, pp. 127–147.

Henson, M. L., and Jergovich, T. A. (2001). Scanning electron microscopy and energy dispersive x-ray spectrometry (SEM/EDS) for the forensic examination of paints and coatings. In: Caddy, B. (Ed.), *Forensic Examination of Glass and Paint: Analysis and Interpretation*. New York: Taylor Francis, pp. 243–272.

Hobbs, A. L., and Almirall, J. R. (2003). Trace elemental analysis of automotive paints by laser ablation–inductively coupled plasma–mass spectrometry (LA–ICP–MS). *Anal. Bioanal. Chem.*, **376**:1265–1271.

Howden, D. R., Dudley, R. J., and Smalldon, K. W. (1977). The non-destructive analysis of single layered household paints using energy dispersive x-ray fluorescence spectrometry. *J. Forensic Sci. Soc.*, **17**:161–167.

Hunter, R. S. (1975). *The Measurement of Appearance*. New York: Wiley.

Infrared Spectroscopy Atlas Working Committee (1991). *An Infrared Spectroscopy Atlas for the Coatings Industry*, Vols. I and II, 4th ed. Blue Bell, PA: Federation of Societies for Coatings Technology.

Infrared Spectroscopy Committee of the Chicago Society for Coatings Technology (1980). *An Infrared Spectroscopy Atlas for the Coatings Industry*. Philadelphia: Federation of Societies for Coatings Technology.

Jenkins, R. (1999). *X-Ray Fluorescence Spectrometry*. New York: Wiley-Interscience.

Johnson, B. B., and Peticolas, W. L. (1976). The resonant Raman effect. *Ann. Rev. Phys. Chem.*, **27**:465–521.

Judd, D. B., and Wyszecki, G. (1975). *Color in Business, Science, and Industry*, 3rd ed. New York: Wiley.

Kanngiesser, B., Manzer, W., Rodriguez, A. F., and Reiche, I. (2005). Three-dimensional micro-XRF investigation of paint layers with a tabletop setup. *Spectrochim. Acta B*, **60**:41–47.

Kendix, E., Nielsen, O. F., and Christensen, M. C. (2004). The use of micro-Raman spectroscopy in architectural paint analysis. *J. Raman Spectrosc.*, **35**:796–799.

Kennedy, R. (2005). Is this a real Jackson Pollock? *New York Times*, May 29, Sec. 2, p. 1.

Kilbourn, J. H., and Marx, R. B. (1994). Polarized light microscopy of extenders in structural paints: forensic applications. *Microscope*, **42**:167–175.

Klockenkamper, R., Knoth, J., Prange, A., and Schwenke, H. (1992). Total-reflection x-ray fluorescence spectroscopy. *Anal. Chem.*, **64**:1115A–1121A.

Klockenkamper, R., von Bohlen, A., and Moens, L. (2000). Analysis of pigments and inks on oil paintings and historical manuscripts using total reflection x-ray fluorescence spectrometry. *X-Ray Spectrom.*, **29**:119–129.

Kopchick, K. A., and Bommarito, C. R. (2006). Color analysis of apparently achromatic automotive paints by visible microspectrophotometry. *J. Forensic Sci.*, **51**:340–343.

Lambourne, R., and Strivens, T. A. (1999). *Paint and Surface Coatings, Theory and Practice*. New York: Woodhead Publishing, William Andrew Publishing.

Learner, T. J. S. (2004). *Analysis of Modern Paints*. Research in Conservation Series. Los Angeles: Getty Publications, The Getty Conservation Institute.

Lewis, P. A. (1995). *Federation Series on Coating Technology: Organic Pigments*, 2nd ed. Blue Bell, PA: Federation of Societies for Coating Technology.

Martin, J. (2008). What materials tell us about the age and attribution of the Matter paintings. *Int. Found. Art Res. J.*, **10(1)**:25–35.

Massonnet, G., and Stoecklein, W. (1999a). Identification of organic pigments in coatings—applications to red automotive topcoats: II. Infrared spectroscopy. *Sci. Justice*, **39**:135–140.

Massonnet, G., and Stoecklein, W. (1999b). Identification of organic pigments in coatings—applications to red automotive topcoats: III. Raman spectroscopy (NIR FT-Raman). *Sci. Justice*, **39**:181–187.

McBane, B. N. (1987). *Federation Series on Coating Technology: Automotive Coatings*. Philadelphia: Federation of Societies for Coating Technology.

McCrone, W. C. (1976). Authenticity of medieval document tested by small particle analysis. *Anal. Chem.*, **48**:676A–679A.

McCrone, W. C. (1979). Particle analysis in the crime laboratory. In: McCrone, W. C., Delly, J. G., and Palenik, S. J. (Ed.), *The Particle Atlas*, Vol. V, 2nd ed. Ann Arbor, MI: Ann Arbor Science Publishers, pp.1379–1401.

McCrone, W. C. (1980). Light microscopical studies of the Turin "shroud": II. *Microscope*, **28**:115–128.

McCrone, W. C. (1981). Light microscopical studies of the Turin "shroud": III. *Microscope*, **29**:19–38.

McCrone, W. C. (1999a). Vinland map 1999. *Microscope*, **47**:71–74.

McCrone, W. C. (1999b). *Judgement Day for the Shroud of Turin*. Buffalo, NY: Prometheus Books.

McCrone, W. C., and Skirius, C. (1980). Light microscopical studies of the Turin "shroud":. *Microscope*, **28**:105–112.

Mirabella, F. M. (1993). *Internal Reflection Spectroscopy: Theory and Applications*. New York: Marcel Dekker.

Morgans, W. M. (1990). *Outlines of Paint Technology*, 3rd ed. New York: Halsted Press.

Nolan, P. J., and Keeley, R. H. (1979). Comparison and classification of small paint fragments by x-ray microanalysis in the SEM. In: *Scanning Electron Microscopy/1979/I*. Chicago: SEM, Inc., AMF O'Hare, pp.449–454.

Norman, E. W. W., et al. (1983). The classification of automotive paint primers using infrared spectroscopy: a collaborative study. *Can. Soc. Forensic Sci. J.*, **16**:163–173.

Nylen, P., and Sunderland, E. (1965). *Modern Surface Coatings*. London: Interscience.

The Oil and Colour Chemists' Association (1983). *Surface Coatings*, Vol. 1, *Raw Materials and Their Usage*, 2nd ed. Randwich, Australia: Tafe Educational Books.

The Oil and Colour Chemists' Association (1984). *Surface Coatings*, Vol. 2, *Paints and Their Applications*, 2nd ed. Randwich, Australia: Tafe Educational Books.

Olin, J. S. (2003). Evidence that the Vinland map is medieval. *Anal. Chem.*, **75**:6745–6747.

Palenik, S. (2007). The contributions of chemical microscopy to the solution of the Green River murders. Presented at the 2007 NIJ/FBI Trace Evidence Symposium, Clearwater Beach, FL, Aug. 13–16.

Palys, B. J., van den Ham, D. M. W., Briels, W., and Feil, D. (1995). Resonance Raman spectra of phthalocyanine monolayers on different supports. a normal mode analysis of zinc phthalocyanine by means of the MNDO method. *J. Raman Spectrosc.*, **26**:63–76.

Panush, S. (1975). Labor pains: the genesis of an automotive color. *Am. Paint Coat. J.*, **60**:2–11.

Percy, R. F. E., and Audette, R. J. (1980). Automotive repaints: just a new look? *J. Forensic Sci.*, **25**:189–239.

Rodgers, P. G., et al. (1976a). The classification of automotive paint by diamond window infrared spectrophotometry: I. Binders and pigments. *Can. Soc. Forensic Sci. J.*, **9**:1–14.

Rodgers, P. G., et al. (1976b). The classification of automotive paint by diamond window infrared spectrophotometry: II. Automotive topcoats and undercoats. *Can. Soc. Forensic Sci. J.*, **9**:49–68.

Rodgers, P. G., et al. (1976c). The classification of automotive paint by diamond window infrared spectrophotometry: III. Case histories. *Can. Soc. Forensic Sci. J.*, **9**:103–111.

Ryland, S. G. (1995). Infrared microspectroscopy of forensic paint evidence. In: Humecki, H. J. (Ed.), *Practical Guide to Infrared Microspectroscopy*. New York: Marcel Dekker, pp.163–243.

Ryland, S. G., et al. (2001). Discrimination of 1990s original automotive paint systems: a collaborative study of black nonmetallic base coat/clear coat finishes using infrared spectroscopy. *J. Forensic Sci.*, **46**:31–45.

Ryland, S. G., and Kopec, R. J. (1979). The evidential value of automotive paint chips. *J. Forensic Sci.*, **24**:140–147.

Ryland, S. G., Jergovich, T. A., and Kirkbride, K. P. (2006). Current trends in forensic paint examination. *Forensic Sci. Rev.*, **18**:97–117.

Schiering, D. W. (1988). A beam condenser/miniature diamond anvil cell accessory for the infrared microspectrometry of paint chips. *Appl. Spectrosc.*, **42**:903–906.

Scientific Working Group for Materials Analysis (2002). Standard guide for using scanning electron microscopy/X-ray spectrometry in forensic paint examination. *Forensic Sci. Commun.*, **4(4)**. http://www.fbi.gov/hq/lab/fsc/backissu/oct2002/bottrell.htm.

Scientific Working Group for Materials Analysis (2007). Standard guide for microspectrophotometry and color measurement in forensic paint analysis. *Forensic Sci. Commun.*, **9(4)**. http://www.fbi.gov/hq/lab/fsc/backissu/oct2007/index.htm.

Scientific Working Group for Materials Analysis (2009). Standard guide for using infrared spectroscopy in forensic paint examinations. On line at http://www.swgmat.org/SWGMAT%20infrared%20spectroscopy.pdf

Snider, A. M., Jr. (1992). X-ray techniques for coatings analysis. In: Golton, W. C. (Ed.), *Analysis of Paints and Related Materials: Current Techniques for Solving Coatings Problems*. West Conshohocken, PA: ASTM, 82–104.

Stachura, S., Desiderio, V. J., and Allison, J. (2007). Identification of organic pigments in automotive coatings using laser desorption mass spectrometry. *J. Forensic Sci.*, **52**:595–603.

Stoecklein, W. (1994). Using the light microscope for analytical procedures: aids for solving cases involving hit-and-run offenses. *Zeiss Inf. Jena Rev.*, **3–4**:19–22.

Stoecklein, W. (2001). The role of colour and microscopic techniques for the characterisation of paint fragments. In: Caddy, B. (Ed.), *Forensic Examination of Glass and Paint: Analysis and Interpretation*. New York: Taylor & Francis, pp.143–164.

Stoecklein, W. (2002). Plate-like pigments in automotive paints: a review. *Paint Coat. Ind.*, May, pp.80–83.

Stoecklein, W., and Fujiwara, H. (1999). The examination of UV-absorbers in two-coat metallic and non-metallic automotive paint. *Sci. Justice*, **39(3)**:188–195.

Stoecklein, W., and Palenik, C. (1998). Forensic analysis of automotive paints: evidential value and the batch problem. *Proceedings of the 4th Meeting of the European Paint Group*, Paris, Oct. 5–6.

Strietberger, H. J., and Dossel, K. F. (Eds.) (2008). *Automotive Paints and Coatings*, 2nd ed. Weinheim, Germany: Wiley-VCH.

Suzuki, E. M. (1996a). Infrared spectra of U.S. automobile original topcoats (1974–1989): I. Differentiation and identification based on acrylonitrile and ferrocyanide C≡N stretching absorptions. *J. Forensic Sci.*, **41**:376–392.

Suzuki, E. M. (1996b). Infrared spectra of U.S automobile original topcoats (1974–1989): II. Identification of some topcoat inorganic pigments using an extended range (4000–220 cm^{-1}) Fourier transform spectrometer. *J. Forensic Sci.*, **41**:393–406.

Suzuki. E. M. (1999a). Infrared spectra of U.S. automobile original topcoats (1974–1989): V. Identification of organic pigments used in red nonmetallic and brown nonmetallic and metallic monocoats—DPP Red BO and Thioindigo Bordeaux. *J. Forensic Sci.*, **44**:297–313.

Suzuki, E. M. (1999b). Infrared spectra of U.S. automobile original topcoats (1974–1989): VI. Identification and analysis of yellow organic automotive paint pigments—Isoindolinone Yellow 3R, Isoindoline Yellow, Anthrapyrimidine Yellow, and miscellaneous yellows. *J. Forensic Sci.*, **44**:1151–1175.

Suzuki, E. M. (2010). Forensic applications of infrared spectroscopy. In: Saferstein, R. (Ed.), *Forensic Science Handbook*, Vol. III, 2nd ed. Upper Saddle River, NJ: Prentice Hall, pp.75–251.

Suzuki, E. M., and Carrabba, M. (2001). In Situ identification and analysis of automotive paint pigments using line segment excitation Raman spectroscopy: I. Inorganic topcoat pigments. *J. Forensic Sci.*, **46**:1053–1069.

Suzuki, E. M., and Marshall, W. P. (1997). Infrared spectra of U.S. automobile original topcoats (1974–1989): III. In Situ identification of some organic pigments used in yellow, orange, red, and brown nonmetallic and brown metallic finishes—benzimidazolones. *J. Forensic Sci.*, **42**:619–648.

Suzuki, E. M., and Marshall, W. P. (1998). Infrared spectra of U.S. automobile original topcoats (1974–1989): IV. Identification of some organic pigments used in red and brown nonmetallic and metallic monocoats—quinacridones. *J. Forensic Sci.*, **43**:514–542.

Suzuki, E. M., and McDermot, M. X. (2006). Infrared spectra of U.S. automobile original finishes: VII. Extended range FT-IR and XRF analyses of inorganic pigments in situ—nickel titanate and chrome titanate. *J. Forensic Sci.*, **51**:532–538.

Thornton, J. I. (2002). Forensic paint examination. In: Saferstein, R. (Ed.), *Forensic Science Handbook*, Vol. 1, 2nd ed. Upper Saddle River, NJ: Prentice Hall, pp.429–478.

Tippett, C. F., Emerson, V. J., Fereday, M. J., Lawton, F., Richardson, A., Jones, L. T., and Lampert, S. M. (1968). The evidential value of the comparison of paint flakes from sources other than vehicles. *J. Forensic Sci. Soc.*, **8(2–3)**:61–65.

Towe, K. M. (2004). The Vinland map is NOT medieval. *Anal. Chem.*, **76**:863–865.

Towe, K. M., Clark, R. J. H., and Seaver, K. A. (2008). Analysing the Vinland map: a critical review of a critical review. *Archaeometry*, **50**:887–893.

Tyler, F. K. (2000). Tailoring TiO_2 chemistry to achieve desired performance properties. *Paint Coat. Ind.*, Feb., pp.32–38.

Wampler, T. P. (2006). *Applied Pyrolysis Handbook*, 2nd ed. Boca Raton, FL: CRC Press.

Willis, S., McCullough, J., and McDermott, S. (2001). The interpretation of paint evidence. In: Caddy, B. (Ed.), *Forensic Examination of Glass and Paint: Analysis and Interpretation*. New York: Taylor & Francis, pp.273–287.

Wilson, R., and Judd, G. (1973). The application of scanning electron microscopy and energy dispersive x-ray analysis to the examination of forensic paint samples. In: Birks, L. S. et al. (Eds.), *Advances in X-Ray Analysis*, Vol. 15. New York: Plenum Press.

Wright, D., Bradley, M., and Mehltretter, A. (2011). Analysis and discrimination of architectural paint samples via a population study. *Forensic Sci. Int.*, **209**:86–95.

Zeichner, A., Levin, N., and Landau, E. (1992). A study of paint coat characterizations produced by spray paints from shaken and nonshaken spray cans. *J. Forensic Sci.*, **37**:542–555.

Zieba, J., and Pomianowski, A. (1981). A statistical criterion for the value of evidence: application to the evaluation of the results of paint spectral analysis. *Forensic Sci. Int.*, **17**:101–108.

Zieba-Palus, J., and Borusiewicz, R. (2006). Examination of multilayer paint coats by the use of infrared, Raman, and XRF spectroscopy for forensic purposes. *J. Mol. Struct.*, **792–793**:286–292.

CHAPTER 6

Analysis Techniques Used for the Forensic Examination of Writing and Printing Inks[†]

GERALD M. LaPORTE

National Institute of Justice, Office of Investigative and Forensic Sciences, Washington, DC

JOSEPH C. STEPHENS

United States Secret Service, Forensic Services Division, Questioned Document Branch, Instrumental Analysis Section, Washington, DC

Summary The analysis and identification of writing and printing inks and toners is generally very important in document examination, especially when used in conjunction with a reference library. Inks can be differentiated based on the chemistry of colorants, solvents, resins, and additives. Instrumental analysis, including GC–MS, HPLC, and FT–IR and Raman spectroscopy, can often be used following visual examination, microscopic observation, and thin-layer chromatography. Analysis of toners can be performed with XRF, SEM–EDS, or pyrolysis GC. Although chemical analysis of materials used to create documents can provide vast amounts of relevant information and strongly support associations between questioned and known materials, in nearly all cases the data obtained will not support a conclusion that identifies a particular writing instrument or printing device.

6.1	Introduction	226
6.2	Ink	226
	6.2.1 Ink composition	227
6.3	Ink analysis	230
	6.3.1 Physical examinations	233
	6.3.2 Optical examinations	236

[†]All references pertaining to manufacturers and their products do not imply endorsement by the United States government, or the authors.

Forensic Chemistry Handbook, First Edition. Edited by Lawrence Kobilinsky.
© 2012 John Wiley & Sons, Inc. Published 2012 by John Wiley & Sons, Inc.

	6.3.3 Chemical examinations	238
	6.3.4 Ink dating	240
6.4	Office machine systems	242
	6.4.1 Inkjet ink	242
	6.4.2 Inkjet ink analysis	243
	6.4.3 Toner printing	245
	6.4.4 Toner analysis	246
	Conclusion	247
	References	248

6.1 INTRODUCTION

Crimes that involve questioned documents can include matters of national security (National Commission on Terrorist Attacks Upon the United States, 2004), homicides, and various types of document fraud. In all cases it is imperative for investigators and forensic scientists to recognize the realm of examinations that can be conducted on documents and the potential information that may be gleaned from their findings. A questioned document (QD) is defined by Lindblom (2006) as "... any material containing marks, symbols, or signs that convey meaning or message." Documents can present themselves in numerous formats, including letters, envelopes, packages, calendars, diaries, currencies, identification cards, financial documents, contracts, wills, and business records. Forensic document examiners (FDEs) can also evaluate a variety of evidence that does not fall into the traditional category of a document, but requires similar expertise, such as written entries found on the body of a victim, or a threat written or painted on a wall.

General acceptance of handwriting analysis to identify a suspected author dates back several decades. While handwriting evaluations are a valuable QD examination, there is a gamut of nonhandwriting examinations that require chemical analyses. Given the variation of materials encountered on a questioned document and the diversity of analytical techniques available, the purpose of this chapter is to provide an overview of the chemical procedures that may be used for document examination. The authors have chosen to highlight some common procedures that may be employed to gather information from documents submitted for forensic analysis. The appropriate scientific procedure is heavily dependent on the composition of the material being analyzed. Often, documents are created using a combination of inks and paper. Correction fluid, adhesives, stains, and various other materials may be encountered on a document as well.

6.2 INK

Ink is one of the most common materials used in the production of a document. Although there are various types of writing and printing inks, chemical and instrumental techniques remain similar when attempting to characterize their basic components. Typically, the goal of such examinations is to determine the source of the manufacturer, compare questioned and known items, and ascertain when the document was produced and/or decipher alterations, erasures, and obliterations.

6.2.1 Ink Composition

All inks, in their basic form, are composed mainly of a colorant(s) suspended in a vehicle (solvents and resins). There are also other organic and inorganic ingredients that may be present in inks, which can include antioxidants, preservatives, wetting agents, lubricants, and trace elements, but these typically form a small fraction of the overall formulation. Nevertheless, their importance should not be discounted because it is possible that a combination of these components allows otherwise similar inks to be discriminated. Before delving into the various analytical approaches for the examination and characterization of inks, it is necessary to gain an understanding of the major components.

Colorants Colorants are a crucial part of all inks because without them, inks would not be discernible under visible light (approximately 380 to 780 nm). The molecular composition of the colorants will dictate how certain wavelengths of visible light are absorbed and reflected, thereby influencing their color. Depending on the vehicle and its interaction with the colorant, two types of colorants can be used: dyes and/or pigments. Dyes are generally considered to be compounds with highly conjugated resonance structures. Their molecular weights can vary from the low hundreds to the high thousands. Dyes can be classified according to their chemical structure or how they are applied to material. It is beyond our scope in this chapter to adequately review the various types of dyes. The most comprehensive volume regarding dye information is *The Colour Index*, published by the Society of Dyers and Colourists and the American Association of Textile Chemists and Colorists (1971). Dyes may be listed by their International Union of Pure and Applied Chemistry (IUPAC) name, Chemical Abstracts Services (CAS) number, or the Colour Index (C.I.) name. *The Colour Index* divides dyes into a series of large groups: acid dyes, azoic dyes (monoazo-, diazo-, triazo-), basic dyes, developers dyes, direct dyes, disperse dyes, fluorescent brighteners, food dyes, ingrain dyes, leather dyes, mordant dyes, natural dyes, oxidation bases, phthalocyanine dyes, reactive dyes, reducing agents, solvent dyes, sulfur dyes, and vat dyes. These colorants are used in many industries to yield colors on an array of substrates; however, solvent dyes and pigments are the most commonly used colorants in writing instruments. Acid dyes and reactive dyes are also used, but less frequently. The molecular structures of a common phthalocyanine dye and a monoazo reactive dye are shown in Figure 6.1.

The major distinguishing feature between dyes and pigments is that the latter consists of fine particles of insoluble material that are suspended in the vehicle. Pigments are generally considered more stable and long-lasting than dyes because pigments are less prone to photodecomposition (lightfast) and are insoluble in

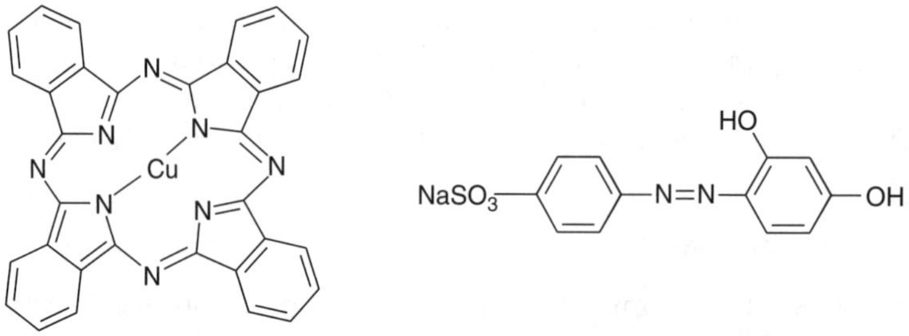

Figure 6.1 Molecular structures of copper phthalocyanine and a monoazo reactive dye that can be found in writing and printing inks.

water (waterfast). Their color can be derived from a metal-centered complex and is generally less vibrant than dyestuffs. Additionally, pigments are more opaque than dyes, so the colorant is more efficient at masking any underlying material. There are five major categories of pigments: organic pigments, toners, lakes, extended pigments, and inorganic pigments. Toners, lakes, and extended pigments are all precipitated from an aqueous solution in conjunction with some catalyst(s).

Solvents The fluid portion of ink that suspends and delivers the colorant to the substrate is known as the *vehicle*. Vehicles are necessary to carry the color from the cartridge to the paper. Once on the paper, the solvent undergoes a series of changes over a fixed period of time, causing the colorant to dry onto the paper. Changes that occur to the solvent–ink mixture over time can include polymerization, evaporation, oxidation, or photodecomposition. These modifications to the original chemical composition of the ink have been the focus of methodologies that are employed when attempting to date a document. A more extensive discussion of ink aging will be covered in the ink analysis section of this chapter.

Although much of the specific information on each company's ink formulations is proprietary, there are some standard chemicals that are used as vehicles. Glycols, alcohols, and water are the most commonly found solvents in use for pens today (Brunelle and Crawford, 2003). 2-Phenoxyethanol (2-PE), ethanol, 1-phenoxypropan-2-ol, benzyl alcohol, and many more solvents can also be used as vehicles. The choice of which solvent or solvents to use often relies on properties related to the writing instrument. The writing instrument type (e.g., fountain pen, ballpoint, felt-tip marker), composition of the ink cartridge, region of sale (e.g., dry, humid), and type of colorants and resins are all considerations when attempting to determine the type of solvent(s) to be used. The desired properties for a single formulation are often achieved through a combination of several vehicles. Other considerations, such as the solubility of the colorants in the vehicles, are also critical, since vehicles are the primary means of delivering the colorant to the substrate.

Resins and Additives Resins, which can be natural or synthetic, are incorporated into inks to provide them with a desired viscosity and a means to bond the ink and the substrate as the ink dries. Normally, the resinous material is dissolved into the vehicle to create a solution in which colorants can be added. Phthalates, ketone resins, styrene-type resins, phenol-type resins, rosin-type resins, and poly(vinylpyrrolidones) are just a few plasticizers that find use in ink formulations as binding resins (U.S. patent 7,462,229). Along with the resins, additives such as biocides, surfactants, lubricants, corrosion inhibitors, preservatives, buffers, and diluting agents are included in ink formulations to enhance properties of the ink (Brunelle and Crawford, 2003). Bügler et al. (2005) used thermal desorption and gas chromatography–mass spectrometry (GS–MS) to characterize ballpoint pen inks. Table 6.1 is a summary of the resins and solvents found in a population of 25 inks.

TABLE 6.1 Summary of Resins and Solvents Found in 25 Different Ballpoint Inks Analyzed Using Thermal Desorption and Gas Chromatography–Mass Spectrometry

Manufacturer	Year of Introduction	Color	Resin Class[a]	Solvents[b]
1	1997	Black	AF	PE, PDE, N-methylpyrrolidone
2	2000	Blue	AF	PE, PDE
1	2000	Black	AF	PE
3	2000	Blue	AF	PE, PDE
	2000	Black	CF	PE, PDE
4	2000	Black	AF	PE, PDE
5	2001	Blue	CF	PE, dipropylene glycol
	1999	Black	AF	PE, PDE, ethyl diethylene glycol
6	2002	Black	Alkyd	PE
7	2002	Black	Alkyd	PE, PDE, propylene glycol
8	2003	Blue	AF	PE, PDE, diisopropylene glycol
	2003	Black	AF	PE, PDE, ethyl diglycol
9	2003	Blue	CF, Alkyd	PE, PDE, hexylene glycol
	2003	Black	AF	PE, PDE, hexylene glycol
10	2003	Blue	Alkyd	PE, butylene glycol
11	2003	Blue	Alkyd	PE, PDE, benzyl alcohol
12	2003	Blue	AF	PE, PDE, butylene glycol
	2003	Black	AF, Alkyd	PE, PDE, butylene glycol
13	2004	Blue	AF	PE, benzyl alcohol
	2004	Black	Unknown	PE, benzyl alcohol
14	2005	Blue	Alkyd	PE, benzyl alcohol
	2005	Black	Alkyd	PE, benzyl alcohol
15	2005	Blue	CF, Alkyd	PE, PDE
	2005	Black	CF, Alkyd	PE, PDE
	2005	Blue	CF	PE, PDE, trioctylphthalate

Source: Adapted, in part, with permission, from the *Journal of Forensic Sciences*, Vol. 50, No. 5, copyright © ASTM International.
[a] AF, acetophenone–formaldehyde resin; CF, cyclohexanone–formaldehyde resin.
[b] PE, 2-phenoxyethanol; PDE, 2-(2-ethoxy) ethoxybenzene.

TABLE 6.2(a) Ink Formulation Containing Glycol

U.S. patent: 4,077,807	Formula
Ethylene glycol monophenyl ether	20.0
1,3-Butylene glycol	20.0
Fatty acid (oleic acid type)	12.0
Solvent Blue 5 dye (such as Victoria Pure 6 Blue BO base)	5.0
Solvent Violet 8 dye (such as Methyl 4 Violet base)	5.0
Resin, ketone condensation type	35.0
Resin, poly(vinylpyrrolidone) type	3.0
Corrosion inhibitor	1.0
Antioxidant	2.0
Parts by weight	103.0

TABLE 6.2(b) Series of Ink Formulas Containing Benzyl Alcohol, Antioxidants, and Corrosion Inhibitors

	Formula			
U.S. patent: 4,077,807	A	B	C	D
Benzyl alcohol	22.0	22.0	22.0	22.0
Ethylene glycol phenyl ether	22.0	22.0	22.0	22.0
Oleic acid	12.0	12.0	12.0	12.0
1,2-Propylene glycol	4.7	4.7	4.7	4.7
Blau Base KG (Solvent Blue 64, BASF)	15.9	15.9	15.9	15.9
Victoria Blue base F4R (Solvent Blue 2, BASF)	7.5	7.5	7.5	7.5
Hexane triol phthalate resin	15.9	15.9	15.9	15.9
2,2-Methylene bis(4-methyl-6-tertiary-)butylphenol (antioxidant)	0.0	2.0	0.0	2.0
Benzotriazole (corrosion inhibitor)	0.0	0.0	0.1	0.1
Parts by weight	100.0	102.0	100.1	102.1

Additional information on specific ink formulations can often be found in patents. Tables 6.2 and 6.3 depict the chemical profiles of several inks obtained from U.S. patents 4,077,807 and 4,097,290.

6.3 INK ANALYSIS

The information obtained from forensic ink examinations can be a critical component of a case, but the analysis process can be extraordinarily complex. We have chosen to generically describe modern techniques that pertain to more commonly encountered inks such as writing instruments and office machine printers (i.e., photocopiers, laser printers, and inkjet printers). Ink analysis revolves around classifying and identifying the various components of ink: colorants, vehicles, and additives. It is very rare to identify a questioned ink as coming from a specific pen or printer, so in most cases, conclusions are reached about the formulation of the ink. Being a mixture, an analytical profile for the ink formulation can be developed using an assortment of instrumental techniques on the various components.

TABLE 6.3 Ink Formulations for the Gillette Company

U.S. Patent 4,097,290	839-41A Black	814-96A Black	814-92C Red	839-41 Green	839-26C Blue
Natural rubber	22.5%	26%	29%	22.5%	22.5%
Aliphatic diluent No. 6	23.5%	—	—	23.5%	23.5%
Aliphatic solvent 360-66	—	17%	—	—	—
VM&P naphtha	—	14%	19%	—	—
Terphenyl (HB40)	30%	26%	—	—	—
DOP	—	—	—	33.5%	27%
Mineral oil	—	—	27%	—	23%
Pigment	20%	17%	15%	16.5%	4%
Stearic and lauric acid	4%	—	—	4%	
	Carbon black (0.08 μm)	Graphite (2–5 μm)	Organic Red (0.015 μm) suspended in mineral oil (35% pigment dry basis)	Phthalocyanine Green (0.015 μm)	Phthalocyanine Blue (0.015 μm) Victoria Blue 10.5% (0.025 μm)

Numerous formulations of writing inks are available to the general public, thereby greatly decreasing the frequency of occurrence of any particular ink formulation. Personal preferences (e.g., fountain, ballpoint, roller, and gel pens), changes in technological features, cost considerations, and the availability of raw materials are some examples of why so many writing inks exist. Furthermore, most ink companies spend substantial resources on research and development. As a result, many of these formulations are considered proprietary and kept secret. It is also important to recognize the significance when two inks are determined to be chemically indistinguishable or "match" each other. As outlined in Section 9.3 in the ASTM's *Standard Guide for Test Methods for Forensic Writing Ink Comparison* (ASTM E1422-05, 2005):

> When the comparison of two or more ink samples by optical or chemical analyses, or both reveals no significant, reproducible, inexplicable differences and there is significant agreement in all observable aspects of the results, it may be concluded that the ink samples match at the level of analysis and that the results indicate that the ink samples are of the same formula or two similar formulas with the same non-volatile components.

Determining the significance of a match can be further interpolated when the analyst has a reference collection for comparison as outlined in the ASTM's *Standard Guide for Writing Ink Identification* (ASTM E1789-11). The U.S. Secret Service (USSS) and the Internal Revenue Service (IRS) jointly maintain the largest known forensic collection of writing inks in the world. The collection includes more than 10500 samples of ink that date back to the 1920s and has been obtained from various manufacturers throughout the world. Pen and ink manufacturers are contacted on an annual basis and requested to submit any new formulations of inks, along with appropriate information, so that the new standards can be chemically tested and added to the reference collection. In addition, writing pens are obtained on the open market and compared with the library of standards to identify additional inks that may not have been formally submitted by a manufacturer. Maintenance of the library is a formidable task that obviously requires significant resources and is not often practical for most forensic laboratories. Consequently, the USSS generally considers offering forensic assistance (made on a case-by-case request) to law enforcement agencies requesting the analyses of writing inks. The Bavarian State Bureau of Investigation in Munich, Germany also maintains a reference collection of over 6000 writing inks (personal communication, Dec. 4, 2009).

Requests for ink analysis typically comprise three types. The first may involve comparing two or more inks to determine if the formulations match each other. This may help to ascertain if any of the written entries were added or altered. The second is to attempt to identify the writing ink on a questioned document and provide investigative information regarding the possible source of the ink. The third and most challenging request is to establish when the written entries were created to help determine if they are authentic with respect to the purported date of preparation of a document(s). This type of request, usually referred to as *ink dating*, is addressed in further detail in a separate discussion. The first two tasks can be achieved by comparing the questioned ink(s) with an adequate collection of standards, as described previously.

Prior to conducting chemical analysis, it is important to determine how the entries on a document were produced (e.g., writing instrument, inkjet printer, toner-based office machine). Without this knowledge it will be difficult to select an appropriate analytical scheme. Moreover, there are physical and optical characteristics of inks that, when in combination with the chemical properties, allow one to form an analytical profile of the ink. It is important to recognize that it is the combination of physical, optical, and chemical results that will allow one to make more accurate and objective conclusions.

6.3.1 Physical Examinations

Physical examinations include the determination of the color and type of ink. Determining the type of ink (e.g., ballpoint, felt tip, fountain pen) often requires training and experience in the evaluation of morphological characteristics using a stereomicroscope (approximately 5× to 50×). It is important to establish the type of writing instrument used to make a questioned entry, as this dictates the extraction solvent necessary for a chemical ink examination. There are various types of writing instruments and writing inks, but generally, inks are separated into three categories: nonaqueous or glycol-based (ballpoint), fluid (porous, felt or fiber tips, gel, and roller pens), and fountain pens. Definitive microscopic differentiation of these three categories of inks is possible in most cases due to their unique morphological characteristics on a substrate. Although there are a variety of writing instruments that utilize fluid inks (e.g., gel pens, roller pens, felt-tip markers), their physical properties are not always definitively discernible. In addition to class characteristics, individual characteristics may be identified, thereby making the microscopic examination a necessary part of the procedure. As an example, Figure 6.2 illustrates a damaged ballpoint casing and the microscopic scratching that appears on

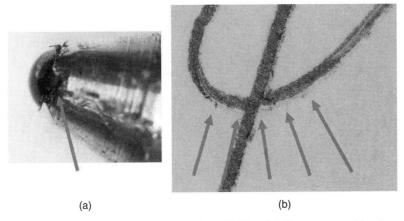

(a)   (b)

Figure 6.2 (a) Damaged ballpoint pen casing. (b) Visual defects generated by damaged writing instrument. *(See insert for color representation.)*

the paper. Such damage to the writing tip may be used to help associate a suspect writing instrument with the questioned document.

Morphological Characteristics in the Written Line When examining a written line, the first step is to make a distinction between absorbent and nonabsorbent (or semiabsorbent) ink. Inks that are water based will absorb into the paper fibers, providing nearly uniform coverage of the substrate over the width of the stroke. Semiabsorbent inks such as ballpoint inks generally do not penetrate the fibers as much but merely rest on top of the paper fibers.

Hesitations, or pauses in the flow of writing, are also indicative of a particular type of writing mechanism. While related to vehicle composition, the absorbing inks tend to saturate and bleed into surrounding paper fibers when the pen is left in contact with the paper for a period of time. To a lesser extent, gel pens exhibit some bleeding, while ballpoint pens yield no bleed when left in contact with a piece of paper for an extended period of time. Figure 6.3 shows ballpoint ink, which exhibits no bleed, and a gel ink with minimal bleed. Both the felt-tipped and roller ball inks bleed significantly on paper.

Strokes of each type of writing instrument are shown in Figure 6.4. Both ballpoint ink, and to a lesser extent gel inks, fail to provide complete coverage of the paper fibers because they are semi- or nonabsorbent. Roller ball inks and felt-tipped (porous) inks are absorbent and therefore fully penetrate most porous substrates, creating uniform coverage.

Additional characteristics observed in the written line can further distinguish writing instruments. For example, fountain pens are often noted for their shading and, when the ink is in short supply, for the appearance of two lines. When there is insufficient ink, the two separate nibs will appear on the substrate as two parallel lines. This phenomenon can also be observed if significant pressure is applied to the fountain pen nib.

Ballpoint inks tend to have a sheen and provide incomplete inking across the paper fibers. Gooping, the release of a large volume of ink in a small area, is unique to ballpoint inks. Striations are voids in the inking that are a result of the ball not rolling through the ink supply prior to being rolled over the paper. Striations that

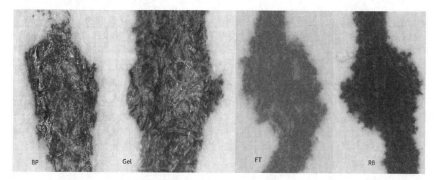

Figure 6.3 Characteristics of the written line from various writing instruments. *(See insert for color representation.)*

INK ANALYSIS   **235**

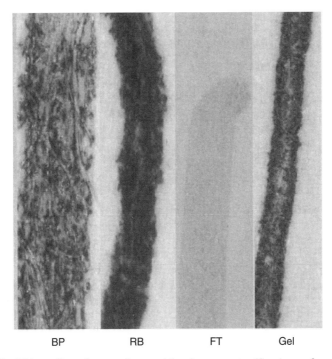

BP   RB   FT   Gel

Figure 6.4 Written lines from various writing instruments. *(See insert for color representation.)*

result from the ball are seen only in ballpoint inks because they do not absorb like other ball mechanism inks (gel, roller ball). The ballpoint mechanism typically leaves a writing groove in the substrate.

Non-ballpoint (porous-tipped) inks are noted for their complete coverage across the width of the inked line. No grooves in the substrate are found and the ink tends to bleed if there is hesitation. There are instances where a felt-tipped pen may become degraded and stray fibers on the tip will leave stray markings off the main inked line.

Gel pens, which utilize a ballpoint mechanism, tend to have heavy saturation at the edge of the written line and less saturation in the center. Since little pressure is needed to apply ink to the substrate, the writing groove present is typically not as deep as one would expect to find from a ballpoint pen. Neither gooping nor striations are visualized.

Writing instruments can often leave information that is pertinent to a handwriting examination. Determining the direction of stroke is a valuable forensic examination. The ink saturation on gel pens, porous-tipped pens, steel pens, and fountain pens makes determination of the direction of stroke challenging. Fountain and steel pens may, in some instances, indicate a direction of stroke, as writers tend to apply more pressure on the downstroke (Hilton, 1982). As pressure is applied to the nib-tipped pens, the nibs separate and the one visible line may become two. The striations in

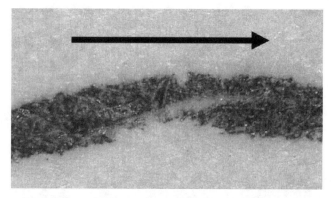

Figure 6.5 Ballpoint pen striations. *(See insert for color representation.)*

ballpoint pens are the best indicator for the direction of stroke. When a ballpoint pen is rotated around the curve of a letter, it often leaves striations on the paper where the inking is incomplete. The striation works its way from the inside of the curve outward. In effect, the thinnest part of the "top" of the curve is always pointing in the direction of the stroke. Figure 6.5 illustrates the direction of stroke of a ballpoint pen. Table 6.4 lists the various physical characteristics that can be used to identify the type of writing instrument.

6.3.2 Optical Examinations

An optical examination, also referred to as a filtered light examination (FLE) or video spectral analysis (VSA), can provide valuable insight regarding the overall composition of an ink. The presence of colorants and other materials will directly affect the manner in which an ink absorbs, reflects, and transmits electromagnetic radiation. Ultraviolet (UV) and infrared (IR) radiation are forms of energy that can be used to examine inks on a document. Near-infrared reflectance (IRR) and infrared luminescence (IRL) properties can help significantly when evaluating properties of an ink. Instruments designed specifically for the purpose of a filtered light examination are best suited for document analysis. Foster & Freeman, Projectina, and other manufacturers provide equipment suited for such analyses.

TABLE 6.4 Summary of the Various Characteristics Typically Found in Writing Inks

	Gooping?	Complete Inking?	Writing Groove?	Ink Bleed?
Fountain pen	—	Some	Yes	Yes
Ballpoint pen	Yes	—	Yes	—
Porous pen	—	Yes	—	Yes
Roller ball pen	—	Yes	Yes	Yes
Gel pen	—	Some	Yes	Some

Drexler and Smith (2002) have described a self-made apparatus that can be manufactured using off-the-shelf equipment, and Richards (1977) provides a basic model for the application of infrared and ultraviolet techniques. Modern commercial equipment typically incorporates the use of a charge-coupled device (CCD), providing users with the opportunity to examine and save images electronically. When illuminating a sample, there are two ways in which the sample can give off energy. The first involves the energy from the illuminant being reflected to the detector. The second way to detect energy from a sample is through luminescence. Luminescence is the absorption of energy at one wavelength and the reradiation of energy at another, typically longer, wavelength. The greatest advantage in examining the luminescence of a sample is the ability to filter out the incident energy on the sample. By filtering out the illuminant, the detector displays only energy that has been reradiated at a longer wavelength. When ink examinations are conducted, both IRL and IRR properties of the ink are recorded. While IRR and IRL properties are often used to determine the regions of interest on a document, they are rarely used to eliminate inks under comparison. Inks that exhibit varying IRR and IRL properties generally contain dyes in the formulation. Typically, dyes do not absorb IR radiation, whereas pigments do. Therefore, pigments will appear black at long wavelengths, whereas dyes often seem to disappear (transmit) in the IR region. When dyes become transparent in the IR region, the underlying substrate often reflects the incident energy, making it seem as though the ink has disappeared. It is important to note that a FLE can be completed before or after chemical analysis is conducted on the ink components. Since ink is a complex combination of chemicals, competing colorants can often reabsorb IRL from other colorants. Some components exhibiting IRL may be visible only after they are separated using chemical analysis, as other components can mask the luminescence. Figure 6.6 demonstrates two inks used for a check alteration with different IRL properties.

An evaluation using optical methods should always be considered, especially when the request is to compare two or more entries, or when evaluating multiple entries on a page. Often, VSA will be used to evaluate if the writing ink on a page is homogeneous. Determining homogeneity may be necessary if further chemical

Figure 6.6 Altered check created using two inks with different IRL properties.

testing is warranted. It is not practical to sample and chemically analyze all entries in most cases, so VSA is used as a screening tool. Additionally, it may be possible to detect if there have been any changes to the document or written entries, such as addition of solvents or materials, alterations, insertions, obliterations, and deletions using a FLE. It is critical to emphasize that if two inks cannot be differentiated via their IRR, IRL, and UV properties, this is not conclusive evidence that they are a match as described in the ASTM's *Standard Guide for Test Methods for Forensic Writing Ink Comparison* (ASTM E1422-05, 2005). It is not uncommon that different inks will exhibit similar or even identical optical properties. Furthermore, when performing such examinations, caution must be taken in rendering conclusions if the evaluations are made on different substrates. The substrate can have an overwhelming effect by masking or absorbing the energy that is reradiated from the ink.

6.3.3 Chemical Examinations

Determining the chemical components present in a writing ink is a critical piece for developing an overall profile. The first stage of analysis typically involves extraction of the ink into the appropriate solution. Often, inks with different extractabilities will have different components. Extracting the same inks that have undergone different aging processes, were stored in different environmental conditions, or were found on different substrates may yield different extractabilities. It is important for an analyst to evaluate ink extractability in order to verify that the ink colorants are being extracted appropriately. Occasionally, some non-ballpoint inks may not extract in an aqueous solvent mixture. These are usually pigment-based fluid inks.

Typically, ballpoint inks are extracted with organic solvents such as pyridine, methanol, or acetonitrile. Fluid inks can be extracted with a mixture of ethanol and water (1 : 1), but this is not always the ideal extracting solvent since some fluid inks are non-aqueous-based (e.g., permanent markers). Recently unpublished research suggests that a tetrahydrofuran (THF) and water mixture (4 : 1) may be suitable for both ballpoint and fluid inks. In some instances, inks may have to be extracted in other solvents, depending on the formulation. While it is necessary to maintain a single solvent or solvent system across a reference collection for searching purposes, there are numerous solvents that can be utilized when conducting an intraplate comparison. Figure 6.7 shows the variations in dyes that are extracted based on the assortment of extraction solvents that were used.

Thin-Layer Chromatography Thin-layer chromatography (TLC) is one of the most widely used and generally accepted scientific methodologies employed to compare and help characterize ink formulations. TLC is an effective and efficient method for separating and identifying colorants. For example, two or more questioned inks can be compared to determine if they are the same, or questioned inks can be associated to a known standard. However, it must be emphasized that TLC is only one portion of an analytical scheme, and the "profile" of an ink is achieved only by using the results from a series of physical, optical, and chemical

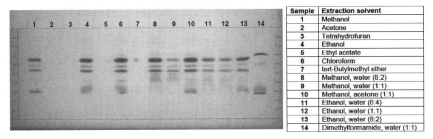

Figure 6.7 Different banding patterns based on changes to extraction solvent. *(See insert for color representation.)*

examinations. In 1999, Roux et al. conducted studies on black and blue ballpoint inks using a FLE, TLC, and reflectance visible microspectrophotometry (MSP). The authors concluded that "the power of the individual techniques to discriminate inks between and within brands, models and batches varied, the most informative techniques being TLC > FLE > MSP." They showed that filtered light examinations of blue inks ($n = 49$) and black inks ($n = 42$) resulted in a discriminating power (DP) of 0.83 and 0.96, respectively. After conducting TLC and in conjunction with FLE, the DP was 0.98 for blue and 0.99 for black, where

$$\text{DP} = \frac{\text{number of discriminated pairs}}{\text{number of possible pairs}}$$

TLC for writing inks has been discussed extensively by Brunelle and Crawford (2003), Witte (1963), Brunelle and Pro (1972), Brunelle and Reed (1984), Kelly and Cantu (1975), and Aginsky (1993a). TLC analysis begins by removing an ink sample from a document and subsequently extracting the ink in an appropriate solvent. The extract is then applied to a silica-coated TLC plate (glass or plastic backing) and placed in a solvent-equilibrated glass chamber containing a solvent or mixture of solvents. A mixture of solvents, often referred to as a solvent system, is often employed to gain timely separation of ink components across the retardation factor range. The sample components then migrate up the plate via capillary action. Typically, the colorants (e.g., dye components) that are present in the ink sample will separate into colored bands or spots.

Other Instrumentation There are numerous research papers regarding the use of instrumental methodologies for the analysis of ink that have been published. The use of gas chromatography–mass spectrometry (GC–MS) and high-performance liquid chromatography (HPLC) for the characterization and/or dating of inks has been discussed by Tebbett et al. (1992), Aginsky (1994, 1996), Andrasko (2001a), Gaudreau and Brazeau (2002), Hofer (2004), LaPorte et al. (2004), Wilson et al. (2004), and Bügler et al. (2005). In addition, spectroscopic techniques such as Fourier transform–infrared (FT–IR) and Raman spectroscopy have been studied for the characterization of inks by Humecki (1985), Merrill and Bartick (1992), and Andermann and Neri (1998). All of these instrumental methods may help further delineate an ink formulation into its individual components, but it is not always necessary. One must take into consideration the environmental changes that can significantly affect the noncolorant components over a period of time immediately following their application to the paper. In cases where a standard reference collection is available, an ink may be determined to match a standard from a database to the exclusion of all others, and further discrimination may not be warranted.

It is critical to understand some fundamental concepts regarding the analysis of inks submitted for forensic examination. First, many inks are composed of volatile and semivolatile components that will evaporate as an ink ages. Therefore, some components that are present in a fresh sample of ink may not be detectable in an aged sample using GC–MS. Although obtaining a supplemental volatile profile may increase the degree of discrimination, limitations may include solvent loss over time or other external factors, such as exposure to high temperatures, light, and/or humidity. Therefore, caution is warranted when making interpretations of qualitative, and especially quantitative, data from instrumental techniques. In addition, there are some commonly utilized dyes that can photodecompose. Triarylmethanes (e.g., methyl violet) are one such group of dyes. They undergo a photochemical reaction that can be "influenced by the concentration of the reactants, the reaction medium, the temperature, the wavelength and intensity of the light" (Weyermann, 2005). Second, when inks are applied to a substrate such as paper, there can be numerous interfering components in the paper that can affect data interpretation. Modern papers can contain optical brighteners, resins, and fillers that can inhibit the analysis of inks. Many inks absorb into the substrate, making FT–IR a difficult methodology, due to the significant contribution of some paper components to IR spectra. Even with spectral subtraction of a paper blank, there may still be an overwhelming effect from some paper ingredients. However, these matrix challenges do not preclude use of the aforementioned technologies. Rather, the implication is that the examiner must consider these variables and the type of request when choosing the appropriate instrumental technique.

In summary, developing a single standardized analytical approach to ink analysis can be challenging, but physical, optical, and TLC examinations should be routine. GC–MS, HPLC, and FT–IR should be considered, but will be dependent on the type of ink, the necessity for further discrimination, and other practical considerations, such as the age of the document.

6.3.4 Ink Dating

Cantu (1995, 1996) has outlined analytical approaches for determining the age of an ink on a questioned document through both static and dynamic methods. The static approach to ink dating generally applies to methods that are based on comparisons with a standard reference collection of inks to determine the first date of production. Numerous ink formulations have been introduced commercially in the past several decades. With the appropriate information and documentation acquired from an ink manufacturer, a database can be used to render more reliable conclusions.

The dynamic approach includes methods that incorporate procedures for the purpose of measuring the physical and/or chemical properties of an ink that change with time. The changes that occur over a given period of time can generally be referred to as *aging characteristics*. Different approaches to measuring the age of an ink once it has been placed on a document have been discussed in the literature by Gaudreau and Brazeau (2002), Hofer (2004), Humecki (1985), Cantu (1987), Brunelle (1992, 1995), Aginsky (1993a, 1998), Andermann and Neri (1998), Brazeau et al. (2000), and Andrasko (2001b, 2003).

When an ink is applied to a substrate and begins "drying," it undergoes numerous changes due to solvent evaporation, oxidation, polymerization, and/or photodecomposition. These processes can change the chemical composition of the ink on the substrate. Given the diversity of inks, substrates, and environmental exposure factors, the complexity of determining what process or processes directly affect the rate at which the ink dries is extremely difficult. The real challenge lies in finding a suitable analytical procedure to isolate the appropriate component for analysis. Brunelle and Crawford (2003) provide a comprehensive review of various methodologies that have been attempted; however, to date, no single ink-dating procedure that has been found to be reliable for all types of inks.

The use of GC–MS for volatile analysis of ballpoint inks in an attempt to determine the age of inks on paper has been studied and reviewed in the literature for more than a decade. Much of the research on volatiles has focused on the analysis of 2-phenoxyethanol (2-PE; also referred to as ethylene glycol monophenyl ether, 1-hydroxy-2-phenoxyethane, β-hydroxyethylphenyl ether, Dowanol EP, and Phenyl Cellosolve). 2-PE is a volatile organic compound commonly used as a principal vehicle in the majority of ballpoint writing inks. Furthermore, it is used in many ink formulations because it is stable in the presence of acids and alkalis. It is nonhygroscopic, nonhazardous, economical, and especially good at dissolving resins and nigrosine (a common solvent black dye used in writing ink formulations). In 1985, Stewart conducted a preliminary study on ballpoint ink aging and the relationship to volatile organic compounds. Beshanishvily et al. (1990) were the first to discuss the identification of 2-PE as it relates to the aging of inks. Since then, Aginsky (1993b) reported that "... significant aging [takes] place over a period of about three months. After this period until the age of fifteen years the extent of the extraction of the volatile component (phenoxyethanol) from the ink entries has been kept at a level of about 20%." Aginsky also describes the ink drying process and surmises that volatile components stop evaporating from a dried sample of ink unless they are freed by heating or a solvent extraction. In 2008, Bügler et al. published a procedure for assessing the age of some ballpoint pen inks using GC–MS coupled with thermal desorption to evaluate the quantity of 2-PE.

Ideally, *ink tags* would be the most reliable method for the dating of inks. Tags can be added to formulations in the form of fluorescent compounds or rare earth elements, and were incorporated into some formulations beginning about 1970 until 1994. Factors have precluded some ink manufacturers from participating in such a program, including, but not limited to, insufficient resources, low priority, and/or disagreement about the type of tag utilized. This is not to say that a widespread tagging agenda is unachievable. On the contrary, efforts continue to convince ink companies to add tags to their formulations. As recently as November 2002, a major ink manufacturer began adding tags to its ink in collaboration with the U.S. Secret Service.

6.4 OFFICE MACHINE SYSTEMS

For the purpose of this discussion, we have chosen to limit the discussion of office machine systems to those that utilize inkjet or toner technology because they are the most commonly encountered printing and copying devices accessible to the general public. Although documents produced via typewriting and dot matrix are not as ubiquitous as some of the other office machine systems mentioned, this does not preclude their importance in some investigations. Occasionally, physical print defects from typewriters, dot matrix, and thermal printers may arise in the printed material and should be examined carefully. Individual and class characteristics from these machines can be important identifying features.

6.4.1 Inkjet Ink

Formulating an inkjet ink to create a high-quality image is an obviously complex task. Inkjet inks for small office/home office (SOHO) machines can have very complex formulations to create acceptable images. In fundamental terms, a well-designed inkjet ink should be able to withstand high heat, dry quickly so as not to smear, have nearly photographic quality, and be light- and water-fast. Since inkjet-printed documents comprise a significant portion of all documents created in the SOHO market and ink formulations are complex, manufacturers utilize a great deal of their resources on the research and development of inks.

Inkjet ink poses several differences as compared to writing inks. Writing ink colorants are mixed together in a single ink formulation to provide a single colored line. Color inkjet, however, is a combination of multiple single-color ink formulations, and each ink formulation can contain one or more colorants. Currently, many SOHO inkjet office machine systems comprise four color printers, incorporating cyan, magenta, yellow, and black (CMYK) into a single print head. The colors can come from either a three-reservoir tricolor cartridge or three single-reservoir cartridges. Typically, the black ink formulation is independent of the colors. Small droplets of ink are jetted onto the substrate to produce a broad range of colors. Figure 6.8 shows microscopic images of inkjet printing on plastic.

Inkjet ink manufacturers typically use acid, direct, and reactive dyes in their formulations. Acid dyes (e.g., azo, anthraquinone, triphenylmethane, azine, and xanthene) are water-soluble anionic dyes that have a wide gamut of hues. Direct dyes (e.g., phthalocyanines) have a variety of bright hues, some of which are fluorescent, as well as possessing relatively good light and water fastness properties. Within the past decade, reactive dyes have been added to some formulations because they contain desired chromophores. This class of dyes link very strongly with

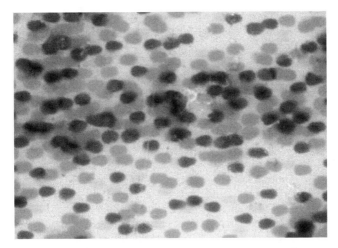

Figure 6.8 Inkjet droplets on a plastic substrate. *(See insert for color representation.)*

cellulosic fibers when heated and they are found in some magentas. Understandably, manufacturers keep their formulations proprietary, which can result in more diverse formulations and thus in greater potential for forensic discrimination. Table 6.5 is a list of various ingredients that are generally used in inkjet ink formulations.

6.4.2 Inkjet Ink Analysis

Many of the same forensic examinations that can be conducted on writing inks may also be completed for inkjet inks. Inkjet inks are typically treated as non-ballpoint pen inks, as they often share similar ink solvents and characteristics. Pagano et al. (2000) presented methodologies for the analysis of jet inks and the different chromatographic systems utilized. Furthermore, LaPorte and Ramotoski (2003) discuss how color inks vary from one manufacturer to the next, making it possible to compare with known standards. LaPorte (2004) also discusses a multifaceted approach to the examination of inkjet printed documents that encompasses physical, optical, and chemical examinations. The U.S. Secret Service maintains the largest known forensic collection of inkjet standards in the world. This collection is composed of print and inkjet ink samples for hundreds of printers and cartridges.

Physical examinations are performed on a printed document to determine the printing process employed. Using at least 10× magnification, inkjet printing appears as a series of irregularly spaced colored dots with some peripheral bleeding. As with writing inks, a FLE is often conducted on inkjet inks. As part of the optical examination, properties of IRR and IRL are noted. Some inkjet inks are formulated using dye-based colorants while others utilize pigments, either in combination or as a replacement to dyes. The selection of colorant (pigment or dye) affects the IRR and IRL properties that are exhibited by the ink. The FDE should exercise caution when evaluating fluorescence and/or absorbance characteristics since inkjet inks may be quenched or enhanced on different types of paper. Optical examinations

TABLE 6.5 Some Common Ingredients That May Be Found in an Inkjet Formulation, Their Purpose, and Approximate Percentage of the Total Weight

Inkjet Ingredient	Purpose	Composition (wt%)
Colorant	Provides the desired color and aesthetic properties	2–8
Solvent	Vehicle that dissolves or suspends the colorant (e.g., water, alcohol)	35–80
Humectant	Inhibits evaporation (e.g., diethylene glycol)	10–30
Surfactant	Lowers the surface tension of the ink to promote wetting (e.g., Surfynol 465)	0.1–2.0
Penetrant	Allows ink to penetrate fibers and promote ambient drying (e.g., pentane-1,5-diol)	1–5
Viscosity modifier	Added to increase viscosity and can be a humectant (e.g., long-chain glycols)	1–3
Dye solubilizer	Helps maintain dye solubility in the solvent because of evaporation in the nozzle (e.g., n-methylpyrrolidone)	2–5
Dispersant	Maintains pigment suspensions [e.g., Derussol carbon black (Degussa)]	3–8
Fixative	Helps maintain image permanence and smear resistance (e.g., water-soluble latex)	1–3
pH buffer	Maintains overall pH to promote colorant stability in the vehicle [e.g., Trizma base (Aldrich Chemical)]	0.1–1.0
Chelation agent	Used to complex metal ions (e.g., ethylenediaminetetraacetic acid)	0.1–0.5
Biocide	Kills bacterial growth in aqueous inks [e.g., Dowicil (Dow Chemical)]	0.1–0.3
UV blocker	Increase lightfastness [e.g., Tinuvin 171 (Ciba-Geigy)]	1–5

can sometimes demonstrate that two or more different inkjet formulations were used to prepare a document(s), but this does not necessarily mean that multiple printers were utilized. It is possible for a single printer to use more that one type of inkjet cartridge or for a cartridge to be refilled with a different ink. Some companies, other than the original manufacturer of an office machine, produce compatible cartridges and ink for aftermarket use that are chemically different from those of the original equipment manufacturer (OEM).

Forensic analysis of inkjet inks centers around TLC. In addition to the physical and optical examinations, chemical analysis is an efficient analytical method for classifying inkjet inks. It is important to note that inkjet individualization based on colorant combinations is extremely difficult, due to the interchangeability of inkjet cartridges. Manufacturers often utilize the same dye combinations through several designs of ink cartridges. Given the complexity of the ink, gross changes in formulation are rare. Investigative research on inkjet inks has shown that plastic-backed

TLC plates seem to be as, if not more, efficient at separating the colorants over a fixed distance. Therefore, most TLC comparisons of inkjet inks are completed on plastic-backed TLC plates. It is critical that TLC plates be visualized with an UV source, minimally short wave (254 nm) and long wave (366 nm), since some inkjet inks incorporate proprietary rhodamine dyes.

TLC is only one of several chemical techniques used for examining inkjet inks. However, there has been limited research focusing on other methods of instrumental analyses. Mazella (1999) used diode array MSP to measure the cyan and magenta components in the range 380 to 760 nm on 70 different brands and models of inkjet printers. Doherty (1998) performed a battery of chemical tests, including tests for waterfastness, solubility, and TLC, on documents produced with various inkjet printers. To date, this has been the most comprehensive study on the examination and classification of office machine systems that utilize inkjet technology.

6.4.3 Toner Printing

Laser printers and photocopiers commonly utilize a resinous, powderlike material referred to as *toner*, which is fused to the paper using heat and/or pressure. Toners consist primarily of organic resins, polymers, and colorants such as carbon black, nigrosines, phthalocyanines, azo-pigments, and quinacrodone (Pagano, 2000). Although manufacturers have a selection of resins to choose from, styrene acrylate copolymer and polyester resins are frequently used. Liquid toners, which are often formulated with a hydrocarbon carrier solvent to deliver extremely fine particles to the substrate surface before the liquid evaporates, can also be utilized. Particle-size ranges are approximately 1 to 3 μm for liquid toner and from 6 to 18 μm for dry toner. The size of the toner particle obviously affects the maximum print resolution.

Laser printers and photocopiers function similarly, such that each contains a rotating drum with a photoconductive coating. The surface of the drum is illuminated by a light source controlled by a series of signals based on the image, causing the exposed areas to be charged. As the drum rotates, oppositely charged toner is deposited, thus adhering to the illuminated areas that have been charged with a positive potential. When the paper passes through the machine, an electrical current imparts a static electrical charge onto the paper. The paper then passes through a series of rollers and comes in contact with the rolling drum, causing the toner to be transferred to the paper, resulting in the creation of text or image(s). Since the nonimage areas of the drum are not charged, they do not attract oppositely charged toner and result in white areas on the paper. Afterward, heat and/or pressure is applied to fuse the toner permanently. For a multicolored process, the same steps are repeated for each color component (i.e., cyan, magenta, and yellow).

Two different types of toner are currently produced: conventional toner and chemically prepared toners (CPTs) (Figure 6.9). Conventional toners are created by mixing the colorant(s) into a large batch of polymer. The fine toner powder is then broken away from the large homogeneous mass. CPTs, on the other hand, are grown chemically from small particles. CPTs have gained favor in recent years

Figure 6.9 (a) Conventional toner. (b) Chemically prepared toner. *(See insert for color representation.)*

due to their more uniform particle-size distributions and shapes. The CPT process allows smaller toner particles to be generated, affording CPTs higher resolutions than conventional toners.

Black toners can be differentiated further based on whether they are mono or dual component. Monocomponent toners incorporate the colorant and the resin into the same large particle. To do so, Fe_xO_y is often used, giving the monocomponent toner a magnetic property. Dual-component toners are typically composed of smaller toner particles and larger developer particles. Although some dual-component particles have magnetic properties, they are not as strong as monocomponent toners. Since dual-component toners have smaller toner particle sizes, they provide better resolution.

6.4.4 Toner Analysis

The analytical scheme for analyzing toners will depend on whether an extensive library is available or if the request is simply to compare a questioned document with other documents or a known printing device. At this time, we are not aware of an extensive collection of black toners that are used for law enforcement purposes, but this does not preclude the possibility that a laboratory has its own internal collection. Various instrumental approaches can be used to analyze toners, and potentially to create a collection of standards, such as pyrolysis GC–MS (Munson, 1989), FT–IR (Andrasko, 1994), x-ray fluorescence (XRF), and SEM–energy-dispersive spectroscopy (SEM–EDS). TLC can also be used to analyze toners and may prove beneficial when conducting comparisons between multiple documents or suspect office machines. Toners can be extracted with chloroform to characterize extractable colorants from the CMY and black components. Given that a number of black toners are carbon black, the forensic value of TLC analysis of black toner can be minimal; however, there are other dyes used in combination with carbon black that may be visualized on a TLC plate. It is highly recommended to visualize developed TLC plates using an UV source. Most important, the results can be used to corroborate other tests and findings.

TLC can be followed with additional methods, but it is critical to understand that the analytical procedures described are not sufficient to individualize a specific printer or copier. It is possible that a manufacturer, or even model, may be distinguished; however, the results are only good enough for discrimination at the class level. XRF analysis can be useful in providing the elemental composition of the toner. The presence and percentage concentration of iron is often a distinguishing factor.

Determining that a toner is iron based (monocomponent) can also be of forensic value, and an excellent area of further research is investigating if quantitative elemental analysis (e.g., Fe) can be used to discriminate toners from different manufacturers or models from the same manufacturer. Scanning electron microscopy (SEM) is also a useful analytical technique, as it provides the high magnification necessary for examining fusing characteristics. Additionally, the method of production used in making the toner can be surmised based on quality images of particle morphology. Given that CPT toners have seen significant use only in the past decade or so, determining that a toner is a CPT is an excellent method for roughly dating a document.

Brandi et al. (1997) offer an extensive discussion and study of toner analysis using diffuse reflectance infrared Fourier transform spectroscopy (DRIFTS), SEM–EDS, and pyrolysis gas chromatography (PyGC). They concluded that DRIFTS was highly discriminating, sensitive, and reproducible after 55 different toners were classified into 15 distinct classes. SEM–EDX was also used to analyze the inorganic contents of 50 toners, subsequently placed into 22 separate classes. The best approach was the combination of the aforementioned techniques, resulting in the characterization of 30 distinct groups of toners. Finally, the authors concluded that "PyGC, with the equipment available, did not appear to provide any greater differentiating power than infrared spectroscopy, gave poor reproducibility, and was a time consuming technique." However, they do acknowledge that a better pyrolysis technique and the use of a mass spectrometer would result in a more discriminating method.

CONCLUSION

The chemical analysis of questioned documents has been relied upon in forensic laboratories for over 50 years. TLC has proven to be a reliable method for the discrimination and identification of some writing and printing inks, especially when used in conjunction with a reference collection. GC–MS is an excellent supplemental procedure and may provide additional information about other components, but the interpretation of the chromatographic and spectral data must be approached with great caution. Inks undergo a drying process that can take place over several months or, possibly, years. Moreover, inks have variable compositions, making the drying processes of different inks nonuniform. This property, in combination with the unknown ink–paper interactions when combined with different substrates and not knowing the storage conditions of a questioned document, result in significant challenges when attempting to characterize and date inks using GC–MS. HPLC and FT–IR can provide additional molecular information, but there has been limited research in the areas of dating inks and characterizing non-ballpoint and printing inks via these methods. Office machines that utilize inkjet and toner technology are ubiquitous and commonly the focus of investigations involving a multitude of crimes. The approach to chemical analyzing inkjets and toners is similar to procedures used for writing inks such that physical, microscopic, and TLC examinations should be a minimum. The gamut of instrumental approaches that follow can vary depending on previous results and the extent of discrimination needed to formulate a reliable conclusion, but it must be recognized that in nearly all cases, the data obtained will not support a conclusion that identifies a particular writing instrument or printing device. Although the information that can be attained from printing and writing inks is based on reliable and generally acceptable analytical methods, it is critical that any person conducting the tests described in this chapter have sufficient training in the physical and microscopic examination of documents. The results from the physical exams help determine how to proceed with chemical methods of analysis, but more important, the combination of all results is necessary to draw relevant and useful conclusions.

REFERENCES

Aginsky, V. N. (1993a). Forensic examination of "slightly soluble" ink pigments using thin-layer chromatography. *J. Forensic Sci.,* **38**:1131–1133.

Aginsky, V. N. (1993b). Some new ideas for dating ballpoint inks: a feasibility study. *J. Forensic Sci.,* **38(5)**:134–150.

Aginsky, V. N. (1994). Determination of the age of ballpoint ink by gas and densitometric thin-layer chromatography. *J. Chromatogr. A,* **678**:117–125.

Aginsky, V. N. (1996). Dating and characterizing writing, stamp pad and jet printer inks by gas chromatography/mass spectrometry. *Int. J. Forensic Doc. Examiners,* **2(2)**:103–115.

Aginsky, V. N. (1998). Measuring ink extractability as a function of age: why the relative aging approach is unreliable and why it is more correct to measure ink volatile components than dyes. *Int. J. Forensic Doc. Examiners*, **4(3)**:214–230.

Andermann, T., and Neri, R. (1998). Solvent extraction techniques: possibilities for dating ballpoint pen inks. *Int. J. Forensic Doc. Examiners*, **4(3)**:231–239.

Andrasko, J. (1994). A simple method for sampling photocopy toners for examination by microreflectance Fourier transform infrared spectrophotometry. *J. Forensic Sci.,* **39(1)**:226–230.

Andrasko, J. (2001a). HPLC analysis of ballpoint pen inks stored at different light conditions. *J. Forensic Sci.,* **46**:21–30.

Andrasko, J. (2001b). Changes in composition of ballpoint pen inks on ageing in darkness. *J. Forensic Sci.,* **47(2)**:324–327.

Andrasko, J. (2003). A simple method for distinguishing between fresh and old ballpoint pen ink entries. *Proceedings of the 3rd Annual European Academy of Forensic Science Meeting*, Istanbul, Turkey, Sept. 22–27.

Beshanishvily, G. S., Trosman, E. A., Dallakian, P. B., and Voskerchian, G. P. (1990). Ballpoint ink age: a new approach. *Proceedings of the 12th International Forensic Scientists Symposium*, Adelaide, Australia, Oct. 15–19.

Brandi, J., James, B., and Gutowski, S. J. (1997). Differentiation and classification of photocopier toners. *Int. J. Forensic Doc. Examiners*, **3(4)**:324–343.

Brazeau, L., Chauhan, M., and Gaudreau, M. (2000). The use of micro-extraction in the development of a method to determine the aging characteristics of inks. *Proceedings of the 58th Annual Conference of the American Society of Questioned Document Examiners*, Ottawa, Ontario, Canada, Aug. 24–29.

Brunelle, R. L. (1992). Ink dating: the state of the art. *J. Forensic Sci.*, **37**:113–124.

Brunelle, R. L. (1995). A sequential multiple approach to determining the relative age of writing inks. *Int. J. Forensic Doc. Examiners*, **1(2)**:94–98.

Brunelle, R. L., and Crawford, K. R. (2003). *Advances in the Forensic Analysis and Dating of Writing Ink*. Springfield, IL: Charles C Thomas.

Brunelle, R. L., and Pro, M. J. (1972). A systematic approach to ink identification. *J. Assoc. Off. Anal. Chem.*, **55**:823–826.

Brunelle, R. L., and Reed, R. (1984). *Forensic Examination of Ink and Paper*. Springfield, IL: Charles C Thomas.

Bügler, J. H., Buchner, H., and Dallmayer, A. (2005). Characterization of ballpoint pen inks by thermal desorption and gas chromatography–mass spectrometry. *J. Forensic Sci.*, **50(5)**:1209–1214.

Bügler, J. H., Buchner, H., and Dallmayer, A. (2008). Age determination of ballpoint ink by thermal desorption and gas chromatography–mass spectrometry. *J. Forensic Sci.*, **53(4)**:982–988.

Cantu, A. A. (1987). On the relative aging of ink: the solvent extraction technique. *J. Forensic Sci.*, **32(5)**:40–51.

Cantu, A. A. (1995). A sketch of analytical methods for document dating: I. The static approach—determining age independent analytical profiles. *Int. J. Forensic Doc. Examiners*, **1(1)**:40–51.

Cantu, A. A. (1996). A sketch of analytical methods for document dating: II. The dynamic approach—determining age dependent analytical profiles. *Int. J. Forensic Doc. Examiners*, **2(3)**:192–208.

Doherty, P. (1998). Classification of ink jet printers and inks. *J. Am. Soc. Questioned Document Examiners,* **1**:88–106.

Drexler, S. G., and Smith, G. (2002). Ink differentiation for the fiscally challenged. *J. Am. Soc. Questioned Document Examiners*, **5(2)**:20–27.

Gaudreau, M., and Brazeau, L. (2002). Ink dating using a solvent loss ratio method. *Proceedings of the 60th Annual Conference of the American Society of Questioned Document Examiners*, San Diego, CA, Aug. 14–18.

Hilton, O. (1982). *Scientific Examination of Questioned Documents*, rev. ed. Washington, DC: CRC Press.

Hofer, R. (2004). Dating of ballpoint pen ink. *J. Forensic Sci.*, **49(6)**:1353–1357.

Humecki, H. J. (1985). Experiments in ballpoint ink aging using infrared spectroscopy. *Proceedings of the International Symposium on Questioned Documents*, FBI Academy, Quantico, VA, pp. 131–135.

Kelly, J. D., and Cantu, A. A. (1975). Proposed standard methods for ink identification. *J. Assoc. Off. Anal. Chem.,* **58**:122–125.

LaPorte, G. M. (2004). Modern approaches to the forensic analysis of inkjet printing: physical and chemical examinations. *J. Am. Soc. Questioned Document Examiners,* **7(1)**:22–36.

LaPorte, G. M., and Ramotowski, R. S. (2003). The effects of latent print processing on questioned documents produced by office machine systems utilizing inkjet technology and toner. *J. Forensic Sci.,* **48(3)**:1–6.

LaPorte, G. M., Wilson, J. D., Cantu, A. A., Mancke, S. A., and Fortunato, S. L. (2004). The identification of 2-phenoxyethanol in ballpoint inks using gas chromatography/mass spectrometry: relevance to ink dating. *J. Forensic Sci.,* **49(1)**:155–159.

Lindblom, B. S. (2006). What is forensic document examination? In: Seaman Kelly, J., and Lindblom, B. S. (Eds.), *Scientific Examination of Questioned Documents*. Boca Raton, FL: CRC Press, Taylor and Francis Group, p. 9.

Mazzella, W. D. (1999). Diode array micro-spectrometry of colour ink-jet printers. *J. Am. Soc. Questioned Document Examiners,* **2**:65–73.

Merrill, R. A., and Bartick, E. G. (1992). Analysis of ballpoint pen inks by diffuse reflectance infrared spectrometry. *J. Forensic Sci.,* **37(2)**:528–541.

Munson, T. O. (1989). A simple method for sampling photocopy toners for examination by pyrolysis gas chromatography. *Crime Lab. Dig.,* **16(1)**:6–8.

National Commission on Terrorist Attacks upon the United States (2004). *The 9/11 Commission Report*. Washington, DC: U.S. Government Printing Office, p. 408.

Pagano, L. W., Surrency, M. J., and Cantu, A. A. (2000). Inks: forensic analysis by thin-layer (planar) chromatography. In: Wilson, I. D. (Ed.), *Encyclopedia of Separation Science*. New York: Academic Press, pp. 3101–3109.

Richards, G. (1977). The application of electronic video techniques to infrared and ultraviolet examinations. *J. Forensic Sci.,* **27(1)**:53–60.

Roux, C., Novotny, M., Evans, I., and Lennard, C. (1999). A study to investigate the evidential value of blue and black ballpoint pen inks in Australia. *Forensic Sci. Int.,* **101**:167–176.

The Society of Dyers and Colourists and the American Association of Textile Chemists and Colorists (1971). Colour Index, 3rd ed. 1971.

Stewart, L. F. (1985). Ballpoint ink age determination by volatile component comparison: a preliminary study. *J. Forensic Sci.,* **30(2)**:405–411.

Tebbett, I. R., Chen, C., Fitzgerald, M. S., and Olson, L. (1992). The use of HPLC with multiwavelength detection for the differentiation of non ball pen inks. *J. Forensic Sci.,* **37(4)**:1149–1157.

Weyermann, C. (2005). Mass spectrometric investigation of the aging processes of ballpoint ink for the examination of questioned documents. Dissertation, Justus-Liebig-University, Giessen, Germany.

Wilson, J. D., LaPorte, G. M., and Cantu, A. A. (2004). Differentiation of black gel inks using optical and chemical techniques. *J. Forensic Sci.,* **49(2)**:364–370.

Witte, A. H. (1963). The examination and identification of inks. In: Lundquist, F. (Ed.), *Methods of Forensic Science*, Vol. II. New York: John Wiley and Sons. pp. 35–77.

CHAPTER 7

The Role of Vibrational Spectroscopy in Forensic Chemistry

ALI KOÇAK

Department of Sciences, John Jay College of Criminal Justice, The City University of New York, New York

Summary Spectroscopy is the study of the interaction of electromagnetic radiation with matter to determine the molecular structure of a solid sample or one dissolved in a specific solvent. This interaction depends on the intrinsic properties of the sample material and can be classified by the energy of the probing electromagnetic radiation. Energy can be in the form of ultraviolet, visible, or infrared light, as well as other forms of energy. One can detect transmitted, absorbed, and/or reflected light. Due to the wide scope of samples typically found in a forensic investigation, a variety of spectroscopic techniques may be needed. Infrared spectroscopy is a good technique to use to identify fibers such as acrylics, nylons, or polyesters, or paints or alkyds, acrylics, or nitrocellulose. The size of the sample may require the use of microscopic infrared spectroscopy, and the nature of the sample may indicate the use of external reflection spectroscopy or attenuated total reflectance spectroscopy. In the chapter we review these techniques as well as related methods of sample identification.

7.1	Introduction to vibrational spectroscopy	252
7.2	Infrared spectroscopy	253
7.3	Infrared sampling techniques	255
	7.3.1 Transmission spectroscopy	255
	7.3.2 External reflection spectroscopy	255
	7.3.3 Attenuated total reflectance	256
	7.3.4 Diffuse reflectance spectroscopy	258
	7.3.5 Infrared microspectroscopy	259
7.4	Raman spectroscopy	260
7.5	Raman spectroscopic techniques	262

Forensic Chemistry Handbook, First Edition. Edited by Lawrence Kobilinsky.
© 2012 John Wiley & Sons, Inc. Published 2012 by John Wiley & Sons, Inc.

7.5.1	Surface-enhanced Raman spectroscopy	262
7.5.2	Resonance Raman scattering	263
7.5.3	Coherent anti-Stokes Raman spectroscopy	263
7.5.4	Confocal Raman spectroscopy	263
7.6	Applications of vibrational spectroscopy in forensic analysis	264
	References	265

7.1 INTRODUCTION TO VIBRATIONAL SPECTROSCOPY

Molecular spectroscopy is the study of interaction of electromagnetic radiation with matter to elucidate information on molecular structure and dynamics, the environment of the sample molecules and their state of association, interactions with solvents, and many other topics (Graybeal, 1988; Colthup et al., 1990; Schrader, 1995). This interaction depends on the intrinsic properties of the material, classified by the frequency range (energy) of the probing electromagnetic radiation, such as ultraviolet–visible (UV–Vis), infrared (Kocak et al., 2006).

The intrinsic properties of a material that govern its interaction with electromagnetic radiation are described by its complex refractive index (Kocak et al., 2006). Usually represented as having a real part, the *refractive index*, and an imaginary part, the *absorption index*. The complex refractive index of a material is a function that describes the response of the material to the electromagnetic radiation applied. The refractive and absorption indices are both functions of the frequency of the electromagnetic radiation applied (Kocak et al., 2006). Different spectroscopic probes are used to observe and measure the interactions between material and the radiation applied. These measured observables (e.g., transmittance, reflectance, emittance), which are in principle expressible in terms of the intrinsic properties of the material, the molecular composition of the substance, and the wavelength of the illuminating radiation, will determine the type and nature of the spectroscopic technique selected. Due to the wide scope of samples under investigation, a variety of spectroscopic techniques are needed. Selection of the appropriate technique will minimize the time needed for analysis and investigation (Diem, 1993; Schrader, 1995; Mirabella, 1998).

Molecules consist of atoms that have a certain mass, which are connected by elastic covalent bonds. As a result, they can perform periodic motions and continual vibrations relative to each other. The energies associated with these vibrations are quantized; within a molecule, only specific vibrational energy levels are allowed (Colthup et al., 1990; Levine, 1991). For a molecule to absorb radiation, the frequency of the illuminating electromagnetic radiation must match the natural frequency of a bond vibration (Colthup et al., 1990). This phenomenon is similar to the example of a child on a swing: energy is transferred to the swing only if the frequency and the phase of the energy delivered match the frequency of swing movement.

This field of spectroscopy in which the vibration of molecules is involved is called *vibrational spectroscopy* (Colthup et al., 1990; Diem, 1993; Schrader, 1995). Vibrational spectroscopy involves different methods, the most important of which in observing the vibration of molecules are infrared and Raman spectroscopy. Depending on the nature of the vibration, which is determined by the symmetry of the molecule (Cotton, 1990); vibrations may be active or forbidden in the infrared or Raman spectrum (Cotton, 1990; Levine, 1991; Diem, 1993).

The infrared and Raman spectra of two molecules are different depending on whether these molecules have, for whatever reason, different constitutions, isotopic distributions, configurations or conformations, or environments (Diem, 1993; Schrader, 1995). Vibrational spectroscopy detects the characteristic vibrational frequencies of molecules, which provide detailed information about chemical bonding and hence the molecular structure. Consequently, its most common application is in the analysis and identification of small particles. As such, the technique lends itself easily to the study of forensic materials such as controlled drug substances, automotive paint chips, inks, fragments of polymers, textile fibers, papers, inorganic compounds, adhesives, and explosives. In this chapter we describe different sampling techniques and methods involved in both infrared and Raman spectroscopy, along with their forensic applications.

7.2 INFRARED SPECTROSCOPY

Infrared spectroscopy is in principle based on arrangements similar to those of any other spectroscopy: a radiation source, such as globar or another infrared-emitting source, is directed at a sample (Kocak, 1998; Griffiths and de Haseth, 2007). This infrared radiation, which is in the mid-infrared-region range (approximately (400 to 4000 cm^{-1}), interacts with a molecule and its chemical bonds (see Figure 7.1). Most organic functional groups have vibrational and rotational frequencies in the mid infrared region. After interaction with the sample, the infrared radiation is collected by an infrared-sensitive detector, such as a deuterated triglycine sulfate (DTGS) or mercury cadmium telluride (MCT) detector (Kocak, 1998; Griffiths and de Haseth, 2007) and in infrared imaging, a focal-plane array (FPA) detector (Lewis and Levin, 1995).

In classical dispersive infrared instrumentation, a Czerny Turner monochromoter (Ingle and Crouch, 1988; Strobel and Heineman, 1989; Skoog et al., 2007) is used to disperse the infrared light into a single wavelength to plot the spectrum of intensity versus wavelength or wavenumber (Ingle and Crouch, 1988). In modern Fourier transform instrumentation, a Michelson interferometer is used. Based on the development of interferometry, Michelson developed the interferometer in 1880 to study the speed of light (Kocak, 1992; Griffiths and de Haseth, 2007). Using the Michelson interferometer, an interferogram is created by the interference of two or more radiation beams (Ingle and Crouch, 1988; Kocak, 1998; Griffiths and de Haseth, 2007). As a result, all frequencies are detected at once in the time or distance domain. The two domains of time or distance and frequency

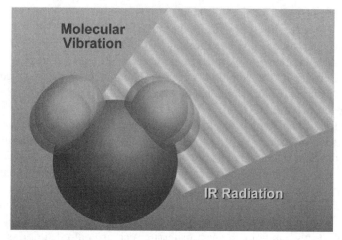

Figure 7.1 Basic principles of IR spectroscopy. *(See insert for color representation.)*

contain the same essential information, and they are interconvertable by fast Fourier transformation (Kocak, 1998; Griffiths and de Haseth, 2007). The spectrum is then plotted as intensity versus wavelength or wavenumber.

Infrared spectroscopy is an important technique for the identification of a class of compound groups or an individual compound present within a specimen. For example, it can be used to identify fibers as acrylics, nylons, or polyesters, or paints as alkyds, acrylics, nitrocellulose, and so on (Kirkbride, 2000). It has a great advantage over other techniques in its ability to obtain structural information that might be difficult to elicit using other techniques. Spectral data representing absorption at specified wavelengths of infrared light characterize chemical and structural groups within the sample, thus providing a molecular "fingerprint" and a tool for sensitive and fast sample composition determination. This allows the technique to be used in the characterization of stereoisomers. For example, it can be a very efficient technique to discriminate between positional isomers of aromatic substances such as various isomers of dimethoxyamphetamine or trimethyl benzene, as well as between other stereoisomers: for example, the isomeric group cocaine, or the group butyl nitrate, isobutyl nitrate, or isoheroin and heroin (Kirkbride, 2000). In drug determination, infrared spectroscopy can be an important tool to distinguish between a free organic and a salt because they have different spectral properties. The N—H stretch in an amine salt is found between 2500 and 3000 cm^{-1}, whereas the N—H stretch in its free base is above 3000 cm^{-1}. Therefore, the spectral properties for drug free bases and their salts are very different within the frequency range of N—H stretch (Kirkbride, 2000; Alzate et al., 2006).

The limit of detection for this technique is usually in the microgram range; this limit may be lowered by using microscope infrared techniques (see below). The technique is relatively rapid and requires minimal sample preparation, which makes it economical and time efficient.

7.3 INFRARED SAMPLING TECHNIQUES

7.3.1 Transmission Spectroscopy

The term *transmission spectroscopy* is used to denote the qualitative and quantitative measurement of the transmittance or absorbance of a material as a function of the wavelength or wavenumber (Ingle and Cronch, 1988; Mirabella, 1998; Griffiths and de Haseth, 2007). The technique involves passing a parallel beam of infrared light through a sample placed in the light beam of the spectrometer, and the *transmittance* of the radiation is measured (see Figure 7.2). This relationship between the incident electromagnetic beam and the pathlength and the concentration of absorbing media is described by Lambert–Beer's law (Swinehart, 1962):

$$A = -\log \frac{I}{I_0} = abc$$

where A is the absorbance, I the intensity of the light, a the absorptivity or extinction coefficient, b the length of the beam in the absorbing medium, and c the concentration of the absorbing species. The resulting spectrum depends on the pathlength or sample thickness, the absorption coefficient of the sample, the reflectivity of the sample, the angle of incidence, the polarization of the incident radiation, and, for particulate matter, on particle size and orientation.

Transmission spectroscopy was the only method in general use until the early 1960s. Thus, large libraries of transmission spectra are readily available. These are frequently used for comparison of spectra obtained by other techniques, even though the frequency response differs between techniques, depending on factors described below. This technique can be used for solid, liquid, and gas sampling. However, to produce useful spectra, in many cases the analyst must engage in a careful and sometimes lengthy sample preparation. The choice of the sample compartment, or cell, is ofen critical. This can cause variations in peak shapes and intensities; for example, a small bubble in a liquid cell will distort intensity ratios (Harrick Scientific Corp., 1987; Mirabella, 1998).

7.3.2 External Reflection Spectroscopy

External reflection spectroscopy is the application of infrared spectroscopy on the characterization of the light reflected from a smooth surface, such as metals, semiconductors, and liquids (Harrick Scientific Corp., 1987; Mirabella, 1998; Humecki,

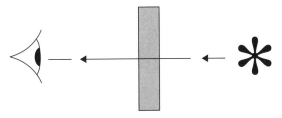

Figure 7.2 Transmitted light as a function of the material and wavelength.

1995). This technique can analyze two classes of samples. The first consists of a highly reflecting substrate (usually, a metal) upon which a material has been deposited (films, contaminants, etc.). The sample beam is reflected from the metal and interacts with film. Since the metal is largely noninteracting, the result is actually an absorption spectrum of the film. This application is sometimes referred to as *reflection–absorption* or *double transmission* (Schrader, 1995; Mirabella, 1998; Griffiths and de Haseth, 2007). A second type of external reflection study is referred to as *specular reflection*, which is the front-surface reflection from the exterior surface of a material. In specular reflection the angle of reflection equals the angle of incidence. These angles are measured from the normal to the reflecting surface. In practice, most external reflection experiments involve a combination of reflection–absorption and specular reflection components from the sample. Since spectral features will differ in the two, it is important to consider these possibilities when interpreting external reflection spectra.

External reflection spectra depend on polarization, angle of incidence, and the substrate's optical properties (see Figure 7.3) (Kortüm, 1969; Harrick Scientific Corp., 1987). The sensitivity of reflected light increases with p-polarized incident light; the phase shift of the reflected light varies with angle of incidence, and hence at large angles of incidence with p-polarized light an optimum sensitivity is obtained. For s-polarized light, the electric field at all angles of incidence has very small values at the interface, resulting in less sensitivity (Harrick Scientific Corp., 1987).

To convert reflectance spectra into absorption-like proportional spectra a Kramers–Kronig transformation (Kocak et al., 2006) should be employed, which is easy to accomplish nowadays, owing to readily available software and computerized spectrometers. Because reflection methods require little or no sample preparation, they are used as frequently as traditional methods. It is a nondestructive method for the measurement of surfaces and coatings without sample preparation. In forensic science it is useful to analyze and distinguish illicit drugs such as cocaine. Koulis et al. (2001) collected and compared the external reflection spectra of cocaine salt and cocaine base.

7.3.3 Attenuated Total Reflectance

The use of internal reflection in infrared (IR) spectroscopy was first suggested by Fahrenfort and Harrick independently in 1959 (Harrick, 1987; Mirabella, 1998). In

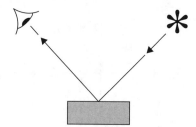

Figure 7.3 External reflection.

the internal reflection mode, an infrared transmissive material with a higher index of refraction than the sample is brought into contact with the sample. The higher-refractive-index material is defined as the internal reflection element (IRE). Typical IREs are ZnSe (refractive index: 2.4), Ge (refractive index: 4.0) and diamond (refractive index: 2.4). In most forensic applications, diamond internal reflection elements are used since they are more durable.

The radiation from the spectrometer's source is directed onto the interface between the sample and the IRE at an incident angle such that the radiation is totally internally reflected (see Figure 7.4). This process causes what is referred to as an *evanescent wave* to be produced a few micrometers beyond the surface of the IRE (Harrick, 1987; Schrader, 1995; Griffiths and de Haseth, 2007). The evanescent wave is absorbed and attenuated by the sample at the characteristic wavelength corresponding to an IR absorption band, producing an attenuated total reflectance (ATR) spectrum. Because the magnitude (the depth of penetration) of the evanescent wave is small, only a few micrometers, ATR is a surface analytical technique and particularly useful for strongly absorbing or thick samples. Several factors can affect the depth of penetration, such as the refractive index of the sample and the crystal, the angle at which the radiation is brought into the crystal, and the wavelength of the radiation (Harrick, 1987; Harrick Scientific Corp., 1987; Griffiths and de Haseth, 2007). Data obtained from ATR measurements are different from the data obtained in transmission measurements. Therefore, they should be compared only for general qualitative purposes. It is worth noting however, that these differences between the two techniques are caused by optical differences rather than compositional ones.

In contrast to external reflection spectroscopy, which requires smooth surface properties, ATR spectroscopy can be used for any surface brought into intimate contact with the internal reflection element. The technique has found a large scope of applications for the analysis of a variety of sample types in a wide spectral range.

For forensic applications, ATR can be utilized in the analysis of pressure-sensitive adhesive tapes (Merrill, 2000). Duct tapes are used in binding victims of violent crimes, and other tapes can be used in packing drugs and illegal materials. Using ATR spectroscopy the major chemical components of these can be determined and leads to the characterization of tapes submitted as evidence to determine if two samples are similar or distinguishable. The brand and the origin

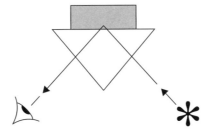

Figure 7.4 Attenuated total reflectance.

of these tapes can also sometimes be determined. It is worthy to note that ATR is a nondestructive technique. However, it may be semidestructive in some cases because the pressure applied to establish a good contact between the sample and the crystal may crush the sample.

7.3.4 Diffuse Reflectance Spectroscopy

Diffuse reflectance is the radiation that penetrates a specimen to a shallow depth and reemerges after scattering off within the sample. It is based not only on reflection and refraction but also on diffraction. These emerging photons from the sample will be collected and analyzed using an infrared detector (see Figure 7.5). Since scattering of the radiation within the sample is an essential phenomenon of the diffuse reflectance technique, it is suitable for inhomogeneous samples that exist as powders or can be converted to powder form. This technique permits the sampling of a solid that has been properly reduced in size, either neat or mixed ($\sim 10\%$) with nonabsorbing powder such as KBr or KCl. The process requires grinding and diluting the sample with KBr or KCl to ensure deeper penetration of the incident light into the sample (Harrick Scientific Corp., 1987; Schrader, 1995; Mirabella, 1998).

The diffuse reflectance technique offers the advantage of having a wide solid sampling range; this in particular is useful when a very wide range of solid samples need to be analyzed, as in forensic laboratories. Suzuki and Gresham (Suzuki et al., 1986b) demonstrated a broad study for the use of the diffuse reflectance technique in forensic analysis. They analyzed the antidepressant drug Alprazolam and obtained the spectral results at two extreme concentrations with a lower limit of 20 μg. In general, the diffuse reflectance technique offers several advantages, among them minimal sample preparation, a high degree of sensitivity (less than 100 ng in some cases), and its suitability for very weakly absorbing samples.

Kubelka and Munk (Suzuki et al., 1986a) were first to describe the effect on paint layers of the pigment particles or other scattering inclusions. This model leads to the well-known Kubilka–Munk equation, which is used for the conversion of the measured reflectance spectrum to an absorption-like proportional representation. Almost all modern Fourier transfer (FT)–IR instrument software packages now contain this conversion program.

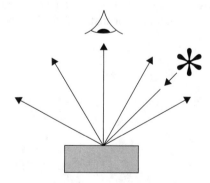

Figure 7.5 Diffuse reflectance.

7.3.5 Infrared Microspectroscopy

The conventional light microscope is an important analytical instrument that is used frequently by forensic scientists, and they are familiar with the technique. It is used in the examination of criminal evidence, including illicit drug crystals, paints, hair and textile fibers, bullets, and documents. These microscopic images are powerful data for the legal evidence, and people without technical backgrounds may have some understanding of the interpretation of the data.

The difficulty that may occur is in distinguishing data obtained from a visual microscope: when, for example, two colored fibers may appear similar under microscopic analysis but are actually stained with two different dyes; or two textile fibers may appear similar under visual microscopical analysis but differ in their molecular composition. A cross section of a polymer may show a uniform microstructure but may have gone through photooxidative degradation.

Combining a conventional light microscope with infrared spectroscopy techniques enables an analyst to obtain an infrared spectrum, or "molecular fingerprint," of a sample and overcome the difficulties of discriminating the molecular differences mentioned above (Katon and Sommer, 1992; Griffiths and de Haseth, 2007). Using infrared microspectroscopy, it is easily possible to obtain high-quality infrared spectra for chemical composition of an unknown substance as small as about micrometers in area. This enables the analyst to obtain the spectral mapping of the specimen and identify the area within. Spectral mapping involves collecting spectra within a region of a sample while maintaining the spatial resolution inherent in infrared microscopy. Spectra collected at a series of positions can be located spatially within the sample. By mapping these data with respect to a particular band of infrared light, the spatial distribution of the specific chemical components in the sample can be obtained.

Application of IR spectral mapping can readily define the presence and distribution of components of a nonhomogeneous specimen (e.g., drug distribution in human hair). This area of research has been the subject of considerable work, and many interesting results have been obtained (Kalasinsky, 1993). Microscopic probing of small areas within longitudinally microtomed hair sections provides a profile of deposition of a drug along a growth line, and thus individual hairs may reveal the hydrophobic or hydrophilic characteristics of a substance.

Infrared microspectroscopy can record specular, diffuse, internal, or ATR spectra in addition to conventional transmission techniques. It is only the sample itself that determines the choice of the type of analysis technique. Infrared microscopy can be employed in the analysis of small samples of paint, as from motor vehicle accidents; in the case of hit-and-run offenses, for example, one may find small paint chips on the clothes of the victim. A paint chip is embedded and a thin section is produced on a microtome and analyzed. This paint chip analysis can lead to identifying the car type, model, and production year by searching the car paint chip library. Other forensic examples could be single textile fibers as evidence of contact and concentrated extracts of low-dosage drugs (Krishnan and Hill, 1990).

7.4 RAMAN SPECTROSCOPY

Like infrared spectroscopy, Raman spectroscopy probes the vibrational states of complex molecules. However, in Raman spectroscopy the excitation mechanism is different from that of infrared spectroscopy. Raman is concerned with the scattering of radiation by the sample rather than absorption of radiation (Herzberg, 1945; Gerrard, 1994; Schrader, 1995). In a Raman spectrometer the sample is irradiated with an intense source of a monochromatic (laser beam) light with much more energy than that required to cause a transition between vibrational energy states. Since the energy of the illuminating radiation does not correspond to a molecular energy level, no absorption transition will occur. This type of interaction between the illuminating light and the molecule, where no absorption takes place, is called *scattering* (see Figure 7.6). This type of scattered light, which produces photons with the same energy, or color, as exciting radiation is called *Rayleigh scattering* (Gerrard, 1994; Schrader, 1995).

In Rayleigh scattering there will be an elastic collision between the incident photon and the molecule. Since as a result of this type of collision the rotational and vibrational energy of the molecule is unchanged, the energy and therefore the frequency of the scattered photon is the same as that of the incident photon. This appears as the strongest component of the scattered radiation. In 1928, C. V. Raman described another type of scattering, now known as the *Raman effect*. Raman scattering, however, occurs when the molecule does not return to the same energy level from which it originated but ends up in the excited state. As a result, the scattered photon no longer has the same energy or frequency as that of the exciting photon. The Raman effect can be described as an inelastic collision between the incident photon and the molecule where, as a result of the collision, the vibrational and rotational energy of the molecule is changed by an amount ΔE, where ΔE is the difference between incident and scattered radiation. This change will cause a small

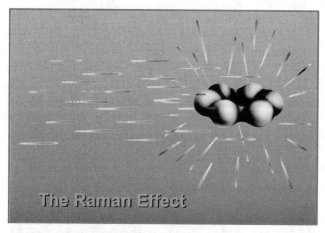

Figure 7.6 The Raman effect: Illuminating light is not absorbed but is scatterred. *(See insert for color representation.)*

fraction of the scattered radiation to exhibit shifted frequencies that correspond to the sample's vibrational transitions. Lines shifted to energies lower than the source are produced by ground-state molecules, while lines at higher frequency are due to molecules in excited vibrational states. These lines, the result of inelastic collision of the light with the sample, are called *Stokes* and *anti-Stokes bands*, respectively. Since molecules of the sample are in the ground state, at room temperature less energy from the source radiation is more favorable than energy gain, and the Stokes band is more intense (see Figure 7.7) (Herzberg, 1945; Diem, 1993; Gerrard, 1994; Schrader, 1995).

The strength of the Raman signal is weak and is inversely proportional to the fourth power of the wavelength of the incident radiation. Therefore, in Raman it is desirable to utilize radiation of relatively short wavelength in the visible region. This may induce fluorescence in the specimen. Fluorescence is likely to block the weak Raman signal and is highly undesirable. A red He–Ne laser (at 632 nm), near-infrared (NIR) diode laser (at about 785 nm), and an Nd:YAG laser (at 1064 nm) can be employed to somewhat reduce this problem and to inhibit fluorescence effect (Turrell and Corset, 1996).

Although both Raman and IR spectroscopy methods probe vibrational states of a specimen, they are both used for qualitative analysis and for molecular structure elucidations, the mechanism of origin of the two techniques is completely different and the two do not produce identical spectra. Infrared absorption can be detected

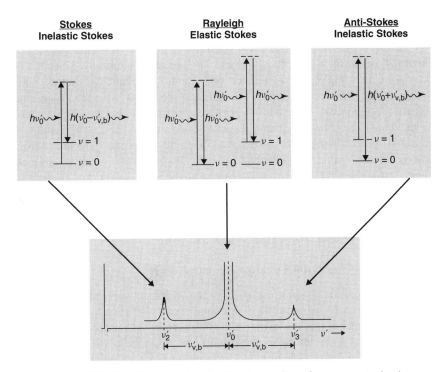

Figure 7.7 The Raman principle. *(See insert for color representation.)*

if the dipole moment in a molecule is changed during the normal vibration. The intensity of an infrared absorption band depends on the change in the dipole moment during this vibration (Colthup et al., 1990).

A Raman active vibration, on the other hand, can be detected if the polarizibility in a molecule is changed during the normal vibration (Colthup et al., 1990; Schrader, 1995). The polarizibility is a measure of how readily the electrons within the molecule are displaced by the electric field of the incident light. The intensity of a Raman band depends on the change of polarizability during the vibration (Colthup et al., 1990; Schrader, 1995). Therefore, infrared spectroscopy is a very useful probe for those molecules that contain functional groups such as C—O, C—N, N—H, and O—H. Infrared spectroscopy provides little information on nonpolar bonds such as C—C, C=C, and N—N. Raman spectroscopy, on the other hand, can provide useful information in relation to polarizability of the bond. Nonpolar bonds, particularly double and triple bonds, yield strong Raman bands. Therefore, Raman is considered to be more useful in fiber identification and comparison, as most fibers are composed of long chains of C—C bonds or aromatic rings. Also, Raman techniques are useful for many dyes and pigments, such as aromatic rings or azo linkage.

Raman spectroscopy has several advantages in the analysis, which can be difficult with IR spectroscopy, particularly when dealing with samples of forensic evidence where the chance of losing or destroying the sample is critical. This is particularly important when analyzing fiber samples mounted on a glass slide for visual microscope analysis. Light microscopy analysis requires embedding the sample on a glass slide. Because glass absorbs strongly in the infrared, the fiber needs to be removed from the slide and cleaned prior to IR analysis. This may involve a high risk of losing a evidentiary sample. In Raman spectroscopy, however, one can analyze the sample directly without removing it from the slide (Bartick, 2002).

Raman spectroscopy has long been utilized in forensic research laboratories. However, recent advanced developments in instrumentation have brought about its use in forensic laboratories. The instrumentation has advanced significantly over the last 10 years compared to what it was 20 to 30 years ago. The spectrum can be acquired more rapidly and the technique has become more sensitive and easier to use. During the last few years, Raman spectroscopic imaging has become one of the most popular topics of spectroscopy.

7.5 RAMAN SPECTROSCOPIC TECHNIQUES

7.5.1 Surface-Enhanced Raman Spectroscopy

Surface-enhanced Raman spectroscopy (SERS) is a Raman spectroscopic technique that provides greatly enhanced Raman signal (typically, 10^3- to 10^6-fold) from Raman-active analyte molecules that have been adsorbed onto colloidal metallic surfaces (e.g., on colloidal Ag prepared with citrate), in the particle-size range 25 to 500 nm (Diem, 1993; Schrader, 1995; Campion and Kambhampati, 1998). The importance of SERS is its surface selectivity and high sensitivity. Conventional

Raman spectroscopy is insensitive for surface studies because the photons of the incident laser light simply propagate through the bulk, and the signal from the bulk overwhelms any Raman signal from the analytes at the surface. SERS selectivity at the surface overcomes this problem.

Surface-enhanced Raman scattering arises from two mechanisms: electromagnetic and chemical enhancement. Usually, the electromagnetic effect is dominant, the chemical effect contributing enhancement only in the range of an order or two of magnitude, but this depends on the surface features of SERS-activated substrate (Campion and Kambhampati, 1998). Electromagnetic enhancement is dependent on the presence of the metal surface's roughness features, while chemical enhancement involves changes to the adsorbate electronic states due to chemisorption of the analyte. The magnitude of the electromagnetic enhancement also depends on the intrinsic characteristics of the metal: Ag, Cu, Au, Pt, and so on. The magnitude of electromagnetic enhancement may vary for each of them. The potential used (i.e., the voltage) will also have an effect on both the electromagnetic and chemical mechanisms of enhancement.

7.5.2 Resonance Raman Scattering

Resonance Raman enhancement occurs if the energy of the photon of the exciting laser beam is close to the energy of an electronic transition of the sample. Intensity enhancement can occur by a factor of 10^2 to 10^6 compared with normal Raman scattering (Diem, 1993). Resonance Raman is used in forensic chemistry to achieve detection limits of 10^{-6} to 10^{-8} M.

7.5.3 Coherent anti-Stokes Raman Spectroscopy

In coherent anti-Stokes Raman spectroscopy (CARS), the sample is illuminated by two lasers, one of them with a fixed wavelength, usually referred to as the *pump laser*, and a second tunable laser beam, referred to as the *Stokes frequency*. The pump beam excites the molecule to the first virtual Raman level. The second laser, a tunable probe beam, stimulates the Stokes transition. The pump then again excites the molecule to a higher virtual level, but starting from the higher vibrational state. Finally, the molecule decays back to the ground state, emitting an anti-Stokes photon (Diem, 1993). CARS produces the same Raman signals as those of conventional Raman spectroscopy, but at levels four to five orders of magnitude greater.

7.5.4 Confocal Raman Spectroscopy

Confocal Raman extracts information about the interior of a sample, pinpointing the exact location of an element embedded inside a sample (Everall, 2004a, 2004b). Laser radiation is focused on a spot with the help of a microscope objective. Advantages of confocal Raman are (1) that it extracts information from a small point within the interior of a large sample, (2) that there is better axial and lateral resolution than with a conventional microscope, (3) that stray light is rejected, which avoids interference and eliminates fake peaks, and (4) superior rejection of fluorescence.

7.6 APPLICATIONS OF VIBRATIONAL SPECTROSCOPY IN FORENSIC ANALYSIS

Vibrational spectroscopy has long played a significant role in both research and applications of forensic science. Indeed, the technique has become one of the important tools in many fields of forensic science. A wide range of applications to physical evidence involved in crime investigations have been developed, including examining everything from drugs and paints, fibers, explosives, polymers, and inks to documents. Vibrational spectra has also been an effective tool for analyzing samples collected from crime scenes, such as blood, fabrics, and soil particles.

De Gelbder et al. (2005) utilized the Raman spectroscopy technique for the analysis of automotive paints to determine the origin of the paint and hence the type of automobile. Determining the origin of the paint is important forensic evidence in many car accidents, particularly in hit-and-run cases.

Several researchers have used vibrational spectroscopic techniques to analyze other types of paints, including white household paints (Bell et al., 2005), which are considered important forensic evidence. However, unlike automotive paints, which can be distinguished by color or type, many household paints cannot be distinguished based on color because they are white. Therefore, the composition of the paint needs to be discriminated, and Raman spectroscopy is a useful tool for this purpose. Bell et al. (2005) have demonstrated that Raman spectroscopy has a considerable potential for the analysis of white paints.

Vibrational spectroscopy is an effective tool for an analysis of evidence samples for the presence of explosive materials, and many vibrational spectroscopy techniques have been used for this purpose (Bartick, 2002). Bartick and Mount (2001) demonstrated the use of Raman and infrared spectroscopy in the analysis of suspect explosive components submitted as case evidence (Bartick and Mount, 2001). Samples of evidence were taken from the home of a suspect who was under suspicion for producing bombs. The evidence included all the components needed to prepare an extremely unstable explosive material, hexamethylenetriperoxideamine (HMTD). One of the major components found was a bottle labeled Welloxide, a liquid stabilizer developer. Welloxide is a hair-coloring developer that contains hydrogen peroxide. Hydrogen peroxide is one of the components used to produce HMTD. Batrick at el. analyzed Welloxide using IR and Raman spectroscopic methods to determine if there was sufficient H_2O_2 to produce HMTD. As a result, they concluded that there was sufficient H_2O_2 in the suspect's possession to produce a bomb. Raman spectroscopy, however, was their method of choice, which showed the distinct presence of hydrogen peroxide's O—O stretching peak at 876 cm^{-1} document. The infrared spectrum, on the other hand, was not observable until the water spectrum was subtracted, and the peak of interest was still weak.

Characterization of drug-packing plastic bags is important forensic evidence in the courtroom to trace the origin of a package. Causin et al. (2006) have utilized the ATR infrared spectroscopic technique to differentiate and discriminate apparently similar plastic shopping bags that were used to pack individual doses. Their study showed that using ATR spectroscopy in combination with other techniques is a useful technique to discriminate mass-produced plastic bags and to trace the source of illicit drug doses with significant results.

ATR is also an important technique in the analysis of pressure-sensitive tapes. Merrill (2000) utilized the ATR technique with single-reflection diamond IRE to analyze four different types of pressure-sensitive adhesive tapes, including duct tapes, electric tape, packing tape, and office tape.

Maynard et al. (2001) used the diffuse reflectance IR spectroscopy technique to study personal lubricants found in sexual assault cases. Using extracts from the lubricant specimen, they were able to define classes and subclasses of these lubricants. The samples used were categorized in groups of three bases: siloxane-, poly(ethylene glycol)-, and glycerin-based lubricants. The infrared results indicated that the base component of each lubricant does not change over the course of three months, and it is possible to identify lubricant traces at least three months after preparing a sample for analysis.

Kimberley et al. (2004) demonstrated the possible use of Raman microspectroscopy to determine the presence of drugs on U.S. dollar bills, which requires minimum sample handling and preserves the sample for further analysis. They were able to measure Raman spectra and identify individual crystals on the currency in a heterogeneous mixture of three compounds: benzocaine, isoxsuprine, and norephedrine.

Documents involving fraud and threatening letters produced on printers and copy machines have been an important concern for forensic investigators. Tracing the model and the origin of the printer or the copy machine is an important hint for crime scene investigators. Merrill et al. (2003) were able to track the origin of a packaged bomb with an address label mailed to a victim's address. Investigators suspected that the bomb was mailed by an employee at a company that had 400 copiers of 61 different models. They requested 61 documents copied from the available 61 models, and they prepared the sample by heating the back of the paper onto aluminum foil attached to a glass microscope slide. Spectra were obtained using microscopic reflection–absorption IR spectroscopy, where the IR beam reflected from the aluminum foil to the detector. Analyzing the documents and comparing the spectra to the ink obtained from the address label allowed the investigators to trace the model of the copier and the suspect.

REFERENCES

Alzate, L. F., et al. (2006). The vibrational spectroscopic signature of TNT in clay minerals. *Vibrat. Spectrosc.*, **42**:357–368.

Bartick, E. G. (2002). Application of vibrational spectroscopy in criminal forensic analysis. In: *Handbook of Vibrational Spectroscopy*. Hoboken, NJ: Wiley.

Bartick, E. G., and Mount, K. H. (2001). Analysis of a suspect explosive component: hydrogen peroxide in hair coloring developer. *Forensic Sci. Commun.*, **3**(4).

Bell, S. E. J., et al. (2005). Forensic analysis of architectural finishes using Fourier transform infrared and Raman spectroscopy: II. White paint. *Appl. Spectrosc.* **59**(11):1340–1346.

Campion, A., and Kambhampati, P. (1998). Surface-enhanced Raman scattering. *Chem. Soc. Rev.*, **27**:241–250.

Causin, V., et al. (2006). A quantitative differentiation method for plastic bags by infrared spectroscopy, thickness measurement and differential scanning caloimetry for tracing the source of illegal drugs. *Forensic Sci. Int.*, **164**:148–154.

Colthup, N. B., Daly, L. H., and Wiberley, S. E. (1990). *Introduction to Infrared and Raman Spectroscopy*. San Diego, CA: Academic Press.

Cotton, F. A. (1990). *Chemical Applications of Group Theory*. New York: Wiley.

De Gelbder, J. et al. (2005). Forensic analysis of automotive paints by Raman spectroscopy. *J. Raman Spectrosc.*, **36**:1059–1067.

Diem, M. (1993). *Introduction to Vibrational Spectroscopy*. New York: Wiley.

Everall, N. (2004a). Depth profiling with confocal Raman microscopy: I. *Spectroscopy*, **19**(10):22–27.

Everall, N. (2004b). Depth profiling with confocal Raman microscopy: II. *Spectroscopy*, **19**(11):16–27.

Fredrick, K. A., et al. (2004). Identification of individual drug crystals on paper currency using Raman microspectroscopy. *Spectrosc. Lett.*, **37**(3):301–310.

Gerrard, D. L. (1994). Raman spectroscopy. *Anal. Chem.*, **66**(12):547–557.

Graybeal, J. D. (1988). *Molecular Spectroscopy*. New York: McGraw-Hill.

Griffiths, P. R., and de Haseth, J. A. (2007). *Fourier Transform Infrared Spectroscopy*. Hoboken, NJ: Wiley.

Harrick, N. J. (1987). *Internal Reflection Spectroscopy*. New York: Harrick Scientific Corporation.

Harrick Scientific Corp. (1987). *Optical Spectroscopy: Sampling Techniques Manual*. New York: Harrick.

Herzberg, G. (1945). *Infrared and Raman Spectra of Polyatomic Molecules*. Princeton, NJ: D. Van Nostrand.

Humecki, H. J. (1995). *Practical Guide to Infrared Microscopy*, Vol. 19. New York: Marcel Dekker.

Ingle, J. D., Jr., and Crouch, S. R. (1988). *Spectrochemical Analysis*. Englewood Cliffs, NJ: Prentice Hall.

Kalasinsky, K. S. (1993). Hair analysis by infrared microscopy for drugs of abuse. *Forensic Sci. Int.*, 63:253–260.

Katon, J. E., and Sommer, A. J. (1992). IR microspectroscopy. *Anal. Chem.*, **64**(19):931–940.

Kirkbride, K. P. (2000). *Analytical Techniques/Spectroscopic Techniques*. San Diego, CA: Academic Press.

Kocak, A. (1998). *Design and Construction of the Step-Scan Fourier Transform Vibrational Circular Dichroism*, Part I. New York: City University of New York.

Kocak, A., et al. (2006). Using the Kramers–Kronig method to determine optical constants and evaluating its suitability as a linear transform for near-normal front-surface reflectance spectra. *Appl. Spectrosc.*, **60**(9):1004–1007.

Kortüm, G. (1969). *Reflectance Spectroscopy*. New York: Springer-Verlag.

Koulis, C. V., et al. (2001). Comparison of transmission and internal reflection infrared spectra of cocaine. *J. Forensic Sci.*, **46**(4):822–829.

Krishnan, K., and Hill, S. L. (1990). *FT-IR Microsampling Techniques: Practical Fourier Transform Infrared Spectroscopy*. San Diego, CA: Academic Press.

Levine, I. N. (1991). *Quantum Mechanics*. Upper Saddle River, NJ: Prentice Hall.

Lewis, E. N., and Levin, I. W. (1995). Fourier transform spectroscopic imaging using an infrared focal-plane array detector. *Appl. Spectrosc.*, **49**(5):672–678.

Maynard, P., et al. (2001). A protocol for the forensic analysis of condom and personal lubricant found in sexual assault cases. *Forensic Sci. Int.*, **124**:140–156.

Merrill, R. A. (2000). Analysis of pressure sensitive adhesive tape: I. Evaluation of infrared ATR accessory advances. *J. Forensic Sci.*, **45**(1):93–98.

Merrill, R. A., et al. (2003). Forensic discrimination of photocopy and printer toners: I. The development of an infrared spectral library. *Anal. Bioanal. Chem.*, **376**:1272–1278.

Mirabella, F. M. (1998). *Modern Techniques in Applied Molecular Spectroscopy*. New York: Wiley.

Schrader, B. (1995). *Infrared and Raman Spectroscopy*. Weinheim, Germany: VCH.

Skoog, D. A., et al. (2007). *Principles of Instrumental Analysis*. Belmont, CA: Thomson Brooks/Cole.

Strobel, H. A., and Heineman, W. R. (1989). *Chemical Instrumentation: A Systematic Approach*. New York: Wiley.

Suzuki, E. M., et al. (1986a). Forensic applications of diffuse reflectance infrared Fourier transform spectroscopy (DRIFTS): I. Principles, sampling methods, and advantages. *J. Forensic Sci.*, **31**:931–952.

Suzuki, E. M., et al. (1986b). Forensic applications of diffuse reflectance infrared Fourier transform spectroscopy (DRIFTS): II. Direct analysis of some tablets, capsule powders, and powders. *J. Forensic Sci.*, **31**:1292–1313

Swinehart, D. F. (1962). The Beer–Lambert law. *J. Chem. Ed.*, **7**:333–335.

Turrell, G., and Corset, J. (1996). *Raman Microscopy: Development and Applications*. San Diego, CA: Academic Press.

CHAPTER 8

Forensic Serology

RICHARD LI

Department of Sciences, John Jay College of Criminal Justice, The City University of New York, New York

Summary Forensic serology is an important component of modern forensic science. The primary activity of forensic serologists is the identification of bodily fluids. Bodily fluid stains are commonly associated with violent criminal cases. Proving the presence of bodily fluids can confirm alleged violent acts for an investigation.

8.1	Introduction	270
8.2	Identification of blood	271
	8.2.1 Oxidation–reduction reactions	272
	8.2.2 Microcrystal assays	275
	8.2.3 Other assays for blood identification	275
8.3	Species identification	278
	8.3.1 Immunochromatographic assays	278
	8.3.2 Ouchterlony assay	280
	8.3.3 Crossed-over immunoelectrophoresis	281
8.4	Identification of semen	282
	8.4.1 Visual examination	282
	8.4.2 Acid phosphatase assays	283
	8.4.3 Microscopic examination of spermatozoa	284
	8.4.4 Immunochromatographic assays	285
	8.4.5 RNA-based assays	286
8.5	Identification of saliva	286
	8.5.1 Visual and microscopic examination	287
	8.5.2 Identification of amylase	287
	8.5.3 RNA-based assays	289
	References	289

Forensic Chemistry Handbook, First Edition. Edited by Lawrence Kobilinsky.
© 2012 John Wiley & Sons, Inc. Published 2012 by John Wiley & Sons, Inc.

8.1 INTRODUCTION

Forensic serology is an important component of modern forensic science. The primary activity of forensic serologists is the identification of bodily fluids. It focuses on identifying the presence of blood, semen, saliva, or other bodily fluids in a questioned sample (DeForest et al., 1983; Gaensslen, 1983; Lee et al., 2001; Shaler, 2002; Greenfield and Sloan, 2005; Jones, 2005; Kobilinsky et al., 2005; Saferstein, 2007; Li, 2008). Bodily fluid stains are commonly associated with violent criminal cases. For example, the identification of blood evidence is often necessary for cases involving homicide, aggravated assault, sexual assault, and burglary. Proving the presence of blood can confirm alleged violent acts for an investigation. The identification of semen and saliva is especially important for the investigation of sexual assault cases. For example, the presence of a suspect's semen stains on a victim's clothing can confirm an alleged sexual activity, and the presence of a suspect's saliva stains on a victim's sex organ can confirm alleged oral copulation. In this chapter we focus on the identification of blood, semen, and saliva evidence.

In certain specific circumstances, the identification of other bodily fluids, although less frequently encountered than blood, semen, and saliva evidence, can have probative value for criminal investigations. The identification of urine can be important for an investigation of alleged assault with urination. The identification of vaginal secretions, menstrual blood, and fecal materials can be important for the investigation of sexual assault cases. For example, the identification of vaginal secretions may be important for the investigation of an alleged vaginal rape with a foreign object. Determining whether bloodstains are attributed to vaginal trauma or menstrual bleeding may be necessary in cases of alleged rape. The identification of fecal samples can be important for cases involving anal rape.

Forensic serology can be considered the process of examining and identifying biological evidence, which occurs prior to the individualization of the biological evidence. *Individualization* of biological evidence is used to determine whether or not a bodily fluid sample has come from a particular person. Today, individualization can be achieved by using forensic DNA analysis. However, the identification of bodily fluid cannot be omitted or replaced by forensic DNA analysis. For example, if a stain from a victim's clothing was processed with forensic DNA analysis for a sexual assault case and the DNA profile of the stain was found to match an alleged suspect, then based on the DNA profiling results alone, one may conclude that the suspect's DNA was found on the victim's clothing, thus, establishing a link between the suspect and the victim. However, if forensic serology testing identified the biological stain as a semen stain, one can conclude that semen was found on the clothing taken from the victim and that the suspect is the source of DNA from the semen stain. Thus, it appears that a sexual act may have occurred.

The identification of bodily fluid can be carried out using presumptive and confirmatory assays to test that the sample is the bodily fluid in question. The advantages of *presumptive assays* are that these assays are sensitive, rapid, and simple. A positive reaction of a presumptive assay indicates the possibility of the presence of the bodily fluid in question. However, presumptive assays are not very specific. Therefore, they should not be considered to be conclusive to the presence of a type of bodily fluid. In contrast, a negative assay suggests that the questioned bodily fluid is absent. Thus, they can be used as a screening method and narrow down biological stains prior to other types of analyses, such as forensic DNA testing. Moreover, these assays can be used as a search method to locate biological stains at the crime scene. Additional assays, such as confirmatory assays, should be conducted afterward if necessary. *Confirmatory assays* are more specific for the bodily fluid in question. These assays are employed to identify bodily fluids with higher certainty than presumptive assays. However, confirmatory assays are more time consuming than presumptive assays.

8.2 IDENTIFICATION OF BLOOD

The identification of blood can be performed using presumptive and confirmative assays (Figure 8.1). The most commonly utilized presumptive assays among forensic laboratories are *oxidation–reduction reaction* assays. The presence of blood can then be confirmed further using confirmatory assays. The most commonly utilized confirmatory assays are *microcrystal assays*.

Currently, since time and budget constraints often exist, confirmatory assays for blood identification are infrequently performed. A reddish-brown stain identified through visual examination is usually tested by using presumptive assays. If the presumptive assay of the alleged blood stain is positive, the stain is then analyzed further by forensic DNA analysis. This approach can only derive a conclusion that the results *indicate* the presence of blood. Therefore, if sufficient amounts of biological materials are available, confirmatory assays should be performed if possible. Thus, if the confirmatory assay is positive, one can conclude that *blood was identified* from the evidence. Additional tests to determine if blood evidence originated from a human or animal source can be performed as necessary and are discussed in Section 8.3.

The blood volume of a normal human is approximately 8% of the body weight. The fluid portion of the blood is called *plasma*. The cellular portion of the blood,

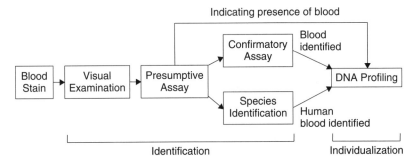

Figure 8.1 Example of the protocol for processing blood samples.

which is suspended in the plasma, consists of *erythrocytes* (also called *red blood cells*), *leucocytes* (also called *white blood cells*), and *thrombocytes* (also called *platelets*). Most presumptive and confirmatory assays of blood identification are based on the detection of hemoglobin. *Hemoglobin* is the protein responsible for the transportation of oxygen and is located in the erythrocytes. The majority of adult human hemoglobin consists of four subunits, two α and two β subunits (designated as $\alpha_2\beta_2$). Each hemoglobin subunit contains a *heme* molecule (also called *ferroprotoporphyrin*). The heme molecule consists of a protoporphyrin IX and a *ferrous* (Fe^{2+}) iron atom (Figure 8.2).

8.2.1 Oxidation–Reduction Reactions

This type of presumptive assay for blood identification is based on the biochemical properties of the heme molecule, which can catalyze an oxidation–reduction reaction. In an oxidation–reduction reaction, changes occur in the oxidation state of chemicals. For example, during the reaction, an oxidant becomes reduced and a reductant becomes oxidized. In the presumptive assays for blood identification, oxidation often coincides with the loss of hydrogen, and reduction often coincides with the gain of hydrogen. An example of an oxidation–reduction reaction for blood identification is shown in Figure 8.3(a).

In an oxidation–reduction reaction for blood identification assays, the most commonly used *reductants* are phenolphthalein, leucomalachite green, tetramethylbenzidine, and luminol (Lee, 1982; Gaensslen, 1983; Sutton, 1999; Greenfield and Sloan, 2005; Laux, 2005; Marie, 2008). Additionally, hydrogen peroxide is usually employed as an *oxidant*. In the presence of heme, a colorless reductant is oxidized, forming a product with color or chemiluminescence. Thus, a positive reaction indicates the possible presence of blood.

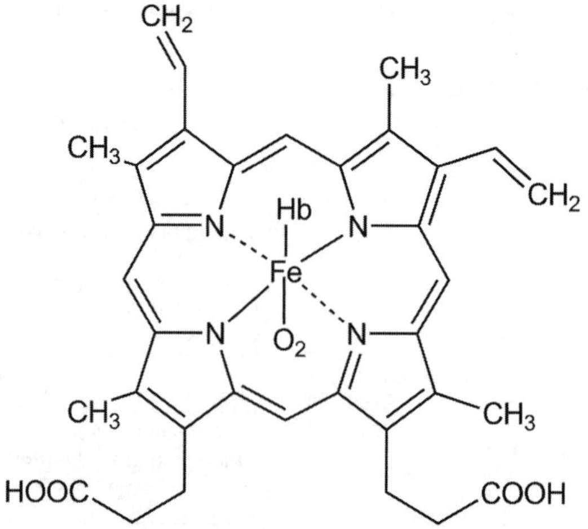

Figure 8.2 Chemical structure of heme. A heme molecule, also called *ferroprotoporphyrin*, consists of a protoporphyrin IX and a ferrous (Fe^{2+}) iron atom.

IDENTIFICATION OF BLOOD 273

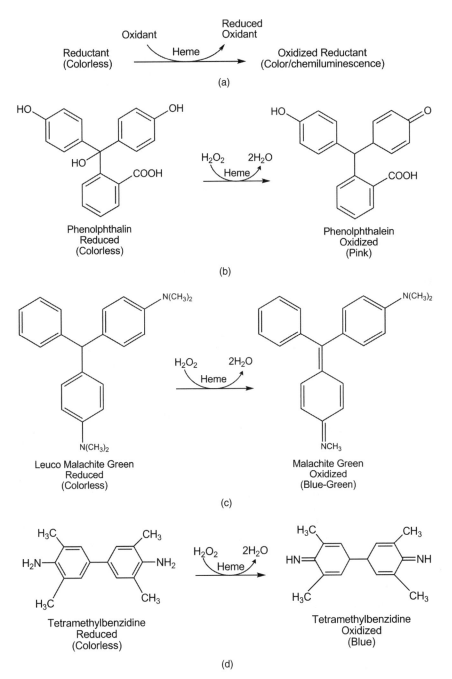

Figure 8.3 Oxidation–reduction reaction and presumptive assays for blood identification: (a) oxidation–reduction reaction as a basis for presumptive assays for blood identification; chemical reactions of (b) Kastle–Meyer assay, (c) leucomalachite green assay, (d) tetramethylbenzidine assay, and (e) luminol assay.

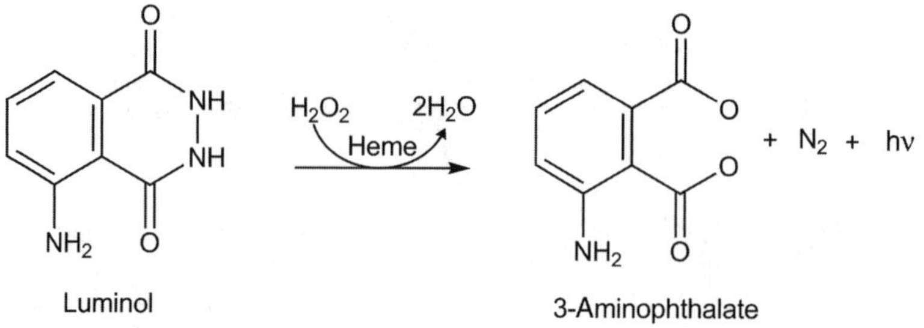

Luminol + H₂O₂ →(Heme, 2H₂O) 3-Aminophthalate + N₂ + hv

Figure 8.3 (*Continued*)

For example, the *Kastle–Meyer assay* utilizes colorless phenolphthalin as a reductant. In an oxidation–reduction reaction catalyzed by heme, phenolphthalin can be oxidized to phenolphthalein. Phenolphthalein shows a pink color in alkaline conditions [Figure 8.3(b)]. The *Leucomalachite green assay* utilizes the leuco base form of malachite green, which is colorless. It can be oxidized in the presence of heme to produce malachite green, which shows a green color [Figure 8.3(c)]. The reaction is carried out under acid conditions. In the *tetramethylbenzidine assay*, the oxidation of tetramethylbenzidine can be catalyzed by heme to produce a green color under acidic conditions [Figure 8.3(d)]. The Hemastix assay (Miles Laboratories) is a portable tetramethylbenzidine assay kit that can be used in laboratories and at crime scenes. In a *luminol assay*, oxidation of luminol (3-aminophthalhydrazide), catalyzed by heme, emits a chemiluminescent light [Figure 8.3(e)]. Thus, the assay can be performed in the dark. Luminol reagents can be sprayed at crime scenes to search for alleged bloodstains. A positive reaction not only locates the blood but also detects the patterns, such as footprints, fingerprints, and splatter patterns. This method can help to pinpoint the location of even small traces of blood. Additionally, it is useful for crime scenes that have been cleaned up and where the blood evidence has been tampered with, leaving no visible stains.

The assays discussed above are very sensitive and able to detect blood samples with up to a 10^{-5}-fold dilution. However, they are not blood specific and possibly lead to false-positive results. For example, a false-positive reaction can be caused by the presence of chemicals that are strong oxidants (such as certain metal salts), even in the absence of heme. Plant peroxidases can also catalyze the oxidation reaction in the absence of heme. Thus, peroxidase-containing plants (such as horseradish) can cause a false-positive result. The presumptive assays can be carried out using the swabbing method. A small amount of the questioned stain is transferred onto a moistened cotton swab. Additionally, a substance control (swabbing a nonstained area adjacent to the stained area) can be prepared. The test is conducted in two steps. First, a drop of the testing reagent (reductant) is added onto the swab with the blood sample. If color develops, it is an indication of a false-positive reaction. If no color develops, the assay can be continued by adding a drop of oxidant reagent (e.g., 3% hydrogen peroxide) onto the swab. The presence of blood results in an immediate development of color, while the absence of blood results in no development of color.

8.2.2 Microcrystal Assays

In microcrystal assays, blood crusts from a bloodstain are treated chemically to convert native heme to heme derivatives. These heme derivatives can form crystals with distinctive morphologies. The crystal morphology of heme derivatives can be examined by using microscopic observation. As a result, the presence of the crystals of heme derivatives confirms the presence of blood. One disadvantage of microcrystal assays is that these assays are less sensitive than the presumptive assays for blood identification. The most commonly utilized microcrystal assays are Takayama and Teichmann crystal assays (Lee, 1982; Gaensslen, 1983). Under normal physiological conditions, the ferrous ion of the heme can form six bonds: four bonds with the nitrogen of protoporphyrin IX, one bond with oxygen, and one bond with a hemoglobin subunit. In the *Takayama crystal assay*, blood crusts are treated in an alkaline condition with pyridine and a reducing sugar, such as glucose, that is capable of reducing other chemicals. Following the treatment, pink-colored pyridine ferroprotophorphyrin crystals form and can be observed under a microscope [Figure 8.4(a)]. In the *Teichmann crystal assay*, blood crusts are treated with glacial acetic acid and salts with heating. Brown-colored prismatic crystals of ferriporphyrin chloride form. Ferriporphyrin chloride is a heme derivative in which the iron is in the ferric (Fe^{3+}) state [Figure 8.4(b)].

8.2.3 Other Assays for Blood Identification

Other confirmatory assays for detecting the presence of hemoglobin are available to confirm the presence of blood. For example, human hemoglobin can be detected by using immunological methods with antihuman hemoglobin antibodies. This type of assays is discussed in Section 8.3. Additionally, hemoglobin can be identified using chromatographic and electrophoretic methods based on the mobility characteristics of hemoglobin, and spectrophotometric methods based on the characteristic light spectrum absorbed by hemoglobin.

Recently, RNA-based assays have been developed and can potentially be utilized for the identification of blood (Juusola and Ballantyne, 2003; Nussbaumer et al., 2006). These assays can detect specific types of *messenger RNA* (mRNA) that are only present in erythrocytes. Messenger RNA is transcribed from a gene and is then translated to produce a protein. The erythrocyte-specific genes utilized for blood identification are *HBA1, PBGD*, and *SPTB* (Table 8.1). *HBA1* encodes for the human hemoglobin α1 subunit, which is abundant in erythrocytes. *PBGD* encodes for porphobilinogen deaminase, an erythrocyte-specific isoenzyme of the heme biosynthesis pathway. *SPTB* encodes β-spectrin, which is a subunit of the major protein of the erythrocyte membrane skeleton. These assays are carried out using the *reverse transcription polymerase chain reaction* (RT-PCR) technique to detect erythrocyte-specific mRNA for blood identification. The RT-PCR usually

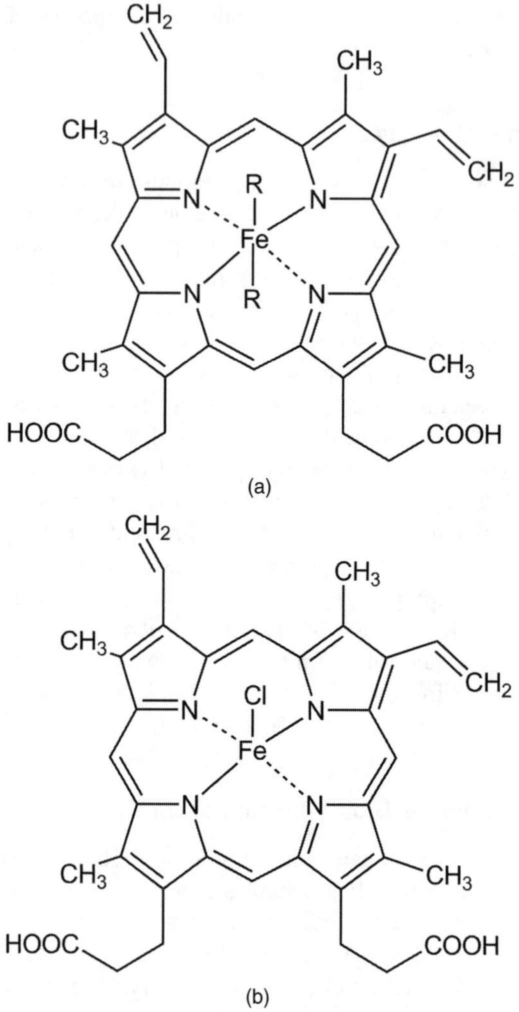

Figure 8.4 Chemical structures of heme derivatives: (a) pyridine ferroprotoporphyrin (R = pyridine) and (b) ferriprotoporphyrin chloride.

consists of two steps (Figure 8.5). In the first step, an mRNA strand is used as a template for the synthesis, carried out by reverse transcriptase, of a DNA strand of complementary sequence, which is referred to as *complementary DNA* (cDNA). In the second step, the resulting cDNA is amplified and the amplified products can then be detected. The advantages of RNA-based assays are that this technique is sensitive and can be used to detect very small quantities of mRNA. Additionally, this technique is specific and can detect erythrocyte-specific mRNA. Moreover, the technique is adaptable to automation. The disadvantage of this technique is that the mRNA present in the sample can be degraded by endogenous ribonucleases.

TABLE 8.1 Use of RNA-Based Assays for Bodily Fluid Identification[a]

Bodily Fluid	Gene Symbol	Gene Product	Description
Blood	HBA1	Hemoglobin α1	Hemoglobin α1 subunit abundant in erythrocytes; transports oxygen
	PBGD	Porphobilinogen deaminase	Erythrocyte-specific isoenzyme of the heme biosynthesis pathway
	SPTB	β-Spectrin	Subunit of the major protein of the erythrocyte membrane skeleton
Semen	KLK3	Kallikrein 3	Also called prostate-specific antigen (PSA)
	PRM1	Protamine 1	DNA-binding proteins involved in condensation of chromatin of spermatozoa
	PRM2	Protamine 2	DNA-binding proteins involved in condensation of chromatin of spermatozoa
Saliva	HTN3	Histatin 3	Histidine-rich protein involved in the nonimmune host defense in the oral cavity
	STATH	Statherin	Inhibitor of the precipitation of calcium phosphate salts in the oral cavity

Source: Adapted from Li (2008).
[a] Tissue-specific genes for identification of bodily fluid are shown.

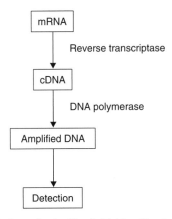

Figure 8.5 RT-PCR technique for bodily fluid identification. The synthesis of cDNA from mRNA can be carried out using reverse transcriptase, and the resulting cDNA is amplified using DNA polymerase.

However, the degree of mRNA degradation, measured by the RT-PCR technique, can potentially be used for determining the age of a bloodstain (Anderson et al., 2005). It is known that mRNA in a blood stain degrades over time after the blood stain is deposited. Thus, the age of a blood stain deposited at a crime scene can be estimated by measuring the detection level of mRNA. This estimation can provide investigative leads, such as when the crime occurred and the length of the postmortem interval.

278 FORENSIC SEROLOGY

8.3 SPECIES IDENTIFICATION

Once a stain is identified as blood, the determination of the origin of a blood stain is critical in forensic casework. The blood sample is tested to determine if it is of human or nonhuman origin. Thus, species identification assays can be useful as a screening method to exclude or eliminate nonhuman stains that are not related to the investigation. Thus, if the blood stain is nonhuman, it is not necessary to analyze the stain further. However, if the case involves animals, it may be necessary to identify the species of the blood sample. Currently, species identification is performed infrequently in forensic laboratories. A blood stain is usually processed for DNA profiling analysis. The DNA isolated is quantified using techniques that detect only higher-primate DNA. Thus, the detection of the DNA can indicate that the donor of a blood stain is of human origin, based on the assumption that the crimes involving primate blood are extremely rare in the United States.

The majority of assays for species identification use antibody-based techniques. *Antibodies* are capable of binding to *antigens* (substances that trigger the generation of antibodies). For example, antihuman antibodies are antibodies bound to human antigens and can be used to determine if the sample is a human blood sample. To produce antihuman antibodies, human antigens are usually introduced into a host animal. The host animal generates specific antibodies against the human antigens. The blood is then removed from the host animal to prepare antibodies. The resulting antibody is a *polyclonal* antihuman antibody containing a mixture of antibodies against various human antigens. Additionally, purified human proteins can be used to generate *monoclonal* antibodies. Monoclonal antibodies are specific, homogeneous, and react with a single determinant site of the antigen. The antihuman antibodies utilized should not cross-react with commonly encountered animals but usually have cross-reactivity with higher primates. However, this is not a great concern, due to the fact that crimes involving primates are very rare. An antibody against animal antigens can also be produced using a similar method for determining the animal species in question.

One commonly employed assay is the immunochromatographic assay (Table 8.2). Additional assays that can be utilized are the Ouchterlony assay and the crossed-over immunoelectrophoresis assay, which relies on the binding of an antigen to an antibody causing the formation of a visible precipitate.

8.3.1 Immunochromatographic Assays

A schematic immunochromatographic membrane device is shown in Figure 8.6. The mechanism of the immunochromatographic assays is illustrated in Figure 8.7. The samples can be prepared by cutting out a portion of a stain or scraping off stains from a surface. The sample is then extracted with a small volume of buffer. The extracted sample is loaded into the sample well of the immunochromatographic device. The loaded sample diffuses across the nitrocellulose membrane. The presence of antigens in the sample results in a colored line at the test zone. The immunochromatographic device also utilizes a control zone to ensure that the

TABLE 8.2 Commonly Used Immunochromatographic Assays for Forensic Application

Forensic Application	Assay	Antigen
Blood identification, species identification	ABAcard HemaTrace (Abacus Diagnostics)	Hemoglobin
	RSID-Blood (Independent Forensics)	Glycophorin A
Semen identification	ABAcard p30 (Abacus Diagnostics)	Prostate-specific antigen
	RSID-Semen (Independent Forensics)	Semenogelin
Saliva identification	RSID-Saliva (Independent Forensics)	Human salivary α-amylase

Source: Adapted from Li (2008).

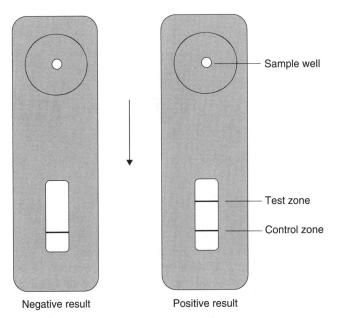

Figure 8.6 Immunochromatographic device. Examples of negative and positive results are shown.

sample has diffused properly along the test strip. Thus, the presence of antigens results in a line at the control zone as well. The assay is considered valid only if the line in the control zone is observed. In contrast, the absence of antigens results in a line in the control zone only. However, a false-negative result may occur in a sample consisting of a very high concentration of antigens. This false-negative result is known as the *high-dose hook effect*. Nevertheless, immunochromatographic assays are rapid, specific, and sensitive and can be used in both laboratory and field tests for species identification.

Most of these assays are based on the detection of human erythrocyte proteins (Lee, 1982; Gaensslen, 1983; Greenfield and Sloan, 2005). For example, purified

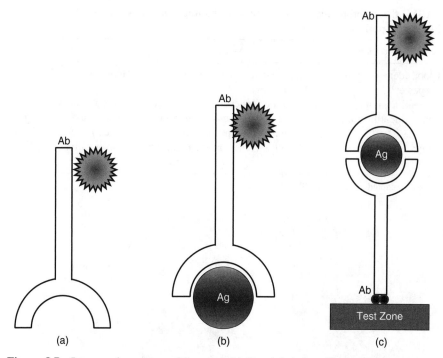

Figure 8.7 Immunochromatographic assay. (a) Dye-labeled antibody (Ab) is placed in the sample well. (b) Sample containing antigen (Ag) is loaded into the sample well and binds to the labeled antibody. This antigen–antibody complex diffuses along the immunochromatographic membrane device. (c) The antigen–antibody complex is captured by an immobilized antibody at the test zone, which forms a colored line for observation.

human hemoglobin can be used to generate antihuman hemoglobin antibodies that detect human hemoglobin. Commercially produced immunochromatographic kits are available, such as the ABAcard HemaTrace kit (Abacus Diagnostics, California). The range of normal blood hemoglobin concentration is 121 to 165 mg/mL. The ABAcard HemaTrace kit can detect very low levels of hemoglobin (as low as 0.07 µg/mL), and specificity studies have shown that the kit is specific for higher primates, including humans. Another immunochromatographic assay utilizes antibodies, recognizing human *glycophorin A* (GPA). Glycophorin A is a human erythrocyte membrane antigen. The RSID-Blood kit (Independent Forensics, Hillside, IL), a commercially produced glycophorin A immunochromatographic kit, is available. The sensitivity of the RSID-Blood kit can be as low as 100 nL of human blood. Species specificity studies have shown that there is no cross reactivity with various animal species, including nonhuman primates.

8.3.2 Ouchterlony Assay

This assay can be performed in an agarose gel supported by a glass slide or polyester film (Lee, 1982; Greenfield and Sloan, 2005). Wells are created by punching holes in the gel layer at desired locations. For example, a pattern with six wells surrounding a center well may be used. An antibody (such as the antihuman antibody) is loaded into the central well with the stain extracts and controls are loaded into the surrounding wells. The diffusion of the antigen and the antibody from the wells is allowed to occur during incubation. In a positive reaction, a line of precipitate forms between each antigen well and the antibody well (Figure 8.8). The gel can then be stained with a dye to enhance the visibility of the line of precipitate. This assay can also analyze more than one antigen in the same assay to determine if the samples in question are from the same or different species of origin.

8.3.3 Crossed-Over Immunoelectrophoresis

Crossed-over immunoelectrophoresis is also known as counterimmunoelectrophoresis (Lee, 1982; Greenfield and Sloan, 2005). In this method, an agarose gel is prepared which contains sample wells opposing each other (Figure 8.9). The antibody and questioned samples are loaded into the wells by pairs. The antihuman antibody is loaded into the wells proximal to the anode, and questioned samples are loaded into the wells proximal to the cathode. The gel electrophoresis is then carried out. During the electrophoresis, the antigen present in a question sample, carrying a negative charge, migrates toward the anode. In contrast, the antibody migrates toward the cathode as a result of flow of fluid. In a positive reaction, a precipitate line forms between the opposing antigen and antibody wells.

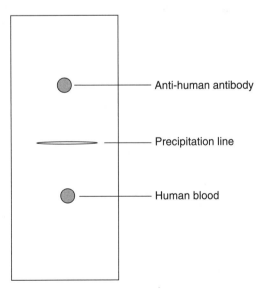

Figure 8.8 Result of an Ouchterlony assay. A precipitate line is observed between a human blood sample and an antihuman antibody.

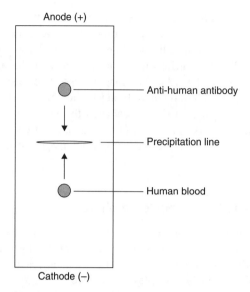

Figure 8.9 Result of a crossover electrophoresis. A precipitate line is observed between a human blood sample and an antihuman antibody.

8.4 IDENTIFICATION OF SEMEN

Normal semen contains seminal fluid and sperm cells (also called *spermatozoa*). The seminal fluid is a mixture of secretions from male accessory glands largely from the seminal vesicles and prostate. Spermatozoa count ranges from 10^7 to 10^8 (spermatozoa/mL of semen) among healthy males. Abnormality in spermatozoa count can be observed among males. For example, *oligospermia* refers to the condition of a male with an abnormally low spermatozoa count. *Aspermia* refers to the condition of a male with no spermatozoa. However, the secretion of the seminal fluid is not affected among these males. A *vasectomy* is a surgical procedure that blocks spermatozoa from reaching the distal portions of the male reproductive tract. A vasectomized male can still produce seminal fluid.

To search for seminal stains on evidence, visual examination of evidence can be facilitated by utilizing alternate light sources. The presence of semen can then be indicated by presumptive assays that detect enzymes present in seminal fluid. The presence of semen can be confirmed by performing confirmatory assays. The most commonly used confirmatory assays are microscopic examination of spermatozoa and immunochromatographic assays that utilize antibody-detecting antigens present in seminal fluid. Recently, RNA-based assays have been developed and are potentially useful for semen identification.

8.4.1 Visual Examination

The visual examination of semen stains can be facilitated by using alternate light sources (Jones, 2005). Semen stains fluoresce when irradiated with alternate light sources. For example, excitation wavelengths from 450 to 495 nm can be used. Fluorescence emitted from a seminal stain can be visualized with colored goggles. Thus, a fluoresced stain indicates the presence of semen and can be tested further using acid phosphatase assays.

8.4.2 Acid Phosphatase Assays

Semen contains high concentrations of *acid phosphatase*, which is secreted primarily from the prostate gland. Thus, acid phosphatase is a useful marker for the identification of semen (Gaensslen, 1983; Greenfield and Sloan, 2005; Jones, 2005). The most commonly used acid phosphatase assays in forensic laboratories are *colorimetric* and *fluororimetric* assays. These assays are based on the principle that acid phosphatase can catalyze a hydrolysis reaction to remove a phosphate group from a substrate.

The acid phosphatase is water soluble and can be transferred from a stain by rubbing the questioned stain area with a moistened cotton swab. The swab can then be used for acid phosphatase assays. The most commonly used colorimetric assay for forensic applications is the use of α-*naphthyl phosphate* as the substrate [Figure 8.10(a)]. The substrate reagent is added to the swab prepared as described above. In the presence of acid phosphatase, the hydrolysis of α-naphthyl phosphate

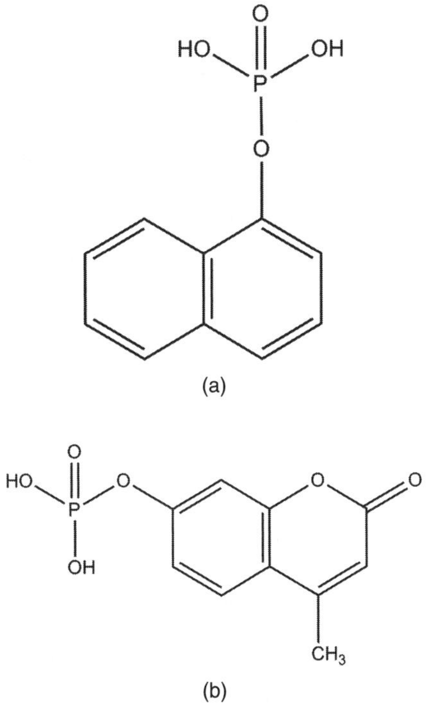

Figure 8.10 Chemical structures of acid phosphatase substrates: (a) α-naphthyl phosphate and (b) 4-methylumbelliferone phosphate.

occurs. By subsequently adding brentamine Fast Blue B solution, a purple precipitate forms at the sites of acid phosphatase activity. The immediate color change to purple indicates the presence of semen.

The fluororimetric assay of acid phosphatase detection is highly sensitive and is generally used for mapping semen stains located on evidence. A piece of moistened filter paper is used to transfer the acid phosphatase from the evidence. The moistened filter paper is overlaid in close contact with the evidence, such as a garment. The filter paper is then lifted from the evidence. Background fluorescence on the filter paper is examined in a dark room with long-wave ultraviolet (UV) light. Any background fluorescence should be marked. The filter paper can then be sprayed with the *4-methylumbelliferone phosphate* reagent [Figure 8.10(b)]. In the presence of acid phosphatase, the hydrolysis of the phosphate residue on 4-methylumbelliferone phosphate occurs, which immediately generates fluorescence under UV light. Semen stains can be visualized as fluorescent areas on the filter paper.

The presumptive assays described above cannot distinguish completely between prostatic acid phosphatase and non-prostatic acid phosphatase *isoenzymes* (multiple forms of acid phosphatase). For example, interference due to contamination with acid phosphatase present in vaginal secretions can occur in specimens collected from victims. Thus, the presence of semen should be confirmed by the performance of confirmatory assays.

8.4.3 Microscopic Examination of Spermatozoa

To prove the presence of semen, light microscopic identification of spermatozoa can be carried out (Jones, 2005). The morphology of human spermatozoa can be characterized as the head, the middle piece, and the tail structures (Figure 8.11). The head contains a nucleus and an acrosomal cap (a membranous compartment at the tip of the head). The head is attached to the midpiece, where the mitochondria are located. The midpiece is attached to the tail, which is a flagellum responsible for the motility of a spermatozoon.

To prepare a sample for microscopic examination, it is necessary to transfer spermatozoa from a questioned stain to a microscope slide. A small portion of a stain is cut and extracted with water followed by gentle vortexing. This suspension containing spermatozoa is then placed onto a slide and is air-dried with low heat. To facilitate the microscopic observation of spermatozoa, a histological staining method, using the *Christmas tree stain*, is utilized to stain the spermatozoa held on the slide. This staining technique utilizes a combination of two dyes. Nuclear fast red is a red dye that stains the nucleus and acrosomal cap of the spermatozoa. Picroindigocarmine is a green dye that stains the neck and tail portions of the sperm. As a result, the nucleus and acrosomal cap are red, while the sperm tails and midpiece are green. In addition to spermatozoa, a victim's epithelial cells can often be observed if the sample is collected from the victim. The epithelial cells are stained blue-green with red nuclei. Today, the spermatozoa from such a slide can be separated from the victim's epithelial cells by utilizing the *laser capture*

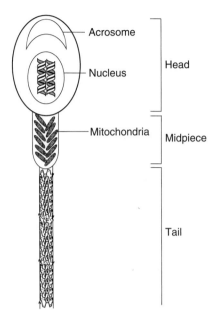

Figure 8.11 Structure of a spermatozoon.

microdissection technique. After the spermatozoa are located on the slide, the head portion of the spermatozoa can be captured. The DNA of the spermatozoa can then be isolated for forensic DNA analysis.

8.4.4 Immunochromatographic Assays

These assays (Jones, 2005) utilize antibodies specific for proteins present in semen such as the *prostate-specific antigen* (PSA) and *seminal vesicle–specific antigens*. PSA (also known as P30 and Kallikrein 3) is a serine protease with a molecular weight of 30 kDa. PSA is produced in the prostate gland and secreted into the seminal fluid. Low levels of PSA can also be found in blood, fecal material, milk, sweat, and urine. Therefore, it can be a concern when a questioned sample is contaminated with those biological materials. Commercially produced immunochromatographic kits are available, such as the ABAcard p30 (Abacus Diagnostics, California). In this device, antihuman PSA antibodies are utilized. The assay is carried out by loading an extracted sample into the sample well. The presence of human PSA results in a pink line at both the test zone and the control zone, while the absence of human PSA results in a pink line in the control zone only. However, the high-dose hook effect, which may cause false-negative results, as described previously, occurs when high quantities of seminal fluid are tested.

Seminal vesicle–specific antigens (also known as *semenogelins*) are the major seminal vesicle–secreted proteins in semen and form a coagulum upon ejaculation. In humans, semenogelin I (SgI) and semenogelin II (SgII) are two major forms of semenogelins. They are present primarily in tissues of the male reproduction system and secreted into seminal fluid. The seminal semenogelin concentration is much higher than seminal PSA concentration. Therefore, the semenogelin test can achieve higher sensitivity than the PSA test. Additionally, semenogelin I and II are present in low amounts in several other tissue types, such as skeletal muscle, kidney, colon, and trachea tissues. However, this is not a great concern since these tissue samples are usually not analyzed for semen detection.

Commercially produced immunochromatographic kits are available, such as the RSID-Semen kit (Independent Forensics, Hillside, IL), in which antisemenogelin antibodies are utilized. The assay is carried out by loading an extracted sample into the sample well. The presence of semenogelins results in a pink line at both the test zone and control zone, while the absence of semenogelins results in a pink line in the control zone only. The RSID-Semen kit can detect seminal fluid with 5×10^4-fold dilution. However, the high-dose hook effect occurs when high amounts of seminal fluid are tested.

8.4.5 RNA-Based Assays

Recently, RNA-based assays have been developed for the identification of semen (Juusola and Ballantyne, 2003; Nussbaumer et al., 2006). These assays can detect specific mRNA that is present exclusively in the spermatozoa and the male accessory glands. For example, the tissue- or cell-specific genes utilized for seminal fluid identification are *KLK3, PRM1*, and *PRM2*. *KLK3* encodes for human PSA. *PRM1* and *PRM2* encode for protamine 1 and protamine 2, respectively. Protamines are DNA-binding proteins involved in the condensation of chromatin in spermatozoa.

8.5 IDENTIFICATION OF SALIVA

The human salivary glands produce 1.0 to 1.5 L of saliva a day. Saliva is a fluid composed mostly of water, but also contains small amounts of electrolytes, proteins, and enzymes. Salivary amylase is one of the enzymes that can be found in saliva. The biochemical function of the amylase is to cleave polysaccharides, such as starch, which is a major component of one's diet. Thus, the identification of the amylase indicates the presence of saliva. Two major amylase isoenzymes (multiple forms of amylase) can be found in humans: human *salivary α-amylase* (HSA) and human *pancreatic α-amylase* (HPA). HSA is produced in the salivary glands and HPA is produced in the pancreas. In addition to saliva, a low level of amylase activity is found in other bodily fluids, such as blood, milk, perspiration, semen, tears, and vaginal secretions.

Visual examinations using alternate light sources can facilitate the search for saliva stains (Jones, 2005). The identification of saliva is carried out by detecting amylase in a questioned sample (Gaensslen, 1983; Greenfield and Sloan, 2005; Jones, 2005). These assays can be divided into presumptive and confirmatory assays. Presumptive assays measure the enzymatic activity of amylase. These assays are not HSA specific and cannot distinguish HSA and its isoenzymes. They are not conclusive as to the specific presence of saliva in a sample. False positives can occur due to contamination by HPA and amylases from plant and bacterial sources. The confirmatory assays, including direct detection of HSA proteins, are more specific than the presumptive assays. Additionally, RNA-based assays (Juusola and Ballantyne, 2003) can potentially be useful as a confirmatory assay for saliva identification.

8.5.1 Visual and Microscopic Examination

The search for saliva stains can be facilitated by using alternate light sources. Saliva stains fluoresce when irradiated with alternate light sources. For example, a 470-nm excitation wavelength can be used. Fluorescence emitted from a saliva stain can be visualized with colored goggles. However, the intensity of fluorescence emitted from saliva stains is less than that of seminal stains. Furthermore, a microscopic examination can be carried out to identify buccal epithelial cells from the stain. The presence of buccal epithelial cells indicates a saliva stain.

8.5.2 Identification of Amylase

Starch–Iodine Assay This assay is based on the principle that iodine (I_2) reacts with starch and develops a dark blue color. In the presence of amylase, starch is cleaved to monosaccharides or disaccharides. The monosaccharides or disaccharides do not react with iodine to develop color. The assay can be carried out in a starch-containing agarose gel with sample wells. The questioned sample is then loaded into the sample well. The gel is incubated and then stained with an iodine solution. The starch-containing gel is stained blue. If amylase is present in the sample, it diffuses out from the sample well and cleaves starch in the gel as it diffuses. A clear area around the sample well indicates amylase activity. The size of the clear area is proportional to the amount of amylase present in the sample. Therefore, the amount of amylase present in a questioned sample can be estimated by comparing it to a standard curve (Figure 8.12).

Phadebas Assay Phadebas (Pharmacia) reagent is an insoluble blue dye–labeled amylase substrate commonly used in forensic laboratories. In the presence of amylase, the dye motif of substrates can be cleaved. The cleaved dye motif is soluble in water and develops a blue color. The amylase reaction can be terminated at an alkaline pH by adding a sodium hydroxide solution. The degree of coloration can be measured (as optical density) at 620 nm using a spectrophotometer and is proportional to the amount of amylase. Then the number of amylase units present in a sample can be determined by comparing the degree of coloration to a standard curve.

Amylase tests can also be used to locate possible saliva stains on evidence through the process of amylase mapping. To perform this process, a piece of filter paper containing Phadebas reagent is placed over the evidence to be tested. The

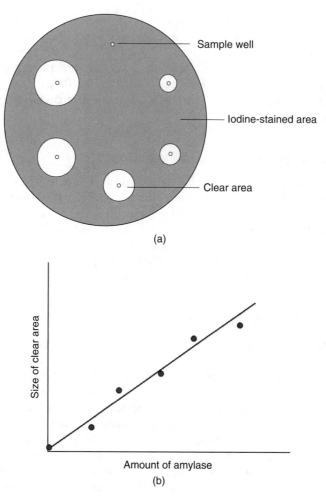

Figure 8.12 Starch–iodine assay for identification of amylase. (a) A series of known concentrations of amylase standards are applied. (b) The size of the clear area resulting from amylase activity is plotted and a liner standard curve (in log scale) can be constructed. The standard curve can be used to determine the amount of amylase present in a questioned sample.

paper is dampened slightly by spraying it with distilled water. The amylase is water soluble and can be transferred from evidence to the moistened filter paper. The amylase transferred to the filter paper reacts with the Phadebas reagent and develops a blue color which corresponds to the site of the saliva stain.

Immunochromatographic Assays Commercially produced immunochromatographic kits are available, such as the RSID-Saliva kit (Independent Forensics). With this device, anti-HSA antibodies are utilized. The assay is carried out by loading an extracted sample into the sample well. The presence of HSA results in a pink line at both the test zone and the control zone, while the absence of HSA results in a pink line in the control zone only. The sensitivity of the RSID-Saliva kit can be as low as 1 μL of saliva.

8.5.3 RNA-Based Assays

Recently, RNA-based assays have been developed for the identification of saliva. These assays can detect specific mRNA that is present exclusively in cells of the oral cavity. Specifically, the tissue- or cell-specific genes utilized for saliva identification are *HTN3* and *STATH*. *HTN3* encodes histatin 3, which is a protein involved in the nonimmune host defense in the oral cavity. *STATH* encodes statherin, which is a salivary protein inhibiting the potentially harmful precipitation of calcium phosphate salts in the salivary glands and the oral cavity.

REFERENCES

Anderson, S., Howard, B., Hobbs, G. R., and Bishop, C. P. (2005). A method for determining the age of a bloodstain. *Forensic Sci. Int.*, **148**:37–45.

DeForest, P., Gaensslen, R., and Lee, H. C. (1983). *Forensic Science: An Introduction to Criminalistics*. New York: McGraw-Hill.

Gaensslen, R. E. (1983). *Sourcebook in Forensic Serology, Immunology, and Biochemistry*. Washington, DC: U.S. Government Printing Office.

Greenfield, A., and Sloan, M. (2005). Identification of biological fluids and stains. In: James, S., and Nordby, J. (Eds.), *Forensic Science: An Introduction to Scientific and Investigative Techniques*, 2nd ed. Boca Raton, FL: CRC Press, Taylor & Francis Group.

Jones, E. (2005). The identification of semen and other body fluids. In: Saferstein, R. (Ed.), *Forensic Science Handbook*, Vol. 2. Englewood Cliffs, NJ: Pearson Prentice Hall.

Juusola, J., and Ballantyne, J. (2003). Messenger RNA profiling: a prototype method to supplant conventional methods for body fluid identification. *Forensic Sci. Int.*, **135**:85–96.

Kobilinsky, L., Liotti, T., and Oeser-Sweat, J. (2005). *DNA: Forensic and Legal Applications*. Hoboken, NJ: Wiley.

Laux, D. L. (2005). The detection of blood using luminol. In: James, S., Kish, P., and Sutton, T. (Eds.), *Principles of Bloodstain Pattern Analysis: Theory and Practice*. Boca Raton, FL: CRC Press, Francis & Talor Group, pp. 369–389.

Lee, H. C. (1982). Identification and grouping of bloodstains. In: Saferstein, R. (Ed.), *Forensic Science Handbook*. Englewood Cliffs, NJ: Prentice Hall.

Lee, H. C., Palmbach, T., and Miller, M. T. (2001). *Henry Lee's Crime Scene Handbook*. San Diego, CA: Academic Press.

Li, R. (2008). *Forensic Biology*. Boca Raton, FL: CRC Press, Taylor & Francis Group.

Marie, C. (2008). Presumptive testing and enhancement of blood. In: Bevel, T., and Gardner, R. (Eds.), *Bloodstain Pattern Analysis*. Boca Raton, FL: CRC Press, Taylor & Francis Group.

Nussbaumer, C., Gharehbaghi-Schnell, E., and Korschineck, I. (2006). Messenger RNA profiling: a novel method for body fluid identification by real-time PCR. *Forensic Sci. Int.*, **157**(2–3):181–186.

Saferstein, R. (2007). *Criminalistics: An Introduction to Forensic Science*, 9th ed. Upper Saddle River, NJ: Prentice Hall.

Shaler, R. C. (2002). Modern forensic biology. In: Saferstein, R. (Ed.), *Forensic Science Handbook*, 2nd ed. Upper Saddle River, NJ: Person Education, pp. 525–614.

Sutton, T. P. (1999). Presumptive blood testing. In: James, S. H. (Ed.), *Scientific and Legal Applications of Bloodstain Pattern Interpretation*. Boca Raton, FL: CRC Press.

CHAPTER 9

Forensic DNA Analysis

HENRIETTA MARGOLIS NUNNO

Department of Sciences, John Jay College of Criminal Justice, The City University of New York, New York

Summary In this chapter we describe how DNA became a valuable forensic tool in identifying the source of physical evidence left at a crime scene. The use of restriction fragment length polymorphism analysis in the mid-1980s was replaced by polymerase chain reaction (PCR) methods, which were more sensitive, requiring far less high-molecular-weight DNA; used less hazardous materials; and were more rapid and more economical. PCR-short tandem repeat–based genetic profile typing methods have improved in sensitivity over the past 20 years and have become a basic tool in the crime lab. Where nuclear DNA is insufficient to generate a full genetic profile, mitochondrial DNA can be used to provide identifying information. The chapter also covers low-copy-number procedures and the typing of single-nucleotide polymorphisms within the human genome.

9.1	Introduction	292
	9.1.1 Background on DNA typing	292
	9.1.2 DNA structure	294
	9.1.3 Nuclear and mitochondrial DNA organization	295
9.2	Methodology	296
	9.2.1 Sample collection and DNA extraction	296
	9.2.2 DNA quantification	297
	9.2.3 Polymerase chain reaction	298
	9.2.4 Short tandem repeats	298
	9.2.5 PCR of STRs	300
	9.2.6 Separation and sizing of STR alleles	301
	9.2.7 Combined DNA index system (CODIS) database	305
	9.2.8 Frequency and probability	306
9.3	Problems encountered in STR analysis	307
	9.3.1 Low-copy-number DNA	307

Forensic Chemistry Handbook, First Edition. Edited by Lawrence Kobilinsky.
© 2012 John Wiley & Sons, Inc. Published 2012 by John Wiley & Sons, Inc.

9.3.2	Degraded DNA and reduced-size (mini) STR primer sets	308
9.3.3	PCR inhibition	310
9.3.4	Interpretation of mixtures of DNA	310
9.3.5	Null alleles and allele dropout	311
9.3.6	Factors causing extra peaks in results observed	312
9.3.7	Stutter product peaks	312
9.3.8	Nontemplate addition (incomplete adenylation)	313
9.3.9	Technological artifacts	313
9.3.10	Single-nucleotide polymorphism analysis of autosomal DNA SNPs	313
9.3.11	Methods used for SNP analysis	314
9.3.12	Mitochondrial DNA analysis	315
9.4	Methodology for mtDNA analysis	316
9.4.1	Preparation of samples	316
9.4.2	MtDNA sequencing methods	316
9.4.3	Reference sequences	317
9.4.4	Screening assays for mtDNA	318
9.4.5	Interpretation of mtDNA sequencing results	319
9.4.6	Statistics: the meaning of a match for mtDNA	320
9.4.7	Heteroplasmy	320
9.4.8	The future of DNA analysis	321
	References	322

9.1 INTRODUCTION

9.1.1 Background on DNA Typing

The structure of DNA was first elucidated by Watson and Crick in 1953. It was shown to take the form of a double helix made by two parallel strands, each of which is composed of four different nucleotides (Figure 9.1) linked together in a specific sequence (Watson and Crick, 1953). While the sequence of the DNA nucleotides is virtually identical in every cell in a person's body, different people differ in portions of their DNA sequences, and each sequence difference is known as a *polymorphism*. In addition, certain sequences of nucleotides are repeated numerous times, frequently one after another, and because people may differ in the number of times that each of these tandem sequence repeats is repeated, there are even more polymorphisms between individuals. It is the examination of such a DNA sequence or repeat number (length) polymorphism that is the basis of DNA identification.

DNA profiling or *typing* (originally called *DNA fingerprinting*) was first used for forensic purposes in the 1980s. The methodology, first described by Alec Jeffreys, used the differences in length of DNA regions created by variations in numbers of repeated sequences to distinguish between individuals (Jeffreys et al., 1985). The repeated regions of DNA that Jeffreys examined are called *variable number tandem repeats* (VNTRs) and the DNA profiling technique that he developed is called

INTRODUCTION

Figure 9.1 DNA structure. The two parallel strands of DNA take the form of a double helix, with each strand composed of four different nucleotides. A is adenine, T is thymine, G is guanine, and C is cytosine.

restriction fragment length polymorphism (RFLP). RFLP uses restriction enzymes, which cut the DNA at specific sequences of bases called *restriction endonuclease recognition sites*, to chop up large pieces of double-stranded DNA in a specific manner. RFLP identifies people by the pattern their DNA segments make when they are separated by size and then hybridized with complementary DNA probes. Because different people may differ in the particular endonuclease recognition sites that are present in their DNA and may also differ in the number of repeats that are present between their recognition sites, the length and therefore the pattern of the various DNA fragments produced by the restriction enzyme digestion will differ from person to person when separated by size.

After the polymerase chain reaction (PCR) was developed in the mid-1980s (Saiki et al., 1985; Mullis, 1990), DNA typing methods began incorporating PCR technology and RFLP was phased out for forensic DNA identification. Because PCR-based methods increase the copy number of DNA fragments to the millions or billions, they greatly reduce the amount of crime scene DNA necessary for DNA identification. In addition, because PCR-based methods are used to amplify relatively small DNA fragments, there is no need for the undegraded DNA required for RFLP. Although the first PCR-based tests used for human identification did not have a high power of discrimination, over the years DNA typing technology has evolved into the current PCR-based methods, which examine short tandem repeat (STR) loci and have powers of discrimination of one in more than a trillion.

9.1.2 DNA Structure

In the Watson and Crick model of DNA structure, each of the two parallel strands forming the double helix is made of linked nucleotides. Each nucleotide contains a phosphate group, a five-carbon sugar (deoxyribose), and one of four nitrogen-containing bases: adenine, thymine, guanine, or cytosine, referred to as A, T, G, and C, respectively. A and G are double-ringed purines and T and C are single-ringed pyrimidines (Figure 9.2). The nucleotides in DNA are linked by bonds (phosphodiester) between the 3'-hydroxyl and 5'-hydroxyl groups of the deoxyribose sugars and the phosphate groups, forming a sugar phosphate backbone on the outside of the double helix [Figure 9.3(a)]. The bases project into the inside and each strand is connected to what is called its *complementary strand* by hydrogen bonding between specific bases. In this complementary base pairing, adenine always pairs with thymine (by forming two hydrogen bonds) and guanine always pairs with cytosine (by forming three hydrogen bonds) (Figure 9.3). There are about 10 nucleotide pairs for each turn of the helix. Polynucleotides are read from the 5' end to the 3' end; since the two polynucleotide chains of the helix run in opposite 5'-to-3' directions, they are said to be *antiparallel* [Figure 9.3(b)].

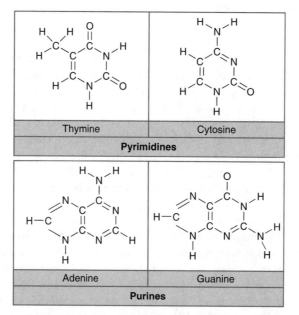

Figure 9.2 Pyrimidines and purines in DNA. The single-ring pyrimidines, thymine (T) and cytosine (C), are shown in the upper panel, and the double-ring purines, adenine (A) and guanine (G), are shown in the lower panel.

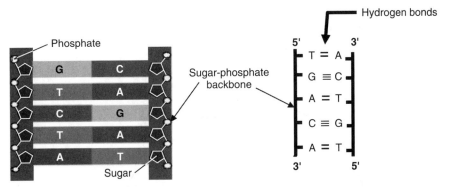

Figure 9.3 Complementary base pairing in DNA structure. (a) A sugar phosphate backbone is on the outside and the four different bases are on the inside of the DNA molecule (left panel). (b) There is base pairing by hydrogen bonding between complementary bases A and T (with two hydrogen bonds) and G and C (with three hydrogen bonds) and both strands run from 5′ to 3′ in opposite directions.

9.1.3 Nuclear and Mitochondrial DNA Organization

Within most human cells DNA is present in both the nucleus and the cytoplasm; cytoplasmic DNA is found in energy-producing organelles called *mitochondria*. Although the DNA found in these two different locations has the same basic double-stranded structure, many differences exist as well.

Nuclear DNA molecules form linear chromosomes and mitochondrial DNA (mtDNA) molecules form circular chromosomes. Whereas nuclear chromosomes are present as pairs (homologs), with one chromosome of each pair received from each parent, mtDNA is inherited uniparentally only through the maternal line. While the nucleus of a typical human cell contains 46 chromosomes made up of 22 pairs of nonsex chromosomes (autosomes) and two sex chromosomes, the total number of mitochondria in a cell varies greatly with the type of cell and its stage of development. Although a set of 23 nuclear chromosomes contains about 3.1 billion nucleotides, the total number of nucleotides in a single mitochondrial chromosome is only 16,569.

Both nuclear and mitochondrial chromosomes contain coding sequences (nucleotides that make up the genes) and noncoding nucleotide sequences whose function is largely unknown. But while the coding and noncoding sequences are distributed intermittently along the length of each nuclear chromosome, in the mitochondrial genome the coding and noncoding areas are separate from each other. The noncoding portion of mtDNA is known as the *control region* (also known as the *D-loop* or *displacement loop*) and it contains about 1100 base pairs (bp), which encompass the two polymorphic hypervariable regions (HV1 and HV2) that are generally used for forensic mtDNA sequencing analysis (Figure 9.4). HV1 occupies positions 16,024 to 16,365, and HV2 occupies positions 73 to 340 in the control region of the mitochondrial chromosome.

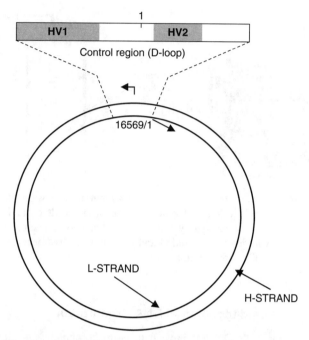

Figure 9.4 Mitochondrial DNA: a schematic representation of the 16,569 bp of the circular mitochondrial DNA genome. Both heavy (H-, outer circle) and light (L-, inner circle) strands are indicated, as are the positions of HV1 (composed of bp 16,024 to 16,365) and HV2 (composed of bp 73 to 340), which are the regions used in forensic mtDNA sequencing.

In addition to differences in organization and number, there are differences in the way that nuclear and mitochondrial DNA molecules are transferred from one generation to the next. Nuclear chromosomes, which are inherited from both parents, undergo genetic recombination, whereas mitochondrial chromosomes, which are inherited from only one parent, do not.

9.2 METHODOLOGY

9.2.1 Sample Collection and DNA Extraction

The isolation of genomic DNA from crime scene evidence samples is complicated by variations in both the quality and quantity of the material that is to be examined. Samples used for DNA analysis may be taken from varied materials, such as fabric, contact swabs, or cigarette butts, and the DNA in these samples may originate from various sources (e.g., blood, saliva, semen, skin). To serve as evidence in court, the biological material must be collected carefully and a chain of custody must be established. In addition, reference DNA samples are necessary to provide material for comparison to crime scene evidence; these can come from convicted offenders, suspects, and/or people who may have left their DNA at the scene merely because they happened to be in the area (elimination specimens).

As methodologies improve, crime scene evidence samples containing less and less DNA (such as those from single fingerprints) are being analyzed. When samples contain very little DNA, purification methods must yield the maximum amount of DNA in the smallest volume possible. Samples containing very small amounts of DNA (usually <100 pg) fall into a category known as low-copy-number DNA (Gill et al., 2000).

The first step in DNA analysis, DNA extraction, involves isolation of the DNA and its separation from other cell components. (Where applicable, cells may first be separated from their substrate swab or fabric by agitation in a minimal amount of liquid.) There are various possible DNA extraction methods, and when dealing with crime scene samples, the type of evidence and the amount of DNA it contains will help determine the extraction method used. Common forensic DNA extraction methods include the use of Chelex beads, solid-phase commercial kits, or organic extraction. [Various methods for manual sample preparation and extraction are described in detail by Butler (2005, Chap. 3).] Whatever extraction method is used, care must be taken to avoid contamination between the samples as well as contamination by extraneous DNA.

For many years most crime labs used the Chelex method for DNA extraction. Chelex is an ion-exchange resin that protects DNA by binding to magnesium ions, thereby inhibiting the function of the magnesium, which requires enzymes that destroy DNA. Cells are first broken open and their DNA released by boiling the cellular material in the presence of Chelex beads. After the centrifugation step that follows, the Chelex and all cellular components except the DNA form a tight pellet at the bottom of the tube. The supernatant liquid containing the extracted DNA is then transferred to a new tube and usually stored frozen at $-20^\circ C$ or $-80^\circ C$ until it is used for analysis.

There is currently a backlog of samples from convicted offenders that must be analyzed and entered into the national database. To ease this backlog and to analyze reference and elimination samples, many laboratories now use commercial kits and/or robotic instruments to prepare these samples, which have abundant DNA and are of a similar type. In addition, new methods and equipment that greatly improve DNA recovery are now available for sample collection, extraction, and quantification, including individualized collection and extraction kits and robotic extraction methodologies (see Brettell et al., 2005, 2007, 2009).

9.2.2 DNA Quantification

Because methods for DNA analysis usually require that the DNA be within a specific concentration range, the amount of human DNA present in an extracted sample must be measured before analysis. In most of the methods used for determining DNA concentration, results for extracted DNA are compared to those for calibrated standards that contain known quantities of DNA. Recent DNA quantification methods have moved from slot blot, to real-time (rt) or quantitative (q)

PCR-based methodology; rtPCR-based methods are more exact than blot-based procedures and can quantify samples with much lower DNA concentrations. Because rtPCR measures the change in fluorescence due to the increase in amplified product that occurs during the initial PCR cycles, it not only quantifies the DNA but indicates how much amplifiable DNA is present in the sample (Butler, 2005, Chap. 4). Kits such as the Quantifiler Human DNA Quantification kit from Applied Biosystems, Inc. are now being used to determine the amount of human DNA present in a sample by rtPCR. In addition, a Quantifiler Duo Quantification kit (Applied Biosystems), which simultaneously quantifies total human and human male DNA, has recently been validated (Barbisin et al. 2009).

9.2.3 Polymerase Chain Reaction

PCR is a process that amplifies DNA and makes millions to billions of copies of a specific DNA sequence through the use of primers. Primers are short segments of DNA (usually 18 to 30 nucleotides in length) that specify which region of the DNA will be copied. For each sequence amplified, two different primers (a forward and a reverse) are used; each primer attaches to a complementary region on one of the two separated DNA strands and designates the beginning or end of the DNA segment to be copied (Figure 9.5). By constructing primers with the appropriate nucleotide sequences, any specific segment of DNA may be amplified. In addition to the primer and the template DNA that is to be copied, a sample undergoing PCR must contain a special type of DNA polymerase which has its optimum function at high temperatures (e.g., *Taq* polymerase, which comes from bacteria native to hot springs), the four building blocks of DNA in a deoxynucleotide triphosphate (dNTP) form (i.e., dATP, dTTP, dGTP, and dCTP), a buffer solution, and salts (e.g., $MgCl_2$) necessary for the DNA polymerase to function.

The PCR process is similar to cellular DNA replication, but with PCR it is temperature changes that accomplish amplification of the DNA. A PCR cycle (see Figure 9.5) consists of three steps. Each cycle starts with a denaturation step where the double-stranded DNA sequence to be copied is separated into two strands by heating to a temperature of about 94°C. The temperature is lowered to about 60°C in the next step, annealing, and the two primers bind by complementary base pairing to each of the single-stranded segments of the target DNA. In the last step, extension, the temperature is raised to about 72°C and the DNA polymerase replicates the DNA by adding the appropriate nucleotide bases (dNTPs) through complementary base pairing. By doubling the amount of DNA with each cycle, the total amount increases exponentially as PCR progresses. The three-step PCR cycle is usually repeated 28 to 30 times for standard forensic DNA analysis.

9.2.4 Short Tandem Repeats

Currently, forensic DNA identification is usually done by PCR-based analysis of short tandem repeat (STR) loci. With PCR of STRs, only small quantities of biological material are necessary and the small PCR product size (about 100 to 500 bp) allows even partially degraded DNA to give full DNA profiles.

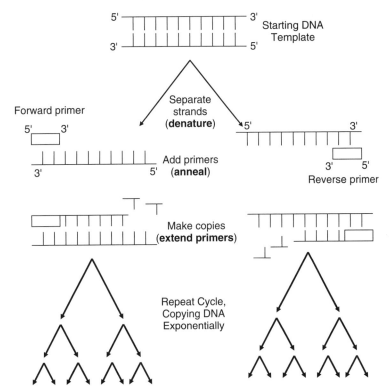

Figure 9.5 DNA amplification by polymerase chain reaction (PCR). A PCR cycle consists of the three steps depicted above, which involve changes in the temperature of the sample. In the first step, denaturation, the two DNA template strands are separated by heat. In the next step, annealing, the sample is cooled to a temperature appropriate for the two primers (forward and reverse) to bind. In the last step, extension, the temperature of the sample is raised to the optimal temperature for the DNA polymerase to extend the primers and produce a copy of each DNA template strand. In each cycle, the number of DNA molecules (containing the sequence between the two PCR primers) doubles and the number of copies increases geometrically. [From Butler (2005), copyright ©2005 Elsevier Academic Press, Fig. 4.2.]

STRs (also known as *microsatellite DNA*) are loci that contain units of short sequences of DNA (two to seven, usually four nucleotides long) that are tandemly repeated (Figure 9.6) anywhere from about a half dozen to several dozen times (see Butler, 2005, Chap. 5). There is a great deal of variation in the human population in the number of times the units of four bases are repeated in an STR, and this length variability is used to distinguish between DNA profiles. Polymorphic STR sequences, which vary throughout the population in the number of repeat units displayed, can be found in all parts of the genome, including introns, exons, and flanking portions of coding areas of the DNA as well as in extragenic areas (Economou et al., 1990; Edwards et al., 1991; Beckman and Weber, 1992). There

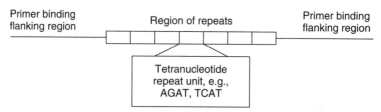

Figure 9.6 Schematic representation of a short tandem repeat (STR) locus. Most STR loci used for forensic DNA analysis contain units of four bases (tetranucleotides) which are repeated one after another. The repeat portion of the locus is surrounded by flanking regions that contain the DNA segments to which the primers bind. One of the primers (usually the forward one) is labeled with a fluorescent dye of a particular color so that the alleles of the STR being amplified can be identified when its size overlaps with that of a different STR.

are probably more than 1 million STRs in the human genome, and among the more than 20,000 tetranucleotide STR loci that have been characterized (Butler, 2006), a set of 13 core loci have been selected for use in forensic DNA profiling in the United States. These 13 STRs are required for a full DNA profile to be uploaded into the FBI's Combined DNA Index System (CODIS) database. The existence of these core loci allows equivalent genetic information to be shared by national, and in many cases international, databases. The 13 core STR loci are used for various forms of human identity testing, including DNA databasing, forensic casework, missing persons/mass disaster victim identification, or parentage testing.

For each of the 13 core STR loci, a person inherits one maternal allele and one paternal allele. Because these loci are either located on different chromosomes or far enough apart on the same chromosome, the 13 core STR loci assort independently of each other during cell division. This allows their population frequencies to be multiplied by each other to obtain a high power of discrimination.

In addition to their presence on autosomal chromosomes, STRs are also found on the sex chromosomes. Y-chromosome STRs, which are present only in males, have been studied for many years as aids in forensic DNA analysis, and commercial kits are now available for their detection. Since Y chromosomes are inherited uniparentally and only one Y chromosome is present in a male individual, Y-STRs do not have the power of discrimination of core autosomal STRs. However, Y-STRs are frequently used to gain additional information when full autosomal STR profiles are lacking and to help distinguish DNA profiles in samples containing both male and female DNA and/or mixtures of DNA from more than one male. STRs can also be found on X chromosomes and, although not used for forensic DNA typing, many of these have been identified and characterized (Gomes et al., 2009; also see Brettell et al., 2005, 2007, 2009).

9.2.5 PCR of STRs

The primers used for PCR-based STR analysis are complementary to the regions that lay outside (flank) the area of the tandem repeats (Figure 9.6), and one of the two PCR primers is labeled with a fluorescent dye. PCR amplification is usually performed on multiple STR loci simultaneously (i.e., by multiplexing). Different fluorescent dyes are used to label the loci that overlap in size so that these STRs can also be distinguished by color. Simultaneous amplification of many loci enables a high power of discrimination to be attained in a single test without the use of too much DNA (e.g., only 1 ng or less of starting material is necessary). Commercial kits containing fluorescent-labeled primers are available for both autosomal STR and Y-STR analysis.

Numerous multiplex STR test kits are available for DNA typing. Newer multiplex kits increase the number of genetic loci that may be amplified and analyzed simultaneously. To analyze all 13 CODIS loci plus the sex-typing locus, amelogenin, many laboratories use a combination of two commercial kits, either the AmpF*l*STR Cofiler and AmpF*l*STR Profiler Plus, or the GenePrint PowerPlex 1.1 and GenePrint PowerPlex 2.1. Some labs use a newer multiplex kit (i.e., the AmpF*l*STR Identifiler or the PowerPlex 16 System), which can amplify the 13 core STR loci plus amelogenin and two extra STR loci in a single reaction. AmpF*l*STR kits are manufactured by Applied Biosystems Inc. (ABI) and PowerPlex kits by Promega Corp. Four different fluorescent dyes are used in the ABI Profiler and Cofiler kits and in all the Promega kits, while the ABI Identifiler kit uses five dyes.

Commercial kits for STR analysis come with solutions of labeled primers whose relative concentrations are appropriate for multiplexing and a "master mix" in which buffers, dNTPs, and necessary salts are already included. The ease and consistency achieved by these kits in generating STR profiles on the same set of core STR loci make national sharing of criminal DNA profiles possible. In addition, STR kits supply fragment sizing standards as well as allelic ladders containing common STR alleles that have been previously sequenced and characterized for the number of repeat units. The sizing standards are used to determine PCR product size, while the allelic ladders calibrate fragment size in relation to STR repeat number for genotyping purposes.

9.2.6 Separation and Sizing of STR Alleles

Electrophoresis exposes DNA to an electrical field that is used to separate DNA fragments. Because of its negative charge in solution, DNA always moves from a negative to a positive electrode when current is applied. Two types of electrophoresis, slab gel and capillary electrophoresis (CE), have been used for forensic DNA analysis. In slab gel electrophoresis, the separation method that was used for RFLP analysis and for early STR analysis, the DNA samples are placed in separate lanes (depressions) on a gel covered with a buffer solution. When current is applied, shorter DNA pieces move faster than longer pieces and the DNA fragments in each lane are separated by size. Bands of DNA are formed by fragments of the same size migrating together. Sample bands are compared to those of DNA fragment sizing standards (and/or a ladder of common STR alleles) and the length of the DNA in each band is estimated by matching the distance it traveled to a ladder standard that traveled a similar distance. The DNA bands formed in RFLP were

302 FORENSIC DNA ANALYSIS

detected by hybridization to labeled probes, while STR gel detection first involved silver staining of bands. More recent methodologies for STR analysis by either gel electrophoresis or CE use detection of PCR primers labeled with different color fluorescent dyes.

In forensic DNA labs today, CE is the method generally used for STR profiling. The separation, detection, and analysis of STR fragments of different sizes is accomplished using CE instruments such as the single-capillary ABI 310 or the 16-capillary ABI 3100 Genetic Analyzers made by Applied Biosystems Inc. (Butler et al., 2004); laboratories with very high throughput may use the ABI 3700 Genetic Analyzer, which has 96 capillaries that operate simultaneously.

CE is fairly similar to gel electrophoresis, except that the separation of fragments takes place in a long narrow glass tube or capillary that is filled with a polymer solution. Samples containing PCR amplified, fluorescently labeled STRs are first added to the capillary and when the current is applied the smaller fragments move faster in the polymer than the larger ones. As they pass through the capillary, the STR fragments are exposed to a laser beam that causes the fluorescent dyes in their primers to emit light of specific wavelengths. Sizing standard fragments labeled with a different dye color are also added to each sample tube. This allows a comparison of test sample fragment movement (measured as migration time) with

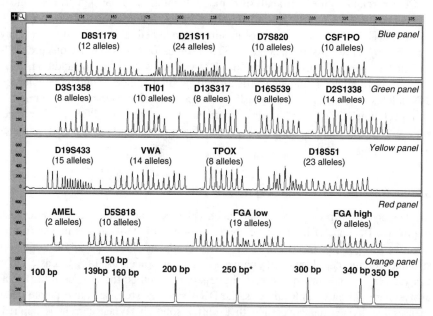

Figure 9.7 Allelic ladders for the AmpF/STR Identifiler kit (Applied Biosystems). This allelic ladder, which includes a total of 205 alleles, is used for genotyping a multiplex PCR reaction involving 15 STRs and the sex-typing test for amelogenin (AMEL). The kit uses five dyes, the colors of which are shown in the upper right portion of each panel. The upper four panels show possible alleles for each locus, and the bottom panel shows the peaks produced by the internal DNA sizing standard that is used with this kit. [From Butler (2005), copyright ©2005 Elsevier Academic Press, Fig. 5.6.]

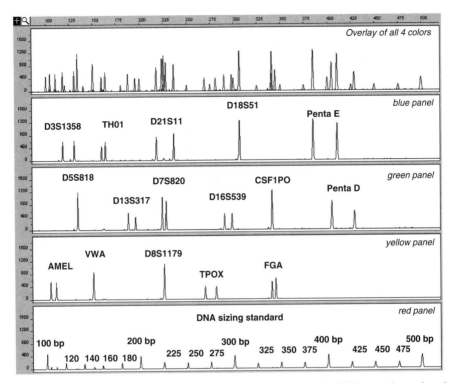

Figure 9.8 Electropherogram showing results for a 1-ng genomic DNA sample produced using the PowerPlex 16 (Promega) kit. This multiplex kit uses four dyes, the colors of which are shown in the upper right portion of each panel, for the labeling of 15 STRs and the amelogenin (AMEL) sex-typing locus. The top panel is an overlay of all four lower panels, including the red lower panel for the internal DNA sizing standard. [From Butler (2005), copyright ©2005 Elsevier Academic Press, Fig. 5.5.]

that of the sizing fragments. A fluorescently labeled allelic ladder sample made up of most of the known alleles for a particular locus is present in the same run and serves as the standard for naming the genotype of the alleles. The allelic ladder for the widely used AmpF/STR Identifiler kit (Applied Biosystems) is shown in Figure 9.7. It contains 205 of the alleles present in the 16 loci that are amplified together. These loci represent 15 STRs plus an amelogenin sex-typing assay.

CE data are presented in the form of an electropherogram (e.g., Figure 9.8) which shows a series of peaks, each of which represents the relative fluorescence of an amplified fragment (allele) plotted as relative fluorescence units (rfu) versus migration time (which is related to size). [Figure 9.8 is an electropheragram of results obtained using the PowerPlex 16 System (Promega), which also amplifies the 13 core loci, two extra STRs, and amelogenin in a single reaction.] In addition, CE instrument software calculates the size and genotype of the various alleles in the electropherograms. A peak is identified as a particular allele based on its fluorescent color emission and its length in base pairs, which should match that

304 FORENSIC DNA ANALYSIS

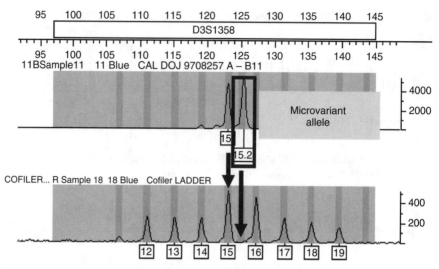

Figure 9.9 Detection of a microvariant allele for D3S1358. When genotyping STRs, alleles are named by a comparison of sample peaks to those of allelic ladders. The sample in the top panel is compared to the allelic ladder for the D3S1358 locus shown in the bottom panel. (Peaks are labeled with the allele number by use of an internal sizing standard.) The allele labeled 15.2 in the top panel does not match any of the ladder alleles. It is a microvariant whose size is two bases larger than allele 15 in the ladder. (www.cstl.nist.gov/STRbase)

of a fragment in the allelic ladder. Alleles that do not match ladder standards are known as *off-ladder* or *microvariant alleles*; these are usually far less common than those in the allelic ladder and therefore may be more statistically significant for human identification. Standard STR allele genotypes are named for the number of repeats they contain. Microvariant alleles contain an incomplete repeat (fewer than four bases) in addition to complete repeats and therefore fall between two allelic ladder standards; they are named by giving the number of full repeats followed by a decimal and then the number of bases in the incomplete repeat (see Figure 9.9).

Like any other experimental, scientific procedure, DNA analysis, requires the inclusion of controls. Negative controls are samples that do not contain any DNA but are subjected to all the same procedures as the samples being analyzed. (Negative controls monitor contamination starting from amplification, whereas reagent blanks, which are frequently used in mitochondrial DNA analysis, are negative controls that monitor contamination starting from extraction.) When no contamination is present, these controls will be negative for DNA. Substrate controls may be run if applicable, and these are specimens taken from an area adjacent to the area (stain) of interest to ensure that the DNA found in the sample did not come from its underlying fabric. Unlike negative controls, a substrate control may contain DNA which may or may not be relevant to the case. To indicate that the test is being done correctly, a known positive control must be included with every analysis and must produce the expected alleles. If negative controls are shown to contain DNA, or if a positive control does not produce the alleles expected, the test is considered inconclusive and the analysis of all samples must be repeated.

DNA fragments of different sizes, whether they come from a heterozygous STR locus or from different STR loci, have different migration times and are therefore distinguishable by CE. For identification purposes, discrimination power increases as more STR loci are studied, and as stated above, at least 13 STRs are required for a DNA profile to be uploaded into the CODIS offenders database. By amplifying multiple STR loci in the same PCR reaction and running them together in the same capillary, evidence is conserved and time and effort expended for DNA analysis are decreased. Many of the alleles of the 13 core STR loci overlap in size, and because of this, primers for each of these overlapping loci must be labeled with different fluorescent dyes so that they may be distinguished from one another.

9.2.7 Combined DNA Index System (CODIS) Database

The 13 core loci that are currently the basis of the U.S. CODIS database were selected in November 1997. The national CODIS database or national DNA index system (NDIS) contains profiles that have been uploaded from various state DNA testing laboratories through their respective systems. A laboratory that wishes to enter data into the CODIS database must meet certain standards to ensure the reliability of its test results.

The national database includes the following 13 core STR loci: D3S1358, D16S539, THO1, TPOX, CSF1PO, D7S820, VWA, FGA, D8S1179, D21S11, D18S51, D5S818, and D13S317. All 13 of these loci are highly polymorphic, and 11 of them are located on different chromosomes. Although CSF1PO and D5S818 are on the same chromosome, they are far enough from each other to assort independently during cell division. None of the core loci are located in coding regions, so they have no functional significance.

Several databases make up CODIS, including a convicted offender database, an evidentiary database, and other specialized databases. DNA obtained from crime scene evidence is included in the evidentiary database. Offenses that require submission of a DNA specimen to the convicted offender database vary from state to state, and the data uploaded to CODIS must meet certain standards. State and local data that differ from those meeting CODIS standards can be used on a local level only. Although most states originally included only those offenders who had been convicted, and usually only those convicted of violent felonies, some states either now include or will soon include profiles of all persons arrested for any serious criminal activity.

Most European countries (including the United Kingdom) uses 10 core STR loci in their DNA databases, and these include eight of the CODIS loci: VWA, FGA, THO1, D8S1179, D21S11, D18S51, D3S1358, and D16S539 (Butler, 2006). The two remaining European core STR loci (D2S1338 and D19S433) can be amplified with the AmpF*l*STR Identifiler kit, which amplifies a total of 15 STR loci in one reaction (Collins et al., 2004).

9.2.8 Frequency and Probability

STR analysis may demonstrate that the genetic profile from the crime scene sample differs from that of the suspect and thus excludes or eliminates the suspect as a potential donor of the evidence. However, if the genetic profiles of the suspect and the crime scene sample are the same, the suspect is included. Thus, DNA evidence can either link a suspect to a crime scene or prove that a person accused incorrectly is innocent of a crime. As of August 2009, 273 prisoners have been exonerated from crimes they did not commit through the use of DNA evidence.

If there is a DNA match, the meaning of that match must be determined. The probability of finding an identical genetic profile in a particular population depends on the frequency of the profile in that population. Because the United States is populated by people from all over the world, it is sometimes difficult to establish the population or populations of origin for a person. The FBI lists the following five large ethnic groups in their CODIS database: (1) African-American, (2) Asian-American, (3) European-American, (4) Southeastern-Hispanic, and (5) Southwestern-Hispanic.

Many data are available giving the frequencies of the core STR alleles in various populations groups, and Table 9.1 shows the frequencies of the most common alleles of some CODIS STR loci in a Caucasian (European-American) population using data from a study conducted by Butler et al. (2003a).

To calculate the probable or expected frequency of a heterozygote in a population, one would multiply the individual allele frequencies by each other and then multiply that product by 2. Since there are two different ways that a heterozygote can be formed (i.e., either one of the two alleles could have come from either the sperm or the egg), the product of the allele frequencies must be doubled. Homozygote probability is calculated differently since only one set

TABLE 9.1 STR Allele Frequencies, U.S. Population Data[a]

Locus	Allele	Caucasian ($n = 302$)	African-American ($n = 258$)	Hispanic ($n = 140$)
CSF1PO	10	0.21689	0.25681	0.23214
	11	0.30132	0.24903	0.29286
	12	**0.36093**	**0.29767**	**0.35714**
TPOX	8	**0.53477**	**0.37209**	**0.47143**
	9	0.11921	0.17829	0.10357
	11	0.24338	0.21899	0.27500
FGA	21	0.18543	0.11628	**0.16786**
	22	**0.21854**	**0.19574**	0.15000
	23	0.13411	0.17054	0.13571
VWA	16	0.20033	**0.24806**	**0.26429**
	17	**0.28146**	0.24225	0.21786

Source: Data from Butler et al. (2003a).

[a] Frequencies for some of the most frequent alleles are given for four CODIS loci. Alleles with highest frequencies are indicated in bold.

of circumstances (i.e., where both the sperm and the egg carry the same allele) leads to homozygote formation. Thus, the frequency of the homozygous allele is multiplied by itself or squared. As an example, in the Caucasian population in Table 9.1, the probability of a person being a 12,11 heterozygote for the CSF1PO locus would be $0.36093 \times 0.30132 \times 2$ (or 0.10876), while the probability that they are an 8,8 homozygote for the TPOX locus would be 0.53477×0.53477 (or 0.28598).

To determine the probability of a "random match" for a genetic profile which contains, for example, two of the 13 STRs, the frequency calculated for each of the two genotypes is multiplied by the other because the two loci are inherited independently. Because the chance of multiple genotypes occurring together in the same person is the product of their individual probabilities, calculations of this type follow what is known as the *product rule*. Thus, the frequency of a person possessing the genotype for CSF1P0 and TPOX described in the paragraph above would be 0.10876×0.28598, or 0.03110. As a genotype for an additional STR is added to a profile, its frequency is multiplied by the product of the frequencies of the other STRs examined.

For the European-American population, the probability of a random match for the genotype that includes all of the most frequent alleles of the 13 CODIS STRs greatly exceeds the entire population on Earth. Thus, one can say with virtual certainty that this genetic profile would be unique on this planet. In addition, all other profiles for the 13 core STRs in the various population groups provide virtual certainty that a particular genetic profile is unique. (There are various web sites that will calculate the random match probability for any core STR profile; see Butler, 2006.)

9.3 PROBLEMS ENCOUNTERED IN STR ANALYSIS

9.3.1 Low-Copy-Number DNA

Blood or semen stains recovered from violent crime scenes usually contain more than enough DNA to perform traditional STR analysis. To obtain amplification of all of the STR alleles present, most commercial STR kits require that starting samples contain from 0.5 ng (500 pg) to 2 ng of DNA. If full profiles could be obtained for samples with much less DNA, a greater range of evidence samples could be analyzed and many more nonviolent crime investigations might be solved by DNA profiling. More and more crime labs are now performing DNA analysis on crime scene samples such as contact swabs and fingerprints which contain very little DNA.

DNA that is present in a very small amount (i.e., less than 100 pg) is known as *low-copy-number* (LCN) *DNA* (Gill et al., 2000; Gill, 2001). Since there are about 6 pg of DNA in a typical human diploid cell, a LCN sample would contain fewer than 17 cells. To analyze samples containing LCN amounts of DNA, the sensitivity of the assay must be increased. This is usually done by increasing to about 34 cycles the number of PCR cycles from the 28 or 30 that are suggested for most commercial STR kits. Since increased sensitivity also increases the potential for analyzing contaminating DNA, extra care should be taken with LCN DNA analysis.

There are also numerous artifacts that frequently arise with LCN DNA analysis that may interfere with interpretation of the DNA profile (Gill et al., 2000). Because such a small quantity of DNA is present in a LCN sample, less than a full complement of the genome may be available for amplification in each PCR reaction; since it is only the loci that are amplified in the first rounds of PCR that will probably show up in the final DNA profile, repeated amplifications of the same LCN sample may give different results. This phenomenon, known as *stochastic variation*, occurs frequently with LCN DNA analysis. Sometimes only one of the two alleles of a heterozygous STR locus will be amplified, and the locus will mistakenly appear in the profile as homozygous (allele dropout). Other times, an entire STR locus may fail to amplify (locus dropout), or extra alleles may appear (allele drop-in). Because of the stochastic variation of these amplification artifacts, PCR of LCN DNA samples is usually done three times, and the only alleles that are reported are those that are present in at least two of the three profiles produced from the same sample (Gill et al., 2000).

Additional complications occur with LCN DNA because the smaller amount of input DNA and the increased sensitivity of the PCR amplification make contamination artifacts such as allele drop-in much more likely. Thus, ultraclean laboratories and DNA-free reagents must be used for LCN DNA analysis. In addition, with LCN DNA it is possible that all the DNA in the original sample was not connected to the crime under investigation; the DNA may have come into the evidence sample by secondary transfer: that is, from people with whom the offender had contact before touching the evidence, or from other people who may have touched the item before the evidence was collected. Thus, LCN DNA analysis results must be interpreted with extreme caution.

Many crime labs in the United States are now analyzing material with LCN amounts of DNA for leads in burglary cases. In England, where LCN DNA analysis was pioneered, forensically relevant DNA profiles have been generated successfully from items such as discarded tools, matchsticks, and weapon handles. Increased use of LCN DNA analysis by crime labs has led to the development of various testing and interpretation protocols (e.g., Caragine et al., 2009; Gill et al., 2009), and there is still controversy surrounding the use of LCN DNA typing techniques (Buckleton, 2009).

9.3.2 Degraded DNA and Reduced-Size (Mini) STR Primer Sets

DNA degradation refers to the breakdown of relatively large fragments of DNA into smaller fragments, and there are various processes (e.g., physical, oxidative, or biological) that may contribute to this degradation. As fragments get smaller in size, the target DNA sequences that are used for the PCR amplification reaction are also broken down and the larger alleles of a locus may decrease in peak height in the electropherogram. If the larger allele for a heterozygous STR is missing,

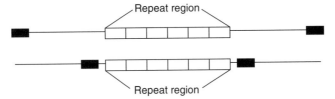

Figure 9.10 Schematic of mini-STR primer positions. Mini STRs are created when primer binding sites (black rectangles) are moved closer to the repeat region in order to shorten the length of the amplified product. The upper portion of the figure shows possible primer positions for a hypothetical traditional STR, while the lower portion shows possible primer positions for a miniSTR with the same number of repeats.

the locus would appear erroneously to be homozygous, and with larger STRs, the entire locus could be missing from the profile. In order to have complete PCR amplification, the full target sequence must be present in the regions where the primers anneal as well as in the portion of DNA between the primer binding sites (Hoss et al., 1996).

Frequently, when the DNA in a forensic sample is present in fragmented form, a full DNA profile cannot be obtained using standard methodology. This problem has been addressed by the development of reduced-size STR primer sets known as *miniplexes*, in which the primer sequences have been moved as close as possible to the repeat region (Figure 9.10) so that smaller STR fragments can be formed (Wiegand and Kleiber, 2001; Butler et al., 2003b). Reduced-size primer sets were first developed for the analysis of STRs by matrix-assisted laser desorption ionization time-of-flight (MALDI-TOF) mass spectrometry, a technique that required shorter DNA fragments for accurate sizing without ladder standards (Butler et al., 1998). Primers for reduced-size STRs also held promise as being better for analyzing degraded DNA and they were used, along with other methods of DNA analysis, to aid in the identification of victims of the World Trade Center disaster. Recent studies comparing mini-STRs with standard STRs and single-nucleotide polymorphism (SNP) assays (see below) found that miniSTRs were usually most effective for the analysis of degraded DNA (Dixon et al., 2006).

Mini-STRs are becoming an important tool for the analysis of degraded DNA in forensic casework samples. Mini primer sets which produce PCR products ranging in size from 50 to 150 bp have been developed for 26 non-CODIS STR markers (Coble and Butler, 2005; Opel et al., 2007; Hill et al., 2008), and a commercial kit for mini-STR amplification of CODIS loci, the AmpFlSTR MiniFiler PCR Amplification kit, is now available from ABI. This kit contains reagents for the mini-STR amplification of the eight largest CODIS loci, which would be the first to drop out in a degraded DNA sample. The MiniFiler kit has recently been validated according to FBI/National Standards and SWGDAM guidelines (SWGDAM, 2000) and was shown to have significant performance improvements over standard STR kits in models of DNA degradation and PCR inhibition (Mulero et al., 2008).

9.3.3 PCR Inhibition

Sometimes substances that interfere with the PCR reaction are present in a forensic sample and coextract with the DNA; the presence of these inhibitors may prevent amplification of some loci or even all the loci. PCR inhibitors may interfere with either the DNA itself or with the polymerase enzyme. Such inhibitors may be present in crime scene DNA samples that were deposited on wood, soil, fabrics, and so on. In addition, hemoglobin in blood, melanin in hair, and certain textile dyes in fabrics, such as that from blue jeans, are inhibitors that coextract with DNA and are frequently present in forensic casework samples (see Besseti, 2007).

The following methods are among those that can be used to decrease PCR inhibition: the DNA sample, and therefore the template, may be diluted so that the inhibitors are also diluted; extra DNA polymerase may be added to overcome inhibition, or bovine serum albumen or betaine may be added to minimize or prevent PCR inhibition; a filtration step may be performed prior to PCR to separate the extracted DNA from the inhibitor [see Raadstom et al. (2004) for a review].

Larger STR loci are more sensitive than smaller loci to PCR inhibition. Because of this, resulting profiles from samples containing inhibitors frequently resemble those of degraded DNA, where larger loci are likely to drop out before smaller ones do. Mini-STR primers, which produce smaller amplified fragments and work well with degraded DNA, may therefore be useful for obtaining full-STR profiles when PCR inhibitors are present.

9.3.4 Interpretation of Mixtures of DNA

If an evidence sample contains a mixture of DNA from more than one person, the profile observed may be confusing and complicated. Mixtures may contain DNA from multiple victims, multiple assailants, victim(s) and assailant(s), or from a single person plus contaminating DNA. If more than two peaks are found for a particular STR locus, the sample is probably a mixture. However, there are cases where three alleles for a single locus (triallelic pattern) may actually be present in a single individual (Butler, 2006), and there are also biological and technological artifacts (described below) that can produce extra peaks. Thus, one must first examine the overall genetic profile in order to determine whether a mixture is present.

Even a two-peak locus that appears to be heterozygous could, in fact, be a mixture from (1) two contributors each of whom is homozygous for a different allele, (2) two contributors both of whom are heterozygous for the same alleles at this locus, or (3) a combination of one contributor who is homozygous and a second who is heterozygous for that locus. If equal amounts of DNA from each person are present, then in the third case one of the two alleles will produce a peak that is three times higher than the other. If the mixture comes from two homozygotes, relative peak heights for the two alleles will depend on how much DNA from each person was present in the mixture.

Since results obtained when DNA mixture components are combined in different ratios are confusing and contain variable peak heights that require further interpretation, it has been suggested that manual interpretation of mixtures not be attempted unless the mixture is the only evidence sample available (Butler 2005, Chap. 7). To aid in mixture interpretation, computer programs are now available to help determine mixture components and ratio; ABI is currently developing GeneMapper ID-X Software Mixture Analysis Tool software for this purpose.

9.3.5 Null Alleles and Allele Dropout

Even with a 1-ng sample of undegraded DNA, there are times when one allele from a person with a heterozygous genotype is not amplified during PCR and the person appears erroneously to be homozygous for that locus. Allele dropout such as this usually occurs because of mutations (sequence polymorphisms) in the flanking region of an STR locus; a mutation in the primer-binding site may prevent one of the primers from annealing, or a mutation close to the primer binding site may block the extension of new DNA during amplification. The presence of alleles with such mutations, known as *null alleles*, can cause errors in the interpretation of STR profiles. However, because the flanking sequence around the tandem repeats is fairly stable and usually not prone to sequence polymorphism, null alleles are rather rare. If a null allele is present and different primers are used for different amplifications, it is possible to obtain a heterozygous genotype with one primer set and an apparent homozygote with another. Since different multiplex kits generally use different primers for the same locus, mutations in or near one of the primer binding sites could produce apparently different genotypes with the different kits. Primer concordance studies have been conducted to examine this phenomenon and, in a comparison of the PowerPlex 16 kit to the Profiler Plus and COfiler kits, allele dropout due to a primer mismatch was found in 22 of 2000 samples in seven of the 13 core STR loci (Budowle and Sprecher, 2001). In addition, because the new MiniFiler STR kit uses new primers with different primer binding sites, it was recently tested for the concordance of its results with those for primers from conventional STR kits (Hill et al., 2007).

Since two identical samples amplified using the same kits will always produce the same genotype results, all samples from the same case should, wherever possible, be tested with the same multiplex primers. However, because DNA database samples (e.g., those of convicted offenders) may have been obtained using a different kit containing primers which differ from those used on the evidence sample, in such a case a database search for an exact match would overlook the true perpetrator if a null allele were present for a primer in one kit but not in the other. To overcome this problem, it has been suggested that fewer stringent match criteria be used in database searches, such as a match of 25 rather than 26 of 26 alleles (Butler, 2005, Chap. 6). In addition, if a primer binds to a region where a polymorphism that produces a null allele is known to occur in some people, the null allele problem may be solved by including "degenerate primers." These are multiple primers for the same locus, each of which will bind to one of the sequence polymorphisms that may be present for that primer binding site. The AmpFlSTR kits contain degenerate primers for the D16S539 locus, which was found to have sequence polymorphism for its primer binding site; degenerate primers for VWA and D8S1179 are also present in some commercial kits (see Butler, 2005, Chap. 6).

9.3.6 Factors Causing Extra Peaks in Results Observed

Electropherograms should always be examined manually to determine that all peaks called are real and not due to artifacts produced during fragment amplification, separation, or detection. Stutter and incomplete adenine addition are two biological artifacts that can cause extra peaks to appear on an electropherograms while pull-up (bleed-through), dye bobs, and spikes are among the technological artifacts that can produce extra peaks (see Figure 9.11 and below).

9.3.7 Stutter Product Peaks

Additional small peaks, several bases shorter than each STR allele peak, are usually present on STR electropherograms. Called *stutter product peaks*, these are formed during PCR when the STR loci are copied by the DNA polymerase. Stutter peaks have been shown to contain one repeat unit less than the true allele peak (Walsh et al., 1996) and they occur because of slipped-strand mispairing during primer extension, which causes either the primer or the template strand to slip, so that one repeat forms a loop that is not base-paired (Hauge and Litt, 1993). (Stutter peaks that are one repeat unit larger than the true allele can occur but are rarely seen.) Stutter peaks can affect DNA profile interpretation, and when mixtures of DNA

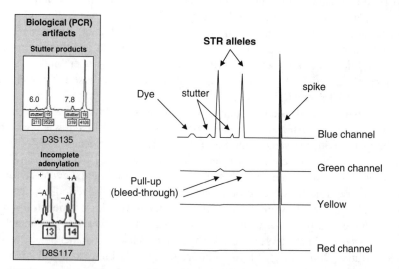

Figure 9.11 Various biology- (left panel) and technology-related artifact peaks that can be observed with STR typing. [From Butler (2005) copyright ©2005 Elsevier Academic Press, Fig. 15.4.]

from two or more sources are present, it can be difficult to determine whether a small peak is a real allele from a minor contributor or a stutter product peak.

Although stutter peak height is proportionally different for each of the different STRs, it is almost always less than 15% of that of the actual allele. Stutter height usually ranges from about 2 to 14% of the adjacent true peak, and the genotyping software does not call peaks in this peak height range.

9.3.8 Nontemplate Addition (Incomplete Adenylation)

The *Taq* polymerase used in PCR frequently adds an extra adenine (A) nucleotide to the 3′ end of a PCR product as it is copying the template strand. This nontemplate addition can lead to the formation of a PCR product that is one base longer than the template DNA. Thus, a peak for an STR may contain two fragments differing in size by only a single adenine base, and this can lead to the formation of electropherograms containing split peaks, broad peaks, or peaks with shoulders. Since sharper peaks increase the accuracy of allele calls by genotyping software, it is preferential to have all peaks either +A or −A rather than a mixture of both. The production of PCR products that all contain an additional adenine can be accomplished by adding an incubation step at 60 or 72°C after the PCR cycles are completed (Kimpton et al., 1993). A final extension step is therefore included in most PCR protocols so that there is enough time for the DNA polymerase to fully adenylate all PCR products. To ensure that accurate calls are made, allelic ladder fragments are also adenylated completely.

9.3.9 Technological Artifacts

Pull-up, also known as *bleed-through* or *matrix failure*, may appear on an electropherogram as a low peak in the same position as a very high allelic peak of a different color. Pull-up is caused by spectral overlap and occurs if the linear range of detection is exceeded due to sample overloading; a peak of a different color pulls up or bleeds through in the same position.

Spikes are sharp peaks that appear as vertical lines. They may be caused by air bubbles in the capillary or by voltage spikes and are easy to distinguish from true peaks since they occur at the same position in all colors.

Dye blobs form peaks that are broader and lower than allele peaks. They are caused by fluorescent dyes coming off their primers and moving through the capillary independently. Dye blobs will possess the spectrum of one of the primer labeling dyes.

9.3.10 Single-Nucleotide Polymorphism Analysis of Autosomal DNA SNPs

A single-nucleotide polymorphism (SNP) is a single base difference between individuals at a particular point in the genome. The human genome contains millions of SNPs (Stoneking 2001), and many have been used for years as markers to track genetic diseases. In addition, SNPs are frequently used together with STR analysis for DNA identification, especially with degraded DNA samples where a complete STR profile cannot be obtained. Both SNP and mini-STR analysis (along with mitochondrial DNA sequencing) were among the methods used in the identification of victims of the World Trade Center disaster (Biesecker et al. 2005).

Whereas STR analysis involves differences in the numbers of multiples of four base repeats, a SNP involves only one differing nucleotide; thus, PCR products from SNPs can be much smaller than those of STRs, making them appropriate for analyzing degraded DNA. However, SNPs are not as informative as STRs since usually only two different nucleotides will be present for a particular SNP, whereas with STRs more than five variations in the number of repeated tetranucleotides can be found at most loci. Thus, to achieve the same random match probability provided by the 13 CODIS STRs, many more SNP markers would have to be analyzed. It has been estimated that 50 to 100 SNPs would probably have to be analyzed to have the same discrimination power as 10 to 16 STRs (Chakraborty et al., 1999; Gill, 2001; Gill et al., 2004).

Because SNPs have lower mutation rates than those of STRs, they are more likely to become fixed in a population and therefore may be potentially useful in predicting phenotypes such as the ethnic origin of an offender. Most of the work on such phenotypic SNPs has, thus far, been related to pigmentation, and certain SNPs have been found to be linked to red hair and fair skin (see Budowle and van Daal, 2008).

9.3.11 Methods Used for SNP Analysis

SNPs are frequently analyzed by *minisequencing*, sometimes called *SNaPshot*, a variation of the typical DNA sequencing method of Sanger et al. (1977) that is described in more detail below for mitochondrial DNA (mtDNA) analysis. In minisequencing the SNP and the DNA region immediately surrounding it are amplified, and this PCR product is then used in a PCR reaction where only one new nucleotide, the one that defines the SNP, is added after the primer binds. In this second PCR, a special type of nucleotide [a dideoxynucleotide triphospate (ddNTP)] is used to stop amplification immediately, and each of the four different ddNTPs is labeled with a different fluorescent dye so that the nucleotide present in the SNP can be identified. In addition, SNP analysis can be performed as a multiplex so that many SNPs can be identified simultaneously. This is done by adding a varying number of extra nucleotides to the 5′ end of the primer (usually, poly T) so that each primer, and therefore each PCR product, differs from the others by several nucleotides, and fragments can then be separated by size (Tully et al., 1996; Vallone et al., 2004).

Other assays used to detect SNPs are variations of techniques used in DNA microarrays (described below in more detail for mtDNA). These arrays consist of many single-stranded DNA probes (fragments representing possible SNPs) attached to a matrix or membrane in a known position in a tightly spaced grid (microarray) or in a linear fashion (linear arrays). The regions of interest are labeled during amplification of the DNA sample (many SNPs can be amplified at once), and the labeled PCR product is then hybridized to the array. Only DNA that is complementary to the various probes will bind to the array and after developing the label, be detectable after washing. By examining which positions on the array are labeled, the sequences (or SNPs) present in the original DNA sample can be determined.

9.3.12 Mitochondrial DNA Analysis

DNA evidence samples may be too limited in amount or too degraded for conventional nuclear DNA analysis or even mini-STR analysis. MtDNA analysis is frequently used to determine the origin of this type of sample. Before miniplex STR primers were developed, mtDNA analysis was the only way to generate any identity information from degraded DNA samples, and it is still used for analysis when there is not enough nuclear DNA or when the nuclear DNA is so degraded that mini-STR analysis is impossible.

Because mitochondrial DNA is inherited from only one parent and therefore does not undergo genetic recombination, mtDNA analysis provides much less genetic identity information than does analysis of nuclear DNA. However, because so many more copies of mtDNA are present in a cell, its analysis can frequently provide information when nuclear DNA cannot. While each cell contains only two copies of each nuclear DNA chromosome (one maternal and one paternal), mtDNA is present in hundreds to thousands of copies per cell. The average cell contains about four or five (with a range of 1 to 15) copies of mtDNA in each mitochondrion, with an estimated average of 500 mtDNA molecules per cell (Satoh and Kuroiwa, 1991).

In addition to the higher number of copies present, other factors contribute to the survival of mtDNA under conditions where nuclear DNA may be highly degraded. The circular nature of the mitochondrial chromosome protects the DNA from enzyme degradation, as does its encapsulation in a double-walled structure within the cell (Butler 2005, Chap. 10). These factors have enabled the success of mtDNA analysis in determining the origin of bones, teeth, and ancient remains.

MtDNA is inherited maternally; thus, all children inherit their mtDNA sequence from their mother and the mtDNA of a male is not passed along. Women have the same mtDNA profile as their siblings, male and female, their children, their grandchildren, and all their maternal relatives. Because the mitochondrial genome does not undergo the type of recombination events that occur in nuclear DNA, unless a mutation occurs, mtDNA is transmitted unchanged in sequence from generation to generation.

Mitochondrial chromosomes are closed circles of double-stranded DNA that contain 16,569 base pairs (bp) numbered in a clockwise direction (Figure 9.4). There is an asymmetric distribution of guanines, the heaviest of the four DNA nucleotides, on the two complementary strands of mtDNA. The strand with the most guanines is known as the heavy (H-) strand and the other as the light (L-) strand. The 16,569 bp of the mitochondrial genome include 37 genes that code for products used in mitochondrial structure and function (exons). There are no introns and very few nucleotides between the coding regions.

The mitochondrial genome also contains a noncoding control region that is 1122 bp in length and contains the origin of replication for one of the two mtDNA strands. The replication of mtDNA begins in the H-strand of the control region. There are two segments of the control region DNA that are highly polymorphic, and these are called *hypervariable* (HV1 and HV2) *regions* (see Figure 4). HV1 is 342 bp long and HV2 is 268 bp long. For forensic mtDNA analysis, all 610 of the HV region bases are sequenced. Unrelated individuals have been shown to vary about 1 to 2% in their hypervariable control region DNA, meaning that 7 to 10 nucleotides of these 610 bases are different (Budowle et al., 1999).

9.4 METHODOLOGY FOR mtDNA ANALYSIS

9.4.1 Preparation of Samples

Because it is present in higher copy number and because its analysis requires more PCR cycles, mtDNA is more sensitive to contamination than nuclear DNA. Thus, care should be taken to avoid contamination by other mtDNA, and the extraction of mtDNA should be performed in a very clean laboratory environment. In addition, the analysis of reference samples should be done after all evidence samples have been processed completely. Reagent blanks and negative controls should also be run to monitor levels of exogenous DNA that may be present in reagents and/or the laboratory environment.

There is usually very little DNA present in the materials that require mtDNA analysis in forensic cases (e.g., hair, teeth, and bones). In the past, mtDNA was frequently quantified indirectly by assuming a fixed ratio between nuclear and mtDNA and taking a fraction of the nuclear DNA; however, there are now methods for the direct measurement of mtDNA using rtPCR (e.g., Meissner et al., 2000; Andreasson et al., 2002). In addition, because the type of sample used for mtDNA analysis frequently contains PCR inhibitors that can be coextracted with the DNA, purification of mtDNA is usually more extensive than that needed for nuclear DNA.

9.4.2 MtDNA Sequencing Methods

The FBI began to examine mtDNA sequences for identification purposes in June of 1996. For forensic analysis, dideoxynucleotide (ddNTP) sequencing (Sanger et al., 1977) is the method generally used to determine the order of the nucleotides in HV1 and HV2, the hypervariable areas of the control region. However, because of the extensive amount of time and labor required for sequencing, more rapid screening methods (described below) are frequently used first, to exclude samples that do not match.

MtDNA sequencing uses the same CE instruments that are used for STR analysis; however somewhat different PCR and CE analysis strategies are used. First, the DNA of each hypervariable region is amplified, usually using 34 to 38 cycles, many more than the 28 to 30 cycles commonly used for nuclear DNA amplification. (Since PCR inhibitors such as melanin can be present in hair samples for mtDNA analysis, extra *Taq* polymerase is sometimes added to overcome this inhibition.) The PCR product for each region is then individually amplified in another PCR reaction. In this PCR, self-terminating nucleotides that stop DNA replication (ddNTPs) are present in addition to the usual dNTP building blocks (Sanger et al., 1977). In the methods currently used with CE sequencing (see Figure 9.12), the four ddNTPs are each labeled with a different color fluorescent dye. When a ddNTP is added to a segment of DNA instead of a dNTP, the extension of the DNA stops immediately and no new nucleotides are added. Because both dNTPs and ddNTPs are present in the PCR reaction, different amplification products are terminated at different points on the DNA template, thereby forming a mixture that contains a series of DNA fragments, each differing by one base pair in length. CE is then used to separate the PCR fragments by length, and because each fragment contains the label of the last base added (in the form of a fluorescent ddNTP), the sequence of bases is obtained for the entire DNA region examined.

9.4.3 Reference Sequences

Following sequencing, the mtDNA sequence of each questioned (evidence) and known (e.g., suspect) sample is compared to a reference sequence and the differences found at specific sites are noted. Human mtDNA was first sequenced in 1981

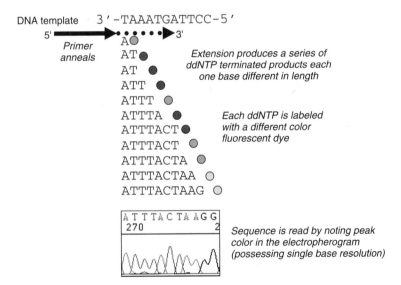

Figure 9.12 DNA sequencing using fluorescent dideoxynucleotide triphosphates (ddNTPs). A primer that anneals next to the specific DNA region to be sequenced is extended by DNA polymerase. Since each sample contains a mixture of dNTPs and ddNTPs for each of the possible nucleotides, some of the extension products continue to be extended, while others are stopped by incorporation of a ddNTP. Each ddNTP is labeled with a different dye, and this allows each extension product to be distinguished by color. The separation of extension products by size allows the DNA sequence to be deduced. [From Butler (2005), copyright ©2005 Elsevier Academic Press, Fig. 10.5.]

(Anderson et al., 1981), and this sequence, known as the Anderson or Cambridge Reference Sequence (CRS), has been used as the reference sequence for forensic mtDNA analysis. Sequencing results are reported in terms of variation from the control region on the L-strand of the CRS. Only the bases that deviate from the reference sequence are reported, and all others are assumed to be the same. For example, a sample differing from the CRS because it has a G instead of an A at position 1620 in the HV1 region would be listed as 1620 G in the sequence report. In the absence of mutation, mtDNA coming from the same person or from a person with the same maternal lineage is expected to have the same DNA sequence and therefore the same differences from the reference sequence. Software packages such as Sequencher (Gene Codes Corp., Michigan) are available to align and compare sequences to the reference sequence and to each other.

The same material that was used to generate the CRS was sequenced again in 1999. Although 11 differences in nucleotides were reported, none of these differences were in the control region (Andrews et al., 1999). Thus, for the HV1 and HV2 regions that are used for forensic mtDNA analysis and reporting, the revised CRS (rCRS) does not differ from the CRS. One difference that was found upon resequencing was the loss of a cytosine at position 3106, making the rCRS standard now used one base shorter than the CRS (16,568 vs.1659 bp). To keep base numbers consistent with those reported previously, a deletion is reported for the rCRS at position 3107. Thus, all numbering is the same as in the original sequence, and each base is assigned a number from 1 to 16,569.

9.4.4 Screening Assays for mtDNA

The regions of the mtDNA genome that are sequenced are those that show the most variability between individuals: the hypervariable control regions, HV1 and HV2. Rapid screening techniques for mtDNA have been developed to determine quickly which samples can be excluded and therefore do not have to be sequenced. Although the variations in mtDNA are scattered throughout HV1 and HV2, there are certain hypervariable sites and regions (hotspots) where most of the variation occurs (Stoneking, 2000), and rapid screening methods concentrate primarily on these hotspots. Mitochondrial SNPs have been examined in these areas (Kline et al., 2005) and minisequencing methods (Tully et al., 1996) similar to that described above for nuclear chromosomal SNPs (e.g., SNaPshot technology) is sometimes used for this purpose (Quintans et al., 2004). In addition, many popular screening methods make use of sequence-specific oligonucleotide (SSO) probes (Stoneking et al., 1991), and these have been used in reverse dot blot and linear array assays (Comas et al., 1999; Gabriel et al., 2003). Rather than sequencing the entire hypervariable regions, SSO probes examine the most polymorphic sites in HV1 and HV2 through hybridization of PCR products to oligonucleotide probes designed to base-pair with the different sequence variants.

Roche Applied Sciences has developed the Linear Array Mitochondrial DNA HVI/HVII Region-Sequence Typing kit. This assay uses SSO probes to distinguish 19 polymorphic sites in 10 different locations within the control region, and it

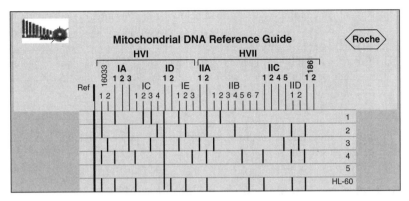

Figure 9.13 Roche linear array mitochondrial DNA HVI/HVII region-sequence typing kit panel results. Samples 1 to 4 are quality control DNA samples; sample 5 is a negative control; sample HL-60 is positive control DNA [HL-60 (NIST) SRM 2392.1]. (Reprinted with permission from Roche Applied Science.)

can do so in only 6 to 8 hours. The primer mix that comes with the kit contains four biotin-labeled primers that amplify the HV1 and HV2 regions simultaneously. Typing of the DNA is accomplished by hybridizing the labeled PCR products to panels containing 33 DNA probes which are striped in 31 lines, and then performing a reaction that produces a color product where fragments containing the biotin-labeled primers have hybridized (see Figure 9.13). In all regions examined, probe 1 is always the Anderson sequence (CRS). This kit has been used successfully for the examination of forensic mtDNA samples (Divne et al., 2005).

9.4.5 Interpretation of mtDNA Sequencing Results

There are FBI guidelines for interpreting mitochondrial sequence comparisons. Known and questioned samples are compared to the rCRS and a comparison of the two samples can result in an exclusion, a failure to exclude, or an inconclusive finding. Following are the guidelines recommended by the Scientific Working Group on DNA Analysis Methods for analyzing mitochondrial sequence data (SWGDAM, 2003).

Exclusion:

> If there are two or more differences in the base sequence of the known and the questioned sequences, one can conclude that there is no common origin between the two specimens and there is an *exclusion*.

Failure to exclude:

> a. If there is no difference in sequence between the questioned and the known samples, one *cannot exclude a common origin*.
> b. If there is heteroplasmy (described below) at the same site for both the known and questioned samples and there are identical bases at every other position, one *cannot exclude a common origin*.

c. If there is heteroplasmy (described below) at one position in one sample but not in the other, and every base at the other positions is identical, one *cannot exclude a common origin*.

d. If there is an ambiguous base (usually denoted as "N") at a specific position in either or both samples and identical bases at every other position, one *cannot exclude a common origin* for the two samples.

Inconclusive:

If there is only one base difference between the known and the questioned sequence, this result is *inconclusive*.

9.4.6 Statistics: The Meaning of a Match for mtDNA

When mtDNA analysis shows a "failure to exclude," a statistical calculation is required to find the meaning of a match. Unlike the 13 core nuclear STR loci whose individual frequencies can be multiplied because they assort independent of each other during cell division, all sites of the mitochondrial genome, and therefore all mtDNA polymorphisms, are linked on the single mitochondrial chromosome. The method currently used to describe the frequency of mtDNA sequences is the counting method; for example, if after the examination of 4000 samples a particular sequence appears for the first time, the frequency of this genetic profile is said to be 1 in 4001. If the sequence had been found once before, the frequency would be indicated as 2 in 4001. Although statistics such as these may not accurately reflect the actual frequency of any sequence because the population studied may be relatively small in size, the counting method is still the primary method used in forensic cases. Another way to express the frequency of mtDNA sequences is by providing 95% confidence limits and providing the upper and lower limits (the highest and lowest frequency estimates) to indicate the range of possible values for the actual frequency of the profile in the relevant population.

A major problem with mtDNA typing occurs because many sequences fall into categories called "most common types." These are particular sequences for HV1 and HV2 that are present in a disproportionately large portion of the population studied. As an example, an analysis of data for more than 1600 persons found that although the sequences for more than 50% of them were unique for the database, the most common type, which has the same sequence as the rCRS, was present 7.1% of the time. In addition, more than 20% of this Caucasian data set was made up of only 18 mtDNA types (Coble et al., 2004). There are numerous databases containing collections of mitochondrial sequence information (Butler, 2005, Chap. 10; Parson and Dür, 2007), and the FBI has compiled published mtDNA sequence data into the mtDNA population database (CODIS).

9.4.7 Heteroplasmy

Heteroplasmy is defined as the presence in a single person of more than one mitochondrial DNA type at a particular base position in the mtDNA sequence. Heteroplasmy can occur in sequence and/or length. Sequence heteroplasmy is detected when two nucleotides appear at a single site as overlapping peaks in a sequencing electropherogram. Length heteroplasmy occurs most often in the hypervariable regions in areas where there are long stretches of cytosines (C-stretches), that is, in positions 16,184 to 16,193 in HV1 and in positions 303 to 310 in HV2, and there are differing numbers of C's within the same sample.

Heteroplasmy can occur in the known and/or the questioned sample. If heteroplasmy at a specific position is found in both samples, this might help indicate an inclusion because heteroplasmy at a specific position can be an important identifying characteristic.

Heteroplasmy has been shown to differ from tissue to tissue in the same person (e.g., hair and blood may show different levels and types of heteroplasmy) (Salas et al., 2001). There may also be heteroplasmy differences in mtDNA populations between a person's other cells, and also differences within a single cell or within one mitochondrion (Butler, 2005, Chap. 6). All persons are thought to be heteroplasmic at some level, even if it is below the level of detection (Comas et al., 1995; Tully et al., 2000). High rates of heteroplasmy are not surprising since the mutation rate in mtDNA is much higher than the rate in nuclear genes; some regions of the mitochondrial genome have been shown to evolve at 6 to 17 times the rate of single-copy nuclear genes (Tully, 1999). In addition, although there is only one copy of a nuclear gene present within a human egg, that egg may contain 100,000 copies of the mitochondrial genome (Chen et al., 1995). Thus, multiple mutations and a good deal of heteroplasmy, some of which may be undetectable, could be present in a person.

9.4.8 The Future of DNA Analysis

As of June 2010, the National DNA Index (NDIS) portion of the FBI CODIS database contained almost 8.5 million profiles from convicted offenders and more than 324,318 forensic (evidentiary) profiles (see http://www.fbi.gov/hq/lab/codis/clickmap.htm for updates). Since STR analyses form the bulk of the profiles already present in CODIS as well as in other DNA databases throughout the world, changing to another type of DNA profiling would probably make current data unusable without reanalysis. Thus, STR typing is expected to continue as the preferred DNA profiling method for many years to come (Gill et al., 2004; Butler et al., 2007). However, since profiles from reduced-size core STRs are fully compatible with existing database information and because mini-STRs can broaden the range of degraded samples that can be analyzed, mini-STR analysis may eventually replace STR typing analysis with currently used primers (Gill et al., 2004). In addition, DNA markers such as SNPs will probably continue to be used along with STRs in order to obtain additional information about particular samples. In addition to the possibility of providing phenotype information (Budowle and van Daal 2008), SNPs are helpful in mtDNA testing and Y-SNPs can serve as lineage markers for Y-chromosome analysis (Butler et al., 2007).

Forensic DNA analysis is a rapidly progressing field, and new instruments are constantly being developed. Most of the advances currently being made in DNA analysis instrumentation are related to increasing the number of samples that can be processed at once (throughput), decreasing processing time and/or decreasing the size of the instruments that are used. To help speed up the analysis of reference samples, robotic instruments that automate DNA extraction and processing are being developed, and many are already in use in high-throughput laboratories. In addition, there is always a need for new user-friendly data analysis software and "expert" software to lessen the manual input required for accurate data interpretation.

The miniaturization of instruments used for forensic DNA analysis is a popular area of current research. Miniature thermal cyclers and small microchip capillary electrophoresis devices are being developed that greatly shorten analysis time and reduce the consumption of sample and of expensive reagents (Yeung et al., 2006; Greenspoon et al., 2007; Liu et al., 2007). A microfabricated capillary array electrophoresis device was recently tested and found to be appropriate for separating STRs (Greenspoon et al., 2008). The ultimate aim of equipment miniaturization is to produce instruments capable of integrating sample preparation, PCR, and DNA analysis which are also small enough and portable enough to be used at the scene of a crime or mass disaster. A prototype of such an instrument was recently developed and tested at a mock crime scene (Liu et al., 2008).

REFERENCES

Anderson, S., Bankier, A. T., Barrell, B. G., de Bruijn, M. H., Coulson, A. R., Drouin, J., Eperon, I. C., Nierlich, D. P., Roe, B. A., Sanger, F., Schreier, P. H., Smith, A. J., Staden, R., and Young, I. G. (1981). Sequence and organization of the human mitochondrial genome. *Nature*, **290**:457–465.

Andreasson, H., Gyllensten, U., and Allen, M. (2002). Real-time DNA quantification of nuclear and mitochondrial DNA in forensic analysis. *Biotechniques*, **33**:402–404, 407–411.

Andrews, R. M., Kubacka, I., Chinnery, P. F., Lightowlers, R. N., Turnbull, D. M., and Howell, N. (1999). Reanalysis and revision of the Cambridge reference sequence for human mitochondrial DNA. *Nat. Genet.*, **23**:147.

Barbisin, M., Fang, R., O'Shea, C. E., Calandro, L. M., Furtado, M. R., and Shewale, J. G. (2009). Developmental validation of the Quantifiler Duo DNA Quantification kit for simultaneous quantification of total human and human male DNA and detection of PCR inhibitors in biological samples. *J. Forensic Sci.*, **54**:305–319.

Beckman, J. S., and Weber, J. L. (1992). Survey of human and rat microsatellites. *Genomics*, **12**:627–631.

Besseti, J. (2007). An introduction to PCR inhibitors. *Profiles DNA*, **10(1)**:9–10.

Biesecker, L. G., Bailey-Wilson, J. E., Ballantyne, J., Baum, H., Bieber, F. R., Brenner, C., Budowle, B., Butler, J. M., et al. (2005). Epidemiology. DNA identifications after the 9/11 World Trade Center attack. *Science*, **310**:1122–1123.

Brettell, T. A., Butler, J. M., and Saferstein, R. (2005). Forensic science. *Anal. Chem.*, **77**:3839–3860.

Brettell, T. A., Butler, J. M., and Almirall, J. R. (2007). Forensic science. *Anal. Chem.*, **79**:4365–4384.

Brettell, T. A., Butler, J. M., and Almirall, J. R. (2009). Forensic science. *Anal. Chem.*, **81**:4695–4711.

Buckleton, J. (2009). Validation issues around DNA typing of low level DNA. *Forensic Sci. Int. Genet.*, **3**:255–260.

Budowle, B., and Sprecher, C. J. (2001). Concordance study on population database samples using the PowerPlex 16 Kit and AmpFlSTR Profiler Plus Kit and AmpFlSTR COfiler Kit. *J. Forensic Sci.*, **46(3)**:637–641.

Budowle, B., and van Daal, A. (2008). Forensically relevant SNP classes. *Biotechniques*, **44**:603–610.

Budowle, B., Wilson, M. R., DiZinno, J. A., Stauffer, C., Fasano, M. A., Holland, M. M., and Monson, K. L. (1999). Mitochondrial DNA regions HVI and HVII population data. *Forensic Sci. Int.*, **103**:23–35.

Butler, J. M. (2005). *Forensic DNA Typing: Biology, Technology, and Genetics of STR Markers*, 2nd ed. New York: Elsevier, Academic Press.

Butler, J. M. (2006). Genetics and genomics of core short tandem repeat loci used in human identity testing. *J. Forensic Sci.*, **51**:253–265.

Butler, J. M. (2007). Short tandem repeat typing technologies used in human identity testing. *BioTechniques*, **43**:Sii–Sv.

Butler, J. M., Li, J., Shaler, T. A., Monforte, J. A., and Becker, C. H. (1998). Reliable genotyping of short tandem repeat loci without an allelic ladder using time-of-flight mass spectrometry. *Int. J. Legal Med.*, **112**:45–49.

Butler, J. M., Schoske, R., Vallone, P. M., Redman, J. W., and Kline, M. C. (2003a). Allele frequencies for 15 autosomal STR loci on U.S. Caucasian, African American, and Hispanic populations. *J. Forensic Sci.*, **48**:908–911.

Butler, J. M., Shen, Y., and McCord, B. R. (2003b). The development of reduced size STR amplicons as tools for analysis of degraded DNA. *J. Forensic Sci.*, **48**:1054–1064.

Butler, J. M., Buel, E., Crivellente, F., and McCord, B. R. (2004). Forensic DNA typing by capillary electrophoresis: using the ABI Prism 310 and 3100 Genetic Analyzers for STR analysis. *Electrophoresis*, **25**:1397–1412.

Butler, J. M., Coble, M. D., and Vallone, P. M. (2007). STRs vs. SNPs: thoughts on the future of forensic DNA testing. *Forensic Sci. Med. Pathol.*, **3**:200–205.

Caragine, T., Mikulasovich, R., Tamariz, J., Bajda, E., Sebestyen, J., Baum, H., and Prinz, M. (2009). Validation of testing and interpretation protocols for low template DNA samples using AmpFlSTR Identifiler. *Croat. Med. J.*, **50**:250–267.

Chakraborty, R., Stivers, D. N., Su, B., Zhong, Y., and Budowle, B. (1999). The utility of STR loci beyond human identification: implications for the development of new DNA typing systems. *Electrophoresis*, **20**:1682–1696.

Chen, X., Prosser, R., Simonetti, S., Jagiello, G., and Schon, E. A. (1995). Rearranged mitochondrial genomes are present in human oocytes. *Am. J. Hum. Genet.*, **57(2)**:239–247.

Coble, M. D., and Butler, J. M. (2005). Characterization of new miniSTR loci to aid analysis of degraded DNA. *J. Forensic Sci.*, **50**:43–53.

Coble, M. D., Just, R. S., O'Callaghan, J. E., Letmanyi, I. H., Peterson, C. T., Irwin, J. A., and Parsons, T. J. (2004). Single nucleotide polymorphisms over the entire mtDNA genome that increase the power of forensic testing in Caucasians. *Int. J. Legal Med.*, **118**:137–146.

Collins, P. J., Hennessy, L. K., Leibelt, C. S., Roby, R. K., Reeder, D. J., and Foxall, P. A. (2004). Developmental validation of a singletube amplification of the 13 CODIS STR loci, D2S1338, D19S433, and amelogenin: the AmpFlSTR Identifiler PCR Amplification Kit. *J. Forensic Sci.*, **49**:1265–1277.

Comas, D., Paabo, S., and Bertranpetit, J. (1995). Heteroplasmy in the control region of human mitochondrial DNA. *Genome Res.*, **5**:89–90.

Comas, D., Reynolds, R., and Sajantila, A. (1999). Analysis of mtDNA HVRII in several human populations using an immobilised SSO probe hybridization assay. *Eur. J. Hum. Genet.* **7**:459–468.

Divne, A. M., Nilsson, M., Calloway, C., Reynolds, R., Erlich, H., and Allen, M. (2005). Forensic casework analysis using the HVI/HVII mtDNA linear array assay. *J. Forensic Sci.*, **50**:548–554.

Dixon, L. A., Dobbins, A. E., Pulker, H. K., Butler, J. M., Vallone, P. M., Coble, M. D., Parson, W., et al. (2006). Analysis of artificially degraded DNA using STRs and SNPs: results of a collaborative European (EDNAP) exercise. *Forensic Sci. Int.*, **164**:33–44.

Economou, E. P., Bergen, A. W., Warren, A. C., and Antoarakis, S. E. (1990). The polydeoxyadenylate tract of Alu repetitive elements is polymorphic in the human genome. *Proc. Natl. Acad. Sci. USA*, **87**:2951–2954.

Edwards, A., Civitello, A., Hammond, H. A., and Caskey, C. T. (1991). DNA typing and genetic mapping with trimeric and tetrameric tandem repeats. *Am. J. Hum. Genet.*, **49**:746–756:

Gabriel, M. N., Calloway, C. D., Reynolds, R. L., and Primorac, D. (2003). Identification of human remains by immobilized sequence-specific oligonucleotide probe analysis of mtDNA hypervariable regions I and II. *Croat. Med. J.*, **44(3)**:293–298.

Gill, P. (2001). Application of low copy number DNA profiling. *Croat. Med. J.*, **42**:229–232.

Gill, P., Whitaker, J., Flaxman, C., Brown, N., and Buckleton, J. (2000). An investigation of the rigor of interpretation rules for STRs derived from less than 100 pg of DNA. *Forensic Sci. Int.*, **2**:1–6.

Gill, P., Werrett, D. J., Budowle, B., and Guerrieri, R. (2004). An assessment of whether SNPs will replace STRs in national DNA databases: joint considerations of the DNA working group of the European Network of Forensic Science Institutes (ENFSI) and the Scientific Working Group on DNA Analysis Methods (SWGDAM). *Sci. Justice*, **44**:51–53.

Gill, P., Puch-Solis, R., and Curran, J. (2009). The low-template-DNA (stochastic) threshold: its determination relative to risk analysis for national DNA databases. *Forensic Sci. Int. Genet.*, **3**:104–111.

Gomes, I., Prinz, M., Pereira, R., Bieschke, E., Mayr, W. R., Amorim, A., Carracedo, A., and Gusmao, L. (2009). X-chromosome STR sequence variation, repeat structure, and nomenclature in humans and chimpanzees. *Int. J. Legal Med.*, **123**:143–149.

Greenspoon, S. A., Yeung, S. H. I., Ban, J. D., and Mathies, R. A. (2007). Microchip Capillary electrophorese: progress toward an integrated forensic analysis system. *Profiles DNA*, **10(2)**:16–18.

Greenspoon, S. A., Yeung, S. H. I., Johnson, K. R., Chu, W. K., Rhee, H. N., McGuckian, A. B., Crouse, C. A., et al. (2008). A forensic laboratory tests the Berkeley microfabricated capillary array electrophoresis device. *J. Forensic Sci.*, **53**:828–837.

Hauge, X. Y., and Litt, M. (1993). A study of the origin of "shadow bands" seen when typing dinucleotide repeat polymorphisms by the PCR. *Hum. Mol. Genet.*, **2**:411–415.

Hill, C. R., Kline, M. C., Mulero, J. J., Lagace, R. E., Chang, C. W., Hennessy, L. K., and Butler, J. M. (2007). Concordance study between the AmpFlSTR MiniFiler PCR amplification kit and conventional STR typing kits. *J. Forensic Sci.*, **52**:870–873.

Hill, C. R., Kline, M. C., Coble, M. D., and Butler, J. M. (2008). Characterization of 26 miniSTR loci for improved analysis of degraded DNA samples. *J. Forensic Sci.*, **53**:73–80.

Hoss, M., Jaruga, P., Zastawny, T. H., Dizdaroglu, M., and Paabo, S. (1996). DNA damage and DNA sequence retrieval from ancient tissues. *Nucleic Acids Res.*, **24(7)**:1304–1307.

Jeffreys, A. J., Wilson, V., and Thein, S. L. (1985). Hypervariable "minisatellite" regions in human DNA. *Nature*, **314**:67–73.

Kimpton, C. P., Gill, P., Walton, A., Urquhart, A., Millican, E. S., and Adam, M. (1993). Automated DNA profiling employing multiplex amplification of short tandem repeat loci. *PCR Methods Appl.*, **3**:13–22.

Kline, M. C., Vallone, P. M., Redman, J. W., Duewer, D. L., Calloway, C. D., and Butler, J. M. (2005). Mitochondrial DNA typing screens with control region and coding region SNPs. *J. Forensic Sci.*, **50**:377–385.

Liu, P., Seo T. S., Beyor, N., Shin, K. J., Scherer, J. R., and Mathies, R. A. (2007). Integrated portable polymerase chain reaction: capillary electrophoresis microsystem for rapid forensic short tandem repeat typing. *Anal. Chem.*, **79**:1881–1889.

Liu, P., Yeung, S. H. I., Crenshaw, K. A., Crouse, C. A., Scherer, J. R., and Mathies, R. A. (2008). Real-time forensic DNA analysis at a crime scene using a portable microchip analyzer. *FSI Genet.*, **2**:301–309.

Meissner, C., Mohamed, S. A., Klueter, H., Hamann, K., von Wurmb, N., and Oehmichen, M. (2000). Quantification of mitochondrial DNA in human blood cells using an automated detection system. *Forensic Sci. Int.*, **113(1–3)**:109–112.

Mulero, J. J., Chang, C. W., Lagace, R. E., Wang, D. Y., Bas, J. L., McMahon, T. P., and Hennessey, L. K. (2008). Development and validation of the AmpFlSTR((R)) MiniFiler(trade mark) PCR Amplification Kit: a MiniSTR multiplex for the analysis of degraded and/or PCR inhibited DNA. *J. Forensic Sci.*, **53**:838–852.

Mullis, K. B. (1990). The unusual origin of the polymerase chain reaction. *Sci. Am.*, **262**:56–61,64–65.

Opel, K. L., Chung, D. T., Drabek, J., Butler, J. M., and McCord, B. R. (2007). Developmental validation of reduced-size STR Miniplex primer sets. *J. Forensic Sci.*, **52(6)**:1263–1271.

Parson, W., and Dür, A. (2007). EMPOP: a forensic mtDNA database. *Genetics*, **1**:88–92.

Quintans, B., Alvarez-Iglesias, V., Salas, A., Phillips, C., Lareu, M. V., and Carracedo, A. (2004). Typing of mitochondrial DNA coding region SNPs of forensic and anthropological interest using SNaPshot minisequencing. *Forensic Sci. Int.*, **140**:251–257.

Raadstom, P., Knutsson, R., Wolffs, P., Lovenklev, M., Lofstrom, C., et al. (2004). Pre-PCR processing: strategies to generate PCR-compatible samples. *Mol. Biotechnol.*, **26**:133–146.

Saiki, R. K., Scharf, S., Faloona, F., Mullis, K. B., Horn, G. T., Erlich, H. A., and Arnheim, N. (1985). Enzymatic amplification of beta-globin genomic sequences and restriction site analysis for diagnosis of sickle cell anemia. *Science*, **230**:1350–1354.

Salas, A., Lareu, M. V., and Carracedo, A. (2001). Heteroplasmy in mtDNA and the weight of evidence in forensic mtDNA analysis: a case report. *Int. J. Legal Med.*, **114**:186–190.

Sanger, F., Nicklen, S., and Coulson, A. R. (1977). DNA sequencing with chain-terminating inhibitors. *Proc. Natl. Acad. Sci. USA*, **74**:5463–5467.

Satoh, M., and Kuroiwa, T. (1991). Organization of multiple nucleoids and DNA molecules in mitochondria of a human cell. *Exp. Cell Res.*, **196(1)**:137–140.

Stoneking, M. (2001). Single nucleotide polymorphisms: from the evolutionary past ... *Nature*, **409**:821–822.

Stoneking, M., Hedgecock, D., Higuchi, R. G., Vigilant, L., and Erlich, H. A. (1991). Population variation of human mtDNA control region sequences detected by enzymatic amplification and sequence-specific oligonucleotide probes. *Am. J. Hum. Genet.*, **48**:370–382.

SWGDAM (2000). Short tandem repeat (STR) interpretation guidelines. Scientific Working Group on DNA Analysis Methods (SWGDAM). *Forensic Sci. Commun.*, **2(3)**.

SWGDAM (2003). Guidelines for Mitochondrial DNA (mtDNA) Nucleotide Sequence Interpretation: Scientific Working Group on DNA Analysis Methods (SWGDAM). *Forensic Sci. Commun.*, **5(2)**.

Tully, G. (1999). Mitochondrial DNA: a small but valuable genome. Presented at the First International Conference on Forensic Human Identification, Forensic Science Service.

Tully, G., Sullivan, K. M., Nixon, P., Stones, R. E., and Gill, P. (1996). Rapid detection of mitochondrial sequence polymorphisms using multiplex solid-phase fluorescent minisequencing. *Genomics*, **34**:107–113.

Tully, L. A., Parsons, T. J., Steighner, R. J., Holland, M. M., Marino, M. A., and Prenger, V. L. (2000) A sensitive denaturing gradient: gel electrophoresis assay reveals a high frequency of heteroplasmy in hypervariable region 1 of the human mtDNA control region. *Am. J. Hum. Genet.*, **67(2)**:432–443.

Vallone, P. M., Just, R. S., Coble, M. D., Butler, J. M., and Parsons, T. J. (2004). A multiplex allele-specific primer extension assay for forensically informative SNPs distributed throughout the mitochondrial genome. *Int. J. Legal Med.*, **118**:147–157.

Walsh, P. S., Fildes, N. J., and Reynolds, R. (1996). Sequence analysis and characterization of stutter products at the tetranucleotide repeat locus vWA. *Nucleic Acids Res.*, **24**:2807–2812.

Watson, J. D., and Crick, F. H. (1953). Genetical implications of the structure of deoxyribonucleic acid. *Nature*, **171**:964–967.

Wiegand, P., and Kleiber, M. (2001). Less is more: length reduction of STR amplicons using redesigned primers. *Int. J. Legal Med.*, **114**:285–287.

Yeung, S. H., Greenspoon, S. A., McGuckian, A., Crouse, C. A., Emrich, C. A., Ban, J., and Mathies, R. A. (2006). Rapid and high-throughput forensic short tandem repeat typing using a 96-lane microfabricated capillary array electrophoresis microdevice. *J. Forensic Sci.*, **51**:740–747.

CHAPTER 10

Current and Future Uses of DNA Microarrays in Forensic Science

NATHAN H. LENTS

Department of Sciences, John Jay College of Criminal Justice, The City University of New York, New York

Summary DNA microarrays have revolutionized basic research in molecular and cellular biology, biochemistry, and genetics. Through hybridization of labeled probes, this high-throughput technology allows the screening of tens or even hundreds of thousands of data points in a single run. The technology is most advanced with nucleic acids, but protein and antibody microarrays are coming of age as well. Because of the unique ability to screen for large numbers of molecules, such as DNA sequences, at once, the potential utility to forensic investigations is tremendous. Indeed, progress has been made demonstrating that microarrays are powerful tools of use to the forensic laboratory. As the technology matures and associated costs come down, the day that microarray analysis becomes a routine part of the forensic toolkit draws nearer.

10.1	Introduction	328
10.2	What is a DNA microarray?	328
	10.2.1 cDNA microarray	329
	10.2.2 Other types of DNA arrays	330
	10.2.3 The birth of "-omics"	331
10.3	DNA microarrays in toxicogenomics	332
	10.3.1 Sharing information	333
	10.3.2 Forensic application	333
10.4	Detection of microorganisms using microarrays	334
	10.4.1 Historical perspective	334
	10.4.2 DNA fingerprinting	335
	10.4.3 DNA fingerprinting by microarrays	336
	10.4.4 DNA sequence-based detection	337
	10.4.5 Where DNA microarrays come in	337

Forensic Chemistry Handbook, First Edition. Edited by Lawrence Kobilinsky.
© 2012 John Wiley & Sons, Inc. Published 2012 by John Wiley & Sons, Inc.

	10.4.6 Looking forward: genetic virulence signatures	338
10.5	Probing human genomes by DNA microarrays	340
	10.5.1 STR analysis	340
	10.5.2 SNP analysis	343
	10.5.3 Exploring an unknown genome?	344
	Conclusion	345
	References	345

10.1 INTRODUCTION

Perhaps once each decade, a technique or technology emerges that revolutionizes scientific research in a way that makes scientists wonder what we ever did without it. In the area of molecular biology, restriction enzymes and molecular cloning, and the polymerase chain reaction, come to mind as examples. More recently, the development of DNA microarray technology is just such an event, allowing molecular research to advance with a rapidity and precision unimaginable just a few years ago (Schena et al., 1995; Shoemaker et al., 2001). Indeed, like polymerase chain reaction (PCR) before it, microarrays have quickly become a seemingly irreplaceable implement in the arsenal of nearly all areas of molecular and genetic research. And as research continues, microarrays are now poised to have a similar impact on molecular forensic investigations.

10.2 WHAT IS A DNA MICROARRAY?

Simply put, DNA microarrays are small pieces of glass, usually the size of a standard microscope slide, onto which tiny spots of DNA have been "printed." The DNA spots on a given microarray, or *chip*, could number in the hundreds, thousands, or even *hundreds of thousands*, allowing a user to probe a given sample quantitatively for the presence of countless specific DNA sequences simultaneously (Schena et al., 1998). In basic research, myriad uses of the DNA microarray platform have evolved, and microarrays printed with other macromolecules, proteins and antibodies especially, have been developed as well (but are not discussed here) (Schena et al., 1998; Hoheisel, 2006; Hughes et al. 2006; Wu and Dewey, 2006; Hall et al., 2007; Werner, 2007; Klenkar and Liedberg, 2008). The information gained in a single cDNA microarray, for example, is equivalent to that of many thousands of gene-specific Northern blots or reverse transcription (RT)-PCR analyses (Allison et al., 2006; Ness, 2006). Thus, by the sheer volume of data obtained in a single run, it is no surprise that microarray technology has had such a revolutionary effect on basic research (Shoemaker et al., 2001). Table 10.1 lists the five major types of DNA microarray and their most common applications.

The use of microarrays in forensics is largely still at the developmental stage, but as research into this area continues, new and powerful possibilities are rapidly

TABLE 10.1 Common Types of DNA Microarrays

Type of Microarray	Printed Spots	Specific Use
cDNA microarray	18 to 30-mers, complementary to coding open reading frames	(a) Analysis of mRNA/miRNA transcription (expression profiling) (b) mRNA splicing analysis
Promoter microarray	PCR products or oligos from known gene promoters	Transcription factor location analysis
Entire chromosome tiling	Varies widely; high resolution: 15–25 bp spanning every ~35 bp of genomic sequence	(a) Transcription factor location analysis (b) Analysis of transcription
Comparative genomic hybridization	60-mers from throughout the genome (coding and noncoding)	Detection of gene deletion or amplification (e.g., during cancer)
SNP microarrays	25-mers	Detection of SNP alleles

emerging. In this chapter we first summarize the basics of microarray technology and the well-established uses in basic research. Then we detail some of the forensic applications of microarrays that are currently under development. Along the way we provide insight into future possibilities for the use of microarrays in forensic science.

10.2.1 cDNA Microarray

The most expansive use of DNA microarrays in scientific laboratories around the world is in the technique of *expression profiling* (Mandruzzato, 2007). Microarrays for this purpose are spotted with DNA molecules, usually 30 to 50-mers, representing complementary DNA sequences from within the coding regions of known or predicted genes (Kreil et al., 2006; Larsson et al. 2006). This platform has been developed so extensively that current high-end cDNA chips bear millions of spots, or *probes*, representing every currently known and predicted coding region in the human genome with multiple spots each. Entire or nearly entire genome coverage has also been achieved for mouse, rat, chicken, sheep, dog, cat, cow, macaque, *Xenopus* (frog), zebrafish, *Drosophila melanogaster, Caenorhabditis elegans*, maize, wheat, rice, barley, grape, cotton, tomato, soybean, sugarcane, *Arabidopsis thaliana*, plasmodium, several species of yeast, *Escherichia coli, Staphylococcus aureus*, and many other microorganisms, plants, animals, and fungi (Affymetrix web site).

cDNA microarrays allow researchers to take a "snapshot" of the gene expression profile, at the level of messenger RNA (mRNA), of a given cell population, as shown in Figure 10.1. Obviously, the utility of this approach is best found in the comparison of two populations: for example, cancer cells vs. tissue-matched normal cells, pre- and postviral infection, or pre- and postexposure to a drug or poison. In this way, the alterations of gene expression that occur as a result of such

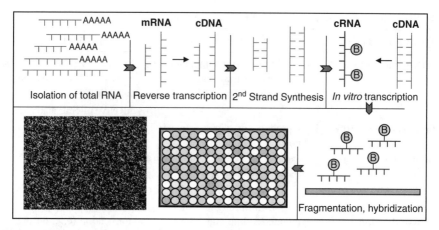

Figure 10.1 Expression profiling by cDNA microarray. Total RNA is isolated from living cells, followed by conversion of mRNA into single-stranded cDNA by reverse transcription using poly(dT) primers. Next, the cDNA is made double-stranded to serve as a template for in vitro transcription using biotinylated NTPs. The cRNA is then fragmented and hybridized onto the microarray platform. cRNA is visualized using a fluorophore-confugated streptavidin moiety and the microarrays are scanned at high resolution to allow quantitation of the fluorescent signal.

treatment conditions can be observed. Following repeated experiments in various settings, contexts, and conditions, a profile of gene expression events emerges. Furthermore, genes for micro-RNAs as well as many known and predicted mRNA splice variants can be discriminated using this same type of chip, and the complexity of the expression profiling data gleaned from microarrays is steadily increasing (Cuperlovic-Culf et al., 2006; Hughes et al., 2006; Kreil et al., 2006; Valencia-Sanchez et al., 2006). It might even be said that our ability to acquire vast amounts of exceedingly complex data is far outpacing our ability to interpret the data.

10.2.2 Other Types of DNA Arrays

Expression profiling is but one application of DNA microarray technology, concerned exclusively with coding regions of the genome. Promoter microarrays are another, displaying DNA sequences representative of gene promoter regions, that is, transcription start sights and proximal sequences of 1 to 10 kb (Ren and Dynlacht, 2004; Blais et al., 2005). Obviously, these types of arrays are not useful for expression studies but, rather, to probe exactly which genes a given transcription factor is physically associated with, at a certain point in time, through a technique called *ChIP-on-chip* or *location analysis* (Buck and Lieb, 2004; Blais et al., 2005; Lee et al., 2006). Other microarrays harbor sequences spanning an entire human chromosome, usually called *tiled arrays* due to the printing pattern of short sequence intervals, or *tiles* (Ishkanian et al., 2004; Graf et al., 2007). These arrays have several uses, including the comprehensive mapping of RNA transcripts and splice variants and chromosome-wide (and eventually true genome-wide) location analysis for transcription factors (Ji and Wang, 2005; Carninci, 2006; Huber et al., 2006). One dramatic example, research using these entire-chromosome arrays, has unexpectedly demonstrated that much more of the "noncoding" regions of the human genome are transcribed than thought previously (Bertone et al., 2004; Johnson et al., 2005). The last major type of DNA array is the comparative genomic hybridization (CGH) array (du Manoir et al., 1993; Kallioniemi et al., 1993). These types of arrays contain DNA sequences, both coding and noncoding, from throughout the genome and are used to detect changes in DNA copy number (gene deletion or gene amplification) that are frequently seen in human cancer and certain genetic diseases (Houldsworth and Chaganti, 1994; Yang et al., 2006; Stjernqvist et al., 2007).

These gene expression arrays, promoter arrays, tiling arrays, and CGH arrays are the most dominantly used microarray platforms in basic molecular research, including toxicogenomics (discussed below) (Ness, 2006). But for most forensic applications, the scrutiny at hand is of small discrete *differences* in closely related DNA sequences that differentiate one individual from another or one strain of bacteria from a close relative. Neither cDNA nor tiling arrays are appropriate for, or capable of, analyzing these subtle DNA differences. Thus, several varieties of microarrays for exactly this purpose have been developed, such as single-nucleotide polymorphism (SNP) arrays and resequencing arrays, which are discussed below. Because they are simply glass slides with spots of DNA, microarrays are easily customized to the application at hand, and countless clever variations have been documented (Mendoza et al., 1999; Lorenz et al., 2003; Dondeti et al., 2004; Thomson et al., 2004; Newton et al., 2005; McGrath et al., 2007). As development costs continually decrease, some laboratories even routinely print their own custom microarrays specific for their area of research and the experimental questions of interest (Ren et al., 2002; Cam et al., 2004; Balciunaite et al., 2005; Blais and Dynlacht, 2005a,b; Acosta-Alvear et al., 2007).

10.2.3 The Birth of "-Omics"

The complete sequencing of the human genome (and that of many other species) has affected nearly every field of biological research (Lander et al., 2001; Venter et al., 2001). Microarray technology is one of the major applications in which these enormous volumes of sequence data are put to direct use (Schena et al., 1998; Shoemaker et al., 2001; Hoheisel, 2006). Indeed, this sequence information, hand-in-hand with the required bioinformatics, was a necessary prerequisite for the development of microarray technology (Schena et al., 1995, 1998; Heller et al., 1997). With this new ability to perform such global genome-wide studies, entirely new scientific disciplines have materialized, and indeed, a new vocabulary has emerged. This is most recognizable by the addition of the suffix "-ome" or "-omics" onto various terms which implies that a global "in total" approach is being applied: for example, genome/genomics, proteome/proteomics, transcriptome, kinome, lipidome, metallome, spliceome, phosphoylome, and glycome (Palsson, 2002; Bishop, 2003, Chaussabel, 2004; Figeys, 2004). Obviously, these new "disciplines" are simply technological approaches to research within existing scientific fields, which are represented by their respective prefixes: for example, metabolomics and toxicogenomics (Aardema and MacGregor, 2002; Dettmer and Hammock, 2004). A new era of molecular research is clearly upon us, the -omics era, and microarrays are at the forefront of this information wave (Chauard et al., 2002; Abu-Issa and Kirby, 2004; Kiechle et al., 2004). In the next few pages we detail how current and future research will bring the -omics wave into the realm of forensic science through the use of DNA microarrays.

10.3 DNA MICROARRAYS IN TOXICOGENOMICS

No area of research has experienced a more profound impact due to advances in microarray technology than the field of genomics (Schena et al., 1998; Bueno-Filho et al., 2006). Stated simply, genomics is the study of the structure and function of DNA and DNA sequences on the genome scale. The field of genomics and, similarly, microarray technology, is intimately connected to the success of the mass sequencing of the human genome (and that of other species) (Schena et al., 1998; Belacel et al., 2006; Lerman et al., 2007). Toxicogenomics is a subdiscipline of genomics that is concerned with deciphering the impact of exogenous molecules such as drugs, poisons, and xenobiotic compounds on gene expression (Lovett, 2000; Aardema and MacGregor, 2002; Gershon, 2002; Hamadeh et al., 2002; Gatzidou et al., 2007). As this name implies, this field was the result of a merger between traditional toxicology research and functional genomics: microarrays are the platform that made this merger possible (Neumann and Galvez, 2002; Pennie, 2002; Ulrich and Friend, 2002; Irwin et al., 2004).

Mechanistic toxicology is the attempt to decipher why and how a drug or poison does what it does. In this way, toxicology is closely related to pharmacology in the search for a "mechanism of action." Predictive toxicology attempts to use this physiological information to make predictions about the toxicity of a related or novel molecule. Toxico*genomics* has greatly accelerated both approaches and deepened our understanding of the biological perturbations caused by chemical compounds (Cunningham, 2006; Fielden and Kolaja, 2006). Drugs and poisons are now grouped in increasingly narrow categories based on common induction of discrete cellular gene expression changes (Oberemm et al., 2005). By aligning physiological response to chemical structures, reliable predictions can be made regarding the toxic effects of a previously untested compound (Raghavan et al., 2005; Fielden and Kolaja, 2006; Martin et al., 2006).

Forensic toxicology is often concerned with the converse situation: the identification of a poisoning agent (which itself may be difficult or impossible to detect), based on the biological effects of that agent. This is often a tenuous process, but a more complete picture of toxin-induced gene expression responses, achieved by microarrays, gives forensic researchers much more to work with in the hunt for specific telltale biomarkers (Lettieri, 2006). Indeed, significant progress has been made in the elucidation of a toxicological "fingerprint" for several drugs and toxins, including arsenic, chromium, nickel, cadmium, methapyriline, clofibrate, hexachlorobenzene, doxorubicin, 5-fluorouracil, and acetaminophen [reviewed by Lettieri (2006)]. The dominant type of microarray technology employed in toxicogenomics is the cDNA microarray for expression profiling (Figure 10.1), although other types of microarrays, such as splicing arrays and CGH arrays, have revealed important pharmacological effects as well (Lee et al., 2005).

10.3.1 Sharing Information

For the voluminous amounts of data gleaned from toxicogenomic studies to be truly useful to the community at large, there must be in place a rigorous common method of analyzing and sharing data. There are several national and international efforts to do just that. The first critical milestone in the creation of public databases for genomic data was the establishment of the MIAME (minimum information about a microarray experiment) guidelines, developed by the European Bioinformatics Institute (EBI) (Brazma et al., 2001). These guidelines provide a common language through which researchers around the world can understand each other's experimental data. Following these guidelines, several publicly accessible repositories for microarray data have been created, the largest being the Gene Expression Omnibus (GEO) (Edgar et al., 2002; Barrett and Edgar, 2006). Within the field of toxicology, the National Center for Toxicology (NCT) within the National Institute of Environmental Health Science (NIEHS) is developing the Chemical Effects in Biological Systems (CEBS) knowledge base (Waters et al., 2003). CEBS, and its counterpart at the EBI, tox-MIAMExpress, aim to compile and integrate vast amounts of genomic data and thus build a complete picture of the complex biological response to toxicological agents and provide that information to the public (Brazma et al., 2003, 2006).

10.3.2 Forensic Application

Our ability to apply what is learned in experimental toxicogenomics to forensic toxicology rests not only on our understanding of the biological response to a given chemical agent, but also on our ability to detect that response in biological tissues, which often present themselves as evidence in very trace amounts or in degraded condition (Nygaard and Hovig, 2006). This presents a serious problem for gene expression analysis, as mRNA is arguably the most labile biological macromolecule of all. But this challenge is certainly not insurmountable and does not apply whatsoever to freshly harvested samples, as is also often possible in forensic practice. But even when access to fresh tissue sample is not possible, conceivable breakthroughs in microarray technology could one day allow recreation of gene expression information. For example, research into chemical modifications of DNA histones has revealed that gene expression is heavily influenced by chromatin architecture, and the precise structure–function relationships of the "histone code" are being teased out methodically (Jenuwein et al., 2001; Turner, 2002).

It seems increasingly likely that scientists will one day be able to infer with relative certainty the level of active transcription of a given gene simply by examining the chromatin structure of the gene promoter (Gilmore and Washburn, 2007). This is relevant to forensic investigations because chromatin is formed by DNA–protein interactions and is far more stable than mRNA. Further, genome-wide scanning of chromatin structure by microarray is already possible through the technology of modified histone-specific ChIP-on-chip (or location analysis) (Roh et al., 2004; Huebert et al., 2006). Regardless of the approach taken to scrutinize gene expression, the unparalleled multiplex capacity of microarrays will be absolutely essential for detecting these residual biological "fingerprints" of drug exposure in a tissue because it will probably involve the altered expression of hundreds of genes. Examining a small number of genes would be far too ambiguous for forensic casework. Without microarrays, the immense labor of examining the expression of hundreds of genes manually makes it difficult to envision the reasonable forensic utility of establishing gene expression signatures in response to drug or toxin exposure.

10.4 DETECTION OF MICROORGANISMS USING MICROARRAYS

DNA microarrays are powerful tools for the rapid detection and identification of microorganisms, and there are countless scenarios in which the identification of a potentially infectious agent is of paramount importance. Indeed, the improved detection of biothreats tropic to humans, livestock, or agricultural plants is a major focus of research efforts at the Centers for Disease Control, the Food and Drug Administration, the National Institutes of Health, the Department of Health and Human Services, and countless nongovernmental research centers. Beyond malicious and bioterrorism-related scenarios, rapid pathogen identification has enormous lifesaving potential in the area of infectious disease diagnostics. Furthermore, and particularly relevant to the current text, precise recognition of microorganisms can be crucial to the investigation of criminal activity. Examples include the detection of microorganisms in deliberate or accidental biopoisoning, and the analysis of bacterial flora to scrutinize or match samples of soil, water, and biological specimens or fluids.

Microarrays are particularly well suited for these purposes for at least three reasons. First, the ability to multiplex with microarrays (i.e., screen for multiple targets simultaneously) far exceeds that of any other method. Second, because there is no need for time-consuming bioassays or decision-tree iterative protocols, microarrays are usually far faster than most currently dominant strategies. Finally, and most powerfully, microarray platforms can be designed intelligently such that little or no prior knowledge of the nature of the microorganism being detected is necessary. All of these advantages have already been demonstrated, as we discuss below. Although microarray technology is not yet commonplace in the forensic analysis of microorganisms, the rapid development and declining costs of the technology will probably lead to increased utilization.

10.4.1 Historical Perspective

Since the invention of the microscope, classifying the various types of microscopic life has challenged and fascinated microbiologists. But the association of various bacteria and fungi with illness brought these classification and identification efforts into the realm of public health. (Because they are even smaller than the maximum resolving power of light microscopes, viruses were not discovered until centuries later.) The Gram stain, which distinguishes bacteria into two classes based on cell wall characteristics, was among the first biochemical techniques in the classification of bacteria and is still in use today. Microscopic size and physical appearance (circular, rod, or corkscrew shape); colony size, shape, and color; growth rate; nutritional requirements; toxicity to laboratory animals; and even traits such as odor once helped scientists tell one species of bacteria from another. Needless to say, all of these methods are laborious and none distinguishes closely related strains or species.

More recently, serological methods have largely supplanted these descriptive techniques. Briefly, *serotyping* is the classification of bacteria based on the presence of certain "markers," usually receptor proteins, that can be detected easily using antibody-based procedures such as ELISA (enzyme-linked immunosorbant assay) (Yolken, 1978). For some pathogenic bacteria, detection assays are designed toward the very receptor that comprises the pathogenic component (Ashkenazi and Cleary, 1989; Munoz et al., 1990; Hearty et al., 2006). This important feature affords these assays a reasonable chance of success in detecting most virulent strains (of that particular species), even, potentially, novel ones. For this reason, along with the rapidity, ease, and negligible expense, serotyping is likely to remain a dominant means of identifying and classifying bacteria (Cebula et al., 2005a).

10.4.2 DNA Fingerprinting

In an effort to distinguish closely related strains of microorganisms, microbiologists and forensic specialists are working to establish a working library of strain-specific genetic fingerprints or *signatures* (Pannucci et al., 2004). These are akin to the DNA fingerprinting that is conducted for identity confirmation of humans using short tandem repeat (STR) or SNP analysis. Knowing the unique genetic features harbored by various strains of pathogenic and benign bacteria, viruses, and fungi will allow investigators to quickly screen biological samples for the presence of these known agents and discriminate between harmless flora and their close virulent relatives.

Several approaches have been taken in an effort to establish DNA fingerprints for individual bacterial strains. RFLP (restriction fragment length polymorphism) analysis of bacterial rRNA genes is one example but fails to distinguish closely related strains or even species (Brindley et al., 1991; Pebay et al., 1992; Alonso et al., 1993; Sriprakash and Gardiner, 1997). RFLP analysis of the entire bacterial chromosome, followed by PFGE (pulse-field gel electrophoresis) is a more dominant technique in bacterial DNA fingerprinting (Linhardt et al., 1992; Gundisch et al., 1993). However, the very large DNA fragments generated by this technique (2 to 5 Mbp) cannot be resolved with great precision. Thus, strain-specific mutations and even larger genomic rearrangements (due to recombination events) could result in no detectable difference in the RFLP fingerprint. In that case, two significantly divergent strains could be indistinguishable by RFLP. Conversely, an otherwise insignificant or even silent mutation that happens to reside within a restriction site could result in a markedly different RFLP fingerprint with little or no phenotypic difference. This slight genetic difference hardly gives rise to a distinct bacterial strain. The point made by both cases is that RFLP analysis simply compares relative sizes of random chunks of the genome. Thus, while perfectly appropriate for identification of humans, RFLP analysis is somewhat clumsy for the identification of rapidly evolving and mutating microorganisms. DNA sequence–based methods would be inherently more discriminating and yield far more information about discrete genetic differences between closely related microbiota.

More recently, amplification-based approaches have emerged that track amplification fragment length polymorphisms (AFLPs) (Janssen et al., 1996; Lin et al., 1996). While some employ random priming and others target specific genomic regions, the readout is again a "fingerprint" of random fragment sizes (Vaneechoutte, 1996). These techniques have proven to be sufficiently discriminating to differentiate closely related strains of bacteria and races of fungi (Keim et al., 1997; Kis-Papo et al., 2003; Hynes et al., 2006). But once again, because the objects of these analyses are (1) chunks of DNA (bands on a gel), not sequences, and (2) totally unrelated to any phenotypic difference, such as pathogenicity, these techniques offer very little genetic insight into the nature of the microorganism of interest. In this new era of high-throughput entire-genome sequencing, sequence-independent techniques for genetic characterization of organisms seem blunt indeed.

10.4.3 DNA Fingerprinting by Microarrays

Microarray technology has now been employed in an effort to establish genetic fingerprints for microorganisms (Willse et al., 2005; Doran et al., 2007). One such example involves the amplification of established repetitive extragenic palindromic (REP) sequences, followed by hybridization onto a small microarray harboring about 200 random nonamers (Willse et al., 2004). This approach was successful in clearly distinguishing 25 closely related strains of *Salmonella enterica* (Willse et al., 2004). A microarray-based genetic fingerprinting system offers distinct advantages over electromobility-based systems. First, the increased complexity of the fingerprint allows much greater precision and statistical confidence, important features for any forensic technology that must withstand scrutiny in a court of law. Second, microarrays afford the potential to screen for hundreds of strains of hundreds of species simultaneously. Third, fingerprints from previously unobserved isolates could be compared to those of known strains and some knowledge of genetic heritage could be inferred. Indeed, bioinformatic methods of designing robust and precise DNA fingerprinting arrays for pathogen detection are maturing (Satya et al., 2008). However, similar to other fingerprinting methods, while microarray fingerprints of repetitive sequences could provide clues into the nature of known or novel microbes, important phenotypic characteristics, such as virulence, could not be predicted with confidence.

10.4.4 DNA Sequence-Based Detection

Single-nucleotide polymorphisms (SNPs) represent the smallest and most discrete genetic difference between two living organisms. These single-base-pair differences can result in a slightly different protein product (a nonsynonymous SNP) or not (a synonymous SNP), based on the degeneracy of the genetic code. Regardless, the comparison of SNPs between closely related strains of bacteria, viruses, or fungi allows a much more thorough appreciation of genetic differences and facilitates the recapitulation of evolutionary lineage (Bochner, 2003). The importance of this cannot be overstated. Consider the hypothetical example of the emergence of a novel strain of *Salmonella enterica* in a poultry packing plant; considering its unique RFLP or AFLP pattern would provide little information other than that this strain is *not* one that has been observed previously. In contrast, a SNP analysis would reveal which known strains of *Salmonella* are the closest genetic relatives to this new strain, and thus a quick and reasonably accurate prediction could be made as to whether or not this new strain is a health hazard. (Prediction of virulence is discussed further in the next section.)

The painstaking process of using SNPs derived from sequencing data to classify strains of bacteria into groups based on hereditary relationships involves statistical analyses, cladisitics, and computer-assisted hierarchical clustering [see Cebula et al. (2005a,b) for two excellent reviews]. Nevertheless, to date, SNP analyses have been used to resolve and identify strains of *Bacillus anthracis, Mycobacterium tuberculosis, Helicobacter pylori, Escherichia coli, Shigella dysenteriae,* and *Salmonella enterica* (Levy et al., 2004; Pearson et al., 2004; Cebula et al., 2005b; Filliol et al., 2006; Van Ert et al., 2007). The most famous example is that of the SNP analysis used to confirm that the multiple *B. anthracis* samples of the fall 2001 anthrax terrorist attack were of the same genetic strain (Hoffmaster et al., 2002).

10.4.5 Where DNA Microarrays Come In

Microarrays afford the possibility of rapid and thorough SNP analyses without having to perform actual sequencing reactions. One such technique, known as *resequencing*, first involves the amplification (and fluorescent labeling) of several specific regions of interest (i.e., those with well-established strain-specific SNPs) from a sample containing an unknown bacterial strain (Hacia, 1999). Target amplicon size is kept relatively small, about 200 bp, and amplicons are denatured and fragmented prior to hybridization. The probe spots on the microarray are small oligonucleotides around 25 bp in length containing a variable SNP at the central position. Each SNP of interest comprises eight spots on the chip: all four possible variations of the SNP, in both DNA strand orientations. Thus, for each biological sample, the two brightest spots will correspond to the SNP version harbored by that particular bacterium, virus, and so on. Of course, the beauty of the microarray is that an essentially unlimited number of SNPs could be analyzed in a single scan (Zwick et al., 2005; Malanoski et al., 2006; Wang et al., 2006). Further, selections of SNPs from many hundreds or even thousands of bacterial species could be printed on a single array. Thus, one could easily envision a single microarray that provides coverage for all known enteric bacteria, for example. It is worth noting that because the resequencing technique involves an amplification step with a fairly small amplicon, only minute amounts of starting material are required.

An approach similar to resequencing, called *minisequencing*, represents another microarray-based method of SNP analysis (Huber et al., 2001; Lindroos et al., 2002). However, the small amplicons containing the SNPs of interest are not fluorescently labeled. Further, the 25-bp probes on the microarray do not harbor the SNP itself. Rather, they are the 25 bp of sequence immediately 5′ of the SNP base. Either prior to hybridization (in-solution minisequencing), or following hybridization (in situ minisequencing), of the targets to the probes, DNA polymerization reactions are performed to extend the probes, but only chain-terminating di-deoxynucleotides (fluorescently labeled) are provided. Thus, the DNA polymerase can only add one base to the probes before chain termination, but this $n + 1$ position will correspond to the SNP base with a coded fluorescent label for each of the four possible bases, as revealed by confocal fluorescent scanning of the chip (see Figure 10.2). The minisequencing approach has two advantages over resequencing. First, it only requires two spots for each SNP rather than eight. (Again, both DNA strands are analyzed as an internal duplicate sample.) Second and more important, the data scan of different color fluorescent spots is unambiguous and far more straightforward to analyze and interpret than the single-color scanning of resequencing. However, the additional chemistry and manipulation of the polymerization step adds cost, labor, and experimental variability, important considerations in the search for a highly automated high-throughput platform for forensic application.

10.4.6 Looking Forward: Genetic Virulence Signatures

Bioinformatics researchers are currently mining the ever-increasing volumes of complete genome sequences from various strains of bacteria in search of "virulence signatures" (Pannucci et al., 2004). The rationale for this search is twofold. First, if a given DNA motif could be established as "necessary" for virulence (and thus present in all pathogenic strains, absent in all innocuous strains), this would represent a potentially momentous breakthrough for the basic scientists investigating mechanisms of microbial pathogenesis. Second, and more germane to the forensics realm, knowing the "virulence signature" of pathogenic strains of *E. coli*, for example, would give investigators a robust tool in detecting the appearance of those pathogens, even, at least in theory, those that have not yet been discovered or are the product of malevolent bioengineering.

It goes without saying that the "training set" of sequence data regarding these virulence signatures would need to be monitored and updated continually as more strains, of both the pathogenic and benign variety, emerge. Experimental validation of the pathogenic mechanisms of virulence-associated DNA sequences should be sought in as timely a manner as possible because of the potential to "overfit" the sequence data, honing in on sequences that are merely correlative and thus not in any way connected with pathogenesis. Examples of "red herring" sequences could

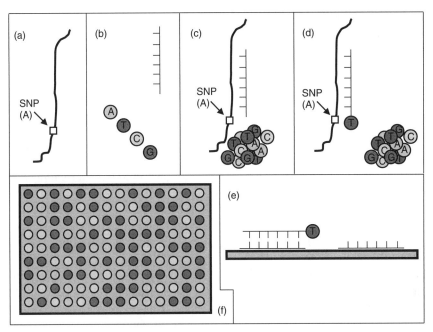

Figure 10.2 SNP analysis by in-solution minisequencing using a four-color microarray. (A) First, the genomic regions containing the target SNPs are amplified by small-amplicon PCR. (B) Then minisequencing reactions are carried out in solution using only dideoxynucleotides (fluorescently labeled). (C) The sequencing primers hybridize with the 3' end oriented one base upstream of the SNPs. (D) The sequencing primers are extended by one labeled base, complementary to each SNP. (E) The mixture is denatured and hybridized to a microarray printed with DNA sequences containing the SNPs. (F) The microarray containing thousands of SNPs is scanned and each spot is called based on fluorescence.

lead to potentially disastrous false-negative readouts during the screening of an unknown. This potential pitfall can only be overcome by experimental rigor and the diligent reporting and sharing of sequence data.

Although traditional sequencing efforts and "shotgun sequencing" had previously been at the forefront of the search for bacterial and viral genetic fingerprints and virulence signatures, the emergence of relatively low-cost whole-genome sequencing is poised to accelerate this research dramatically (Lorenz, 2002; Read et al., 2003; Pannucci, et al., 2004; Dorrell et al., 2005; Afset et al., 2006; Burrack and Higgins, 2007). Comparative genomics and bioinformatics, followed by experimental validation, will probably illuminate the key genomic elements that confer the pathogenic phenotype(s) for each organism. Indeed, in a spectacularly successful example, a detailed virulence signature for *Staphylococcus aureus* has been established, corresponding to genetic elements in 84 different genes, including those for antibioitic resistance (Spence et al., 2008). Such information is immediately applicable toward building a robust and extremely sensitive microarray that could be used to detect the presence of a virulent form of *S. aureus* with relative certainty.

Once established and validated, virulence signatures for multiple pathogens (each in tens or hundreds of naturally occurring degenerate permutations) can be spotted onto DNA microarrays. Because the spot density capability of DNA chips is so high and increasing (currently >1,000,000 unique probes per chip is possible), it is conceivable that a single chip could harbor the virulence signatures of thousands of different pathogens, allowing broad but precise screening of biological samples in a single pass.

Although the use of microarrays in the forensic identification of microorganisms is still in its infancy, the enormous advantages justify further development of this technology. As costs continue to decline and access increases, it is not inconceivable that microarrays will, one day soon, become indispensables tools for the rapid and reliable identification of microbial strains.

10.5 PROBING HUMAN GENOMES BY DNA MICROARRAYS

The identification of human beings by "DNA fingerprint" has, in a relatively short time, revolutionized forensic science and criminal proceedings (Higuchi et al., 1988; Kobilinsky and Levine, 1988; Lee et al., 1994). This revolution cannot possibly be overstated. However, recent advances in molecular biology and genomics have made possible even further development of forensic DNA technology. Despite this, the extent of DNA technology in most forensic laboratories is currently limited to "DNA fingerprinting" for matching an unknown biological sample to that of a known person. In that regard, DNA microarrays offer several advantages over traditional STR analysis, including adding significant automation, quantitative analysis, and statistical power, as well as further extending the technique to even smaller and more degraded DNA samples than is currently possible. In addition, looking beyond, DNA microarrays may one day allow forensic investigators to transform an unknown and therefore unhelpful DNA sample into a crucial clue for investigating crime.

10.5.1 STR Analysis

Short tandem repeat (STR) analysis is the current technique of choice for identifying unique individual markers of a given DNA sample, and in most cases provides such immense discriminating power that the U.S. and British judicial systems fully accept STR results into a court of law, provided that the technical execution is within established protocols. This technique exploits the polymorphic nature of some repetitive sequences within the human genome, called *microsatellite DNA*. These regions consist of a simple DNA sequence, 3 to 7 bp in length, that is repeated anywhere from 3 to 15 times. [There exists examples of both shorter and longer repeats (satellite or minisatellite DNA) and regions where many hundreds or thousands of repeats exist in tandem, but these are not typically part of forensic STR analysis.] Biallelic in nature (one each on the paternal and maternal chromosome), these microsatellites, or STRs, are highly polymorphic and stably inherited. That is, a given person will have two alleles for a given STR (e.g., 12 and 8 repeats on each of a pair of homologous chromosomes, one inherited from her biological mother, the other from her father), and one will be passed on to each of her children at random. Thus, two nonrelated individuals will have a very low probability of having the same two alleles for this given STR. Even siblings have only a 25% chance of inheriting the same two numbers of repeats for each and every STR. When examining multiple STRs, the likelihood that two different individuals will harbor the same alleles drops dramatically with each additional STR included in the analysis. Accordingly, the standard U.S. Department of Justice's CODIS protocol calls for an examination of 13 established STR loci, which places the statistical odds of two nonidentical individuals having the same two alleles for all 13 loci at less than 2×10^{-15}, a number whose reciprocal is more than 10,000 times more than the number of human persons who have ever walked the Earth (Lee et al., 1994; McEwen, 1995; Budowle et al., 1999; Ruitberg et al., 2001). (Government officials are currently deliberating an increase in the number of CODIS STR loci to 20.)

Clearly, the discovery, development, and careful propagation of STR analysis represented an unparalleled triumph in the application of molecular biology to criminal justice, and it may be tempting to think that current abilities of DNA analysis already meet or exceed their usefulness, with no further improvement necessary. However, a critical examination of current techniques in STR analysis reveals several compelling reasons to pursue further development in this area. First, the approach of STR analysis itself could benefit with increased automation, multiplexing, and high-throughput capabilities, as current protocols call for extensive sample handling and manipulation and significant technician labor. In addition to the costs of time and labor, these manipulations bring opportunities for experimental error which could potentially jeopardize the trustworthiness of a given DNA match. Although the PCR amplifications can now be multiplexed routinely and efficiently (reducing labor and handling), traditional STR analysis still requires forensic researchers to electro-resolve the PCR products and make judgments regarding relative mobility rates of these DNA fragments, as compared to known DNA standards. The result is a DNA "fingerprint" that is somewhat qualitative, rather than strictly quantitative, as is most desirable when objectivity is of such supreme importance.

Without altering the basic premise, small DNA microarrays can be used in the execution of traditional STR analysis. A major challenge in any microarray-based, and thus hybridization-based, approach to forensic STR analysis is the detection of subtle differences in sequence repeat number. For example, while discriminating between 3 and 15 repeats would be relatively straightforward, it is exceedingly difficult to discriminate between 9 and 10 repeats using differential hybridization onto a standard microarray with the necessary degree confidence. Accordingly, researchers have devised two creative solutions to this obstacle. First, electronically active microarrays can be utilized to establish and fine-tune an extremely high stringency hybridization environment (Radtkey et al., 2000). In this way, even the subtlest differences in hybridization energies can be exaggerated and the results are clear spot calls on the microarray, indicating the exact repeat number for each

STR with precision. For this reason, the technology of electronic microarrays has found its way into several microarray platforms (including forensic SNP analysis, discussed below) and development continues (Heller, 2002).

More recently, a second approach to array-based STR analysis has been reported that involves the use of standard glass microarrays and a sequence of enzymatic steps (Kemp et al., 2005). In this somewhat complicated method, called variable-length probe array (VLPA), the microarray spots harbor single-stranded probes containing STRs of various lengths (repeat numbers) bound to the microarray surface at their 5′ end. Following amplification and 5′ fluorescent labeling of STR regions in the target DNA sample, the labeled amplicons are hybridized to the microarray. Obviously, the targets will hybridize to probes with both more and fewer numbers of STR repeats than those of the target itself. To ensure proper alignment at each STR set, complementary *clamp sequences* are included at the 5′ end of the probe DNAs and the 3′ end of the target sequences. Thus, at probe spots containing *fewer* repeats than the hybridized target sequence, the 5′ end of the target DNA (which contains the label) will overhang. At probe spots containing *more* repeats than the target DNA, the spotted 3′ end of the probe DNA will overhang the target DNA. Next, the slides are treated with S1 endonuclease, which digests single-stranded DNA. Thus, all overhangs are removed and, importantly, the fluorescent label will be released from any target DNA that is bound to a probe spot containing an STR sequence that is shorter (fewer repeats) than the target DNA. Conversely, the fluorescent label will remain on all target DNAs that are bound at spots containing probes of equal or greater length, despite the digestion of the 3′ end of probe DNAs that are longer than the target DNA. For example, if a target DNA sample contains 11 repeats of a given STR, fluorescent signals of approximately equal intensities would be detected at probe spots harboring 11, 12, 13, 14,... repeats, but no signal whatsoever would be detected at probe spots harboring 10, 9, 8, 7,... repeats. Although it is not yet clear how this system can delineate heterozygous alleles, a major advantage of this approach is that the spot calls are unambiguous (signal or no signal) and thus easy to automate in a high-throughput manner. Creative use of the clamp sequence could also allow a great deal of multiplexing of many samples simultaneously.

A third microarray-based approach toward STR analysis that has been reported involves the use of the branch migration capabilities of double-stranded DNA. This assay begins similarly to VLPA with hybridization of labeled target DNA to probes of varying length using a clamp sequence to ensure alignment. Next, a succession of two competing probes in opposite orientations initiate branch migration to displace probe–target interactions that are not perfect matches for the number of tandem repeats. Thus, array scanning will elicit signal from only one spot per STR allele, and heterozygosity can be detected and delineated using an algorithm specially designed for this system (Pourmand et al., 2007). In summary, several options for array-based STR analysis have been developed, each with distinct advantages and caveats. The major advantage of any microarray-based platform is the enormous potential for rapid high-throughput screening of thousands of samples simultaneously. And as microarray costs continue to decline, the feasibility of their routine use in forensic applications increases.

10.5.2 SNP Analysis

Although the power of STRs in forensic investigations is truly remarkable, there are limitations to STR analysis that warrant continued research in alternative technologies. Specifically, because each STR locus is first amplified in a PCR reaction, there is a lower limit for required template, beneath which amplification is too inefficient and unreliable for forensic purposes. This minimum sample size is indeed minute, allowing complete DNA profiles to be collected from scant physical evidence such as one intact hair follicle. However, biological material that serves as a template in a PCR reaction is consumed in the process, and when the amount of sample is limiting, as is often true in forensic investigations, there is little or no margin for error. Currently, the lower limit of input DNA to achieve a reliable CODIS (FBI's Combined DNA Index System) DNA profile is between 100 and 200 pg of pristine human genomic DNA (Lorente et al., 1997; Krenke et al., 2002; Leclair et al., 2003). With 6 pg of DNA in each human nucleus, this means that in the best of circumstances and with flawless technique, a biological sample must contain at least 15 to 30 intact cells with undamaged DNA. Although this is certainly an impressive level of detection, the collection of trace forensic evidence often falls below this standard in quantity or quality of DNA. Following the death of a human cell, DNA begins the steady process of decay, and trace DNA specimens from hair or dead skin cells collected at a crime scene are often not of adequate quality for any type of STR analysis.

Single-nucleotide polymorphism (SNP) analysis provides an alternative to STR analysis with much lower requirements of quantity and quality of input DNA, thus allowing further reduction of the threshold requirements of biological material from which a quality DNA profile can be extracted (Walsh, 2004; Nygaard and Havig, 2006). This increased sensitivity arises due to the amplification of significantly smaller amplicons. Unavoidably, forensic STR analysis involves the amplification of repeat sequences as large as 400 bp, and there is a direct relationship between the size of an amplicon and the amount of intact template DNA required for efficient amplification. Because SNPs are a single base in length, amplicons can be designed that are less than 50 bp in length. Thus, SNP analysis is far more robust than STR analysis with trace and/or degraded DNA samples. However, because there are only four possible alleles for each SNP (A, T, C, or G), rather than 7 to 22 possible alleles for each CODIS STR, a far greater panel of SNPs must be examined in order to achieve discriminating power similar to that of forensic STR analysis. For the level of certainty required of a criminal prosecution, at least 50 SNPs would be needed for a positive identification, and a panel of about 100 SNPs would probably be appropriate for a universal forensic protocol. It is this large number of individual loci that need to be examined that has probably discouraged more widespread application of SNP analysis in forensic DNA matching. However, microarrays circumvent this challenge completely, and hundreds or even thousands of SNPs could be analyzed simultaneously with relative ease. Further, clamping sequences or a similar mechanism could facilitate complete SNP screening for multiple (even hundreds) of different biological samples on a single chip.

Human SNP analysis by microarray is already a well-established technology, although application is limited primarily to basic research and clinical diagnosis of genetic diseases. Further, the current dbSNP database at the National Institutes of Health contains millions of documented SNPs in the human population (NCBI web site, http://www.ncbi.nlm.nih.gov/projects/SNP/). Thus, it will not require great strides to bring the technology of SNP microarrays to the forensic realm. Indeed, electronically active arrays have already proven their utility in forensic applications (Borsting et al., 2004, 2005; Huang et al., 2004; Tsang et al., 2004). Exactly which cadre of SNPs will perform best for forensic purposes remains to be seen, but among the most important considerations are high variability of alleles in the human population and ease of amplification in a large multiplex PCR. The latter factor is determined by the genomic "neighborhood" of each SNP and must be tested empirically. A microarray-based strategy for detecting SNPs involves either resequencing or minisequencing (see Section 10.4.4 and Figure 10.2). Although to date, current reports of forensic SNP analysis by microarray have utilized in-solution minisequencing prior to hybridization, resequencing microarrays could prove just as powerful and would involve less labor and manipulation (Allen and Divne 2005; Divne and Allen, 2005).

10.5.3 Exploring an Unknown Genome?

Consider the example, as happens often, when a DNA sample collected from a crime scene does not match with a known DNA profile in any database or from a possible suspect. Currently, the unknown DNA sample would then be considered a "dead end," useful only for screening additional suspect(s), if they are somehow identified by further criminal investigation. However, the DNA sample itself is an entire genome, containing much more information than the simple DNA fingerprints that are used to match samples. Within the DNA sequence of a genome is all the genetic information necessary to build that person, and *only* that person. Thus, a DNA sample with no known matches could be far from a dead-end. It could potentially tell us much about the perpetrator him- or herself, including such genetic traits as eye color, hair color, race and ethnic origin, approximate height/weight/build, and genetic diseases or predispositions. When investigating a crime, such information could be extremely valuable for narrowing down overwhelmingly large pools of suspects into just a few people or one person. Focusing on a smaller suspect pool would greatly reduce the disturbance of innocent people, as well increase the likelihood that detectives could discover additional evidence implicating the guilty party. Further, if a suspect pool was narrowed to a sufficiently small number of people, a court of law could then compel the collection of DNA samples to facilitate a comprehensive match.

Forensic research into the precise genetic basis for physical traits such as race/ethnic origin and eye color is already proceeding rapidly (Frudakis et al.,

2003; Rowold and Herrera, 2003; Sturm and Frudakis, 2004; Amorim and Pereira, 2005; Ray et al., 2005; Sobrino and Carracedo, 2005; Wetton et al., 2005; Lao et al., 2006; Kayser et al., 2008). However, information of this type will be most valuable to forensic science when many traits are analyzed at once; for example, "the DNA sample left at the crime scene was from a woman of mixed Central European and East Asian descent, with green eyes, blonde natural hair color, medium build, projected height between 5′6″ and 5′9″ and strong predisposition for asthma." Even accepting the caveat that many of these characteristics will fluctuate and are influenced by environment in combination with genetics, it is easy to imagine the power of such information when investigating a crime. However, this area of forensic research is still in its infancy, awaiting rigorous research, and must be accompanied by careful consideration of the social and ethical concerns that are raised by such an approach. Notwithstanding, there exists no technological or scientific barrier to the development of this approach.

CONCLUSION

The genomics revolution, led by microarray technology, has already informed various subdisciplines in forensic science, most notably toxicogenomics and SNP analysis, and further microarray-catalyzed innovation is surely afoot. One of the hallmark features of the genomics era is the staggering scale and speed with which research and technology development is now progressing. Thus, the wait might not be as long as one would think. It can now be said that the revolutionary potential of microarray technology rivals that of PCR. However, the potential application to forensics is much broader than just DNA fingerprinting and human ID. It will be interesting to observe how quickly potential becomes reality and innovative research becomes the gold standard. A currently limiting factor, both in terms of research and in application, is the seemingly exorbitant cost of microarray printing, fluidics, and high-resolution scanners. However, these costs are steadily decreasing and the ability to strip and reuse microarrays would dramatically slash research costs (although it would not be appropriate for casework). Further, the need to purchase and maintain such microarray instrumentation, including array printers, fluidics stations, and even scanners, is being alleviated by a more "full service" approach, for both in-stock and custom arrays, from the major microarray companies. As declining costs bring microarray technology within reach for more research laboratories, increased numbers of thorough forensic validation studies can be performed and published. Hopefully, this will persuade forensic practitioners that the technology is not just powerful but *reliable* as a bona fide tool for forensic casework.

REFERENCES

Aardema, M. J., and MacGregor, J. T. (2002). Toxicology and genetic toxicology in the new era of "toxicogenomics": impact of "-omics" technologies. *Mutat. Res.*, **499**:13–25.

Abu-Issa, R., and Kirby, M. L. (2004). Take heart in the age of "omics." *Circ. Res.*, **95**:335–336.

Acosta-Alvear, D., et al. (2007). XBP1 controls diverse cell type- and condition-specific transcriptional regulatory networks. *Mol. Cell*, **27**:53–66.

Affymetrix web site. http://www.affymetrix.com.

Afset, J. E., et al. (2006). Identification of virulence genes linked with diarrhea due to atypical enteropathogenic *Escherichia coli* by DNA microarray analysis and PCR. *J. Clin. Microbiol.*, **44**:3703–3711.

Allen, M., and Divne, A. M. (2005). Universal tag arrays in forensic SNP analysis. *Methods Mol. Biol., (Clifton, N.J.)* **297**:141–154.

Allison, D. B., Cui, X., Page, G. P., and Sabripour, M. (2006). Microarray data analysis: from disarray to consolidation and consensus. *Nat. Rev. Genet.*, **7**:55–65.

Alonso, R., et al. (1993). Comparison of serotype, biotype and bacteriocin type with rDNA RFLP patterns for the type identification of *Serratia marcescens*. *Epidemiol. Infect.*, **111**:99–107.

Amorim, A., and Pereira, L. (2005). Pros and cons in the use of SNPs in forensic kinship investigation: a comparative analysis with STRs. *Forensic Sci. Int.*, **150**:17–21.

Ashkenazi, S., and Cleary, T. G. (1989). Rapid method to detect Shiga toxin and Shiga-like toxin I based on binding to globotriosyl ceramide (Gb3), their natural receptor. *J. Clin. Microbiol.*, **27**:1145–1150.

Balciunaite, E., et al. (2005). Pocket protein complexes are recruited to distinct targets in quiescent and proliferating cells. *Mol. Cell. Biol.*, **25**:8166–8178.

Barrett, T., and Edgar, R. (2006). Gene expression omnibus: microarray data storage, submission, retrieval, and analysis. *Methods Enzymol.*, **411**:352–369.

Belacel, N., Wang, Q., and Cuperlovic-Culf, M. (2006). Clustering methods for microarray gene expression data. *Omics*, **10**:507–531.

Bertone, P., et al. (2004). Global identification of human transcribed sequences with genome tiling arrays. *Science*, **306**:2242–2246.

Bishop, M. (2003). Omics research and bioinformatics: joined-up thinking or anarchy? *Brief. Bioinf.*, **4**:313.

Blais, A., and Dynlacht, B. D. (2005a). Devising transcriptional regulatory networks operating during the cell cycle and differentiation using ChIP-on-chip. *Chromosome Res.*, **13**:275–288.

Blais, A., and Dynlacht, B. D. (2005b). Constructing transcriptional regulatory networks. *Genes Dev.*, **19**:1499–1511.

Blais, A., et al. (2005). An initial blueprint for myogenic differentiation. *Genes Dev.*, **19**:553–569.

Bochner, B. R. (2003). New technologies to assess genotype–phenotype relationships. *Nat. Rev. Genet.*, **4**:309–314.

Borsting, C., Sanchez, J. J., and Morling, N. (2004). Multiplex PCR, amplicon size and hybridization efficiency on the NanoChip electronic microarray. *Int. J. Legal Med.*, **118**:75–82.

Borsting, C., Sanchez, J. J., and Morling, N. (2005). SNP typing on the NanoChip electronic microarray. *Methods Mol. Biol. (Clifton, N.J.)*, **297**:155–168.

Brazma, A., et al. (2001). Minimum information about a microarray experiment (MIAME): toward standards for microarray data. *Nat. Genet.*, **29**:365–371.

Brazma, A., et al. (2003). ArrayExpress: a public repository for microarray gene expression data at the EBI. *Nucleic Acids Res.*, **31**:68–71.

Brazma, A., Kapushesky, M., Parkinson, H., Sarkans, U., and Shojatalab, M. (2006). Data storage and analysis in ArrayExpress. *Methods Enzymol.*, **411**:370–386.

Brindley, P. J., Heath, S., Waters, A. P., McCutchan, T. F. and Sher, A. (1991). Characterization of a programmed alteration in an 18S ribosomal gene that accompanies the experimental induction of drug resistance in *Schistosoma mansoni*. *Proc. Natl. Acad. Sci. USA*, **88**:7754–7758.

Buck, M. J., and Lieb, J. D. (2004). ChIP-chip: considerations for the design, analysis, and application of genome-wide chromatin immunoprecipitation experiments. *Genomics*, **83**:349–360.

Budowle, B., Moretti, T. R., Baumstark, A. L., Defenbaugh, D. A., and Keys, K. M. (1999). Population data on the thirteen CODIS core short tandem repeat loci in African Americans, U. S. Caucasians, Hispanics, Bahamians, Jamaicans, and Trinidadians. *J. Forensic Sci.*, **44**:1277–1286.

Bueno-Filho, J. S., Gilmour, S. G., and Rosa, G. J. (2006). Design of microarray experiments for genetical genomics studies. *Genetics*, **174**:945–957.

Burrack, L. S., and Higgins, D. E. (2007). Genomic approaches to understanding bacterial virulence. *Cur. Opin. Microbiol.*, **10**:4–9.

Cam, H., et al. (2004). A common set of gene regulatory networks links metabolism and growth inhibition. *Mol. Cell*, **16**:399–411.

Carninci, P. (2006). Tagging mammalian transcription complexity. *Trends Genet.*, **22**:501–510.

Cebula, T. A., et al. (2005a). Molecular applications for identifying microbial pathogens in the post-9/11 era. *Expert Rev. Mol. Diagn.*, **5**:431–445.

Cebula, T. A., Jackson, S. A., Brown, E. W., Goswami, B., and LeClerc, J. E. (2005b). Chips and SNPs, bugs and thugs: a molecular sleuthing perspective. *J. Food Prot.*, **68**:1271–1284.

Chaussabel, D. (2004). Biomedical literature mining: challenges and solutions in the "omics" era. *Am. J. Pharmacogenom.*, **4**:383–393.

Chouard, T., Weiss, U., and Dhand, R. (2002). Good "omics" for the poor? *Nature*, **419**:489.

Cunningham, M. L. (2006). Putting the fun into functional toxicogenomics. *Toxicol. Sci.*, **92**:347–348.

Cuperlovic-Culf, M., Belacel, N., Culf, A. S., and Ouellette, R. J. (2006). Microarray analysis of alternative splicing. *Omics*, **10**:344–357.

Dettmer, K., and Hammock, B. D. (2004). Metabolomics: a new exciting field within the "omics" sciences. *Environ. Health Perspect.*, **112**:A396–A397.

Divne, A. M., and Allen, M. (2005). A DNA microarray system for forensic SNP analysis. *Forensic Sci. Int.*, **154**:111–121.

Dondeti, V. R., Sipe, C. W., and Saha, M. S. (2004). In silico gene selection strategy for custom microarray design. *BioTechniques*, **37**:768–770, 772, 774–766.

Doran, M., et al. (2007). Oligonucleotide microarray identification of *Bacillus anthracis* strains using support vector machines. *Bioinformatics (Oxford)*, **23**:487–492.

Dorrell, N., Hinchliffe, S. J., and Wren, B. W. (2005). Comparative phylogenomics of pathogenic bacteria by microarray analysis. *Curr. Opin. Microbiol.*, **8**:620–626.

du Manoir, S., et al. (1993). Detection of complete and partial chromosome gains and losses by comparative genomic in situ hybridization. *Hum. Genet.*, **90**:590–610.

Edgar, R., Domrachev, M., and Lash, A. E. (2002). Gene Expression Omnibus: NCBI gene expression and hybridization array data repository. *Nucleic Acids Res.*, **30**:207–210.

Fielden, M. R., and Kolaja, K. L. (2006). The state-of-the-art in predictive toxicogenomics. *Curr. Opin. Drug Discov. Dev.*, **9**:84–91.

Figeys, D. (2004). Combining different "omics" technologies to map and validate protein–protein interactions in humans. *Brief. Funct. Genom. Proteom.*, **2**:357–365.

Filliol, I., et al. (2006). Global phylogeny of *Mycobacterium tuberculosis* based on single nucleotide polymorphism (SNP) analysis: insights into tuberculosis evolution, phylogenetic accuracy of other DNA fingerprinting systems, and recommendations for a minimal standard SNP set. *J. Bacteriol.*, **188**:759–772.

Frudakis, T., et al. (2003). Sequences associated with human iris pigmentation. *Genetics*, **165**:2071–2083.

Gatzidou, E. T., Zira, A. N., and Theocharis, S. E. (2007). Toxicogenomics: a pivotal piece in the puzzle of toxicological research. *J. Appl. Toxicol.*, **27**:302–309.

Gershon, D. (2002). Toxicogenomics gains impetus. *Nature*, **415**:4–5.

Gilmore, J. M., and Washburn, M. P. (2007). Deciphering the combinatorial histone code. *Nat. Methods*, **4**:480–481.

Graf, S., et al. (2007). Optimized design and assessment of whole genome tiling arrays. *Bioinformatics (Oxford)*, **23**:i195–i204.

Gundisch, C., Kirchhof, G., Baur, M., Bode, W., and Hartmann, A. (1993). Identification of *Azospirillum* species by RFLP and pulsed-field gel electrophoresis. *Microb. Releases*, **2**:41–45.

Hacia, J. G. (1999). Resequencing and mutational analysis using oligonucleotide microarrays. *Nat. Genet.*, **21**:42–47.

Hall, D. A., Ptacek, J., and Snyder, M. (2007). Protein microarray technology. *Mech. age. Dev.*, **128**:161–167.

Hamadeh, H. K., Amin, R. P., Paules, R. S., and Afshari, C. A. (2002). An overview of toxicogenomics. *Curr. Issues Mol. Biol.*, **4**:45–56.

Hearty, S., Leonard, P., Quinn, J., and O'Kennedy, R. (2006). Production, characterisation and potential application of a novel monoclonal antibody for rapid identification of virulent *Listeria* monocytogenes. *J. Microbiol. Methods*, **66**:294–312.

Heller, M. J. (2002). DNA microarray technology: devices, systems, and applications. *Ann. Rev. Biomed. Eng.*, **4**:129–153.

Heller, R. A., et al. (1997). Discovery and analysis of inflammatory disease-related genes using cDNA microarrays. *Proc. Nat. Acad. Sci. USA*, **94**:2150–2155.

Higuchi, R., von Beroldingen, C. H., Sensabaugh, G. F., and Erlich, H. A. (1988). DNA typing from single hairs. *Nature*, **332**:543–546.

Hoffmaster, A. R., Fitzgerald, C. C., Ribot, E., Mayer, L. W., and Popovic, T. (2002). Molecular subtyping of *Bacillus anthracis* and the 2001 bioterrorism-associated anthrax outbreak, United States. *Emerg. Infect. Dis.*, **8**:1111–1116.

Hoheisel, J. D. (2006). Microarray technology: beyond transcript profiling and genotype analysis. *Nat. Rev. Genet.*, **7**:200–210.

Houldsworth, J., and Chaganti, R. S. (1994). Comparative genomic hybridization: an overview. *Am. J. Pathol.*, **145**:1253–1260.

Huang, Y., et al. (2004). Multiple sample amplification and genotyping integrated on a single electronic microarray. *Electrophoresis*, **25**:3106–3116.

Huber, M., et al. (2001). Detection of single base alterations in genomic DNA by solid phase polymerase chain reaction on oligonucleotide microarrays. *Anal. Biochem.*, **299**:24–30.

Huber, W., Toedling, J., and Steinmetz, L. M. (2006). Transcript mapping with high-density oligonucleotide tiling arrays. *Bioinformatics (Oxford)*, **22**:1963–1970.

Huebert, D. J., Kamal, M., O'Donovan, A., and Bernstein, B. E. (2006). Genome-wide analysis of histone modifications by ChIP-on-chip. *Methods (San Diego)*, **40**:365–369.

Hughes, T. R., Hiley, S. L., Saltzman, A. L., Babak, T., and Blencowe, B. J. (2006). Microarray analysis of RNA processing and modification. *Methods Enzymol.*, **410**:300–316.

Hynes, S. S., Chaudhry, O., Providenti, M. A., and Smith, M. L. (2006). Development of AFLP-derived, functionally specific markers for environmental persistence studies of fungal strains. *Can. J. Microbiol.*, **52**:451–461.

Irwin, R. D., et al. (2004). Application of toxicogenomics to toxicology: basic concepts in the analysis of microarray data. *Toxicol. Pathol.*, **32(Suppl. 1)**:72–83.

Ishkanian, A. S., et al. (2004). A tiling resolution DNA microarray with complete coverage of the human genome. *Nat. Genet.*, **36**:299–303.

Janssen, P., et al. (1996). Evaluation of the DNA fingerprinting method AFLP as an new tool in bacterial taxonomy. *Microbiology (Reading)*, **142(Pt. 7)**:1881–1893.

Jenuwein, T., and Allis, C. D. (2001). Translating the histone code. *Science*, **293**:1074–1080.

Ji, H., and Wong, W. H. (2005). TileMap: create chromosomal map of tiling array hybridizations. *Bioinformatics (Oxford)*, **21**:3629–3636.

Johnson, J. M., Edwards, S., Shoemaker, D., and Schadt, E. E. (2005). Dark matter in the genome: evidence of widespread transcription detected by microarray tiling experiments. *Trends Genet.*, **21**:93–102.

Kallioniemi, O. P., et al. (1993). Comparative genomic hybridization: a rapid new method for detecting and mapping DNA amplification in tumors. *Semin. Cancer Biol.*, **4**:41–46.

Kayser, M., et al. (2008). Three genome-wide association studies and a linkage analysis identify HERC2 as a human iris color gene. *Am. J. Hum. Genet.*, **82**:411–423.

Keim, P., et al. (1997). Molecular evolution and diversity in *Bacillus anthracis* as detected by amplified fragment length polymorphism markers. *J. Bacteriol.*, **179**:818–824.

Kemp, J. T., Davis, R. W., White, R. L., Wang, S. X., and Webb, C. D. (2005). A novel method for STR-based DNA profiling using microarrays. *J. Forensic Sci.* **50**:1109–1113.

Kiechle, F. L., Zhang, X., and Holland-Staley, C. A. (2004). The -omics era and its impact. *Arch. Pathol. Lab. Med.*, **128**:1337–1345.

Kis-Papo, T., Kirzhner, V., Wasser, S. P., and Nevo, E. (2003). Evolution of genomic diversity and sex at extreme environments: fungal life under hypersaline Dead Sea stress. *Proc. Natl. Acad. Sci. USA*, **100**:14970–14975.

Klenkar, G., and Liedberg, B. (2008). A microarray chip for label-free detection of narcotics. *Anal. Bioanal. Chem.*, **391(5)**:679–1688.

Kobilinsky, L., and Levine, L. (1988). Recent application of DNA analysis to issues of paternity. *J. Forensic Sci.*, **33**:1107–1108.

Kreil, D. P., Russell, R. R., and Russell, S. (2006). Microarray oligonucleotide probes. *Methods Enzymol.*, **410**:73–98.

Krenke, B. E., et al. (2002). Validation of a 16-locus fluorescent multiplex system. *J. Forensic Sci.*, **47**:773–785.

Lander, E. S., et al. (2001). Initial sequencing and analysis of the human genome. *Nature*, **409**:860–921.

Lao, O., van Duijn, K., Kersbergen, P., de Knijff, P., and Kayser, M. (2006). Proportioning whole-genome single-nucleotide-polymorphism diversity for the identification of geographic population structure and genetic ancestry. *Am. J. Hum. Genet.*, **78**:680–690.

Larsson, O., Wennmalm, K., and Sandberg, R. (2006). Comparative microarray analysis. *Omics*, **10**:381–397.

Leclair, B., et al. (2003). STR DNA typing: increased sensitivity and efficient sample consumption using reduced PCR reaction volumes. *J. Forensic Sci.*, **48**:1001–1013.

Lee, H. C., Ladd, C., Bourke, M. T., Pagliaro, E., and Tirnady, F. (1994). DNA typing in forensic science: I. Theory and background. *Am. J. Forensic Med. Pathol.*, **15**:269–282.

Lee, K. M., Kim, J. H., and Kang, D. (2005). Design issues in toxicogenomics using DNA microarray experiment. *Toxicol. Appl. Pharmacol.*, **207**:200–208.

Lee, T. I., Johnstone, S. E., and Young, R. A. (2006). Chromatin immunoprecipitation and microarray-based analysis of protein location. *Nat. Protocols*, **1**:729–748.

Lerman, G., et al. (2007). Functional genomics via multiscale analysis: application to gene expression and ChIP-on-chip data. *Bioinformatics (Oxford)*, **23**:314–320.

Lettieri, T. (2006). Recent applications of DNA microarray technology to toxicology and ecotoxicology. *Environ. Health Perspect.*, **114**:4–9.

Levy, D. D., Sharma, B., and Cebula, T. A. (2004). Single-nucleotide polymorphism mutation spectra and resistance to quinolones in *Salmonella enterica* serovar *enteritidis* with a mutator phenotype. *Antimicrob. Agents Chemother.*, **48**:2355–2363.

Lin, J. J., Kuo, J., and Ma, J. (1996). A PCR-based DNA fingerprinting technique: AFLP for molecular typing of bacteria. *Nucleic Acids Res.*, **24**:3649–3650.

Lindroos, K., Sigurdsson, S., Johansson, K., Ronnblom, L., and Syvanen, A. C. (2002). Multiplex SNP genotyping in pooled DNA samples by a four-colour microarray system. *Nucleic Acids Res.*, **30**:e70.

Linhardt, F., Ziebuhr, W., Meyer, P., Witte, W., and Hacker, J. (1992). Pulsed-field gel electrophoresis of genomic restriction fragments as a tool for the epidemiological analysis of *Staphylococcus aureus* and coagulase-negative staphylococci. *FEMS Microbiol. Lett.*, **74**:181–185.

Lorente, M., et al. (1997). Sequential multiplex amplification: utility in forensic casework with minimal amounts of DNA and partially degraded samples. *J. Forensic Sci.*, **42**:923–925.

Lorenz, M. C. (2002). Genomic approaches to fungal pathogenicity. *Curr. Opin. Microbiol.*, **5**:372–378.

Lorenz, M. G., Cortes, L. M., Lorenz, J. J., and Liu, E. T. (2003). Strategy for the design of custom cDNA microarrays. *BioTechniques*, **34**:1264–1270.

Lovett, R. A. (2000). Toxicogenomics. Toxicologists brace for genomics revolution. *Science*, **289**:536–537.

Malanoski, A. P., Lin, B., Wang, Z., Schnur, J. M., and Stenger, D. A. (2006). Automated identification of multiple micro-organisms from resequencing DNA microarrays. *Nucleic Acids Res.*, **34**:5300–5311.

Mandruzzato, S. (2007). Technological platforms for microarray gene expression profiling. *Adv. Exp. Med. Biol.*, **593**:12–18.

Martin, R., Rose, D., Yu, K., and Barros, S. (2006). Toxicogenomics strategies for predicting drug toxicity. *Pharmacogenomics*, **7**:1003–1016.

McEwen, J. E. (1995). Forensic DNA data banking by state crime laboratories. *Am. J. Hum. Genet.*, **56**:1487–1492.

McGrath, P. T., et al. (2007). High-throughput identification of transcription start sites, conserved promoter motifs and predicted regulons. *Nat. Biotechnol.*, **25**:584–592.

Mendoza, L. G., et al. (1999). High-throughput microarray-based enzyme-linked immunosorbent assay (ELISA). *BioTechniques*, **27**:778–780, 782–776, 788.

Munoz, M. L., et al. (1990). Antigens in electron-dense granules from *Entamoeba histolytica* as possible markers for pathogenicity. *J. Clin. Microbiol.*, **28**:2418–2424.

Ness, S. A. (2006). Basic microarray analysis: strategies for successful experiments. *Methods Mol. Biol. (Clifton, N.J.)*, **316**:13–33.

Neumann, N. F., and Galvez, F. (2002). DNA microarrays and toxicogenomics: applications for ecotoxicology? *Biotechnol. Adv.*, **20**:391–419.

Newton, S. S., Bennett, A., and Duman, R. S. (2005). Production of custom microarrays for neuroscience research. *Methods (San Diego)*, **37**:238–246.

Nygaard, V., and Hovig, E. (2006). Options available for profiling small samples: a review of sample amplification technology when combined with microarray profiling. *Nucleic Acids Res.*, **34**:996–1014.

Oberemm, A., Onyon, L., and Gundert-Remy, U. (2005). How can toxicogenomics inform risk assessment? *Toxicol. Appl. Pharmacol.*, **207**:592–598.

Palsson, B. (2002). In silico biology through "omics." *Nat. Biotechnol.*, **20**:649–650.

Pannucci, J., et al. (2004). Virulence signatures: microarray-based approaches to discovery and analysis. *Biosens. Bioelectron.*, **20**:706–718.

Pearson, T., et al. (2004). Phylogenetic discovery bias in *Bacillus anthracis* using single-nucleotide polymorphisms from whole-genome sequencing. *Proc. Natl. Acad. Sci. USA*, **101**:13536–13541.

Pebay, M., et al. (1992). Detection of intraspecific DNA polymorphism in *Streptococcus salivarius* subsp. *thermophilus* by a homologous rDNA probe. *Res. Microbiol.*, **143**:37–46.

Pennie, W. D. (2002). Custom cDNA microarrays; technologies and applications. *Toxicology*, **181–182**:551–554.

Pourmand, N., et al. (2007). Branch migration displacement assay with automated heuristic analysis for discrete DNA length measurement using DNA microarrays. *Proc. Natl. Acad. Sci. USA*, **104**:6146–6151.

Radtkey, R., et al. (2000). Rapid, high fidelity analysis of simple sequence repeats on an electronically active DNA microchip. *Nucleic Acids Res.*, **28**:E17.

Raghavan, N., Amaratunga, D., Nie, A. Y., and McMillian, M. (2005). Class prediction in toxicogenomics. *J. Biopharm. Stat.*, **15**:327–341.

Ray, D. A., et al. (2005). Inference of human geographic origins using Alu insertion polymorphisms. *Forensic Sci. Int.*, **153**:117–124.

Read, T. D., et al. (2003). The genome sequence of *Bacillus anthracis* Ames and comparison to closely related bacteria. *Nature*, **423**:81–86.

Ren, B., and Dynlacht, B. D. (2004). Use of chromatin immunoprecipitation assays in genome-wide location analysis of mammalian transcription factors. *Methods Enzymol.*, **376**:304.

Ren, B., et al. (2002). E2F integrates cell cycle progression with DNA repair, replication, and G(2)/M checkpoints. *Genes Dev.*, **16**:245.

Roh, T. Y., Ngau, W. C., Cui, K., Landsman, D., and Zhao, K. (2004). High-resolution genome-wide mapping of histone modifications. *Nat. Biotechnol.*, **22**:1013–1016.

Rowold, D. J., and Herrera, R. J. (2003). Inferring recent human phylogenies using forensic STR technology. *Forensic Sci. Int.*, **133**:260–265.

Ruitberg, C. M., Reeder, D. J., and Butler, J. M. (2001). STRBase: a short tandem repeat DNA database for the human identity testing community. *Nucleic Acids Res.*, **29**:320–322.

Satya, R. V., Zavaljevski, N., Kumar, K., and Reifman, J. (2008). A high-throughput pipeline for designing microarray-based pathogen diagnostic assays. *BMC Bioinformatics*, **9**:185.

Schena, M., Shalon, D., Davis, R. W., and Brown, P. O. (1995). Quantitative monitoring of gene expression patterns with a complementary DNA microarray. *Science*, **270**:467–470.

Schena, M., et al. (1998). Microarrays: biotechnology's discovery platform for functional genomics. *Trends Biotechnol.*, **16**:301–306.

Shoemaker, D. D., et al. (2001). Experimental annotation of the human genome using microarray technology. *Nature*, **409**:922–927.

Sobrino, B., and Carracedo, A. (2005). A. SNP typing in forensic genetics: a review. *Methods Mol. Biol. (Clifton, N.J.)*, **297**:107–126.

Spence, R. P., et al. (2008). Validation of virulence and epidemiology DNA microarray for identification and characterization of *Staphylococcus aureus* isolates. *J. Clin. Microbiol.*, **46**:1620–1627.

Sriprakash, K. S., and Gardiner, D. L. (1997). Lack of polymorphism within the rRNA operons of group A streptococci. *Mol. Gen. Genet.*, **255**:125–130.

Stjernqvist, S., Ryden, T., Skold, M., and Staaf, J. (2007). Continuous-index hidden Markov modelling of array CGH copy number data. *Bioinformatics (Oxford)*, **23**:1006–1014.

Sturm, R. A., and Frudakis, T. N. (2004). Eye colour: portals into pigmentation genes and ancestry. *Trends Genet.*, **20**:327–332.

Thomson, J. M., Parker, J., Perou, C. M., and Hammond, S. M. (2004). A custom microarray platform for analysis of microRNA gene expression. *Nat. Methods*, **1**:47–53.

Tsang, S., et al. (2004). Development of multiplex DNA electronic microarrays using a universal adaptor system for detection of single nucleotide polymorphisms. *BioTechniques*, **36**:682–688.

Turner, B. M. (2002). Cellular memory and the histone code. *Cell*, **111**:285–291.

Ulrich, R., and Friend, S. H. (2002). Toxicogenomics and drug discovery: Will new technologies help us produce better drugs? *Nat. Rev.*, **1**:84–88.

Valencia-Sanchez, M. A., Liu, J., Hannon, G. J., and Parker, R. (2006). Control of translation and mRNA degradation by miRNAs and siRNAs. *Genes Dev.*, **20**:515–524.

Vaneechoutte, M. (1996). DNA fingerprinting techniques for microorganisms: a proposal for classification and nomenclature. *Mol. Biotechnol.*, **6**:115–142.

Van Ert, M. N., et al. (2007). Strain-specific single-nucleotide polymorphism assays for the *Bacillus anthracis* Ames strain. *J. Clin. Microbiol.*, **45**:47–53.

Venter, J. C., et al. (2001). The sequence of the human genome. *Science* **291**:1304–1351.

Walsh, S. J. (2004). Recent advances in forensic genetics. *Expert Rev. Mol. Diagn.*, **4**:31–40.

Wang, Z., et al. (2006). Identifying influenza viruses with resequencing microarrays. *Emerg. Infect. Dis.*, **12**:638–646.

Waters, M., et al. (2003). Systems toxicology and the chemical effects in biological systems (CEBS) knowledge base. *EHP Toxicogenomics*, **111**:15–28.

Werner, T. (2007). Regulatory networks: linking microarray data to systems biology. *Mech. Age. Devel.*, **128**:168–172.

Wetton, J. H., Tsang, K. W., and Khan, H. (2005). Inferring the population of origin of DNA evidence within the UK by allele-specific hybridization of Y-SNPs. *Forensic Sci. Int.*, **152**:45–53.

Willse, A., et al. (2004). Quantitative oligonucleotide microarray fingerprinting of *Salmonella enterica* isolates. *Nucleic Acids Res.*, **32**:1848–1856.

Willse, A., et al. (2005). Comparing bacterial DNA microarray fingerprints. *Stat. Appl. Genet. Mol. Biol.*, **4**, Art. 19.

Wu, X., and Dewey, T. G. (2006). From microarray to biological networks: analysis of gene expression profiles. *Methods Mol. Biol. (Clifton, N.J.)*, **316**:35–48.

Yang, T. P., Chang, T. Y., Lin, C. H., Hsu, M. T., and Wang, H. W. (2006). ArrayFusion: a Web application for multi-dimensional analysis of CGH, SNP and microarray data. *Bioinformatics (Oxford)*, **22**:2697–2698.

Yolken, R. H., et al. (1978). Enzyme-linked immunosorbent assay for identification of rotaviruses from different animal species. *Science*, **201**:259–262.

Zwick, M. E., et al. (2005). Microarray-based resequencing of multiple *Bacillus anthracis* isolates. *Genome Biol.*, **6**:R10.

CHAPTER 11

Date-Rape Drugs with Emphasis on GHB

STANLEY M. PARSONS

Department of Chemistry and Biochemistry, Program in Biomolecular Science and Engineering, Neuroscience Research Institute, University of California, Santa Barbara, California

Summary The date-rape drugs γ-hydroxybutyrate (GHB), 3,4-methylenedioxymethamphetamine, flunitrazepam, and ketamine are discussed with an emphasis on GHB. Recreational, predatory, and lethal doses, why polydosing is dangerous, metabolism including membrane transport, and diagnostic metabolites are covered. Similarities to and differences from the effects and metabolism of ethanol are also discussed. The advantages of field tests to detect date-rape drugs, and limitations of antibodies and advantages of enzymes for field tests, are described. Development of a rapid enzymatic test for GHB is described.

11.1	Introduction		357
11.2	Molecular mechanisms of action		357
	11.2.1	Receptors and transporters	357
	11.2.2	Real GHB receptors	359
11.3	Societal context of date-rape agents		361
	11.3.1	Acute effects of date-rape agents on cognition and behavior	361
	11.3.2	Medicinal uses of date-rape drugs	361
	11.3.3	Self-abuse	362
	11.3.4	Date rape, death, and regulation	363
11.4	Metabolism fundamentals		363
	11.4.1	Complexity in unraveling metabolism of GHB-related compounds	363
	11.4.2	Isozymes in GHB-related metabolism	364
	11.4.3	Subcellular compartmentalization of enzymes, transporters, and substrates	364
	11.4.4	Dynamics and equilibria for enzymes and transporters	365
	11.4.5	Thermodynamics-based analysis of metabolic flux	366

Forensic Chemistry Handbook, First Edition. Edited by Lawrence Kobilinsky.
© 2012 John Wiley & Sons, Inc. Published 2012 by John Wiley & Sons, Inc.

	11.4.6	Metabolism of endogenous GHB versus ingested GHB and prodrugs	367
	11.4.7	Directionality of in vivo and in vitro enzymatic activity	367
	11.4.8	Transporters and enzymes mediating GHB-related metabolism	367
11.5	Biosynthesis of endogenous GHB		368
	11.5.1	First step for GHB biosynthesis in the known pathway	368
	11.5.2	Second step for GHB biosynthesis in the known pathway	368
	11.5.3	Third step for GHB biosynthesis in the known pathway	371
	11.5.4	Which step in GHB biosynthesis is rate limiting?	373
	11.5.5	Are there other biosynthetic pathways to endogenous GHB?	374
11.6	Absorption and distribution of ingested GHB		376
	11.6.1	Gastrointestinal tract	376
	11.6.2	Blood	377
11.7	Initial catabolism of GHB		377
	11.7.1	Transport into mitochondria	377
	11.7.2	Iron-dependent alcohol dehydrogenase ADHFe1	377
	11.7.3	Poorly characterized catabolism of GHB	379
11.8	Chemistry of GHB and related metabolites not requiring enzymes		380
11.9	Experimental equilibrium constants for redox reactions of GHB		380
11.10	Estimated equilibrium constants for redox reactions of GHB in vivo		381
11.11	Different perspectives on turnover of endogenous GHB are consistent		384
11.12	Disposition of succinic semialdehyde		385
11.13	Conversion of prodrugs to GHB and related metabolites		386
	11.13.1	γ-Butyrolactone	386
	11.13.2	1,4-Butanediol	387
11.14	Subcellular compartmentalization of GHB-related compounds		388
11.15	Comparative catabolism of ethanol, 1,4-butanediol, fatty acids, and GHB		389
11.16	Catabolism of MDMA, flunitrazepam, and ketamine		390
11.17	Detection of date-rape drugs		390
	11.17.1	Compounds diagnostic for dosing by synthetic date-rape drugs	390
	11.17.2	Compounds diagnostic for dosing by GHB	390
	11.17.3	Gold-standard testing	391
	11.17.4	Many applications for reliable field tests	392
	11.17.5	Hospital emergency department example	392
	11.17.6	Preparation of a sample for delayed analysis	393
	11.17.7	Time window available to detect dosing	393
	11.17.8	Extending the time window	394
11.18	Special circumstances of GHB		395
	11.18.1	Industrial connection	395
	11.18.2	Enzymes acting on GHB in bacteria, yeast, and plants	395
	11.18.3	Possible accidental intoxication by GHB in the future	395
11.19	Considerations during development of field tests		396
	11.19.1	Shortcomings of antibody-based screens for simple analytes	396
	11.19.2	Advantages of enzyme-based screens for simple natural analytes	397
11.20	Development of an enzymatic test for GHB		399
	11.20.1	Sensitivity required for the hospital emergency department	399

11.20.2	Choice of enzyme	399
11.20.3	Reliable field test for GHB	400
Conclusion		402
Notes		404
References		406

11.1 INTRODUCTION

Prominent examples of *date-rape drugs* are the natural compound γ-hydroxybutanoic acid (GHB) and the synthetic compounds 3,4-methylenedioxymethamphetamine (MDMA), 5-(2-fluorophenyl)-1,3-dihydro-1-methyl-7-nitro-2H-1,4-benzodiazepin-2-one (flunitrazepam), and 2-(2-chlorophenyl)-2-(methylamino)cyclohexanone (ketamine) (Figure 11.1; Scott-Ham and Burton, 2005; Saint-Martin et al., 2006; Abanades et al., 2007a; Juhascik et al., 2007). These compounds are grouped in ignominy because they can be administered surreptitiously in the relatively small amounts required to make a person compliant with sexual assault. Ethanol has similar effects, but generally it is not called a date-rape drug because it is legal and knowingly consumed. The term *agent* as used here refers to ethanol and date-rape drugs, whereas the term *drug* is restricted to date-rape drugs.

In this chapter we discuss primarily date-rape drugs.[1] We also touch on ethanol just enough to point out some similarities, differences, and interactions between ethanol and date-rape drugs. As GHB is a research topic of special interest to the author, it is emphasized throughout. Much important progress has been made recently in understanding the physiology, biochemistry, cell biology, and molecular biology relevant to these agents. Important progress has also been made in developing better methods to detect dosing. The goals of this chapter are to unify the biological and chemical perspectives of date-rape drugs and to improve our ability to detect dosing. The wide range of subjects covered has required the use of footnotes to explain specialized terminology. References are numerous, but to keep the number of them reasonable, they are not comprehensive (apologies to authors not cited). A number of recent excellent reviews of individual date-rape drugs and forensic methods to detect them may be consulted for more citations: Maitre et al., 2000; Palmer, 2004; Wong et al., 2004a,b; Britt and McCance-Katz, 2005; Gonzalez and Nutt, 2005; Drasbek et al., 2006; Pardi and Black, 2006; Snead and Gibson, 2006; Wedin et al., 2006; Wolff and Winstock, 2006; Colado et al., 2007; Kintz, 2007; Seamans, 2008; Sinner and Graf, 2008.

11.2 MOLECULAR MECHANISMS OF ACTION

11.2.1 Receptors and Transporters

To cause cognitive and behavioral effects, psychotropic agents must interact with neuroreceptors or neurotransporters in the central nervous system. This

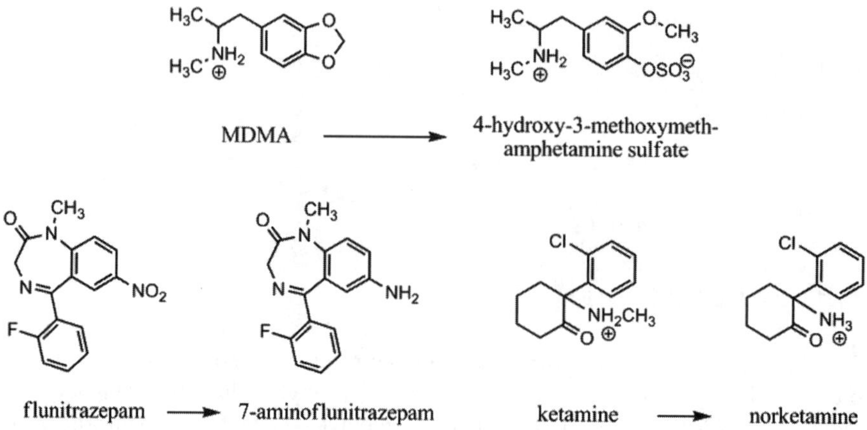

Figure 11.1 Structures of some synthetic date-rape drugs and their major metabolites found in human urine.

requirement means that the agents must pass through the blood–brain barrier to come into contact with neurons. A neuroreceptor generally spans the cytoplasmic membrane of a neuron, and the active site faces the extracellular space so that it can bind to a neurotransmitter coming from there. A neurotransporter can span either the cytoplasmic membrane or the membrane of a synaptic vesicle[2] inside a neuron.

The blood–brain barrier is breached by nonspecific diffusion for ethanol, flunitrazepam, and ketamine (Adachi et al., 2005; Cuadrado et al., 2007). It is breached by specific transport for MDMA, metabolites of MDMA, and GHB (Drewes et al., 2001; Bhattacharya and Boje, 2004; Escobedo et al., 2005). Thus, access by the latter compounds to brain potentially is saturable and inhibitable. More will be said later about transport versus nonspecific diffusion of GHB and related molecules. Some date-rape agents have actions in the periphery, but such actions won't be discussed much, except for metabolism.

Once they have reached neurons in the brain, the agents have many effects. Recent research indicates that ethanol acts primarily by means of a limited number of molecular mechanisms (Vengeliene et al., 2008). Prominent among them is a three-way, synergistic interaction of ethanol, neurosteroids[3], and the inhibitory neurotransmitter γ-aminobutanoic acid (GABA) on separate sites in the ionotropic $GABA_A$ receptor. Together these agents promote opening of the chloride channel in the receptor at concentrations of GABA lower than otherwise required (Herd et al., 2007; Mody et al., 2007; Mody, 2008).[4] Open chloride channels make most neurons resistant to electrical stimulation. This three-way synergy is especially effective at a subtype of the $GABA_A$ receptor located *outside* of synapses, where extracellular GABA concentrations are low but do not fluctuate much. As $GABA_A$ receptors are distributed throughout the brain, the result of ethanol consumption is widespread tonic (i.e., persistent) neural depression.

GHB is present naturally in brain at concentrations that depend on the region (Doherty et al., 1978; Vayer et al., 1988). Values range up to about 10 μM, averaged across tissue compartments. At the >100-fold-higher concentrations reached in sexual assault and self-abuse situations, GHB is a partial agonist for $GABA_B$, the metabotropic receptor (Gobaille et al., 1999; Lingenhoehl et al., 1999; Wong et al., 2004a,b).[5] Activation of $GABA_B$ receptor inhibits the release of neurotransmitters from nerve terminals by several intracellular second-messenger mechanisms (Ulrich and Bettler, 2007). GHB does not produce the full range of its behavioral effects in mice that have one of the two subunit types in $GABA_B$ receptors deleted from the genome (Bettler and Braeuner-Osborne, 2004; Gassmann and Bettler, 2007). The observation confirms involvement of $GABA_B$ receptors in GHB effects in vivo.

GHB increases neurosteroid levels three- to fivefold when it activates $GABA_B$ receptors (Barbaccia et al., 2002; Biggio et al., 2007). Increased neurosteroid levels in turn potentiate opening of chloride channels in $GABA_A$ receptors. Thus, high doses of GHB directly or indirectly activate both major classes of GABA receptor. These actions produce molecular, cellular, and behavioral responses partially overlapping those produced by ethanol and GABAergic drugs[6] such as flunitrazepam, another date-rape drug discussed in this chapter (see below; Baker et al., 2008; Helms et al., 2008; Carter et al., 2009).

Activation of GABA receptors by GHB is entangled with the possibility that a small fraction of ingested GHB or GHB prodrug is converted to GABA itself by reversal of the biosynthetic pathway from GABA to GHB (Figure 11.2; Sections 11.5.2 and 11.5.3) (Collier and De Feudis, 1970; Vayer et al., 1985a; Hechler et al., 1997; Gobaille et al., 1999). Such reversal is thermodynamically feasible. GABA derived from GHB potentially could act on all GABA receptors, including recently discovered $GABA_A$-ρ (also known as $GABA_C$; Schmidt, 2008). However, reversal is controversial (Mohler et al., 1976; Snead et al., 1989; Crunelli et al., 2006; Ren and Mody, 2006). It might not occur when membrane barriers are intact, and the issue is unresolved. Action of GHB on recently identified GHB receptors is described in Section 11.2.2.

Other date-rape drugs have different molecular mechanisms of action. Flunitrazepam binds to an allosteric site in both synaptic and nonsynaptic $GABA_A$ receptors, where it potentiates GABA action to depress the central nervous system (Brenneisen and Raymond, 2001). MDMA (and/or an MDMA metabolite) interacts with several types of transporter to cause release of the neurotransmitter 5-hydroxytryptamine (also known as serotonin) and the neurotransmitter-hormone oxytocin.[7] In addition, extensive use of MDMA can cause permanent neuronal damage, probably by promoting the formation of reactive oxygen species such as hydrogen peroxide and hydroxyl radical (de la Torre et al., 2004)! Ketamine is an antagonist at the NMDA type of receptor for the excitatory neurotransmitter glutamate (Raeder, 2003).[8]

11.2.2 Real GHB Receptors

The mechanisms described above do not involve a receptor selective for GHB. Two forms of the human GHB receptor (GHBh1 and C12K32) have been cloned,

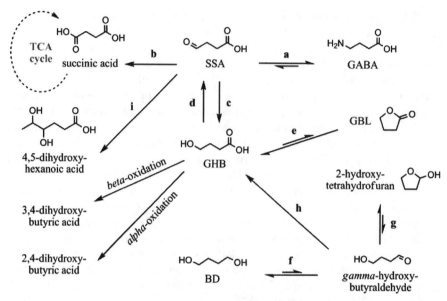

Figure 11.2 Known metabolism of GHB and its prodrugs. Chemical structures are shown above or next to the names or abbreviations. Reversible steps between metabolites are shown with paired half-head arrows and irreversible steps with a single arrow. An arrow longer than a paired arrow indicates the favored direction. Enzymes catalyzing the reactions alphabetically labeled are listed in Table 11.2. Reaction a: GABA transaminase makes SSA in mitochondrial matrix. Reaction b: ALDH5A1 in mitochondrial matrix oxidizes SSA to succinic acid that enters the tricarboxylic acid cycle. Reaction c: AKR1A1 and AKR7A2 reduce SSA in cytoplasm. This step requires transport of SSA from mitochondrial matrix (Figure 11.3). Reaction d: ADHFe1 oxidizes GHB in mitochondrial matrix after transport of GHB into matrix. The reaction probably is irreversible because reaction b is very fast. Untested is the possibility that irreversible ω-oxidation of GHB by cytochrome P450 and catalase also occurs (Section 11.15). Reaction e: γ-Lactonase 1 in blood and liver hydrolyzes GBL. Reaction f: One or more unidentified ADHs oxidize BD in the periphery. A different, irreversible oxidative mechanism that might involve cytochrome P450 and catalase occurs in brain (Section 11.15). Reaction g: Cyclization is spontaneous. Reaction h: Unidentified ALDH in the periphery. A different oxidative mechanism also might exist by analogy to the metabolism of ethanol (Section 11.15). Reaction i: An intermediate in the pyruvate decarboxylase reaction in mitochondria probably forms 5-keto-4-hydroxyhexanoic acid that then is reduced. α- and β-oxidation of GHB also occur, possibly in mitochondria, peroxisomes, and smooth endoplasmic reticulum (Section 11.15).

expressed, and characterized preliminarily (Kemmel et al., 2006; Andriamampandry et al., 2007). Both forms bind to physiologically relevant, sub-μM concentrations of GHB to regulate intracellular second messengers. The receptors also bind to the specific antagonist NCS-382, but not to GABA. GHBh1 expressed in a heterologous host quickly desensitizes after exposure to a saturating concentration of

GHB, whereas C12K32 does not.[9] A putative GHB receptor, which is of uncertain significance because it has aberrant pharmacological properties, has also been cloned from rat (Andriamampandry et al., 2003).

The cloning of the human receptors, combined with prior discoveries of neurotransmitter-like properties of GHB, such as its storage by synaptic vesicles, establishes GHB as a classical neurotransmitter (Kemmel et al., 1998; Muller et al., 2002). The full significance of neuro-signaling by endogenous GHB is not known (Ticku and Mehta, 2008). The terms *de novo* and *endogenous* are used here as synonyms. They mean not arising from ingested GHB or GHB prodrug. Whether GHB receptors in vivo desensitize during dosing with exogenous GHB, and what the physiological consequences of such desensitization would be, are not known. Characterization of these properties is just beginning (Coune et al., 2010).

11.3 SOCIETAL CONTEXT OF DATE-RAPE AGENTS

11.3.1 Acute Effects of Date-Rape Agents on Cognition and Behavior

Ethanol reduces social inhibition, increases suggestibility, and at large doses induces sedation and amnesia (Vengeliene et al., 2008). GHB increases libido and suggestibility, causes euphoria, relaxation, and reduction of social inhibitions, and at large doses induces sedation and amnesia (Drasbek et al., 2006; Kueh et al., 2008). Flunitrazepam induces hypnosis, sedation, anxiolysis, and at large doses, amnesia (Lane et al., 2008). MDMA causes mild euphoria and stimulates feelings of love, empathy, and connection to other people (Colado et al., 2007). Ketamine produces analgesia, anesthesia, and hallucinations (Rowland, 2005). Despite differences in the molecular mechanisms of action and required doses (Sections 11.2.1, 11.2.2, and 11.3.3), the agents are similar to each other in producing mental states that make victims compliant with sexual assault (Elliott, 2008; Lawyer et al., 2010).

11.3.2 Medicinal Uses of Date-Rape Drugs

Date-rape drugs have been used to treat clinical indications. For example, GHB is used for sleep disorders and alcohol-withdrawal syndrome (Agabio and Gessa, 2002; Johnston et al., 2003; Pardi and Black, 2006; Robinson and Keating, 2007; Castelli, 2008; Caputo et al., 2009). GHB might also be useful in cancer treatment and protection from or functional recovery after focal cerebral ischemia (Sadasivan et al., 2006; Emi et al., 2007; Gao et al., 2008). Flunitrazepam has been used for insomnia and anxiety (Matsuo and Morita, 2007), and ketamine has been used for pain (Okon, 2007; Morton, 2008). MDMA is being tested for treatment of posttraumatic stress disorder (Parrott, 2007; Vastag, 2010). Accordingly, the compounds are manufactured by reputable pharmaceutical companies and are available through prescription. Unfortunately, they also are available from illicit sources.

11.3.3 Self-Abuse

Date-rape agents frequently are ingested voluntarily without medical supervision (http://www.drugabuse.gov/Infofacts/clubdrugs.html). Some people engage in this behavior because they perceive the cognition- and behavior-altering properties of the agents to be compatible with recreation (Camacho et al., 2005; Griffiths and Johnson, 2005; Abanades et al., 2007a; Kim et al., 2007b; Horowitz et al., 2008; Lee and Levounis, 2008). Common oral doses for recreational use by people who have not developed tolerance are given in Table 11.1. Effective doses differ over a 50,000-fold range for the various agents!

In addition to having cognitive and behavioral effects, GHB stimulates release of human growth hormone (Volpi et al., 2000; Gonzalez and Nutt, 2005). Some bodybuilders consume it in huge amounts (18 to 250 g/day divided into multiple doses) in an effort to increase muscle mass. The practice results in tolerance and addiction, and severe withdrawal symptoms occur when addicted individuals attempt to discontinue use (Bennett et al., 2007; LeTourneau et al., 2008; Wojtowicz et al., 2008). In rats, the addiction decreases membrane fluidity and increases GABA reuptake from extracellular space in brain (Bhattacharya et al., 2006). GHB greatly alters the levels of some brain proteins (van Nieuwenhuijzen et al., 2010). It increases the acetylation of histone H3, thereby causing epigenetic modification of gene expression (Klein et al., 2009). The latter observations demonstrate physiological adaptation consistent with addiction.

TABLE 11.1 Effective Oral Doses for Human Beings[a]

Compound	LD_{50} (mg/kg)	Recreation (mg/kg)	$\frac{LD_{50}}{Recreation}$
Ethanol	5000[b]	500[c]	10
GHB	400	50	8
GBL/BD[d]	200	25	8
MDMA	30	2	16
Flunitrazepam	0.4	0.01	30[e]
Ketamine	40	1	>38

Source: Data from Miller (1979), Wang and Bai (1998), Gable (2004), Carai et al. (2006), and http://toxnet.nlm.nih.gov/. Effective doses for rodents are higher.

[a]The numbers apply to normal, healthy, and otherwise undrugged human adults of either gender who have not developed tolerance.
[b]This amount corresponds to about 0.6% ethanol in physiological fluids.
[c]This amount corresponds to consumption of about one-half bottle of wine.
[d]GBL and 1,4-butane diol (BD) are quickly converted to GHB in the body (Sections 11.13.1 and 11.13.2). Effective doses are lower than for GHB because the compounds are absorbed faster (Table 11.5). Safety factors are similar, however.
[e]Rounding errors create the apparent discrepancy relative to the ratio computed from numbers on the left.

11.3.4 Date Rape, Death, and Regulation

In chemical assault for the purpose of date rape, the perpetrator attempts to deliver a high recreational dose that incapacitates the victim. However, date-rape agents can be lethal. The estimated oral LD_{50} values for human beings who have not developed tolerance are given in Table 11.1. The ratio of the LD_{50} to the recreational dose (i.e., the safety factor) is also given for each agent. Some agents have small safety factors, particularly ethanol, GHB, and its prodrugs γ-butyrolactone (GBL) and 1,4-butanediol (BD) (Sections 11.13.1 and 11.13.2). Because different agents have mostly different molecular targets that can interact with each other physiologically, these agents have the potential to synergize and push the central nervous system into an extremely depressed state. Deaths occur at doses of combined agents that are difficult to predict and often much lower than those required for single agents (Jonasson et al., 2000; Lalonde and Wallage, 2004; Mazarr-Proo and Kerrigan, 2005; Jones et al., 2007; Verschraagen et al., 2007; Hovda et al., 2008; Knudsen et al., 2008; Akins et al., 2009). To impede drug-assisted date-rape and possible death, most national governments restrict access to date-rape drugs. Proscription has created a need for reliable analytical procedures to detect the compounds and their metabolites.

11.4 METABOLISM FUNDAMENTALS

11.4.1 Complexity in Unraveling Metabolism of GHB-Related Compounds

Because it is a neurotransmitter, GHB is degraded (catabolized) and biosynthesized (anabolized) at more-or-less constant low rates in mammals. Conceptually, such turnover can be divided into two phases, one starting with and one ending with GHB. In contrast, when GHB is ingested, catabolism starts with a very high concentration and proceeds at a high rate. The cascade of steps that ensues can be conceptualized as parallel to that followed in the catabolism of endogenous GHB. However, is it precisely the same or partially different when much higher concentrations are involved? When prodrug GBL or BD is ingested, rapid metabolic activation generates a higher concentration of GHB. Activation can be conceptualized as being parallel to the de novo biosynthesis of GHB, except that the steps are different. Overall, a full understanding of the actions of GHB and GHB-related compounds in vivo requires descriptions of at least four distinguishable phases of metabolism. The intellectual and experimental overlay of these phases onto each other creates much more complexity than exists for other date-rape drugs, each of which undergoes only one catabolic phase.

To provide the reader with the concepts needed to resolve uncertainties encountered in GHB metabolism, the rest of Section 11.4 covers some relevant fundamentals. In addition to its utility in defining metabolic pathways with certainty, information at the substantial level of detail presented in this chapter is important because it will probably reveal new ways to detect dosing and to mitigate the adverse effects of an overdose. Indeed, the potential payoff motivates the comprehensive approach taken.

11.4.2 Isozymes in GHB-Related Metabolism

Before the tools of recombinant DNA, cell biology, and genomic sequences were available, research on GHB metabolism was extraordinarily difficult. In part, this occurred because of the existence of *isozymes*, which catalyze the same type of reaction on the same type of chemical functional group in different substrates. For example, isozymic alcohol dehydrogenases catalyze the oxidation of alcohols (GHB is an alcohol) to carbonyl compounds. Isozymes are encoded by different genes.[10] Moreover, enzymes catalyze reversible reactions in both directions. In the case of biosynthesis and catabolism of GHB, it can be unclear in which direction a reversible reaction proceeds in vivo.

Isozymes often resemble each other in molecular weight, net charge, and other physicochemical properties due to evolutionary kinship. Isolation of an enzymatic activity from a natural source such as brain, kidney, or liver can produce a mixture of isozymes. Mixtures are especially likely when isozymes form heterooligomers with each other. Unfortunately, some in vitro data are compromised by lack of convincing isozymic identity and/or homogeneity. Many of the isozymes discussed in the chapter now have been expressed in and isolated from a heterologous host such as the bacterium *Escherichia coli*, which is methodology that produces a single isozyme. Where the issue is critical, we cite only references not suffering from isozyme ambiguity.

Genomic and mRNA sequences reveal that more forms of many enzymes exist than formerly realized, including the enzymes that catalyze reduction and oxidation (redox) reactions in GHB-related pathways. There are nine functional human genes for NAD^+-dependent alcohol dehydrogenases (ADHs), 14 for NADPH-dependent aldo-keto reductases (AKRs), and 23 for $NAD(P)^+$-dependent aldehyde dehydrogenases (ALDHs) (http://www.ihop-net.org/; http://www.med.upenn.edu/akr; Hyndman et al., 2003; Hoffmann and Valencia, 2004; Vasiliou and Nebert, 2005; Jin and Penning, 2007). ADHs generally bind NAD^+ in preference to $NADP^+$, and thus they catalyze oxidation of alcohols using primarily NAD^+ rather than $NADP^+$ in vivo. AKRs generally bind NADPH in preference to NADH, and thus they catalyze reduction of aldehydes and ketones primarily using NADPH rather than NADH in vivo. In contrast, many aldehyde dehydrogenases catalyze oxidation of aldehydes to carboxylic acids using either NAD^+ or $NADP^+$.[11]

11.4.3 Subcellular Compartmentalization of Enzymes, Transporters, and Substrates

An isozyme can be located in the same or different cellular and subcellular compartments as another isozyme of the same superfamily. It can have multiple natural substrates, some better than others. Also, a metabolite of interest might be a good substrate for several isozymes when characterized in vitro. When complexity in substrate–isozyme relationships exists in vitro, one cannot choose on the basis of substrate selectivity alone which isozyme functions in a particular pathway in vivo.[12] Rather, the choice can require knowledge of organ, cellular, and subcellular locations for both substrates and isozymes. Also, the levels of expression and cofactors, and the values for Michaelis–Menten parameters, can be important.[13]

If a pathway flows through a noncytosolic compartment such as the mitochondrion, a candidate isozyme and its substrate must be in that compartment. Delivery of a protein to the compartment requires an appropriate intracellular trafficking signal. If the signal is coded in the amino acid sequence, which it usually is, it is coded in the mRNA sequence also. Delivery of the substrate is spontaneous if it is hydrophobic enough to pass through membranes rapidly by nonspecific diffusion. However, if a substrate is highly polar or ionic (e.g., GHB; Section 11.14), delivery requires an appropriate transporter in the compartmental membrane. Isoporters also exist.[14]

11.4.4 Dynamics and Equilibria for Enzymes and Transporters

A transporter and an enzyme are similar to each other in that both are catalysts and both exhibit Michaelis–Menten constants for reaction in the forward and reverse directions. However, they can respond to substrate analogs very differently. For an enzyme, a binding analog of a substrate is always an inhibitor with respect to the substrate, as it blocks binding by substrate. For a transporter, membrane sidedness opens an additional possibility (Figure 11.3). Suppose that the rate at which the empty binding site for substrate reorients across the membrane is slower than the rate for the filled binding site. Such behavior is common. The steady-state rate for transport is limited by the rate of the slowest step, so it is limited in this case by reorientation of the empty binding site to begin a new transport cycle. Binding of an analog to the empty binding site facing the *trans* side of the membrane can speed the steady-state rate for transport of substrate by creating fast exchange between the substrate and the analog. Alternatively, if an analog binds but is not a substrate, it will slow the rate of substrate transport.

A proposed reaction in a metabolic pathway, whether catalyzed by an enzyme or a transporter, cannot have a very unfavorable equilibrium constant in order to function. The constant is important in part because of the Haldane relationship, which for a saturable, single-substrate single-product reaction is $K_{eq} = V_{maxforward} K_{Mproduct} / V_{maxreverse} K_{Msubstrate}$ (Cleland, 1982). The symbols have the conventional meanings defined in Michaelis–Menten kinetics. The Haldane relationship is valid even for crude preparations of enzyme or transporter. When $K_{eq} < 1$, $V_{maxforward}$ or $K_{Mproduct}$ (or both terms) is small relative to the corresponding term in the denominator. When $V_{maxforward}$ is small, the reaction is intrinsically slow. When $K_{Mproduct}$ is small (which means tight binding), product inhibition can prevent the enzyme or transporter from combining with more substrate. In either case, the reaction proceeds slowly, or not at all, in the direction written. This conclusion is important when analyzing incompletely characterized enzymes and transporters, as the K_{eq} value often can be estimated even when Michaelis–Menten values are unknown.

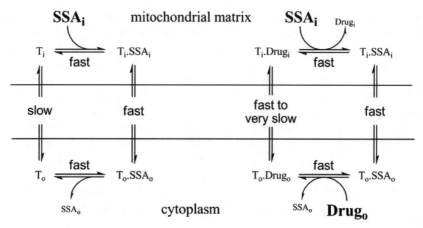

Figure 11.3 *Trans*-stimulation and inhibition of succinic semialdehyde efflux from mitochondria. The transporter for mitochondrial efflux of SSA is unknown, but it might be a member of the MCT family, which as a group has low substrate selectivity (Section 11.6.1). A hypothetical transport cycle starts with an empty binding site for substrate facing mitochondrial matrix (T_i in the cycle on the left) or a drug-loaded binding site facing mitochondrial matrix ($T_i \cdot Drug_i$ in cycle on right). The cycle on the left is rate limited by slow reorientation of an empty transporter to generate T_i. The cycle on the right runs faster than the cycle on the left if a drug molecule in cytoplasm ($Drug_o$) binds to T_o, and reorientation of the $T_o \cdot Drug_o$ complex and exchange to $T_i \cdot SSA_i$ are fast. The cycle would not run at all if a drug molecule binds and stops reorientation of the transporter. Both transport cycles are driven clockwise by net flow of substrates thermodynamically downhill.

11.4.5 Thermodynamics-Based Analysis of Metabolic Flux

But what minimal value for K_{eq} makes a reaction favorable enough for metabolism? Thermodynamics-based analysis of metabolic networks incorporating mass and flux balances provides insights (Beard and Qian, 2005; Henry et al., 2007). The concentrations of substrates and products common to many steps in metabolism, such as [H^+], [NAD^+], [$NADP^+$], [NADH], and [NADPH], are set by the network as a whole. Homeostasis often maintains them constant even in the face of large metabolic stress. If this condition applies, relevant concentrations can be substituted into the overall expression for K_{eq}. Substitution produces a modified K_{eq} value equal to the concentration of reaction-specific product divided by the concentration of reaction-specific substrate that would be present at equilibrium. If the modified K_{eq} is > 1, robust flux in the forward direction occurs because the reaction seeks equilibrium for which [product]/[reactant] > 1. If the modified K_{eq} value is < 1, flux in the reverse direction would occur *if* there is a source of reactant (which is the product of forward-direction flux) from the network. The principles apply whether reactions in vivo reach equilibrium or are in steady state.

A cell can often overcome a single modified $K_{eq} < 1$ in a multistep pathway in order to produce flux in the forward direction. It does so by adjusting the steady-state concentrations of the reaction-specific substrate and product to make the free-energy change at that step negative (think LeChâtelier's principle). However, when modified $K_{eq} \ll 1$, mass and flux constraints in the network preclude adjustments sufficient to obtain forward-direction flux (Feist et al., 2007). In this chapter, modified $K_{eq} \leq 10^{-4}$ is assumed to be too unfavorable to support normal in vivo metabolism in the direction written. This value is a low threshold requirement, as it is smaller than the modified K_{eq} values for all dehydrogenase and reductase reactions of the tricarboxylic acid (TCA) cycle and glycolysis. In a similar but inverse manner, modified $K_{eq} \geq 10^4$ will be taken as irreversible in vivo.

11.4.6 Metabolism of Endogenous GHB versus Ingested GHB and Prodrugs

Metabolism of endogenous GHB occurs within a metabolic network, and it is subject to the modified K_{eq} threshold and mass and flux balances outlined in Section 4.5. In contrast, because concentrations of ingested GHB (and its prodrugs) are not limited by anabolism, the catabolism of ingested GHB is not subject to the network restrictions. A large in vivo concentration of exogenous GHB might overcome high K_M and subthreshold values for modified K_{eq} at steps that endogenous GHB cannot overcome. Thus, catabolism of ingested GHB and prodrugs might follow multiple pathways (Sections 11.7.3 and 11.15), whereas catabolism of endogenous, low-concentration GHB probably follows mostly a single high-affinity pathway.

11.4.7 Directionality of in Vivo and in Vitro Enzymatic Activity

Reversible reactions producing no significant flux in the forward direction in vivo often can be made to run in that direction in vitro. This phenomenon occurs because constraints on [H^+], [oxidant], [reductant], and pathway-specific [reactant] and [product] are much less restrictive in vitro than they are in vivo. Typically, the initial velocity for an enzymatic reaction is characterized in nearly saturating concentrations of substrates, the absence of products, and optimal nonphysiological pH. Demonstration of an enzymatic reaction under such conditions does not mean that the reaction functions in vivo. On the other hand, barring an unknown activation requirement, an enzyme that does not catalyze a reaction of interest under otherwise optimized conditions in vitro is unlikely to function in vivo.

11.4.8 Transporters and Enzymes Mediating GHB-Related Metabolism

Transporters carrying GHB across plasma membranes of human beings for absorption by the body and metabolism by organs are listed in Table 11.2. Enzymes mediating GHB-related metabolism in human beings are listed in Table 11.3. Some of the listings are certain and some are speculative, as explained in the text. The table includes accession numbers for annotated amino acid sequences at the National Center for Biotechnology Information (NCBI). It also includes

TABLE 11.2 Some Monocarboxylate Transporters for GHB in Cytoplasmic Membranes[a]

Abbreviation	Synonym	NCBI Accession
MCT1[b]	SLC16A1	NP_003042
MCT2[b]	SLC16A7	O60669
MCT4[b]	SLC16A4	NP_004687
SMCT1[c]	SLC5A8[d]	NP_666018
SMCT2[c]	SLC5A12	Q1EHB4

[a] All of these transporters have low substrate affinity and selectivity.
[b] H^+-dependent monocarboxylate transporters 1, 2, and 4.
[c] Sodium-coupled monocarboxylate transporters 1 and 2.
[d] Also known as the sodium iodide–related cotransporter.

identification numbers for three-dimensional atomic-resolution structures in the Protein Data Bank (PDB) when they are known (http://www.ncbi.nlm.nih.gov/; http://www.rcsb.org/pdb/home/home.do). Inspection of Tables 11.2 and 11.3 confirms that impressive progress in knowledge of the molecular and cellular biology of GHB-related pathways has been made, but much more remains to be done. The structures of known and probable chemical intermediates in the pathways are shown in Figure 11.2.

11.5 BIOSYNTHESIS OF ENDOGENOUS GHB

11.5.1 First Step for GHB Biosynthesis in the Known Pathway

A well-documented de novo pathway for the biosynthesis of GHB in brain originates primarily with glutamate. The pathway is shown in Figure 11.2. Glutamate decarboxylases 1 and 2 yield GABA in mitochondrial matrix. The enzymes are isozymes of each other, and the relative frequency of occurrence of cDNA clones reveals that the mRNAs are highly to well expressed in all human tissues (http://www.ncbi.nlm.nih.gov/IEB/Research/Acembly/). Minor amounts of GABA also arise by deamination of the polyamines spermine, ornithine, and putresine (Tillakaratne et al., 1995).

11.5.2 Second Step for GHB Biosynthesis in the Known Pathway

GABA transaminase reacts GABA with 2-oxoglutarate to produce succinic semialdehyde (SSA) plus glutamate in mitochondrial matrix (Figure 11.2). The reaction transfers the pro-(S)-hydrogen from GABA (Santaniello et al., 1978). Only one gene exists for human GABA transaminase. The mature polypeptide is 472 amino acids long, M_r 53.3 kDa, and a soluble homodimer (Schousboe et al., 1980; Kugler, 1993; Rao and Murthy, 1993). The relative frequency of cDNA clones has revealed that the mRNA is expressed at high to moderate levels in essentially all human tissues (http://www.ncbi.nlm.nih.gov/IEB/Research/Acembly/).[15] GABA

TABLE 11.3 Enzymes Mediating GHB-Related Metabolism[a]

Common Name	Isozyme	Probable Compartment[b]	Substrate	Product	K_M (mM)	NCBI Accession	Isoform[c]	PDB
GABA transaminase	Only one	Mitochondria	GABA	Succinic semialdehyde (SSA)	0.4	P80404	—	1OHW[d]
Low K_M SSA reductase[e]	AKR7A2	Cytoplasm[f]	SSA	GHB	0.020	O43488	AAH07352, AAH10852	2BP1
Aldehyde reductase	AKR1A1	Cytoplasm	SSA	GHB	0.231	P14550	—	2ALR
2-Oxoglutarate-dependent GHB transhydrogenase[g]	ADHFe1	Mitochondria	GHB	SSA	0.07	NP_653251		
SSA dehydrogenase	ALDH5A1	Mitochondria	SSA	Succinate	0.006	P51649	NP_733936, NP_001071	2W8N, 2W8O[h]
α-Oxidation (cytochrome P450)	—	Endoplasmic reticulum/peroxisomes	GHB	2,4-Dihydroxy-butanoate	1.4 ?	—	—	—
β-Oxidation	Enzymes of fatty acid oxidation	Endoplasmic reticulum/peroxisomes	GHB	3,4-Dihydroxy-butanoate	—	—	—	—
ω-Oxidation (cytochrome P450)	—	Endoplasmic reticulum	GHB	SSA	1.4 ?	—	—	—
ω-Oxidation (catalase)	Only one	Peroxisomes	GHB	SSA	—	P04040	—	1DGB
γ-Lactonase[i]	1	Blood and liver	γ-Butyrolactone (GBL)	GHB	10	NP_000437	P27169	1V04[j]
β-Oxidation (cytochrome P450)	—	Endoplasmic reticulum	GBL	3-Hydroxy-GBL	—	—	—	—

(continued)

TABLE 11.3 (Continued)

Common Name	Isozyme	Probable Compartment[b]	Substrate	Product	K_M (mM)	NCBI Accession	Isoform[c]	PDB
γ-oxidation (cytochrome P450)	—	Endoplasmic reticulum	GBL	4-Hydroxy-GBL → SSA	—	—	—	—
Unidentified NAD$^+$-dependent ADH	—	—	BD	γ-Hydroxy-butyraldehyde[k]	—	—	—	1HSZ, 1U3V, 1U3U[l]
Cytochrome P450	2E1	Endoplasmic reticulum	BD	γ-Hydroxy-butyraldehyde[k]	—	—	—	—
Unidentified ALDH	—	—	γ-Hydroxy-butyraldehyde[m]	GHB	—	—	—	—
NAD$^+$-dependent GHB dehydrogenase	Only one	Bacterial cytoplasm	GHB	SSA	2	AAT18823	—	—

[a] The enzymes correlate with some of the pathways in Figure 11.2.
[b] In human beings, except for bacterial NAD$^+$-dependent GHB dehydrogenase.
[c] NCBI accession for mRNA sequences of alternatively spliced variants. Other polymorphisms are not considered.
[d] Pig enzyme.
[e] Also known as aflatoxin B1 aldehyde reductase.
[f] A few reports support mitochondrial and Golgi localization.
[g] Also known as hydroxyacid-oxoacid transhydrogenase (HOT) and ADH8. However, the term ADH8 is also used for retinol dehydrogenase. ADHFe1 uses added 2-oxoglutarate instead of added NAD$^+$ as the oxidant. No operative NAD$^+$-dependent GHB dehydrogenase has been identified in human beings.
[h] Oxidized and reduced forms.
[i] The enzyme is also known as paraoxonase.
[j] Chimeric human, rabbit, mouse, and rat γ-lactonase enzyme (NCBI accession AAR95986).
[k] Presumed but based on indirect evidence.
[l] These are structures of ADH1B, which is only very weakly implicated.
[m] Assumed reactant on basis of indirect evidence.

and 2-oxoglutarate (a TCA cycle intermediate) are present in essentially all organs (Tanaka, 1985; Zambonin et al., 1991; Abe et al., 1998). Thus, SSA is probably produced in all organs, although this speculation has not been tested. Because SSA is the immediate precursor to GHB (below), it would be of interest to determine SSA levels in all organs of the mammalian body now that a quantitative analytical method has been developed (Struys et al., 2005a).

The overall equilibrium constant from GABA to SSA is 0.04 (Van Bemmelen et al., 1985). As [2-oxoglutarate]/[glutamate] in mitochondrial matrix is ~15, the modified K_{eq} value is ~0.6 (Wiesner et al., 1988). Although modified K_{eq} < 1, the reaction can readily be made to run robustly in the forward direction by lowering the SSA concentration. Lowering indeed occurs by irreversible and rapid conversion of SSA to succinate by SSA dehydrogenase (isozyme ALDH5A1, Section 11.12).

Other than from catabolism of ingested GHB or GHB prodrug, the only significant source of SSA known in mammals is GABA by means of the transaminase reaction. A very minor source is catabolism of vitamin B_6 (http://www.genome.ad.jp/kegg/pathway.html). Another theoretically possible source is 2-oxoglutarate, which in turn is derived from many metabolites. The enzyme 2-oxoglutarate decarboxylase is part of the 2-oxoglutarate dehydrogenase complex in the TCA cycle. It makes SSA bind covalently to thiamine pyrophosphate, which then transfers the SSA to dihydrolipoyl acyltransferase, also in the dehydrogenase complex. The transfer reaction is a form of substrate channeling in which a newly made product molecule is passed directly into the active site of another enzyme without dissociation of free product into solution. There has been no report of the SSA intermediate escaping from the 2-oxoglutarate dehydrogenase complex, although the possibility appears not to have been tested at the level of accuracy required to detect the small amount of free SSA used for the production of endogenous GHB (Section 11.5.3). Human beings and other mammals contain a gene and mRNA coding for 2-oxoglutarate decarboxylase-like hypothetical mitochondrial protein (i.e., OGDHL) of uncertain function that could possibly release SSA (Bunik and Degtyarev, 2008). Formation of free SSA by 2-oxoglutarate decarboxylase not part of a 2-oxoglutarate dehydrogenase complex is a robust reaction in some microbial species.

11.5.3 Third Step for GHB Biosynthesis in the Known Pathway

In brain, a very small fraction (~0.1%) of the SSA made from GABA is reduced to GHB (Gold and Roth, 1977; Rumigny et al., 1981). The remaining 99.9% is oxidized to succinate (Section 11.12). The only enzymes known to be capable of reducing aldehydes to alcohols in mammals are AKRs acting in their forward (reductive) direction and ADHs acting in their reverse (reductive) direction (http://www.genome.ad.jp/kegg/pathway.html). Among human AKRs tested, only AKR1A1 and AKR7A2 reduce SSA at a significant rate in vitro (Cash et al., 1979). Among tested mammalian ADHs, only ADHFe1 reduces SSA at a significant rate in vitro (Section 11.7.2). Reduction of SSA by an enzyme that prefers NADH over NADPH has not been demonstrated to exist in mammals. Nevertheless, it can be inferred potentially to exist from a single observation of apparent NAD^+-dependent oxidation of GHB by a mammalian extract (Lyon et al., 2007). For reasons discussed in Sections 11.7.2, 11.7.3, and 11.10, however, only AKR1A1 and AKR7A2 currently are realistic candidates for GHB biosynthesis.

The AKR1A1 polypeptide is 325 amino acids long, M_r 36.6 kDa, and a soluble monomer. The atomic-resolution structure has been determined (El-Kabbani et al., 1994). The AKR7A2 polypeptide is 359 amino acids long, M_r 39.6 kDa, and a soluble dimer of 79 kDa. It also is known as aflatoxin B1 aldehyde reductase (AFAR; Ireland et al., 1998; Schaller et al., 1999).[16] The atomic-resolution structures of AKR7A2 and the mouse ortholog (Akr7a5) have been determined (Debreczeni et al., 2005; Zhu et al., 2006). AKR1A1 has a k_{cat} value for SSA fivefold higher than that of AKR7A2, but AKR7A2 has a K_M value for SSA 11-fold lower (20 μM) than that of AKR1A1 (230 μM) (O'Connor et al., 1999).[17] Thus, the catalytic efficacies (k_{cat}/K_M) are very similar to each other. However, AKR7A2 is twofold faster at low concentrations of SSA, and AKR1A1 is fivefold faster at high concentrations. The analysis does not take into account different levels of expression of the enzymes in vivo.

Which of the isozymes is likely to dominate GHB biosynthesis? To begin the analysis, consider the distribution of endogenous GHB. It is found in brain, heart, kidney, liver, lung, skeletal muscle, and brown fat of human beings and rodents (Nelson et al., 1981; Maitre, 1997; Moriya and Hashimoto, 2005; Richard et al., 2005; Zacharis et al., 2008). GHB is not restricted to the nervous system, but its nonnervous role is not understood. In brain, GHB is located primarily in cytosol and synaptic vesicles, as indicated by subcellular fractionation of homogenates (Snead, 1987; Muller et al., 2002). Brown fat contains an especially high concentration (37 nmol/g), which suggests the possibility that GHB is catabolized slowly in this tissue. A question regarding detection of dosing presents itself. Does the brown fat of dosed human adults take up ingested GHB (or converted prodrug) and retain it longer than other tissues do? If so, is biopsy of brown fat an acceptable forensics technique for living human beings (van Marken Lichtenbelt et al., 2009)?

Only trace levels of endogenous GHB occur in blood and cerebrospinal fluid. The very low levels mean that transfer of GHB from organs that make it to organs that do not is unlikely. Such transfer would require high-affinity ($K_M \leq 1$ μM) uptake of GHB with high selectivity to prevent competitive inhibition of uptake by other common monocarboxylic acids present in blood and cerebrospinal fluid. However, a transporter having these properties has not been described (Section 11.6.1). Tissues that contain endogenous GHB thus make it. They must contain both the biosynthetic enzymes and the initial, pathway-specific substrate, which in the only well-documented de novo pathway to GHB is SSA generated from mitochondrial GABA. However, if another significant pathway to GHB exists in human beings, it is possible that an intermediate of that pathway could be passed from one organ to another through blood or cerebrospinal fluid (below).

Immunostaining for human AKR1A1 and AKR7A2 proteins and their rodent orthologs indicates that AKR7A2 and its orthologs are present in essentially all tissues but that AKR1A1 and its orthologs are not (Ireland et al., 1998; O'Connor et al., 1999; Kelly et al., 2000; Grant et al., 2001; Hedberg et al., 2001; Picklo et al., 2001a,b; Hinshelwood et al., 2002, 2003; Fung et al., 2004; Cui et al., 2009).[18] Assays for mRNA and relative frequencies for cDNA clones coding for AKR1A1 and AKR7A2 reveal that both mRNAs are present at very high to significant levels in essentially all human tissues (http://genome.ucsc.edu/; http://bioinfo2.weizmann.ac.il/cgi-bin/genenote/home_page.pl; http://www.ncbi.nlm.nih.gov/IEB/Research/Acembly/).[19] Why do the cellular distribution maps for mRNA and protein corresponding to AKR1A1 disagree? It is possible that immunostaining of AKR1A1 in some tissues does not occur because the major epitopes are masked (Shi et al., 1995).[20] On the other hand, comparisons between the transcriptome and the proteome of human beings often exhibit poor correlations. The disjunction suggests that control mechanisms responding to different physiological imperatives operate at the transcriptional and proteomic levels of expression (Rogers et al., 2008). More research would be required to determine whether substantially more AKR7A2 than AKR1A1 truly is present in vivo. This situation presumably would mean AKR7A2 rather than AKR1A1 dominates GHB biosynthesis, as in fact has been concluded by some researchers (Rumigny et al., 1981). The results reported in Section 11.5.4 support this possibility. They also indicate, however, that it probably does not matter which of these AKRs dominates in different cell types.

11.5.4 Which Step in GHB Biosynthesis is Rate Limiting?

In cultured human SH-SY5Y neuroblastoma cells, suppression of AKR7A2 expression with siRNA causes about a 90% decrease in AKR7A2 protein and SSA reductase activity, but only a 35% decrease in the amount of GHB (Lyon et al., 2007).[21] The compound zopolrestat, a selective inhibitor of AKR1A1 activity, blocks only about one-half of the 10% residual SSA reductase activity in an extract from suppressed cells, even when the SSA concentration is saturating for AKR1A1. Apparently, SH-SY5Y cells express only a small amount of AKR1A1 ($\leq 5\%$ of the total SSA reductase activity), although they make substantial amounts of GHB whether or not AKR7A2 is fully expressed.[22] The results are consistent with rate-limiting AKR7A2 activity in normal GHB biosynthesis and induction of an unknown AKR1A1-independent pathway when AKR7A2 is suppressed.

However, there is a simpler hypothesis. The results are *also* consistent with neither AKR7A2 nor AKR1A1 activity being rate limiting to GHB biosynthesis. Instead, the rate-limiting step could be presentation of SSA for reduction. In this scenario, about 10% residual SSA reductase activity in SH-SY5Y cells could be sufficient to support 65% of the normal rate of GHB biosynthesis.

To assess the feasibility of this model, subcellular compartmentalization of SSA reduction must be analyzed. AKR7A2 and AKR1A1, and their rodent orthologs, have been shown to be in cytosol or adsorbed to cytosolic surfaces of nuclear, Golgi, and mitochondrial membranes (Claros and Vincens, 1996; Andriamampandry et al.,

1998; Witzmann et al., 1998; O'Connor et al., 1999; Picklo et al., 2001a; Kelly et al., 2002, 2004; Keenan et al., 2006; Lyon et al., 2007).[23,24] Moreover, reduction by NADPH is nearly 1000-fold more favorable in cytosol than in mitochondrial matrix (Section 11.10). Both factors argue that SSA is reduced in cytosol. This conclusion requires that SSA made in mitochondria be exported to cytosol before it is reduced to GHB.

Presentation of SSA to AKR1A1 and AKR7A2 could be a complex process. It might require three steps, and none of them has been well characterized.

1. Assuming that the GABA transaminase reaction is the only significant source of SSA (Section 11.5.2), a possible rate-limiting step is escape of newly made SSA from substrate channeling in a complex formed between GABA transaminase and ALDH5A1, as discussed in Section 11.12. Such potential channeling should not be confused with the known channeling that occurs in the glutamate dehydrogenase complex (Section 11.5.2).
2. A definitely required and possibly rate-limiting step is transport of SSA out of mitochondrial matrix. The transport might be performed by the same transporter that takes exogenous GHB into mitochondria for catabolism (Section 11.7.1). Alternatively, it might be performed by an unknown transporter.
3. Another possible rate-limiting step is dehydration of SSA hydrate. The fraction of SSA that hydrates spontaneously to form *gem*-diol during steady-state metabolism in vivo is not known. It could be as much as one-half (Section 11.8). Hydrated and nonhydrated forms of SSA probably are transported out of mitochondrial matrix at different rates, and such discrimination could influence the fraction of the SSA in cytoplasm that is unhydrated and available for reduction.

Overall, because the hypothesis is parsimonious, it seems likely that the rate-limiting step in biosynthesis of endogenous GHB is presentation of reducible SSA to AKR1A1 and AKR7A2 in cytoplasm. However, the slow molecular process within the presentation requirement is unidentified.

11.5.5 Are There Other Biosynthetic Pathways to Endogenous GHB?

Some experimental observations have been interpreted to support the existence of a GABA-independent pathway to endogenous GHB:

1. Substantial amounts of endogenous GHB are found in organs containing ≤ 1% as much GABA as brain has (Drummond and Phillips, 1974; Nelson et al., 1981). The implication is that such organs do not contain enough GABA to explain their GHB contents.
2. Anticonvulsants and inhibitors of glutamate decarboxylase and GABA transaminase affect levels of GABA and GHB in rat brain paradoxically (Snead et al., 1982).
3. Exogenous BD is converted quickly to GHB (Section 11.13.2) by a pathway that does not involve GABA (Snead et al., 1989). Free and chemically bound

forms of BD have been reported to be present in undosed human beings and rats at about one-tenth the amount of GHB, as required if BD were part of a biosynthetic pathway to GHB (Barker et al., 1985; Poldrugo et al., 1985).

4. NAD^+-dependent ADH activity that is able to oxidize exogenous BD increases in SH-SY5Y cells treated with siRNA suppressing AKR7A2, as if ADH activity were up-regulated by a feedback mechanism sensing levels of GHB or AKR7A2 (above).

Although many of them have nagged the field for years, the observations and their interpretations are ambiguous or uncertain. Observation 1 is not compelling because GHB biosynthesis in the "reference" organ brain is very curious. Only 1 part in 1000 of GABA-derived SSA is transported to cytosol to become GHB (Sections 11.5.3 and 11.12). Low levels of GABA characteristic of peripheral organs could support the formation of relatively large amounts of GHB if (1) the presentation of reducible SSA to AKR1A1 and AKR7A2 in cytoplasm is rate limiting, and (2) a larger fraction of mitochondrial SSA is presented for reduction in those organs than in brain. Even a 100-fold increase would decrease mitochondrial flux of SSA to succinate only 10%!

The results of an experiment monitoring metabolism of $[2,3,3,4,4-^2H_5]$glutamic acid to SSA and GHB are informative (Niwa et al., 1983). A high concentration of unlabeled GHB and about one-third additional $[^2H_4]$GHB are present in spent culture medium after incubation of primary kidney cells in a minimum essential medium containing 2 mM DL-$[^2H_5]$glutamic acid. GHB is barely detectable after similar incubation of primary liver cells. Instead, high concentrations of SSA are found. The observation is consistent with the approximately 17-fold greater amount of GHB in intact kidney (22 to 28 nmol/g) compared to intact liver (1.4 to 1.6 nmol/g), even though these organs contain similar amounts of GABA (21 and 15 nmol/g, respectively) (Drummond and Phillips, 1974; Nelson et al., 1981). As there are many metabolic sources of glutamic acid, the presence of excess deuterium atoms in about one-fourth of the GHB indicates that most GHB secreted by kidney cells comes from glutamic acid. Different partitioning ratios at metabolic branch points in the conversion of glutamate to GHB probably account for different GHB/GABA ratios in different organs.

The experiment also observed small amounts of 4-hydroxy-2-butenoic acid in both spent media. The compound does not arise from GHB or glutamic acid, as it did not contain excess deuterium. Brain and kidney contain the compound at levels one-sixth to one-half of the amounts of GHB (Niwa et al., 1982; Vayer et al., 1985b). Possibly, 4-hydroxy-2-butenoic acid can be reduced in vivo to GHB, but the conversion has not been demonstrated. This possibility should be tested.

Observation 2 used drugs and inhibitors thought at the time to act primarily on ion channels and enzymes. However, the analysis did not consider the role of the mitochondrial exporter for SSA. Aminooxyacetic acid, valproic acid, and 3-mercaptopropionic acid are three of the compounds utilized in the study, and they are monocarboxylic acids. As discussed in Section 11.4.3 and illustrated in Figure 11.3, depending on how they interact with the unknown exporter for SSA, monocarboxylic acids could either stimulate or inhibit the rate of SSA export and thus GHB biosynthesis. Unfortunately, there is no information available on this issue for the compounds used. All of the "inhibitors" utilized are now known to be much less selective than once thought (Meldrum and Rogawski, 2007), and their effects on the known pathway for GHB biosynthesis cannot be predicted.

Regarding observation 3, recent mass spectrometric analyses have not found BD in human beings, rats, or mice (Sakurada et al., 2002; Richard et al., 2005; http://www.metabolibrary.ca/). Extracts of plants contain minor amounts of long-chain diesters of BD, so trace BD in mammals might arise from diet, which could change through time (Bergelson, 1969). However, a trace dietary source is unlikely to support a robust alternative pathway to GHB, and such a source certainly cannot do so in cultured SH-SY5Y cells. A report that GBL is present in the brains of undosed rats at about one-tenth the amount of GHB (Snead et al., 1989) also has not been confirmed by more recent literature.

Finally, observation 4 is not compelling either, as no demonstration has been reported that the ADH activity induced in SH-SY5Y cells is selective for BD. Overall, available data do not make a persuasive case that a second pathway for biosynthesis of endogenous GHB exists in human beings.

11.6 ABSORPTION AND DISTRIBUTION OF INGESTED GHB

11.6.1 Gastrointestinal Tract

The first signs of intoxication by GHB can appear in as little as 10 min, which suggests that the stomach absorbs some GHB (Tanaka et al., 2003). Strong behavioral disturbances and peak concentrations of GHB in blood occur between 0.5 and 2 h, which is $\geq$ the time required for gastric emptying (Poggioli et al., 1999; Carter et al., 2006b; Abanades et al., 2007b; Goodwin et al., 2009). Thus, the small intestine absorbs most of each ingested dose of GHB.

Intestinal absorption probably is mediated by H^+-dependent monocarboxylate transporters (MCTs) and sodium-coupled monocarboxylate transporters (SMCTs) located in apical and basolateral membranes of epithelial cells in the stomach and small intestine (Table 11.2; Goddard, 1998; Iwanaga et al., 2006; Ganapathy et al., 2008). SMCTs and MCTs transport a wide range of metabolites involved in energy metabolism. Substrates include lactic acid, pyruvic acid, β-D-hydroxybutanoic acid, acetoacetic acid, and butanoic acid. The transporters are found throughout the human body, including in the blood–brain barrier and nerve terminals (Benavides et al., 1982; Halestrap and Price, 1999; Tunnicliff, 2002; Enerson and Drewes, 2003; Nishimura and Naito, 2005; Pierre and Pellerin, 2005; Simpson et al., 2007; Wang and Morris, 2007; Morris and Felmlee, 2008; Cui and Morris, 2009). SMCTs and MCTs exhibit moderate to low affinity and low selectivity for substrates. The K_M values for GHB are 30 to 700 μM for different SMCTs and 1 to 10 mM for different MCTs.

11.6.2 Blood

Liver, kidney, lung, heart, brain, brown fat, white fat, skeletal muscle, and pancreas catabolize exogenous GHB (Kleinzeller, 1944; Roth and Giarman, 1966; Khizhnyak, 1976; Lettieri and Fung, 1976; Shumate and Snead, 1979; Struys et al., 2006b; Lenz et al., 2008). Because they express different amounts and mixes of SMCTs and MCTs, different organs probably dominate uptake (and catabolism) of different doses of GHB. However, when fatal dosing occurs with GBL, the postmortem amounts of GHB per gram of tissue are within a factor of 2 of each other for all organs (Lenz et al., 2008).

11.7 INITIAL CATABOLISM OF GHB

11.7.1 Transport into Mitochondria

After GHB is taken into cells, where is it catabolized? Mitochondria isolated from rat liver and kidney convert GHB to SSA, and the rate increases when the mitochondria are disrupted (Gibson and Nyhan, 1989). The observations suggest that (1) metabolizing enzyme(s) is(are) in mitochondrial matrix, and (2) *transport* of GHB into mitochondrial matrix *is capacity limited*. The latter term is synonymous with the existence of rate limitation at high doses of GHB. Mitochondrial uptake probably is mediated by an isoform of the MCT family (Gladden, 2007; Yoshida et al., 2007; Hashimoto and Brooks, 2008; Hashimoto et al., 2008). Comprehensive characterization of the roles that intracellular transporters have in GHB catabolism would be welcomed.

11.7.2 Iron-Dependent Alcohol Dehydrogenase ADHFe1

An atypical alcohol dehydrogenase called ADHFe1 initiates one pathway for mitochondrial catabolism of GHB (Kaufman et al., 1988).[25] Antibody against recombinantly tagged ADHFe1 heterologously expressed in a mammalian cell line stains mitochondria intensely (Kim et al., 2007a).[26] mRNA for ADHFe1 is expressed at very high or significant levels in essentially all human and mouse tissues and in SH-SY5Y cells (Deng et al., 2002; Kardon et al., 2006; Kim et al., 2007a; Lyon et al., 2009; http://www.genecards.org/cgi-bin/carddisp.pl?gene=ADHFE1&search=adhfe1; http://www.ncbi.nlm.nih.gov/IEB/Research/Acembly/). Four alternatively spliced mRNAs encoding different putative isoforms of the enzyme (M_r = 50.3, 45.1, 32.5, and 27.4 kDa) have been detected (Deng et al., 2002; http://www.uniprot.org/uniprot/Q8IWW8#Q8IWW8-2). Two of the isoforms encode an N-terminal targeting sequence for import by mitochondria, and two do not, suggesting that some ADHFe1 is in matrix and some in cytosol. The two-compartment hypothesis has not been tested carefully, however, and it could have important consequences for the location of GHB catabolism. Because the dehydrogenase activity characterized requires unique oxidants (not a nicotinamide, o-quinone, or flavin cofactor; see below), it can be detected unambiguously. It has been found in homogenates of liver, kidney, heart, brain, skeletal muscle, and lung of rats.

The active site of ADHFe1 probably contains tightly bound iron in the +2 oxidation state [i.e., Fe(II) or Fe^{2+}] (Gibson and Nyhan, 1989; Kaufman and Nelson, 1991; Struys et al., 2004; 2006a; Kardon et al. 2006). However, this statement is an inference based on sequence homology with iron-dependent alcohol dehydrogenases isolated from microorganisms. No chemical analysis of mammalian ADHFe1 for metal ions has been reported, as the enzyme has not been purified to homogeneity. The enzyme prefers to use 2-oxoglutarate as the oxidant, from which stoichiometric D-2-hydroxyglutarate [also known as (R)-2-hydroxyglutarate] results. The activity does not require added nicotinamide cofactor. Nevertheless, the sequence contains the NAD(P)-binding motif G-X-G-X-X-G, where G is glycine and X is any amino acid (Wierenga et al., 1985). Presumably, a nicotinamide cofactor is very tightly bound, but again this statement is an inference. Such a cofactor would have to cycle between its oxidized and reduced states as the enzyme binds to and reacts with GHB or 2-oxoglutarate in alternating fashion (Figure 11.4). This kinetic mechanism couples the thermodynamics of GHB oxidation to free [2-oxoglutarate]/[D-2-hydroxyglutarate] and not free [NAD(P)$^+$]/[NAD(P)H].[27]

ADHFe1 has interesting substrate selectivity, as GHB, D-2-hydroxyglutarate, L-3-hydroxybutanoic acid, and L-lactic acid are substrates for the oxidative reaction. SSA, 2-oxoglutaric acid, 2-oxoadipic acid, oxaloacetic acid, pyruvic acid, and 2-oxobutanoic acid are substrates for the reductive reaction. The substrates are listed here in order of decreasing preference (Kaufman et al., 1988; Struys et al., 2005b). Although the reaction couple between GHB and SSA presumably proceeds best in vitro, it produces no net chemical change. The next best reaction couple is GHB/2-oxoglutarate, which is why we assume here that ADHFe1 in vivo uses 2-oxoglutarate preferentially to oxidize GHB. The reverse reaction between the SSA/D-2-hydroxyglutarate couple can be made to proceed in vitro. However, it is very unlikely to occur in vivo, as the modified equilibrium constant probably is $\ll 1$ (Section 11.10) and the concentration of SSA is low (Section 11.12).

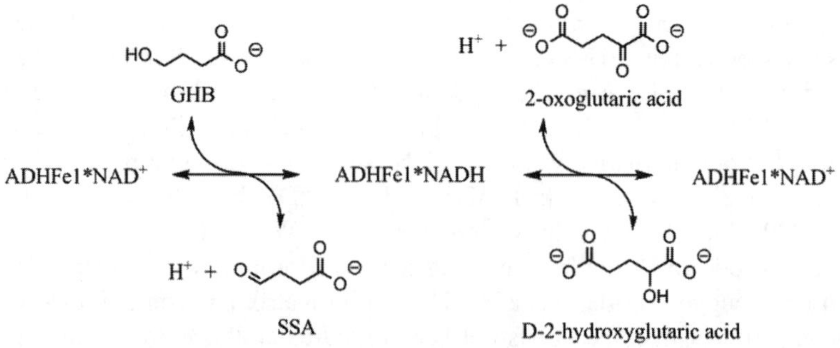

Figure 11.4 ADHFe1-catalyzed reaction. The asterisks indicate very tight complexes that do not release NAD$^+$ or NADH. Substrate-binding steps are implicit. GHB and 2-oxoglutarate are preferred substrates, but they are not the only substrates.

The known substrates differ in the length of the carbon chain, the terminal or internal position of the redox site, and the number of carboxylate groups. The great diversity in structures suggests that unidentified substrates exist. However, 2-hydroxyisovalerate, 3-hydroxyisovalerate, malate, 3-hydroxyglutarate, L-2-hydroxyglutarate, lactate, glycolate, and 2-hydroxybutanoic acid are not substrates. K_M values for 2-oxoglutaric acid differ up to 70-fold when the enzyme is prepared from liver, fibroblasts, or kidney (Struys et al., 2005b). A smaller variation occurs for SSA (Kaufman et al., 1988). This observation, combined with the existence of alternatively spliced mRNAs (above), raises the possibility that different organs predominantly express isoforms of ADHFe1 that have different substrate selectivities (and possibly cofactor requirement).

11.7.3 Poorly Characterized Catabolism of GHB

Additional pathways for catabolism of GHB *do* exist, but they are poorly characterized. When the GHB load in rodents and human beings is large, some of it is converted to 2,4-dihydroxybutanoic acid (also known as 3-deoxytetronic acid) and 3,4-dihydroxybutanoic acid (also known as 2-deoxytetronic acid) by α- and β-hydroxylations, respectively (Figure 11.2; Walkenstein et al., 1964; Lee, 1977; Jakobs et al., 1981; Niwa et al., 1982; Vayer et al., 1985b; Brown et al., 1987; Vamecq and Poupaert, 1990; Gibson et al., 1998; Shinka et al., 2002, 2003). These metabolites are also derived from sugars and fatty acids, so they are not distinctive for dosing by GHB (http://www.metabolibrary.ca/). Enzymes and subcellular compartments that might mediate α- and β-hydroxylations of GHB are discussed in Section 11.15.

An NAD^+-dependent GHB dehydrogenase activity has been reported in a crude preparation of mitochondria from human SH-SY5Y cells (Lyon et al., 2007). Unfortunately, the activity was not otherwise characterized. Could it be due to an alternatively spliced ADHFe1 that requires NAD^+?

"$NADP^+$-dependent dehydrogenase" was suggested to catalyze oxidation of GHB using tightly coupled reduction of D-glucuronate (Kaufman et al., 1979; Kaufman and Nelson, 1991). The name is confusing, as it names an oxidizing agent rarely used by the ADH superfamily in vivo (Section 11.10). The reaction proposed is similar in concept to obligatorily coupled oxidation of GHB and reduction of 2-oxoglutarate catalyzed by ADHFe1. In vitro, D-glucuronate does stimulate oxidation of GHB under certain conditions. However, the effect apparently is an artifact of optimized conditions that efficiently capture binary $NADP^+$ enzyme or NADPH·enzyme complexes with GHB or D-glucuronate, respectively. Under in vivo-like conditions for substrate concentrations and pH, bound NADPH has time to dissociate and join the pool of free NADPH. Such dissociation couples GHB oxidation to free $[NADP^+]/[NADPH]$ and not free [D-glucuronate]/[L-gulonate]. Ironically, $NADP^+$-dependent dehydrogenase is biosynthetic AKR1A1 discussed in Section 11.5.3. Moreover, the suggested coupling between the oxidation of GHB and the reduction of D-glucuronate has been tested comprehensively in rats and found wanting (Bhattacharya and Boje, 2007). Because critical information was not available when it was proposed, the hypothesis should be gently retired.

11.8 CHEMISTRY OF GHB AND RELATED METABOLITES NOT REQUIRING ENZYMES

A surprisingly large number of spontaneous reactions affects the stability, enzymatic kinetics, equilibrium constants, or subcellular compartmentalization of GHB-related compounds and reactions. GHB cyclizes to GBL with elimination of water, and GBL hydrolyzes to GHB over a period of several days at pH 2 to form an equilibrium mixture consisting of about one-third GHB and two-thirds GBL (Ciolino et al., 2001; Perez-Prior et al., 2005). At pH 7.4, the reactions require months to form an equilibrium mixture containing 1.5% GBL (Fishbein and Bessman, 1966). They are too slow to be of physiological significance, but they can become important during long-term storage of urine and blood for forensic analysis.

At 37°C and neutral pH, SSA covalently hydrates in about 1 min to form an equilibrium mixture consisting of approximately equal amounts of the free aldehyde and *gem*-diol (Le Henaff, 1967; Pocker and Dickerson, 1969). Dehydration occurs at a similar rate. Enzymes catalyzing reduction or oxidation of SSA act only on the free aldehyde (Miller and Pitts, 1967; O'Connor et al., 1999). Thus, dehydration can be the rate-limiting step for reduction or oxidation of SSA (Deetz et al., 1984; Marchitti et al., 2008). Hydrated SSA is likely to form in vivo primarily when newly synthesized SSA must be transported to a different compartment before the next covalent change to SSA occurs. This circumstance probably arises during biosynthesis of endogenous GHB (Section 11.5.4).

In tris(hydroxymethyl)aminomethane (Tris) buffer, SSA rapidly forms the imine (also known as the Schiff base) (Hauptmann and Gabler, 1968; Ogilvie and Whitaker, 1976). In organic solvent and thus possibly in fat droplets of adipocytes, SSA rapidly cyclizes to form an equilibrium mixture consisting of about one-third 4-hydroxybutyrolactone and two-thirds open-chain SSA (Jaeger et al., 2008). The cyclization does not occur in water at pH 7. A cyclic trimer of SSA called 1,3,5-trioxane–2,4,6-tripropanoic acid forms in organic solvent but not in dilute aqueous solution (Carriere, 1921). If 4-hydroxy-GBL is formed by cytochrome P450–mediated hydroxylation of GBL, it will undergo spontaneous ring opening to SSA. The presumed γ-hydroxybutyraldehyde intermediate of BD catabolism cyclizes in water to form the hemiacetal (Section 11.13.2).

11.9 EXPERIMENTAL EQUILIBRIUM CONSTANTS FOR REDOX REACTIONS OF GHB

The reactions considered in this section reversibly interconvert GHB and SSA. Irreversible oxidations (e.g., to form 2,4-dihydroxybutanoic acid, 3,4-dihydroxybutanoic acid, or succinic acid; see Section 11.15) are not considered. A reversible reaction exhibits an equilibrium constant that is independent of the enzyme used to establish the equilibrium or even whether an appropriate enzyme exists. The experimental constant (K_{eq}) obtained from literature sources has been adjusted in this section to pH 7.0 to yield K'_{eq}.[28] Experimental constants were obtained at 25°C and not 37°C, but the temperature effect is small.

A K_{eq} value for the reaction catalyzed by bacterial NAD^+-dependent GHB dehydrogenase in the direction toward SSA has been determined (Nirenberg and Jakoby, 1960). Unfortunately, Tris buffer was used, and spontaneous formation of the imine made oxidation of GHB much more favorable than it would be in vivo. Nevertheless, there is a way to estimate a relevant value. As the structures of the alcohol functional groups in GHB and other simple primary alcohols (except methanol) are essentially the same as each other, the K_{eq} values for oxidation to aldehydes by NAD^+ must be essentially the same. They are available for ethanol, 3-hydroxypropionate, and 1-butanol in phosphate or pyrophosphate buffer (Goldberg et al., 1993). Adjusted to pH 7.0, the values are $9.97 \times 10^{-5}, 9 \times 10^{-5}$, and 9.1×10^{-5}, respectively, which as expected are essentially the same. Accordingly, the average value (9.5×10^{-5}) is assumed for K'_{eq} in the NAD^+-dependent oxidation of GHB. Formation of aldehyde hydrate is ignored, as the reaction occurs to the same extent for all of these aldehydes and is physiological. Also, the reciprocal of K'_{eq} for the NAD^+-dependent oxidation of GHB gives K'_{eq} for the NADH-dependent reduction of SSA to GHB, or 1.1×10^4.

Because the reduction potentials for $NADP^+$ and NAD^+ are essentially the same, the K'_{eq} values for the NADP-dependent reactions are essentially the same as for the NAD-dependent reactions, namely 9.5×10^{-5} for $NADP^+$-dependent oxidation of GHB and 1.1×10^4 for NADPH-dependent reduction of SSA to GHB at pH 7.0.[29]

K_{eq} for the reaction between GHB and 2-oxoglutarate catalyzed by ADHFe1 has not been determined. However, the sum of the NAD^+-dependent GHB dehydrogenase-catalyzed and NAD^+-dependent D-2-hydroxyglutarate dehydrogenase-catalyzed reactions in opposite directions is the ADHFe1-catalyzed reaction. K'_{eq} for the half-reaction to form D-2-hydroxyglutarate in phosphate buffer is 6.8×10^4 (Buckel and Miller, 1987). K'_{eq} for the reaction catalyzed by ADHFe1 is the product of the constants for the half-reactions, or $(9.5 \times 10^{-5})(6.8 \times 10^4) = 6.5$. The prime in K'_{eq} is not strictly needed because the equilibrium is independent of pH. For simplicity of nomenclature it will be retained.

K'_{eq} for this reaction can also be calculated using published Michaelis–Menten parameters and the Haldane relationship for a two-substrate, two-product enzyme ($K'_{eq} = V_{maxforward} K_{Mproduct1} K_{Mproduct2} / V_{maxreverse} K_{Msubstrate1} K_{Msubstrate2}$). For ADHFe1 from human liver, the estimate is 7.5, and for ADHFe1 from rat kidney, it is 12 (Kaufman et al., 1988; Struys et al., 2005b). All three of the values calculated are reassuringly similar. The average of 8.7 is taken here for K'_{eq} in the ADHFe1 reaction.

11.10 ESTIMATED EQUILIBRIUM CONSTANTS FOR REDOX REACTIONS OF GHB IN VIVO

This section incorporates into K'_{eq} expressions the pH and [oxidant]/[reductant] values found in vivo. The resulting numbers are modified equilibrium constants that predict $[SSA]_{eq}/[GHB]_{eq}$ values for cytosolic and mitochondrial compartments. They differ $>10^{10}$-fold, depending on the direction of reaction, type of oxidant or reductant, and the subcellular compartment. The values are instructive in determining which class of redox enzyme in which compartment can function at a particular metabolic step. Modified equilibrium constants do not inform among isozymes utilizing the same [oxidant] and [reductant] reaction couple, however.

In mitochondrial matrix, pH = 7.7, free $[NAD^+]/[NADH] \approx 7.7$ (coincidentally the same number as the pH), and free $[NADP^+]/[NADPH] \approx 12$.[30] In cytosol, the values are pH = 7.4, free $[NAD^+]/[NADH] \approx 1164$, and free $[NADP^+]/[NADPH] \approx 0.014$ (Veech et al., 1969; Zhang et al., 2002; Balut et al., 2008). Free $[NAD^+]/[NADH]$ and free $[NADP^+]/[NADPH]$ exclude enzyme-bound forms, which substantially exceed the free forms and distort many published ratios. The uncertainty in the ratios is about twofold, which in the current discussion is small (Lin and Guarente, 2003). Moreover, dosing of rats with GHB does not affect cytosolic $[NAD^+]/[NADH]$ or $[NADP^+]/[NADPH]$, although minor alkalosis occurs (MacMillan, 1979). Thus, metabolic homeostasis is nearly maintained. Although extensive analyses of NAD and NADP levels in different organs are not available, mitochondrial redox states in normal brain, heart, kidney, liver, and testis are similar to each other (Mayevsky and Chance, 2007). Published values for pH, free $[NAD^+]/[NADH]$, and free $[NADP^+]/[NADPH]$ in cytosol and mitochondrial matrix should be applicable to all major organs. Adjustment of K'_{eq} values to incorporate pH, free $[NAD^+]/[NADH]$, and free $[NADP^+]/[NADPH]$ yields $K^{mito}_{red}, K^{cyto}_{red}, K^{mito}_{ox}$, and K^{cyto}_{ox}[31] values in the reducing or oxidizing direction in mitochondrial matrix or cytosol (each as indicated) (Table 11.4). No reversible GHB metabolism is known to occur in other subcellular compartments.

Reactions catalyzed by NAD^+-dependent ADHs are addressed first. For oxidation of GHB, $K^{mito}_{ox} \approx 3.7 \times 10^{-3}$ and $K^{cyto}_{ox} \approx 2.8 \times 10^{-1}$. The reaction is unfavorable in both compartments, but not so unfavorable that similar reactions do not occur. For example, NAD^+ oxidizes malate to oxaloacetate in the mitochondrial matrix ($K^{mito}_{ox} \approx 4 \times 10^{-4}$, or about 10-fold smaller than estimated here for GHB) and NAD^+ oxidizes ethanol to acetaldehyde in cytosol ($K^{cyto}_{ox} \approx 2.8 \times 10^{-1}$, or equal to that estimated here for GHB). It is clear that NAD^+ *could* oxidize GHB in both cytosol and mitochondrial matrix. Whether it *does* depends on additional factors discussed in Sections 11.11 and 11.15.

The reciprocals of the constants for oxidation of GHB by NAD^+ give the constants for reduction of SSA by NADH. They are $K^{mito}_{red} \approx 2.7 \times 10^2$ and $K^{cyto}_{red} \approx 3.6 \times 10^0 = 3.6$. The reaction is favorable in both compartments. However, of six human ADH isozymes that have been tested in vitro under conditions even more favorable to reduction than exist in vivo, none catalyzes the reduction of SSA (Deetz et al., 1984). Why? Apparently, none of the isozymes forms a productive complex with SSA. Also, no other well-documented mammalian NADH-dependent ADH reduces SSA. Not only must the modified equilibrium constant be acceptable

TABLE 11.4 Modified Equilibrium Constants for Reactions Interconverting GHB and Succinic Semialdehyde[a]

Reduction Reaction	Enzyme[b]	K_{red}^{mito}	K_{red}^{cyto}
SSA + NADH + H$^+$ ↔ GHB + NAD$^+$	NAD$^+$-dependent GHB dehydrogenase (reverse)	2.7×10^2	3.6×10^0
SSA + NADPH + H$^+$ ↔ GHB + NADP$^+$	SSA reductase	1.8×10^2	1.5×10^5

Oxidation Reaction	Enzyme	K_{ox}^{mito}	K_{ox}^{cyto}
GHB + NAD$^+$ ↔ SSA + NADH + H$^+$	NAD$^+$-dependent GHB dehydrogenase	3.7×10^{-3}	2.8×10^{-1}
GHB + NADP$^+$ ↔ SSA + NADPH + H$^+$	NADP$^+$-dependent GHB dehydrogenase or SSA reductase (reverse)	5.7×10^{-3}	6.7×10^{-6}
GHB + 2-oxoglutarate ↔ SSA+ D-2-hydroxyglutarate	2-Oxoglutarate-dependent GHB transhydrogenase	$> 8.7^c$	-

[a] All equilibrium constants are unitless. The experimentally determined, pH-independent, overall equilibrium constant at 25°C from the literature was adjusted to pH 7.0 (footnote 28). Then the constants were adjusted to mitochondrial and cytoplasmic conditions to give K_{red}^{mito}, K_{red}^{cyto}, K_{ox}^{mito}, and K_{ox}^{cyto} in the reducing and oxidizing directions, respectively, by the reductant and oxidant couple indicated (footnote 31). The values apply to most or all organs. Values less than 10^{-4} probably cannot support metabolism in the direction written, and values greater than 10^4 probably produce irreversible reaction (Section 11.4.5).
[b] The enzyme names are phenomenological descriptions and not assigned isozyme names.
[c] This value is a lower boundary based on [2-oxoglutarate]/[D-2-hydroxyglutarate] > 1. The actual value probably is ≥ 870 (Section 11.10).

in vivo, but an isozyme that binds SSA productively must exist for the reaction to occur.

Reactions catalyzed by AKRs are addressed next. For reduction of SSA and other simple aldehydes by NADPH, $K_{red}^{mito} \approx 1.8 \times 10^2$ and $K_{red}^{cyto} \approx 1.5 \times 10^5$. The reaction is strongly favorable in both compartments. Cytosolic reduction probably is irreversible, as K_{red}^{cyto} is well above the threshold of 10^4 for irreversibility (Section 11.4.5)! But is a human AKR known that catalyzes this reaction? Yes, AKR1A1 and AKR7A2 do so (Section 11.5.3).

The reciprocals of K_{red}^{mito} and K_{red}^{cyto} for AKRs give the constants for oxidation of GHB and other simple alcohols by NADP$^+$. They are $K_{ox}^{mito} \approx 5.7 \times 10^{-3}$ and $K_{ox}^{cyto} \approx 6.7 \times 10^{-6}$. The reaction is unfavorable in both compartments. Nevertheless, as K_{ox}^{mito} is ≈ 10-fold more favorable than for oxidation of malate (although by NAD$^+$), oxidation of GHB by NADP$^+$ in mitochondrial matrix probably *could* occur. Whether it *does* occur depends on additional factors discussed in Section 11.11. In contrast, NADP$^+$-dependent oxidation of GHB in cytosol is nearly 1000-fold less favorable than in mitochondrial matrix and well below the 10^{-4} threshold for function (Section 11.4.5). Cytosolic NADP$^+$-dependent GHB dehydrogenase has very little chance of functioning in vivo.

The oxidation of GHB by 2-oxoglutarate catalyzed by ADHFe1 in mitochondrial matrix is addressed next. Free [2-oxoglutarate]/[D-2-hydroxyglutarate] controls K_{ox}^{mito}. The ratio is not known. Nevertheless, a useful boundary on its value can be set. FAD-dependent (*not* NAD$^+$-dependent) D-2-hydroxyglutarate dehydrogenase is present in human and rat mitochondrial inner membrane facing the matrix (Rzem et al., 2004). Mutations in the gene lead to a human disorder called D-2-hydroxyglutaric aciduria, which presents with up to 30-fold increase in the concentration of D-2-hydroxyglutarate in urine (Struys, 2006; Wickenhagen et al., 2009). The phenotype tells us that FAD-dependent D-2-hydroxyglutarate dehydrogenase controls free [2-oxoglutarate]/[D-2-hydroxyglutarate]. Moreover, the ratio in normal people is > 1, as otherwise a reservoir of 2-oxoglutarate sufficient to produce the observed large increase in [D-2-hydroxyglutarate] when the dehydrogenase is blocked would not exist. The physiological observations are consistent with biochemical determinations that oxidations by FAD-dependent ADHs are $\geq$ 100-fold more favorable than oxidations by NAD$^+$-dependent ADHs (Goldberg and Alberty, 2008). Because [2-oxoglutarate]/[D-2-hydroxyglutarate] surely is > 1, K_{ox}^{mito} for ADHFe1 surely is > 8.7, and it probably is ≥ 870 (Table 11.4). Thus, the reaction of GHB with 2-oxoglutarate catalyzed by ADHFe1 probably is very favorable in vivo. However, it is reversible in vitro (Section 11.7.2).

If isoforms of ADHFe1 exist that prefer an oxidant other than 2-oxoglutarate or are present in cytoplasm, perhaps as a result of alternative splicing of the mRNA (Section 11.7.2), the calculation just presented could change. Reduction of SSA catalyzed by ADHFe1 in vivo has not been proposed, and as already indicated, it is very unlikely to occur due to a low SSA concentration. It will not be addressed.

11.11 DIFFERENT PERSPECTIVES ON TURNOVER OF ENDOGENOUS GHB ARE CONSISTENT

Cytosolic reduction of SSA to GHB catalyzed by NADPH-dependent AKRs, and oxidation of GHB to SSA in mitochondrial matrix catalyzed by ADHFe1, are strongly favored among reactions interconverting SSA and GHB. This conclusion is consistent with most of what is known about substrate selectivity by AKR and ADH isozymes, subcellular distributions of isozymes, and the presence of MCTs in mitochondrial inner membrane. If one hypothesizes that transport of SSA out of mitochondrial matrix is rate limiting for biosynthesis of endogenous GHB, and transport of GHB into mitochondrial matrix is rate limiting for degradation of exogenous but not endogenous GHB, a generally consistent picture emerges. Endogenous GHB is made by excess AKR1A1 and AKR7A2 activity in cytosol. Because its concentration is low, cycling endogenous GHB probably is degraded almost entirely by ADHFe1. In contrast, exogenous GHB probably saturates mitochondrial uptake and/or ADHFe1 activity. Under these conditions, other pathways for catabolism of GHB set in, as evidenced by the formation of products other than SSA. Several possible alternative catabolic pathways are discussed in Section 11.15.

11.12 DISPOSITION OF SUCCINIC SEMIALDEHYDE

Most SSA is irreversibly and rapidly oxidized to succinate by SSA dehydrogenase in mitochondrial matrix (Ryzlak and Pietruszko, 1988).[32] GABA aminotransferase, which in the absence of ingested GHB or GHB prodrug makes most and perhaps all SSA, might form a complex with SSA dehydrogenase (Hearl and Churchich, 1984). Such a complex could allow most newly made SSA to be transferred directly into the active site of SSA dehydrogenase without dissociation into free solution. In this event, escape from channeling would be required for export of GABA-derived SSA from mitochondria (Section 11.5.2). The idea is untested, but such an escape could be the rate-limiting step in GHB biosynthesis and a regulatory control point. More should be learned about the reported complex between GABA aminotransferase and SSA dehydrogenase.

Even though the selectivity of ALDH isozymes for SSA has not been studied systematically, ALDH5A1 can be assigned to SSA dehydrogenase activity in vivo (Marchitti et al., 2008). The primary reason is that rare genetic defects map to the ALDH5A1 gene in human beings, cause GHB aciduria, and block SSA dehydrogenase activity (Knerr et al., 2007; Malaspina et al., 2009). The enzyme has been expressed in *E. coli*, purified to homogeneity, and characterized as a soluble homotetramer of 58 kDa monomers 544 residues long (Kang et al., 2005). It strongly prefers NAD^+ over $NADP^+$, and the K_M value for SSA is ~6 µM. The protein and mRNA are present in essentially all organs at levels that correlate approximately with those for GABA transaminase, which is consistent with the possible existence of substrate channeling between GABA transaminase and ALDH5A1 (Miller and Pitts, 1967; Chambliss et al., 1995a,b; Tillakaratne et al., 1995; http://www.ncbi.nlm.nih.gov/IEB/Research/Acembly/). Two isoforms of ALDH5A1 presumably exist, although they have not been observed directly, as alternative splicing in the coding region of the mRNA occurs (Blasi et al., 2002). Both isoforms are predicted to contain N-terminal mitochondrial targeting sequences (Claros and Vincens, 1996). Succinate resulting from ALDH5A1 activity enters the tricarboxylic acid cycle for final disposition, which includes conversion to α-amino acids and CO_2 (Figure 11.2).

Mice carrying deletion of the orthologous gene (Aldh5a1) have been created as a model for human GHB aciduria (Hogema et al., 2001; Pearl et al., 2009). The block causes levels of GHB and GABA in urine and organ extracts to increase as much as about 20- and threefold, respectively (Gibson et al., 2002; Gupta et al., 2004; Chowdhury et al., 2007). The different percent increases support the notion that different GHB/GABA ratios can arise from causes not related to the existence of a GABA-independent pathway to endogenous GHB (Section 11.5.5). Metabolic profiles of people and mice lacking ALDH5A1 activity are similar to those of genetically normal people who ingest large amounts of GHB (Brown et al., 1987; Gibson et al., 1997; Shinka et al., 2002, 2003; Struys et al., 2006a,b; Sauer et al., 2007).

Human beings and mice lacking SSA dehydrogenase activity produce a small amount of 4,5-dihydroxyhexanoic acid and its γ-lactone [both compounds occur as

(4R, 5R) and (4S, 5R) stereoisomers] (Brown et al., 1987; Gibson et al., 2002). The reaction probably occurs by condensation of SSA with the nucleophilic thiamine-bound acetyl intermediate created by pyruvate decarboxylase in the pyruvate dehydrogenase complex (Shaw and Westerfeld, 1968; Brown et al., 1987). Resulting 5-oxo-4-hydroxyhexanoic acid [in a mixture of (4R) and (4S)] could then be reduced by an AKR at the 5-oxo center. Spontaneous cyclization would produce the γ-lactone. The condensates are not found in baboons acutely dosed with GHB, however, which suggests that they might not be satisfactory markers for acute dosing of people (Struys et al., 2006b). Moreover, the γ-lactone, which is also known as solerole, is found in many foods and beverages, further indicating that these compounds are not satisfactory markers for GHB dosing (Krajewski et al., 1995; Aubert et al., 2003; Rocha et al., 2005).

The atomic-resolution structure of human ALDH5A1 has been obtained (Kim et al., 2009). The enzyme is controlled by redox modulation. It is active when two cysteine residues in the active site are reduced and is inactive when the cysteines are oxidized to the disulfide. Reactive oxygen species such as H_2O_2 cause formation of the disulfide. This phenomenon links GHB catabolism to other oxidative physiology (Mamelak, 2007). $GABA_A$ and NMDA receptors are regulated by redox modulation as well, thereby creating the potential for high-order physiological synergism from abuse of GHB, flunitrazepam, ketamine, and ethanol (Pan et al., 2000; Ikonomidou et al., 2000)! Indeed, GHB recently was shown to increase oxidative stress in vivo by decreasing defenses to oxidative insults (Sgaravatti et al., 2007, 2009).

11.13 CONVERSION OF PRODRUGS TO GHB AND RELATED METABOLITES

11.13.1 γ-Butyrolactone

Exogenous GBL is rapidly converted to GHB in human beings, baboons, cats, and rodents (Roth and Giarman, 1966; Richard et al., 2005; Lenz et al., 2008; Goodwin et al., 2009). It is thus a prodrug that is hydrolyzed by γ-lactonase present primarily in blood and liver (Figure 11.2; Ingels et al., 2000; Teiber et al., 2003; Draganov et al., 2005). There are three γ-lactonase isozymes in human beings, but GBL is hydrolyzed overwhelmingly by γ-lactonase 1 in vitro (Billecke et al., 2000). The enzyme is very nonselective and targets many xenobiotic compounds, as it hydrolyzes phosphotriesters, aryl carboxylate esters, carbonates, and lactones. There is very little γ-lactonase activity in hen, rat, and rabbit brains (Monroy-Noyola et al., 2007). The low activity in brain, combined with the good ability of low-polarity GBL to permeate membranes nonspecifically (Section 11.14) and pass into the periphery (Sgaragli and Zilletti, 1970; Abraham et al., 1995), probably explains why GBL injected into a cerebral ventricle has no behavioral effect in cat, rabbit, and rat (Gessa et al., 1967; Carter et al., 2006a).

GBL is predicted by a computer-based, expert toxicological system to undergo hydroxylations by cytochrome P450 to give 3-hydroxy-GBL and 4-hydroxy-GBL (Rosenkranz, 2001). Although not tested, both of these compounds probably would be hydrolyzed by a γ-lactonase. Also, 4-hydroxy-GBL will undergo spontaneous ring opening to SSA. Hydrolysis of 3-hydroxy-GBL produces 3,4-dihydroxybutanoic acid, which is made as well by β-oxidation of GHB and metabolism of carbohydrates and fatty acids. It would be of interest to determine whether 3,4-dihydroxybutanoic acid arising from GBL and exogenous GHB differs in chirality from 3,4-dihydroxybutanoic acid arising from carbohydrates and fatty acids (Sakata, 1990).

11.13.2 1,4-Butanediol

Similar to GBL, exogenous BD is rapidly converted to GHB in human beings, rodents, and baboons. It thus is a prodrug (Figure 11.2; Maxwell and Roth, 1972; Vree et al., 1978; Poldrugo and Snead, 1984; Snead et al., 1989; Ingels et al., 2000; Richard et al., 2005; Irwin, 2006; Kapadia et al., 2007; Goodwin et al., 2009). Metabolism of BD in the periphery is inhibited by pyrazole and 4-methylpyrazole, which are inhibitors of many Zn^{2+}-dependent ADHs (Bessman and McCabe, 1972; Quang et al., 2004; Carai et al., 2006). Partially purified ADHs, probably containing mixtures of isozymes obtained from human and horse livers, oxidize BD in vitro (Pietruszko et al., 1978). In contrast, metabolism of BD injected into a cerebral ventricle of rat is not inhibited by pyrazole (Poldrugo and Snead, 1986; Snead et al., 1989). Ethanol competitively inhibits BD metabolism in both the periphery and brain. The findings suggest that BD is metabolized by (1) ADH in the periphery, similarly to ethanol, and (2) a non-ADH pathway in brain that ethanol also follows (Gerak et al., 2004; Nishimura and Naito, 2005). As 4-methylpyrazole blocks the behavioral effects of peripherally administered BD in mice, most BD metabolism occurs in the periphery (Carai et al., 2002; Quang et al., 2002). The metabolic competition between BD and ethanol has been simulated numerically, and metabolism of BD has been shown to be capacity limited (Fung et al., 2008). A weak correlation between slower metabolism of BD in human beings and a common allele of ADH1B has been reported (Thai et al., 2007).[33] Nevertheless, no compelling evidence supports assignment of BD dehydrogenase to a particular ADH isozyme. A moderate rate of conversion of BD to GHB in brain, combined with the good ability of low-polarity BD to permeate membranes nonspecifically (Section 11.14) and pass into the periphery (Abraham et al., 1995), probably explains why BD injected into a cerebral ventricle has no behavioral effect in rat (Carter et al., 2006a).

The first step in conversion of BD to GHB, no matter what the enzymatic mechanism, presumably produces γ-hydroxybutyraldehyde. About 86% of this compound spontaneously cyclizes in aqueous solution to the hemiacetal called 2-hydroxytetrahydrofuran (Hurd and Saunders, 1952). However, neither the open-chain nor the hemiacetal form has been identified in human beings or rodents. A search for these compounds, or a distinctive condensate of them (as occurs for SSA, Section 11.12), could be rewarding. Such a metabolite might be more metabolically stable than are BD and GHB and serve as an indicator of dosing after BD has disappeared from blood and urine. The open-chain form of γ-hydroxybutyraldehyde is probably oxidized to GHB by an ALDH, as the ALDH inhibitor disulfiram partially blocks the behavioral effects of BD in mice (Carai et al., 2002). The cyclized hemiacetal form could be oxidized to GBL, but whether this occurs in vivo (e.g., catalyzed by a cytochrome P450 or catalase; see Section 15) is not known.

11.14 SUBCELLULAR COMPARTMENTALIZATION OF GHB-RELATED COMPOUNDS

The logarithm to the base 10 of the distribution coefficient D ($\log D$) measures partitioning of a compound between octanol and water at a given pH and 25°C. At a pH of about 7, the GHB-related compounds GABA, SSA, GHB, 2,4-dihydroxybutanoic acid, 3,4-dihydroxybutanoic acid, and 4,5-dihydroxyhexanoic acid each carry one or two electrical charges, and they have $-\log D$ values of ≥ 2.65 (Table 11.5). In vivo these compounds are subject to compartmentalization controlled by membrane transporters. This restriction in turn means that only certain isozymes can access the compounds.

In contrast, ingested GHB is exposed to intragastric pH between 1 and 4, depending on the person, time of day, and presence or absence of food in the stomach (van Herwaarden et al., 1999). Because its pK_a value is 4.72, GHB is predominantly protonated and uncharged while it is in the stomach (Tao et al., 2008). Protonated GHB has a $-\log D = 0.70$, which is 200-fold less polar than unprotonated GHB is (Table 11.5). This value approaches that for ethanol ($-\log D = 0.19$), which the stomach absorbs nonspecifically (Levitt et al., 1997). Thus, the stomach probably absorbs some ingested GHB nonspecifically. Ingested GHB is not exposed to the acid in the stomach long enough to cyclize to GBL (Sections 11.6.1 and 11.8).

TABLE 11.5 Octanol–Water Partition Coefficients

GHB Related	$-\log D^a$	Prodrug Related	$-\log D^a$
GABA	3.14	GBL	0.76
SSA	2.65	BD	1.02
GHB	3.01, 0.70*	4-Hydroxybutyraldehyde	0.57
2,4-Dihydroxybutanoate	4.18	2-Hydroxytetrahydrofuran[b]	0.93
3,4-Dihydroxybutanoate	4.79		
4,5-Dihydroxyhexanoate	3.63		

[a]-Log D values are for pH 7 and 25°C, except for the GHB value marked with *, which applies to pH 1–3 (GHB is mostly protonated and uncharged in this pH range in the stomach). The acids are named to indicate they are charged. Estimates were done by SciFinder Scholar with Advanced Chemistry Development (ACD/Labs) Software V8.14 for Solaris. Larger positive values mean less solubility in octanol, which correlates approximately with less nonspecific permeation through membranes. Compare these values with the values for ethanol (0.19) and acetaldehyde (0.16), which quickly and nonspecifically permeate biological membranes by partitioning into and across the lipid bilayer.
[b]Cyclized 4-hydroxybutyraldehyde in the hemiacetal form.

GBL, BD, and the probable BD metabolites 4-hydroxybutyraldehyde and 2-hydroxytetrahydrofuran are uncharged and have small $-\log D$ values at all physiological values of pH (Table 11.5). These molecules probably permeate membranes nonspecifically (Roth and Giarman, 1966; Lenz et al., 2008). BD has been shown to permeate barrier epithelia resistant to polar but uncharged molecules (Wright and Pietras, 1974). Thus, ingested GBL and BD are absorbed rapidly and efficiently by the stomach (Arena and Fung, 1980; Perez-Prior et al., 2005; Thai et al., 2007; Fung et al., 2008; Hicks and Varner, 2008; Goodwin et al., 2009). Poor compartmentalization is especially relevant to BD, which consequently has access to many, perhaps all, isozymes that could act on it during its presumed two-step conversion to GHB (Section 11.13.2).

11.15 COMPARATIVE CATABOLISM OF ETHANOL, 1,4-BUTANEDIOL, FATTY ACIDS, AND GHB

Most ethanol consumed by human beings is catabolized in the stomach and liver, which use primarily NAD^+-dependent cytosolic ADH1B and ADH1C for the initial step to acetaldehyde (Lee et al., 2006). However, these organs also use hydrogen peroxide (H_2O_2)-dependent catalase located in peroxisomal matrix and NADPH- and oxygen (O_2)-dependent cytochrome P450 2E1 located on the cytosolic face of smooth endoplasmic reticulum for the initial step to acetaldehyde. Some ethanol is catabolized by the brain, which uses primarily catalase and cytochrome P450 2E1 (Zimatkin et al., 2006; Hipolito et al., 2007). In the catalase reaction, H_2O_2 is reduced to two water molecules as the alcohol group is oxidized to the aldehyde group. In the cytochrome P450 reaction, one of the O atoms from O_2 attacks C1 of ethanol to form hydrated acetaldehyde, and the other O atom is reduced to H_2O by NADPH. These reactions are irreversible. Human beings have one gene for catalase but ≥ 58 genes for cytochrome P450s. Overall, human beings use *three* types of enzymatic reactions to oxidize ethanol to acetaldehyde. The prominence of each type depends on the organ, and due to different K_M values, probably the dose of ethanol. Does similar parallelism occur in the catabolism of BD and GHB?

Observations presented in Section 11.13.2 strongly suggest that metabolism of BD in the periphery and brain proceeds along the same pathways that ethanol uses. Consistent with this hypothesis, BD is oxidized in vitro to formaldehyde and uncharacterized products by cytochrome P450 in rat liver microsomes (Clejan and Cederbaum, 1992).[34] The BD isomer 2,3-butandiol is oxidized in vitro to 3-hydroxy-2-butanone by catalase (Magner and Klibanov, 1995). Unidentified isozymes of NAD^+-dependent ADH also act on BD in vitro (Section 11.13.2). Thus, all three types of enzymes acting on ethanol in human beings also act on BD or BD isomer in vitro.

Similar studies should be carried out on GHB, which can be conceptualized as either a primary alcohol such as ethanol or a small fatty acid. As a primary alcohol, catalase and cytochrome P450 could convert GHB to SSA. As a fatty acid, catalase and cytochrome P450 could convert GHB to 2,4-dihydroxybutanoic acid by α-hydroxylation (Poulos et al., 1993; Alderson et al., 2004; http://pfam.sanger.ac.uk/family?acc=PF00067). Also as a fatty acid, mitochondria and peroxisomes could convert GHB to 3,4-dihydroxybutanoic acid by β-hydroxylation (Visser et al., 2007). In addition, a cytochrome P450 might convert GHB to SSA by ω-oxidation (Orrenius, 1969). As discussed in Section 11.7.3, large amounts of 2,4-dihydroxybutanoic acid and 3,4-dihydroxybutanoic acid, respectively, are formed after dosing by GHB. No in vitro test of whether catalase or cytochrome P450 acts on GHB has been reported. However, GHB competitively inhibits demethylation of N-nitrosodimethylamine by rat cytochrome P450 2E1 with an IC_{50} value of 1.4 mM, which implies that it probably is bound to the active site (Wang et al., 1995). The IC_{50} value is similar to the concentrations of GHB reached during dosing.

Overall, there are ample hints that exogenous BD and GHB follow multiple catabolic pathways in human beings. It would be of great interest to characterize them systematically in vitro and in vivo. GBL probably will follow GHB pathways, as it is rapidly hydrolyzed in vivo (Section 11.13.1).

11.16 CATABOLISM OF MDMA, FLUNITRAZEPAM, AND KETAMINE

Because MDMA, flunitrazepam, and ketamine are not natural compounds, there are no biosynthetic pathways for them. The constraints of a metabolic network do not apply (Section 11.4.5). Even the Haldane relationship does not apply, as most of the steps are irreversible and do not have a K_{eq}. For MDMA the major end metabolite is 4-hydroxy-3-methoxymethamphetamine sulfate ester, for flunitrazepam it is 7-aminoflunitrazepam, and for ketamine it is norketamine (Figure 11.1) (Feng et al., 2007).

11.17 DETECTION OF DATE-RAPE DRUGS

11.17.1 Compounds Diagnostic for Dosing by Synthetic Date-Rape Drugs

Because MDMA, flunitrazepam, and ketamine, and their metabolites are not part of natural metabolism, the presence of any one of them in a physiological sample is per se evidence for ingestion of the drug. However, as flunitrazepam and ketamine are used occasionally in medicine, evidence of dosing is not necessarily evidence of drug-assisted sexual assault or self-abuse. The context must be considered in reaching a conclusion.

11.17.2 Compounds Diagnostic for Dosing by GHB

D-2-hydroxyglutaric acid, 4,5-dihydroxyhexanoic acid, and SSA have strong, direct metabolic links to GHB (Figure 11.2; Sections 11.7.2 and 11.12). Nevertheless, none of these metabolites currently is a good marker of dosing, as the endogenous levels in undosed human beings are only partially characterized. Variance in endogenous levels of 2,4-dihydroxybutanoic acid and 3,4-dihydroxybutanoic acid, and of their breakdown products (formic acid, 3-hydroxypropanoic acid, glycolic acid, and acetic acid), is very large because of person-to-person differences in metabolism and diet. Elevated levels provide little evidence for dosing by GHB (Chalmers et al., 1976). GBL and BD are excreted at detectable levels only if they per se have been ingested and not after GHB has been ingested (McCusker et al., 1999; Blanchet et al., 2002; Lora-Tamayo et al., 2003; Wood et al., 2004; Shima et al., 2005).[35]

The amount of GHB present in blood and urine taken from undosed normal people has been comprehensively characterized, and it is always less than 5 and 10 µg/mL, respectively (Elliott, 2003; LeBeau et al., 2006). Persons suffering from uncontrolled diabetes excrete large amounts of the GHB isomers α- and β-hydroxybutanoic acid, but they excrete no excess GHB (Shima et al., 2005). Persons suffering from SSA dehydrogenase deficiency excrete large amounts of the GHB, but the frequency of this disorder is so low that it can be ignored in testing of the public for dosing (Koelker et al., 2008). Even though 96 to 99% of ingested GHB or prodrug typically undergoes catabolism, the current best physiological marker for dosing by these substances is a value of GHB above 5 or 10 µg/mL in serum or urine, respectively (Villain et al., 2003; Yeatman and Reid, 2003; Crookes et al., 2004; Kasprzak et al., 2006; LeBeau et al., 2006). As discussed above for other date-rape drugs, evidence of dosing is not necessarily evidence for drug-assisted sexual assault or self-abuse, as GHB is occasionally used in medicine. The context must be considered.

11.17.3 Gold-Standard Testing

Most regional forensic laboratories and many large clinical laboratories conduct *gold-standard tests* for date-rape drugs and their metabolites. Such tests include, but are not limited to, such techniques as spiking of a urine sample with deuterated drug or metabolite, extraction into organic solvent, chemical derivatization to improve physical properties, high-performance liquid or gas chromatography to reduce background signals, and mass spectrometry to quantitate isotope dilution (Kastrissios et al., 2005; Rendle, 2005; de Oliveira et al., 2007; Kraemer and Paul, 2007; Langman, 2007; Maurer, 2007; Musshoff and Madea, 2007a,b; Pizzolato et al., 2007; Pragst, 2007; Samyn et al., 2007; Smith et al., 2007; Snow, 2007; Tagliaro et al., 2007; de Paoli and Bell, 2008; Gong et al., 2008). Such sophisticated methods can provide superior sensitivity as well as quantitative and unambiguous results that constitute evidence of dosing suitable to systems of justice. However, the certitude that these methods provide is gained at high cost. Expensive instrumentation, highly trained persons to maintain the instruments and perform the tests, relatively slow turnaround, and substantial operating costs are characteristic. For these reasons, gold-standard testing must be strongly justified by circumstances. It is usually not used for routine screening and not available outside high-end laboratories.

11.17.4 Many Applications for Reliable Field Tests

Rapid, reliable, inexpensive, and semiquantitative testing by responsible principals could clarify many situations in which illicit drugs are suspected. Such test results assist in making rapid decisions and taking immediate appropriate action. The principals include staff persons working in hospitals and medical clinics, narcotics enforcement, prisons, and probation departments. The public also has an interest in reliable and easy detection of illicit drugs in bars, schools, and households. These and similar circumstances are termed "field" situations in this chapter to distinguish them from laboratory situations.

An international panel has recommended that at least two uncorrelated test methods be used to document presence of an illicit drug when the evidence is to be presented to a judicial system. The result of a validated field test can qualify for one of the methods if it is supported by the result from a gold standard, category A test (Santos, 2007). An opportunity for synergism exists here, as gold-standard testing can be made to be faster, more efficient, and less expensive by targeting the illicit drug putatively identified in a field test (Cone and Huestis, 2007; Peters, 2007).

11.17.5 Hospital Emergency Department Example

Because their molecular mechanisms of action are generally different, each agent causes some distinctive symptoms after an overdose. However, when agents are co-ingested, as often occurs, mixed symptoms are produced (Ricaurte and McCann, 2005; Wu et al., 2006; Van Sassenbroeck et al., 2007; Munir et al., 2008). A medical doctor can find it difficult to reach a correct diagnosis based on symptoms and oral history alone when dealing with possible polydosing (Bjornaas et al., 2006; Liechti et al., 2006; Devlin and Henry, 2008).

A hypothetical scenario that illustrates the need for both field and gold-standard testing is the following. A disheveled and comatose woman is brought into a hospital emergency department by a friend who does not know what happened. Staff members are aware that recent cases of sexual assault in the city involved date-rape drugs, and they want to test immediately for candidate drugs and ethanol. Rapid discovery of specific agents has many benefits. A correct diagnosis would be more likely, and appropriate treatment could be selected quickly. Expensive and unnecessary tests for head trauma, metabolic disorders, central nervous system infection, and increased intracranial pressure might be avoided. Also, physiological samples could be collected and preserved for gold-standard testing if warranted: for example, when sexual assault was suspected (Zvosec et al., 2007). Field and gold-standard testing are not competitive; they are complementary.

Rapid tests for ethanol are available, of course. Field tests based on antibody recognition of MDMA, ketamine, flunitrazepam, and other date-rape drugs are also available. They require 10 to 20 min to complete (Kunsman et al., 1996; Salamone and Tsai, 2002; Negrusz et al., 2005), and many suppliers are available (http://www.varianinc.com/cgi-bin/nav?products/dat/index&cid=KJOQJOKKFJ; http://www.biocompare.com/jump/2831/Immunoassays.html). Antibody test kits typically are cross-reactive for the major metabolites of a drug. The ketamine kit, for example, picks up both ketamine and norketamine. Cross-reactivity for the metabolites extends the window of time available for detection of dosing, as metabolites can linger in persons longer than the original drug does.

Unfortunately, no reliable field test for GHB is currently available. Purported tests exist, but they are very insensitive, based on unreliable or poorly selective chemistry, or require unrealistically ideal conditions. They should not be used by a responsible principal (Meyers and Almirall, 2004; Grossman et al., 2005; Beynon et al., 2006; Child and Child, 2007; Quest and Horsley, 2007). A sensitive and reliable field test for GHB is described in Section 11.20.3.

11.17.6 Preparation of a Sample for Delayed Analysis

If a physiological fluid is to be tested for a date-rape drug, it should be obtained as soon as possible after a suspected incident, as any drug will be subject to metabolism and excretion (Kastrissios et al., 2005; Drummer, 2007). However, for many reasons that include short-term amnesia caused by the drug, victims often do not report suspected assault until one or more days after an incident. A victim usually will have ingested only a single dose of the drug. The combination of a single dose and significant delay before the report greatly increases the sensitivity required to detect dosing. If testing is to be done days to months after sampling, such as often occurs for gold-standard analysis, the sample must be stored properly to avoid degradation, sequestration, or artifactual generation of analyte (El Mahjoub and Staub, 2000; Clauwaert et al., 2001; Hijazi et al., 2001; Elliott et al., 2004; Holmgren et al., 2004; Berankova et al., 2006; LeBeau et al., 2007; Saude and Sykes, 2007). When a physiological fluid is prepared properly and stored no more than several days in a refrigerator, large errors generally do not occur. However, "proper" depends on both the matrix and analyte. Also, enzymatic treatment to release covalently conjugated analyte can be important for some types of analytes in order to increase sensitivity. The ability of staff members to properly prepare a physiological fluid for gold-standard analysis would benefit from the knowledge gained in a positive field test.

11.17.7 Time Window Available to Detect Dosing

The window of time available to detect dosing depends on the drug. After single effective doses, MDMA, flunitrazepam, and ketamine are routinely detectable in urine for 1.5 to 4 days. Extremely sensitive gold-standard detection methods can extend the time window significantly. Also, metabolites often are detectable for even longer (Verstraete, 2004; Adamowicz and Kala, 2005).

In contrast, single doses of GHB are detectable above the threshold values of 5 and 10 μg/mL in serum or urine, respectively, for as little as 3 h and as long as 12 h in different persons (Section 11.17.2; Palatini et al., 1993; Kavanagh et al., 2001; Borgen et al., 2003; Brenneisen et al., 2004; Haller et al., 2006; Bodson et al., 2008). The time window is quite variable in part because large doses can be detected for *disproportionately* longer time periods, as absorption, metabolism, and elimination of GHB are limited by physiological capacity (Palatini et al., 1993; Scharf et al., 1998; Bhattacharya and Boje, 2007; Odujebe et al., 2007; Jones et al., 2009). Also, metabolism rates are affected by diet and ethanol (Poldrugo et al., 1989; Kaufman and Nelson, 1991; Karch et al., 2001; Borgen et al., 2003; Haller et al., 2006; Thai et al., 2007; Wang et al., 2008). Time courses have been simulated numerically (Bhattacharya and Boje, 2006). In summary, physiological fluids must be sampled in ≤ 12 h of possible consumption in order to detect GHB above threshold. Merely increasing the sensitivity of testing does not solve the problem of a limited time window for GHB.

11.17.8 Extending the Time Window

Several relatively new gold-standard technologies extend the time available to detect dosing. One technology utilizes hair, which incorporates a circulating drug when the strands are being formed. In this technique, full-length hairs are obtained from a person, cleaned, and cut into 3-mm segments 4 to 5 weeks after putative dosing. The segments are dissolved, and the solubilized hair is analyzed by tandem mass spectrometry. If the person was dosed, the protocol produces a peak in drug concentration for the hair segment formed when the dosing occurred, and baseline values for hair segments that flank the segment that was formed during dosing. A single GHB dosing of a human being has been detected in this manner (Kintz, 2007).

Another applicable technology is *gene expression profiling* by DNA microarray analysis. Blood mRNAs for the proteins epiregulin (member of the epidermal growth factor family) and PEA-15 (phosphoprotein enriched in astrocytes of 15 kDa) were significantly elevated 48 h after mice were given a single large dose of GHB (Larson et al., 2007). Also, the Fos transcription factor is greatly elevated in rat brain after a high dose of GHB, although samples of brain normally would not be tested in living human beings (van Nieuwenhuijzen et al. 2009). Because Fos is widely distributed, Fos mRNA in the periphery is also probably elevated, although this hypothesis has not been tested. The effects of MDMA and ketamine on mRNA levels in rodent brain have been studied by gene expression profiling (Thiriet et al., 2002; Funada et al., 2005; Lowe et al., 2007). The method has great promise for forensic science. However, there remain the hurdles of (1) identifying suitable sources of mRNA, (2) identifying at least one mRNA that is diagnostic for a target drug, (3) establishing baseline values for mRNAs in the person tested, and (4) reducing statistical variation enough to make results accurate.

Gene expression profiling illustrates one reason why detailed knowledge of metabolic pathways might be valuable to forensic science. The RNAs coding for or regulating the expression of isoceptors, isozymes, and isoporters linked to a target drug are the most likely to respond to dosing. For example, mRNAs for MCTs (Section 6.1) and Fos in white blood cells are available from fresh blood, and one of them might be responsive to dosing by GHB.

11.18 SPECIAL CIRCUMSTANCES OF GHB

11.18.1 Industrial Connection

GHB differs from other date-rape drugs in one more important respect. Because of their rapid and efficient conversion to GHB in the body, GBL and BD have psychotropic effects very similar to those of GHB. Yet because GBL and BD are used by industry for multiple legitimate purposes, they are inexpensive, widely available, and weakly controlled. Chemical mixtures containing $\leq 70\%$ GBL recently were exempted (http://openregs.com/regulations/view/103806/exempt_chemical_mixtures_containing_gamma-butyrolactone). As a result, GBL and BD increasingly are being adopted as alternatives to GHB in drug-assisted sexual assault and self-abuse (Kapadia et al., 2007; Thai et al., 2007; Wood et al., 2008). The rapid test for GHB described below will not detect GBL or BD. However, it will flag their ingestion by detecting excess GHB in serum or urine.

11.18.2 Enzymes Acting on GHB in Bacteria, Yeast, and Plants

Many physiologically stressed bacteria and yeast biosynthesize polymers containing GHB. The polymers are a carbon- and energy-storage material (Moore et al., 2005; Song et al., 2005, Bach et al., 2009).[36] Polymerization requires activation of the GHB carboxyl group, which can be done by butyrate kinase, 4-hydroxybutyrate-CoA transferase, or 4-hydroxybutyryl-CoA synthetase (AMP forming) (Cary et al., 1988; Valentin et al., 2000; Huber et al., 2008). When environmental conditions become favorable to growth, microorganisms hydrolyze stored polymer to mobilize the carbon. Some species of bacteria also consume tetrahydrofuran, which is an environmental pollutant and a common solvent, by oxidizing it to GBL and then hydrolyzing the GBL to GHB (Kohlweyer et al., 2000). In all characterized cases, GHB is brought into central metabolism by an Fe(II)-containing ADH that uses dissociable NAD^+ as the hydride acceptor. These enzymes have sequences similar to human ADHFe1 (Section 11.7.2). Also, some plants use GHB as a redox ballast to survive periods of anoxia, and they make it with SSA reductase (Breitkreuz et al., 2003; Hoover et al., 2007). Only one isozyme is known for each of these enzymes.

11.18.3 Possible Accidental Intoxication by GHB in the Future

The possibility of accidental intoxication by GHB is increasing from an ironic source. Researchers have engineered microorganisms and plants to make large amounts of polymers containing GHB and/or BD monomers. Such polymers can resemble polyethylene and polypropylene in their rheological properties and are termed *bioplastics* (Van der Walle et al., 2001). Bionolle and Ecoflex are commercial examples (Gross and Kalra, 2002).[37]

Although currently minor, the fraction of plastics containing GHB and/or BD monomers could increase dramatically. The possibility arises because "synbio" organisms can make them from renewable low-value inputs such as corn stover and sewage sludge (Zhang and Wang, 1994; Tsunemasa, 1998; Poirier, 2001; Rabetafika et al., 2006). Bioplastics require less input of energy and valuable feedstocks than petroleum-based plastics do (Kurdikar et al., 2000; Efe et al., 2008). Moreover, bioplastics are hydrophilic polyesters that are biodegradable by esterases found in soil-dwelling microorganisms. Some bioplastics are even compostable, meaning that they are degraded completely in only a few months (Tokiwa and Pranamuda, 2001; Deguchi et al., 2002; Abou-Zeid et al., 2004). Composting reduces the volume of waste that must be accommodated by landfills, which are increasingly scarce (Davis and Song, 2006). The combined virtues of renewability and biodegradability might encourage much greater use of bioplastics containing GHB and/or BD in the future (Dornburg et al., 2004; Okino, 2005a,b; Kale et al., 2007; http://www.bioplastics24.com/).

If disposal facilities were to contain large amounts of bioplastics, leachate from them will contain monomers, including GHB and BD. The concentrations in primary leachate will probably be low and nontoxic, assuming that both the bioplastics and drainage meet professional specifications (http://www.astm.org/Standard/index.shtml). However, concentrations could increase to intoxicating levels if leachate were allowed to form puddles that evaporate, as GHB and BD are not very volatile and are chemically stable. Bacteria in the puddles might not efficiently scavenge GHB and BD, due to lack of micronutrients or preferential use of better carbon sources, such as glucose derived from plasticized starch also present in the disposal facility (Greer, 2006). This scenario is most likely to occur around residential compost piles. Dosing of neighborhood wildlife and omnivorous children by decomposing bioplastics might become a significant problem, thanks to green products. Regular environmental monitoring for GHB and BD could become a necessity.

11.19 CONSIDERATIONS DURING DEVELOPMENT OF FIELD TESTS

11.19.1 Shortcomings of Antibody-Based Screens for Simple Analytes

Many field tests for date-rape drugs are based on binding of free drug and its metabolites to an antibody preparation (Singh et al., 2006). Although the method works well for complex synthetic molecules having multiple points of contact between the analyte and the antibody combining site, it often does not work for simple analytes.[38] The reasons are easy to identify. To make a small molecule immunogenic, it must be covalently "conjugated" to a large carrier such as a protein. Conjugation in this case causes a big perturbation in analyte structure. Thus, antibodies often bind to the conjugated analyte better than to the free analyte (Benacerraf and Paul, 1970). Moreover, most antibodies used in forensic science are polyclonal, which means that preparation begins with the blood obtained from an immunized animal. Many different amino acid sequences and three-dimensional structures are present in the antibody combining sites. Polyclonal heterogeneity has adverse consequences on analyte selectivity, as discussed below.

Another type of antibody is monoclonal, which means that the preparation is obtained from an amplified clonal line of an antibody-producing cell. One amino acid sequence and one three-dimensional structure are present in all of the antibodies made by the clonal line. A different clonal line generated against the same analyte will in general secrete an antibody having a different amino acid sequence exhibiting a different three-dimensional structure in the antibody combining site. Thus, cocaine has been found by x-ray crystallography to bind in different conformations and orientations to different monoclonal antibodies (Pozharski et al., 2005). The result implies that an analyte binds to individual members of a polyclonal preparation in many different ways.

It follows that within the diversity of complexes that form with an analyte, a preparation of polyclonal antibodies can also bind many analogs of the analyte. From highest to lowest affinity, the potential complexes will include conjugated analyte, structural relatives of conjugated analyte, free analyte, and metabolites and other analogs of free analyte. Indeed, the ability of a polyclonal preparation to recognize a drug *and* its metabolites depends on antibody heterogeneity. Moreover, if a polyclonal preparation is purified with an immunoaffinity step, which will of necessity use conjugated analyte, the final preparation is biased even more away from binding free analyte and toward binding conjugated analyte (Choi et al., 1999). It is not surprising that complex mixtures of antibody combining sites exposed to complex mixtures of small molecules, *none* of which were the actual immunogen, often produce the worst possible outcome for a small-molecule screen: namely, a high rate of false positives and negatives.

11.19.2 Advantages of Enzyme-Based Screens for Simple Natural Analytes

Enzymes can provide a solution to the shortcomings of antibodies in the detection of simple analytes. First, there is no need to conjugate the analyte covalently. Thus, misdirection toward analyte conjugates and their analogs does not happen. Second, an enzyme has a single amino acid sequence and three-dimensional structure, thereby eliminating binding-site heterogeneity that decreases substrate selectivity. Third, most enzymes have evolved to discriminate against metabolites similar in structure to the substrate desired. Such discrimination is important to minimize crosstalk between pathways (Hartl et al., 1985; Hartl, 1989; Galperin et al., 1998).[39] In contrast, immunization has no mechanism to discriminate against metabolites resembling the conjugated analyte. Fourth, substrate bound in an enzymatic active site must form the near attack conformation (NAC) before it can react (Bruice, 2002). The NAC is accessed by librational dynamics, which are optimized by evolution. The NAC theory is complementary to the better known theory that an enzyme binds to the transition state for a reaction tighter than to the ground state. In both theories, protein dynamics contribute to selectivity in catalysis. In contrast, protein dynamics play no consequential role in an ordinary antibody–analyte complex. In summary, enzymes possess numerous mechanisms for molecular discrimination that antibodies in general and polyclonal antibodies in particular do not have.

Enzyme-based screens also have a time advantage that can be important in field applications. Positive results start to become known within seconds of mixing reagents and a test sample when a high concentration of analyte is present. In contrast, antibody-based screens require some sort of previsualization processing such as lateral flow chromatography or blocking of nonspecific protein adsorption to suppress background before the results can be read. An immunochemical method such as enzyme-linked immunosorbant assay (ELISA), which uses an enzyme to amplify the signal arising from analyte, does not negate the requirement for previsualization processing. In addition, enzymes often have a cost advantage. The complementary DNA sequence for an enzyme is cloned into an expression vector only one time, and the vector is used to transform bacteria only one time. Transformed bacteria can be propagated indefinitely and used to produce the enzyme in a genetically "tagged" form established by the vector. The tag is a polypeptide or protein chosen to allow facile purification of the resulting "fusion protein" by affinity chromatography at moderate cost. In contrast, polyclonal antibodies usually are obtained from immunized animals in a relatively time-consuming and expensive process. Monoclonal antibodies are even more expensive to produce.

The preceding observations suggest rules of thumb for choosing between an antibody and an enzyme for the development of a field test. An enzyme should be considered for a small natural metabolite, as a cognate enzyme *must* exist and an antibody has a high probability of failure. GHB is an example of this type of analyte. To avoid false positives the enzyme must be highly selective for analyte in the samples to be tested. Because of low selectivity, an enzyme that acts on xenobiotics usually would not be appropriate. Also, some readily implemented method must exist to visualize test results outside a laboratory.

An antibody should be considered for a structurally complex or large analyte, as the problem of immunological perturbation by conjugation is lessened. If the analyte is manufactured by synthetic organic chemistry, an enzyme selective for it will not exist in nature. MDMA, flunitrazepam, and ketamine are examples of this type of analyte. Also, even if the analyte is natural, the cloning, expression, and isolation of a cognate enzyme might not be practical, or the enzymatic reaction might not be easy to visualize in the field. An antibody might then be the best recourse. Cocaine is an example of this type of analyte. The pathway for its biosynthesis is not known with certainty, and thus an enzyme possibly having good selectivity for cocaine cannot be identified with certainty. Human carboxylesterase 1 and cytochrome P450 act on cocaine, but they are low-selectivity enzymes targeting many xenobiotic compounds (Ladona et al., 2000; Humphrey and O'Hagan, 2001; Bencharit et al., 2003).

Of course, no rule of thumb is infallible. Some analytes are large natural products, and new tools such as diabodies, aptamers, and directed evolution are being developed continually in the search for new types of biosensors (Fickert et al., 2006; Poyot et al., 2007).

11.20 DEVELOPMENT OF AN ENZYMATIC TEST FOR GHB

11.20.1 Sensitivity Required for the Hospital Emergency Department

Patients presenting to a hospital emergency department because of possible drug overdose are in crisis (Chin et al., 1998). Only moderate sensitivity is required to detect a drug if it is a major cause of the distress, as the amount in the patient's system *at that moment* must be large. In the case of GHB, levels of 30 to 600 μg GHB/mL of serum and 400 to 6000 μg GHB/mL of urine have been found by gold-standard analyses of samples that have been prepared and stored correctly (Louagie et al., 1997; Sporer et al., 2003; Couper et al., 2004; Elliott, 2004; Van Sassenbroeck et al., 2007). These levels are well within the capabilities of typical enzyme-based tests. *In the potentially life-and-death context of a hospital emergency department, a rapid test with good sensitivity is better than a slow test with superior sensitivity.*

11.20.2 Choice of Enzyme

As footnoted above, attempts to develop a field test based on antibody recognition of GHB have failed. An enzyme-based test is an attractive alternative. Three enzymes that act on GHB have been characterized well enough to consider. They are vertebrate NADPH-dependent SSA reductase (AKR1A1 or AKR7A2), vertebrate 2-oxoglutarate–dependent ADHFe1, and bacterial Fe(II)-containing NAD^+-dependent GHB dehydrogenase (Sections 11.5.3, 11.7.2, and 11.18.2). To use a reductase, the reaction would have to be run in the reverse direction using $NADP^+$ as oxidant.

What considerations can help us decide which of these enzymes might be suitable? The oxidation catalyzed by ADHFe1 is thermodynamically favorable, but ADHFe1 has only moderate selectivity for GHB among common metabolites. It would be susceptible to false-positive results when testing physiological fluids and other complex matrices. This possibility is decisively unfavorable, leaving vertebrate SSA reductase and bacterial ADH to consider. The overall equilibrium constants are the same, providing no thermodynamic basis to choose between them (Section 11.9). A search of online databases containing DNA and protein sequences reveals many putative SSA reductases and GHB dehydrogenases, as deduced by sequence homology (http://www.ncbi.nlm.nih.gov/; http://www.uniprot.org/). However, sequence homology is much better at predicting the chemistry carried out than the preferred substrate (Hulo et al., 2006). Unless substrate selectivity has been studied by direct biochemical methods, one cannot be sure which substrate is preferred.

A search of online databases for enzymatic substrates, followed by thorough reading of the primary literature, reveals that only a few SSA reductases and GHB dehydrogenases have been characterized biochemically. Examples of these databases are BRENDA (http://www.brenda-enzymes.info/) and KEGG (http://www.genome.ad.jp/kegg/pathway.html). Abstracts of primary literature are freely accessible in PubMed (http://www.ncbi.nlm.nih.gov/) or by subscription to SciFinder Scholar (http://www.cas.org/products/sfacad/index.html). Most SSA reductases have high affinity for SSA and low affinity for GHB, which promotes fast and complete reduction of moderate concentrations of SSA but slow and incomplete oxidation of moderate concentrations of GHB, a decidedly unfavorable property (Keenan et al., 2006). This characteristic narrows the choice of enzyme to bacterial GHB dehydrogenase. Moreover, because a single isozyme is likely, and cloning and expression are usually easier, a bacterial enzyme is preferred to an enzyme from a eukaryotic organism such as an animal, plant, or yeast. The air-stable Fe(II)-containing NAD^+-dependent GHB dehydrogenase from the bacterium *Ralstonia eutropha* meets all criteria (Valentin et al., 1995). The properties of this family of enzymes have been summarized (http://pfam.sanger.ac.uk/family?acc=PF00465). GHB dehydrogenase from the bacterium *Clostridium kluyveri* has also been characterized biochemically. However, it is unacceptable because it is rapidly inactivated by oxygen, which would be a major liability in a field test (Wolff and Kenealy, 1995).

The GHB dehydrogenase reaction is not visible to the unaided eye. This circumstance means that a secondary reaction must be coupled to the reaction to visualize the results of the test. Fortunately, the electron pair stored in NADH can be used by a well-known "turnkey" reaction that reduces a large number of colorless prodyes to intensely colored products (Berridge et al., 2005). All required reagents are commercially available. Such a coupled secondary reaction also helps solve another problem. Oxidation of GHB with NAD^+ is significantly unfavorable at neutral pH (Section 9). Unless something is done to compensate, only a fraction of the GHB in a test sample will react, and it will do so slowly because of the asymptotic approach to equilibrium. Both properties are decidedly undesirable. However, the turnkey reactions are so favorable that the resulting two-step reaction is favorable overall.

11.20.3 Reliable Field Test for GHB

To develop the test, the gene for GHB dehydrogenase in *R. eutropha* was copied from total genomic DNA using appropriate primers and polymerase chain reaction. The resulting DNA was purified, end-trimmed with restriction enzymes, and fused to the gene for the affinity-tag glutathione *S*-transferase, which is under the control of a lactose promoter in the commercial expression vector pGEX-2T. *E. coli* was transformed with the recombinant vector to produce and isolate large quantities of fusion protein that links glutathione *S*-transferase to GHB dehydrogenase. About 150 mg of purified fusion protein typically is obtained from 1 L of bacterial culture in a fast procedure utilizing an affinity matrix containing covalently bound glutathione (Bravo et al., 2004).

A solution assay at pH 8.5 in 2-amino-2-methyl-1,3-propanediol buffer was devised for quantitative research using a laboratory spectrometer. The higher pH value decreases the concentration of product H^+ and makes the reaction more favorable (Fig. 5). The buffer is Tris-like, resulting in formation of the imine of SSA as it is formed. Formation of imine (Section 11.8) pulls GHB oxidation thermodynamically and prevents product inhibition due to SSA buildup. The NADH produced by the GHB dehydrogenase reaction is used to reduce (with the

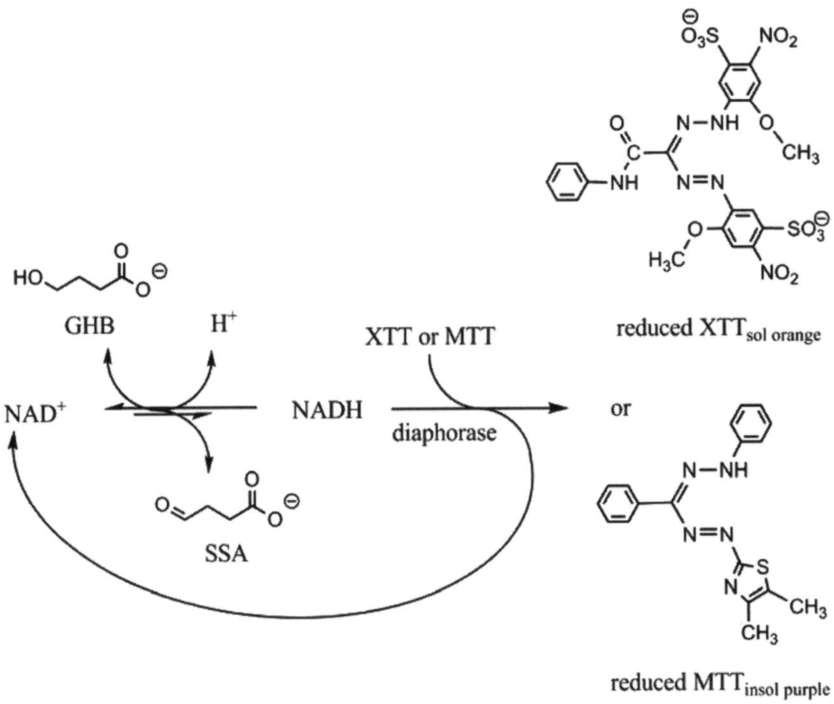

Figure 11.5 Coupled assay for GHB. NAD$^+$ and GHB dehydrogenase from *Ralstonia eutropha* oxidize GHB to SSA in the first step. Because it is otherwise unfavorable, the reaction is run under pulling conditions. A Tris-like buffer (2-amino-2-methyl-1,3-propanediol) that also forms the imine of SSA is utilized at pH 8.5. In addition, diaphorase uses NADH made in the first step to reduce nearly colorless XTT to a soluble orange dye for a laboratory assay using a spectrometer or MTT to an insoluble purple dye deposited on a dipstick for a field assay. Different prodyes that yield different colors and solubility are available.

aid of the electron transfer enzyme diaphorase) a prodye called XTT [sodium 3,3-[(phenylamino)carbonyl]-3,4-tetrazolium-bis(4-methoxy-6-nitro)benzenesulfonic acid hydrate] to form a soluble orange dye (λ_{max} 450 nm) and regenerate NAD$^+$ (Figure 11.5). This reaction also pulls GHB oxidation thermodynamically and removes product inhibition due to NADH buildup. Thus, the assay implements five strategies to make it favorable, faster, and visible.

Michaelis–Menten parameters were determined in 1 mM NAD$^+$. The K_M value for GHB is 2.19 ± 0.13 mM and the V_{max} value is 0.049 ± 0.001 absorbance/min per microgram of fusion protein. The lowest concentration of GHB reported in the urine of emergency department patients ($\sim$400 µg/mL) corresponds to about 4 mM. The K_M value is well matched to this concentration, thus providing a rapid and sensitive detection of GHB.

Ethanol is a weakly binding alternative substrate of the enzyme, having an estimated K_M value of 413 ± 14 mM, or 200 times higher than for GHB. The affinity is low enough that ethanol concentrations realistically found in physiological fluids do not produce false-positive results. Recent research has shown that other simple alcohols, such as methanol, 1-propanol, 2-propanol, 1-butanol, and 1-pentanol, are significantly worse substrates than ethanol is (Vinckier et al., 2011).

Multiple studies agree that GHB arising from ingested GHB or GHB prodrug is concentrated about 10-fold in urine relative to blood (Liu et al., 2006). Thus, urine rather than serum is preferred for dose testing. If allowed to react for about 2 h in 1 mL of coupled assay buffer, all of the GHB in a small sample of spiked normal human urine (10 μL) is reacted. The amount of orange color produced is linearly dependent on the amount of GHB. No substances in urine contribute significantly to color development, even after 2 h of reaction. This variation of the test, called an *endpoint assay*, has a limit of detection of 37 μg GHB/mL urine (total A_{450} to background $A_{450} = 2$ to 1). The same procedure using urine containing 0.47% (w/v) ethanol gives the same results, confirming that a realistic concentration of ethanol in urine does not interfere with the endpoint assay.

The solution assay uses a spectrophotometer and is not as fast or convenient as a dipstick assay that can be used in the field. To develop a prototype for a dipstick, formation of an insoluble purple dye by reduction of the soluble prodye MTT (3-[4,5-dimethylthiazol-2-yl]-2,5-diphenyltetrazolium bromide) was used (Figure 11.5). Small samples (10 μL) of water, normal human urine, or normal human urine containing 0.63% (w/v) ethanol and having different concentrations of spiked GHB were spotted onto paper, after which assay reagent (10 μL) was spotted. The test is developed for no longer than 2 min to avoid false-positive results. No significant color develops in the absence of GHB, and strong purple color develops within a few seconds for samples containing concentrations of GHB typically present in urine during an overdose. Color development is fast because both the test sample and reagents are diluted very little. The three tested matrices gave the same results, demonstrating that neither urine nor a physiologically high concentration of ethanol interfere (Figure 11.6). The precision around the cutoff was also assessed. Of 20 samples for each condition, none were positive in the absence of GHB, 25% were positive (and 75% were negative) at 20 μg GHB/mL, 75% were positive at 50 μg GHB/mL, and all were positive at 100 μg GHB/mL and more. Thus, a concentration of GHB four times lower than the lowest concentration found in the urine of patients suffering from GHB overdose is detected every time. A dipstick test strip not requiring spotting of reagents before use is under development for long shelf life and use in hospital emergency departments (Harper et al., 2008).

CONCLUSION

Impressive progress in understanding the fundamental molecular and cellular biology of date-rape drugs has been made in recent years. Nevertheless, many important questions remain. The roles of RNA and regulation of enzymatic activity and gene expression have been studied only a little. For example, does protein phosphorylation regulate the metabolism of date-rape drugs? Use of cultured human cells

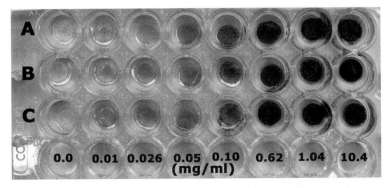

Figure 11.6 Prototype of a dipstick assay for GHB. Ten-microliter samples containing the concentrations of GHB indicated were spotted onto white-paper circles lying in wells of a microtiter plate. Ten microliters of assay reagent containing buffer, GHB dehydrogenase, NAD^+, diaphorase, and MTT prodye was spotted onto a circle of filter paper in each well, and color was allowed to develop for 2 min at 23° C. Row A, GHB in water; row B, GHB in normal human urine; row C, GHB in normal human urine containing 0.63% (w/v) ethanol. [From Bravo et al. (2004); used with permission.] *(See insert for color representation.)*

and genetically modified mice should be especially effective tools to answer these questions. In the case of GHB and its prodrugs, the cellular, subcellular, and enzymatic mechanisms for catabolism are not fully understood. Even when a particular isozyme or isoporter has been implicated in GHB catabolism (e.g., mitochondrial ADHFe1), we do not know whether it dominates other isozymes or isoporters under all relevant conditions in all relevant cell types, and whether it is rate limiting to the pathway. Neuronal signaling by GHB under normal and overdose conditions is not understood. Thus, potential consequences of GHB overdose that manifest in subtle, delayed, or synergistic damage have not been identified.

Impressive progress in the forensic science of date-rape drugs has also been made in recent years. Especially sensitive analytical methods have been developed. A surrogate for drugs and their metabolites in the form of perturbed mRNA levels has been discovered. The window of time available to detect dosing has been greatly extended. In the case of GHB, new metabolites have been found that are potentially diagnostic, and more probably remain to be found. An enzyme-based test for GHB has been developed that fills a gap in field tests for date-rape drugs. Moreover, the demand for forensic expertise in date-rape drugs will probably increase in the future due to (1) the possible need for comparison of two uncorrelated test methods in judicial proceedings, (2) development of new tests based on increased knowledge of biology, and (3) greater need for environmental testing. It is a rewarding and exciting time to be working in forensic science!

Acknowledgments

The author thanks fellow researchers David Harris and Dawn Bravo for their insights and hard work, Lawrence Kobilinsky for his invitation to contribute to this book, Karl Sporer, Elizabeth Ellis, Toshimitsu Niwa, and Aaron Rowe for comments on the manuscript draft, Zachary Golden for alerting me to new statuatory law, Don Aue for ab initio calculations of stabilities for GHB metabolites in water, Tetsuya Kawamura for translation services, and Leslie D. Odell for extraordinary editorial assistance.

NOTES

1. GHB also is known as Xyrem and "grievous bodily harm"; MDMA as ecstasy; flunitrazepam as Rohypnol and roofies; and ketamine as Ketanest and special K.
2. Synaptic vesicles are small and spherical or ellipsoidal and reside inside nerve terminals. They store concentrated neurotransmitter for evoked release as a quantum.
3. The central and peripheral nervous systems can synthesize neurosteroids at higher levels in females than in males. This difference might help explain why women often are more susceptible to ethanol at equivalent body burdens. An example of a neurosteroid is allopregnanolone.
4. Subtypes of $GABA_A$ receptor located inside synapses are not as responsive to ethanol as the subtype located outside synapses, which contains the δ-subunit.
5. The $GABA_B$ receptor modulates intracellular second messengers. An agonist activates a receptor. In this case, a high concentration of GHB activates the $GABA_B$ receptor similarly to a low concentration of GABA.
6. GABAergic means GABA-like or GABA-potentiating.
7. All neurotransmitters are agonists, even if transported instead of secreted out of a neuron.
8. An antagonist does not activate a receptor when it binds, but it inhibits binding of agonist, thus preventing activation by agonist.
9. Desensitization means that the receptor becomes unresponsive to agonist, even though no antagonist is present. Many different molecular and cellular mechanisms can mediate desensitization by receptors.
10. Isozymes (also known as *paralogs*) usually, but not always, have partially conserved sequences. The following examples illustrate nomenclature for the dehydrogenases and reductases discussed here. AKR1A1 is listing 1 (the last "1") in family 1 (the first "1", having $\geq$ 40% sequence identity among members) and subfamily A ($\geq$60% sequence identity among members) of the AKR superfamily. This enzyme is also commonly called aldehyde reductase, as it reduces a wide range of aldehydes. Similar nomenclature is used for ADHs and ALDHs by replacing "AKR" with "ADH" or "ALDH," as appropriate. AKR1A1 has no necessary relationship to ADH1A1. An isozyme could have several isoforms that arise from alternative splicing of the pre-mRNA.
11. This occurs because oxidation of an aldehyde to a carboxylate is irreversible. The reduction potential of an $NAD^+/NADH$ or $NADP^+/NADPH$ couple is not very important in an irreversible oxidation. A few mammalian alcohol dehydrogenases use non-nicotinamide redox cofactors such as flavins.
12. For example, SSA is a good substrate for AKR1A1 *and* AKR7A2 (Section 11.5.3). AKR1A1 also prefers the natural substrates 4-hydroxynonenal, methylglyoxal, 16-oxoestrone, D-glyceraldehyde, and D-glucuronic acid. AKR7A2 also prefers the natural substrates aflatoxin B1 aldehyde and 16-oxoestrone.

13. When two enzymes act on the same substrate, one of them can dominate at a low concentration of substrate and the other one at a high concentration if the Michaelis–Menten parameters for the two enzymes (and potentially also for an intervening transporter) have appropriate relationships. This possibility is relevant for GHB and its metabolites, as concentrations in vivo change > 100-fold between normal and dosed conditions.
14. Similar to isozymes, isoporters are transporters coded by different genes, and they often exhibit conservation of sequence and transport mechanism. Different isoporters usually have different preferred substrates.
15. However, mRNA levels often do not correlate quantitatively with protein levels, due in part to regulation of translation and differences in protein stability.
16. Amazingly, both SSA and aflatoxin B1 are preferred substrates for AKR7A2/AFAR1, even though the structures of SSA and aflatoxin B1 differ greatly. The observation warns us that in the absence of experimental evidence, one cannot eliminate the possibility that an isozyme having a preferred substrate very different in structure from SSA nevertheless also prefers SSA.
17. AKR7A2 sometimes is called the low-K_M SSA reductase.
18. An ortholog is the same enzyme in a different species. It will have a somewhat different sequence, due to species divergence, but its functional properties are usually very similar.
19. DNA microarray analysis determines the amount of specific mRNAs in total mRNA isolated from a tissue. Single-stranded synthetic DNA is anchored to thousands of microscopic spots in an indexed two-dimensional pattern on the surface of a "chip." The DNA on each spot is complementary to a different mRNA. A preparation of total mRNA (or its DNA complement) is hybridized to the chip-bound DNA, after which each spot optically reveals the amount of hybridization.
20. Epitope masking can occur when tissue fixatives react with and destroy an epitope or when the epitope is dependent on native conformation destroyed by denaturing electrophoresis, as occurs in Western blots.
21. Small interfering RNA (siRNA) blocks effective expression of a specific gene by inducing cellular degradation of the mRNA transcribed from that gene.
22. A low level of AKR1A1 in SH-SY5Y cells is not necessarily indicative of the amounts in vivo, as transformed cells are disregulated.
23. Golgi is a subcellular organelle that among other tasks forms synaptic vesicles that secrete neurotransmitters from nerve terminals.
24. Isoforms can have different amino acid sequences even though they arise from the same gene. This often occurs by alternative splicing of the mRNA. The total number of isoforms for ADHs, AKRs, and ALDHs thus exceeds the 46 genes for these enzymes by an unknown margin.
25. ADHFe1 is also known as HOT (hydroxyacid-oxoacid transhydrogenase) and ADH8. However, the latter designation has been used by some authors for retinol dehydrogenase, which is a completely different enzyme from ADHFe1.
26. Genetic tagging attaches a peptide to a protein to make a fusion protein that is to be expressed heterologously. In this case, an epitope tag recognized by a commercially available antibody was attached to ADHFe1.
27. Square brackets around a chemical name indicate the activity of the chemical species, which we approximate as molar concentration.

28. The ratio of the multiplicative concentrations of all chemical products divided by the multiplicative concentrations of all chemical reactants present at equilibrium is equal to K_{eq}. Thus, for ADH-catalyzed oxidation of an alcohol, $K_{eq} =$ [aldehyde]$_{eq}$[NADH]$_{eq}$[H$^+$]$_{eq}$/([alcohol]$_{eq}$[NAD$^+$]$_{eq}$), which has net units of molar. The apparent value of K_{eq} for the reaction at pH 7.0 is given by $K'_{eq} = K_{eq}/[10^{-7.0}\text{M}]$, which is unitless. Similar equations apply to other reactions considered in this chapter.

29. A reported equilibrium constant of $K'_{eq} = 5.8$ for reduction of SSA must be in error (Hearl and Churchich, 1985).

30. At the pH and [NAD$^+$]/[NADH] prevailing in mitochondria, [aldehyde]$_{eq}$/[alcohol]$_{eq}$ = K_{ox}^{mito} = $K'_{eq}([10^{-7.0}\text{M}]/[10^{-7.7}\text{M}])$([NAD$^+$]$_{mito}$/[NADH]$_{mito}$) = $(9.5 \times 10^{-5})(10^{0.7})(7.7) = 3.7 \times 10^{-3}$. The free-energy change for the reaction is given by $\Delta G_{ox}^{mito} = -2.3RT \log K_{ox}^{mito}$, where the symbols have their usual meanings. Analogous expressions apply to NADP$^+$-dependent reactions, reverse reactions, and cytoplasm.

31. K_{ox}^{cyto}, equilibrium constant for oxidation of GHB modified for the pH and ratio of oxidant to reductant prevailing in cytosol; K_{red}^{cyto}, equilibrium constant for reduction of SSA modified for the pH and ratio of reductant to oxidant prevailing in cytosol; K_{ox}^{mito}, equilibrium constant for oxidation of GHB modified for the pH and ratio of oxidant to reductant prevailing in mitochondrial matrix; K_{red}^{mito}, equilibrium constant for reduction of SSA modified for the pH and ratio of reductant to oxidant prevailing in mitochondrial matrix.

32. Oxidation of an aldehyde to a carboxylic acid is irreversible in part because three driving forces are present that are not present in the oxidation of alcohols: neutralization of the carboxylic acid proton, electronic delocalization of the negatively charged carboxylate, and strong hydration of the carboxylate.

33. An allele is a mutant (or wild-type) form of a gene.

34. Microsomes are vesicles created from shearing of endoplasmic reticulum by homogenization of an organ such as liver, which is rich in endoplasmic reticulum. They are rich in P450 isozymes.

35. From this point onward in the chapter, any time that consequences of dosing by GHB are discussed, it should be assumed that similar, and perhaps even more extreme, consequences attach to dosing by GBL or BD.

36. Such species include *Ralstonia eutropha*, *Azotobacter vinelandii*, *Escherichia coli*, *Clostridium kluyveri*, and *Pseudomonas putida*. Poly(GHB) is also known as poly(γ-butyrolactone).

37. The trademark Bionolle is registered to Showa Highpolymer Co., Ltd. The chemical name is poly(butylene succinate). Butylene refers to 1,4-butanediol diester. Ecoflex is registered to BASF Group. The chemical name is poly(butylene adipate).

38. Several companies have attempted to raise antibodies that are selective for GHB, which is a small molecule in the immunogenic context, but they failed (personal communications to SMP).

39. Enzymes that act on xenobiotics are exceptions.

REFERENCES

Abanades, S., Farre, M., Barral, D., Torrens, M., Closas, N., Langohr, K., Pastor, A., and de la Torre, R. (2007a). Relative abuse liability of gamma-hydroxybutyric acid, flunitrazepam, and ethanol in club drug users. *J. Clin. Psychopharmacol.*, **27**:625–638.

Abanades, S., Farre, M., Segura, M., Pichini, S., Pastor, A., Pacifici, R., Pellegrini, M., and de la Torre, R. (2007b). Disposition of gamma-hydroxybutyric acid in conventional and nonconventional biologic fluids after single drug administration: issues in methodology and drug monitoring. *Ther. Drug Monit.*, **29**:64–70.

Abe, T., Kurozumi, Y., Yao, W.-B., and Ubuka, T. (1998). High-performance liquid chromatographic determination of β-alanine, β-aminoisobutyric acid and γ-aminobutyric acid in tissue extracts and urine of normal and (aminooxy)acetate-treated rats. *J. Chromatogr. B*, **712**:43–49.

Abou-Zeid, D.-M., Mueller, R.-J., and Deckwer, W.-D. (2004). Biodegradation of aliphatic homopolyesters and aliphatic–aromatic copolyesters by anaerobic microorganisms. *Biomacromolecules*, **5**:1687–1697.

Abraham, M. H., Chadha, H. S., and Mitchell, R. C. (1995). Hydrogen bonding. 36. Determination of blood–brain barrier distribution using octanol–water partition coefficients. *Drug Des. Discov.*, **13**:123–131.

Adachi, H., Inagaki, Y., Okazaki, N., and Ishibe, Y. (2005). Racemic ketamine and S(+)-ketamine concentrations in cerebrospinal fluid after epidural and intravenous administration in rabbits. *Yonago Acta Med.*, **48**:33–40.

Adamowicz, P., and Kala, M. (2005). Urinary excretion rates of ketamine and norketamine following therapeutic ketamine administration: method and detection window considerations. *J. Anal. Toxicol.*, **29**:376–382.

Agabio, R., and Gessa, G. L. (2002). Therapeutic uses of γ-hydroxybutyrate. In: Tunnicliff, G., and Cash, C. D. (Eds.), *Gamma-Hydroxybutyrate*. London: Taylor & Francis, pp. 169–187.

Akins, B. E., Miranda, E., Lacy, J. M., and Logan, B. K. (2009). A multi-drug intoxication fatality involving Xyrem (GHB). *J. Forensic Sci.*, **54**:495–496.

Alderson, N. L., Rembiesa, B. M., Walla, M. D., Bielawska, A., Bielawski, J., and Hama, H. (2004). The human *FA2H* gene encodes a fatty acid 2-hydroxylase. *J. Biol. Chem.*, **279**:48562–48568.

Andriamampandry, C., Siffert, J.-C., Schmitt, M., Garnier, J.-M., Staub, A., Muller, C., Gobaille, S., Mark, J., and Maitre, M. (1998). Cloning of a rat brain succinic semialdehyde reductase involved in the synthesis of the neuromodulator γ-hydroxybutyrate. *Biochem. J.*, **334**:43–50.

Andriamampandry, C., Taleb, O., Viry, S., Muller, C., Humbert, J. P., Gobaille, S., Aunis, D., and Maitre, M. (2003). Cloning and characterization of a rat brain receptor that binds the endogenous neuromodulator gamma-hydroxybutyrate (GHB). *FASEB J.*, **17**:1691–1693.

Andriamampandry, C., Taleb, O., Kemmel, V., Humbert, J.-P., Aunis, D., and Maitre, M. (2007). Cloning and functional characterization of a gamma-hydroxybutyrate receptor identified in the human brain. *FASEB J.*, **21**:885–895.

Arena, C., and Fung, H.-L. (1980). Absorption of sodium γ-hydroxybutyrate and its prodrug γ-butyrolactone: relationship between in vitro transport and in vivo absorption. *J. Pharm. Sci.*, **69**:356–358.

Aubert, C., Guenata, Z., Ambid, C., and Baumes, R. (2003). Changes in physicochemical characteristics and volatile constituents of yellow- and white-fleshed nectarines during maturation and artificial ripening. *J. Agric. Food Chem.*, **51**:3083–3091.

Bach, B., Meudec, E., Lepoutre, J.-P., Rossignol, T., Blondin, B., Dequin, S., and Camarasa, C. (2009). New insights into γ-aminobutyric acid catabolism: evidence for γ-hydroxybutyric acid and polyhydroxybutyrate synthesis in *Saccharomyces cerevisiae*. *Appl. Environ. Microbiol.*, **75**:4231–4239.

Baker, L. E., Searcy, G. D., Pynnonen, D. M., and Poling A. (2008). Differentiating the discriminative stimulus effects of gamma-hydroxybutyrate and ethanol in a three-choice drug discrimination procedure in rats. *Pharmacol. Biochem. Behav.*, **89**:598–607.

Balut, C., vande Ven, M., Despa, S., Lambrichts, I., Ameloot, M., Steels, P., and Smets, I. (2008). Measurement of cytosolic and mitochondrial pH in living cells during reversible metabolic inhibition. *Kidney Int.*, **73**:226–232.

Barbaccia, M. L., Colombo, G., Affricano, D., Carai, M. A. M., Vacca, G., Melis, S., Purdy, R. H., and Gessa, G. L. (2002). GABA(B) receptor-mediated increase of neurosteroids by gamma-hydroxybutyric acid. *Neuropharmacology*, **42**:782–791.

Barker, S. A., Snead, O. C., Poldrugo, F., Liu, C. C., Fish, F. P., and Settine, R. L. (1985). Identification and quantitation of 1,4-butanediol in mammalian tissues: an alternative biosynthetic pathway for gamma-hydroxybutyric acid. *Biochem. Pharmacol.*, **34**:1849–1852.

Beard, D. A., and Qian, H. (2005). Thermodynamic-based computational profiling of cellular regulatory control in hepatocyte metabolism. *Am. J. Physiol. Endocrinol. Metabol.*, **288**:E633–E644.

Benacerraf, B., and Paul, W. E. (1970). Hapten-carrier relation. *Ann. NY Acad. Sci.*, **169**:93–104.

Benavides, J., Rumigny, J. F., Bourguignon, J. J., Wermuth, C. G., Mandel, P., and Maitre, M. (1982). A high-affinity, sodium ion-dependent uptake system for γ-hydroxybutyrate in membrane vesicles prepared from rat brain. *J. Neurochem.*, **38**:1570–1575.

Bencharit, S., Morton, C. L., Xue, Y., Potter, P. M., and Redinbo, M. R. (2003). Structural basis of heroin and cocaine metabolism by a promiscuous human drug-processing enzyme. *Nat. Struct. Biol.*, **10**:349–356.

Bennett, W. R. M., Wilson, L. G., and Roy-Byrne, P. P. (2007). Gamma-hydroxybutyric acid (GHB) withdrawal: a case report. *J. Psychoactive Drugs*, **39**:293–296.

Berankova, K., Mutnanska, K., and Balikova, M. (2006). Gamma-hydroxybutyric acid stability and formation in blood and urine. *Forensic Sci. Int.*, **161**:158–162.

Bergelson, L. D. (1969). Diol lipids. *Prog. Chem. Fats Other Lipids*, **10**:241–286.

Berridge, M. V., Tan, A. S., and Herst, P. M. (2005). Tetrazolium dyes as tools in cell biology: new insights into their cellular reduction. *Biotechnol. Annu. Rev.*, **11**:127–152.

Bessman, S. P., and McCabe, E. R. B., III (1972). 1,4-Butanediol, a substrate for rat liver and horse liver alcohol dehydrogenases. *Biochem. Pharmacol.*, **21**:1135–1142.

Bettler, B., and Braeuner-Osborne, H. (2004). The GABA$_B$ receptor: from cloning to knockout mice. In: Schousboe, A., and Braeuner-Osborne, H. (Eds.), *Molecular Neuropharmacology*. Totowa, NJ: Humana Press, pp. 129–144.

Beynon, C. M., Sumnall, H. R., McVeigh, J., Cole, J. C., and Bellis, M. A. (2006). The ability of two commercially available quick test kits to detect drug-facilitated sexual assault drugs in beverages. *Addiction*, **101**:1413–1420.

Bhattacharya, I., and Boje, K. M. K. (2004). GHB (γ-Hydroxybutyrate) carrier-mediated transport across the blood–brain barrier. *J. Pharmacol. Exp. Ther.*, **311**:92–98.

Bhattacharya, I., and Boje, K. M. K. (2006). Potential γ-hydroxybutyric acid (GHB) drug interactions through blood–brain barrier transport inhibition: a pharmacokinetic simulation-based evaluation. *J. Pharmacokinet. Pharmacodyn.*, **33**:657–681.

Bhattacharya, I., and Boje, K. M. K. (2007). Feasibility of D-glucuronate to enhance γ-hydroxybutyric acid metabolism during γ-hydroxybutyric acid toxicity: pharmacokinetic and pharmacodynamic studies. *Biopharm. Drug Dispos.*, **28**:1–11.

Bhattacharya, I., Raybon, J. J., and Boje, K. M. K. (2006). Alterations in neuronal transport but not blood–brain barrier transport are observed during gamma-hydroxybutyrate (GHB) sedative/hypnotic tolerance. *Pharm. Res.*, **23**:2067–2077.

Biggio, G., Concas, A., Follesa, P., Sanna, E., and Serra, M. (2007). Stress, ethanol, and neuroactive steroids. *Pharmacol. Ther.*, **116**:140–171.

Billecke, S., Draganov, D., Counsell, R., Stetson, P., Watson, C., Hsu, C., and La Du, B. N. (2000). Human serum paraoxonase (PON1) isozymes Q and R hydrolyze lactones and cyclic carbonate esters. *Drug Metab. Dispos.*, **28**:1335–1341.

Bjornaas, M. A., Hovda, K. E., Mikalsen, H., Andrew, E., Rudberg, N., Ekeberg, O., and Jacobsen, D. (2006). Clinical vs. laboratory identification of drugs of abuse in patients admitted for acute poisoning. *Clin. Toxicol.*, **44**:127–134.

Blanchet, B., Morand, K., Hulin, A., and Astier, A. (2002). Capillary gas chromatographic determination of 1,4-butanediol and γ-hydroxybutyrate in human plasma and urine. *J. Chromatogr. B*, **769**:221–226.

Blasi, P., Boyl, P. P., Ledda, M., Novelletto, A., Gibson, K. M., Jakobs, C., Hogema, B., Akaboshi, S., Loreni, F., and Malaspina, P. (2002). Structure of human succinic semialdehyde dehydrogenase gene: identification of promoter region and alternatively processed isoforms. *Mol. Genet. Metab.*, **76**:348–362.

Bodson, Q., Denooz, R., Serpe, P., and Charlier, C. (2008). Gamma-hydroxybutyric acid (GHB) measurement by GC–MS in blood, urine and gastric contents, following an acute intoxication in Belgium. *Acta Clin. Belg.*, **63**:200–208.

Borgen, L. A., Okerholm, R., Morrison, D., and Lai, A. (2003). The influence of gender and food on the pharmacokinetics of sodium oxybate oral solution in healthy subjects. *J. Clin. Pharmacol.*, **43**:59–65.

Bravo, D. T., Harris, D. O., and Parsons, S. M. (2004). Reliable, sensitive, rapid and quantitative enzyme-based assay for gamma-hydroxybutyric acid (GHB). *J. Forensic Sci.*, **49**:379–387.

Breitkreuz, K. E., Allan, W. L., Van Cauwenberghe, O. R., Jakobs, C., Talibi, D., Andre, B., and Shelp, B. J. (2003). A novel γ-hydroxybutyrate dehydrogenase: identification and expression of an *Arabidopsis* cDNA and potential role under oxygen deficiency. *J. Biol. Chem.*, **278**:41552–41556.

Brenneisen, R., and Raymond, L. (2001). Pharmacology of flunitrazepam and other benzodiazepines. In: Salamone, S. J. (Ed.), *Benzodiazepines and GHB*. Totowa, NJ: Humana Press, pp. 1–16.

Brenneisen, R., Elsohly, M. A., Murphy, T. P., Passarelli, J., Russmann, S., Salamone, S. J., and Watson, D. E. (2004). Pharmacokinetics and excretion of gamma-hydroxybutyrate (GHB) in healthy subjects. *J. Anal. Toxicol.*, **28**:625–630.

Britt, G. C., and McCance-Katz, E. F. (2005). A brief overview of the clinical pharmacology of "club drugs." *Subst. Use Misuse*, **40**:1189–1201.

Brown, G. K., Cromby, C. H., Manning, N. J., and Pollitt, R. J. (1987). Urinary organic acids in succinic semialdehyde dehydrogenase deficiency: evidence of alpha-oxidation of 4-hydroxybutyric acid, interaction of succinic semialdehyde with pyruvate dehydrogenase and possible secondary inhibition of mitochondrial beta-oxidation. *J. Inherit. Metab. Dis.*, **10**:367–375.

Bruice, T. C. (2002). A view at the millennium: the efficiency of enzymatic catalysis. *Acc. Chem. Res.*, **35**:139–148.

Buckel, W., and Miller, S. L. (1987). Equilibrium constants of several reactions involved in the fermentation of glutamate. *Eur. J. Biochem.*, **164**:565–569.

Bunik, V. I., and Degtyarev, D. (2008). Structure-function relationships in the 2-oxo acid dehydrogenase family: substrate-specific signatures and functional predictions for the 2-oxoglutarate dehydrogenase-like proteins. *Proteins*, **71**:874–890.

Camacho, A., Matthews, S. C., Murray, B., and Dimsdale, J. E. (2005). Use of GHB compounds among college students. *Am. J. Drug Alcohol Abuse*, **31**:601–607.

Caputo, F., Vignoli, T., Maremmani, I., Bernardi, M., and Zoli, G. (2009). Gamma-hydroxybutyric acid (GHB) for the treatment of alcohol dependence: A review. *Int. J. Environ. Res. Public Health*, **6**:1917–1929.

Carai, M. A. M., Colombo, G., Reali, R., Serra, S., Mocci, I., Castelli, M. P., Cignarella, G., and Gessa, G. L. (2002). Central effects of 1,4-butanediol are mediated by GABA$_B$ receptors via its conversion into γ-hydroxybutyric acid. *Eur. J. Pharmacol.*, **441**:157–163.

Carai, M. A. M., Colombo, G., Quang, L. S., Maher, T. J., and Gessa, G. L. (2006). Resuscitative treatments on 1,4-butanediol mortality in mice. *Ann. Emerg. Med.*, **47**:184–189.

Carriere, E. (1921). The acid aldehydes of the succinic series. *Ann. Chim. Appl.*, **17**:38–132.

Carter, L. P., Koek, W., and France, C. P. (2006a). Lack of effects of GHB precursors GBL and 1,4-BD following i.c.v. administration in rats. *Eur. J. Neurosci.*, **24**:2595–2600.

Carter, L. P., Richards, B. D., Mintzer, M. Z., and Griffiths, R. R. (2006b). Relative abuse liability of GHB in humans: a comparison of psychomotor, subjective, and cognitive effects of supratherapeutic doses of traizolam, pentobarbital, and GHB. *Neuropsychopharmacology*, **31**:2537–2551.

Carter, L. P., Koek, W., and France, C. P. (2009). Behavioral analyses of GHB: receptor mechanisms. *Pharmacol. Ther.*, **121**:100–114.

Cary, J. W., Petersen, D. J., Papoutsakis, E. T., and Bennett, G. N. (1988). Cloning and expression of *Clostridium acetobutylicum* phosphotransbutyrylase and butyrate kinase genes in *Escherichia coli*. *J. Bacteriol.*, **170**:4613–4618.

Cash, C. D., Maitre, M., and Mandel, P. (1979). Purification from human brain and some properties of two NADPH-linked aldehyde reductases which reduce succinic semialdehyde to 4-hydroxybutyrate. *J. Neurochem.*, **33**:1169–1175.

Castelli, M. P. (2008). Multi-faceted aspects of gamma-hydroxybutyric acid: a neurotransmitter, therapeutic agent and drug of abuse. *Mini Rev. Med. Chem.*, **8**:1188–1202.

Chalmers, R. A., Healy, M. J. R., Lawson, A. M., and Watts, R. W. E. (1976). Urinary organic acids in man: II. Effects of individual variation and diet on the urinary excretion of acid metabolites. *Clin. Chem.*, **22**:1288–1291.

Chambliss, K. L., Zhang, Y.-A., Rossier, E., Vollmer, B., and Gibson, K. M. (1995a). Enzymic and immunologic identification of succinic semialdehyde dehydrogenase in rat and human neural and nonneural tissues. *J. Neurochem.*, **65**:851–855.

Chambliss, K. L., Caudle, D. L., Hinson, D. D., Moomaw, C. R., Slaughter, C. A., Jakobs, C., and Gibson, K. M. (1995b). Molecular cloning of the mature NAD^+-dependent succinic semialdehyde dehydrogenase from rat and human: cDNA isolation, evolutionary homology, and tissue expression. *J. Biol. Chem.*, **270**:461–467.

Child, A. M., and Child, P. (2007). Ability of commercially available date-rape drug test kits to detect gamma-hydroxybutyrate in popular drinks. *J. Can. Soc. Forensic Sci.*, **40**:131–141.

Chin, R. L., Sporer, K. A., Cullison, B., Dyer, J. E., and Wu, T. D. (1998). Clinical course of gamma-hydroxybutyrate overdose. *Ann. Emerg. Med.*, **31**:716–722.

Choi, J., Kim, C., and Choi, M. J. (1999). Influence of the antibody purification method on immunoassay performance: hapten–antibody binding in accordance with the structure of the affinity column ligand. *Anal. Biochem.*, **274**:118–124.

Chowdhury, G. M. I., Gupta, M., Gibson, K. M., Patel, A. B., and Behar, K. L. (2007). Altered cerebral glucose and acetate metabolism in succinic semialdehyde dehydrogenase-deficient mice: evidence for glial dysfunction and reduced glutamate/glutamine cycling. *J. Neurochem.*, **103**:2077–2091.

Ciolino, L. A., Mesmer, M. Z., Satzger, R. D., Machal, A. C., McCauley, H. A., and Mohrhaus, A. S. (2001). The chemical interconversion of GHB and GBL: forensic issues and implications. *J. Forensic Sci.*, **46**:1315–1323.

Claros, M. G., and Vincens, P. (1996). Computational method to predict mitochondrial proteins and their targeting sequences. *Eur. J. Biochem.*, **241**:779–786.

Clauwaert, K. M., Van Bocxlaer, J. F., and De Leenheer, A. P. (2001). Stability study of the designer drugs "MDA, MDMA and MDEA" in water, serum, whole blood, and urine under various storage temperatures. *Forensic Sci. Int.*, **124**:36–42.

Clejan, L. A., and Cederbaum, A. I. (1992). Structural determinants for alcohol substrates to be oxidized to formaldehyde by rat liver microsomes. *Arch. Biochem. Biophys.*, **298**:105–113.

Cleland, W. W. (1982). An analysis of Haldane relationships. *Methods Enzymol.*, **87**:366–369.

Colado, M. I., O'Shea, E., and Green, A. R. (2007). MDMA and other "club drugs." In: Sibley, D. R., Hanin, I., Kuhar, M., and Skolnick, P. (Eds.), *Handbook of Contemporary Neuropharmacology*, Vol. 2. Hoboken, NJ: Wiley, pp. 613–658.

Collier, B., and De Feudis, F. V. (1970). Conversion of gamma-hydroxybutyrate to gamma-aminobutyrate by mouse brain in vivo. *Experientia*, **26**:1072–1073.

Cone, E. J., and Huestis, M. A. (2007). Interpretation of oral fluid tests for drugs of abuse. *Ann. NY Acad. Sci.*, **1098**:51–103.

Coune, P., Taleb, O., Mensah-Nyagan, A. G., Maitre, M., and Kemmel, V. (2010). Calcium and cAMP signaling induced by gamma-hydroxybutyrate receptor(s) stimulation in NCB-20 neurons. *Neuroscience*, **167**:49–59.

Couper, F. J., Thatcher, J. E., and Logan, B. K. (2004). Suspected GHB overdoses in the emergency department. *J. Anal. Toxicol.*, **28**:481–484.

Crookes, C. E., Faulds, M. C., Forrest, A. R. W., and Galloway, J. H. (2004). A reference range for endogenous gamma-hydroxybutyrate in urine by gas chromatography–mass spectrometry. *J. Anal. Toxicol.*, **28**:644–649.

Crunelli, V., Emri, Z., and Leresche, N. (2006). Unravelling the brain targets of γ-hydroxybutyric acid. *Curr. Opin. Pharmacol.*, **6**:44–52.

Cuadrado, M. U., Ruiz, I. L., and Gomez-Nieto, M. A. (2007). QSAR models based on isomorphic and nonisomorphic data fusion for predicting the blood–brain barrier permeability. *J. Comput. Chem.*, **28**:1252–1260.

Cui, D., and Morris, M. E. (2009). The drug of abuse gamma-hydroxybutyrate is a substrate for sodium-coupled monocarboxylate transporter (SMCT) 1 (SLC5A8): characterization of SMCT-mediated uptake and inhibition. *Drug Metab. Dispos.*, **37**:1404–1410.

Cui, Y., Tian, M., Zong, M., Teng, M., Chen, Y., Lu, J., Jiang, J., Liu, X., and Han, J. (2009). Proteomic analysis of pancreatic ductal adenocarcinoma compared with normal adjacent pancreatic tissue and pancreatic benign cystadenoma. *Pancreatology*, **9**:89–98.

Davis, G., and Song, J. H. (2006). Biodegradable packaging based on raw materials from crop and their impact on waste management. *Ind. Crop Prod.*, **23**:147–161.

de la Torre, R., Farre, M., Roset, P. N., Pizarro, N., Abanades, S., Segura, M., Segura, J., and Cami, J. (2004). Human pharmacology of MDMA: pharmacokinetics, metabolism, and disposition. *Ther. Drug Monit.*, **26**:137–144.

de Oliveira, C. D. R., Roehsig, M., de Almeida, R. M., Rocha, W. L., and Yonamine, M. (2007). Recent advances in chromatographic methods to detect drugs of abuse in alternative biological matrices. *Curr. Pharm. Anal.*, **3**:95–109.

de Paoli, G., and Bell, S. (2008). A rapid GC-MS determination of gamma-hydroxybutyrate in saliva. *J. Anal. Toxicol.*, **32**:298–302.

Debreczeni, J. E., Lukacik, P., Kavanagh, K., Dubinina, E., Bray, J., Colebrook, S., Haroniti, A., Edwards, A., Arrowsmith, C., Sundstrom, M., Von Delft, F., Gileadi, O., and Oppermann, U. (2005). Structure of the aflatoxin aldehyde reductase in complex with NADPH. *Protein Data Bank*, 2BP1.

Deetz, J. S., Luehr, C. A., and Vallee, B. L. (1984). Human liver alcohol dehydrogenase isozymes: reduction of aldehydes and ketones. *Biochemistry*, **23**:6822–6828.

Deguchi, J., Maruyama, K., Tsukahara, T., Shirahama, R., inventors; Mitsubishi Chemical Corp, assignee. (2002). Quick landfill-disposal of biodegradable mulch films. Japan patent JP 2002348406, Dec. 4.

Deng, Y., Wang, Z., Gu, S., Ji, C., Ying, K., Xie, Y., and Mao, Y. (2002). Cloning and characterization of a novel human alcohol dehydrogenase gene (ADHFe1). *DNA Seq.*, **13**:301–306.

Devlin, R. J., and Henry, J. A. (2008). Clinical review: major consequences of illicit drug consumption. *Crit. Care*, **12**:202.

Doherty, J. D., Hattox, S. E., Snead, O. C., and Roth, R. H. (1978). Identification of endogenous γ-hydroxybutyrate in human and bovine brain and its regional distribution in human, guinea pig, and rhesus monkey brain. *J. Pharmacol. Exp. Ther.*, **207**:130–139.

Dornburg, V., Lewandowski, I., and Patel, M. (2004). Comparing the land requirements, energy savings, and greenhouse gas emissions reduction of biobased polymers and bioenergy: an analysis and system extension of life-cycle assessment studies. *J. Ind. Ecol.*, **7**:93–116.

Draganov, D. I., Teiber, J. F., Speelman, A., Osawa, Y., Sunahara, R., and La Du, B. N. (2005). Human paraoxonases (PON1, PON2, and PON3) are lactonases with overlapping and distinct substrate specificities. *J. Lipid Res.*, **46**:1239–1247.

Drasbek, K. R., Christensen, J., and Jensen, K. (2006). Gamma-hydroxybutyrate: a drug of abuse. *Acta Neurol. Scand.*, **114**:145–156.

Drewes, L. R., Gerhart, D. Z., Leino, R. L., and Enerson, B. E. (2001). Expression and modulation of blood–brain monocarboxylate transporters. In: Kobiler, D., Lustig, S., and Shapira, S. (Eds.), *Blood–Brain Barrier: Drug Delivery and Brain Physiology*. New York: Kluwer Academic/Plenum, pp. 9–17.

Drummer, O. H. (2007). Requirements for bioanalytical procedures in postmortem toxicology. *Anal. Bioanal. Chem.*, **388**:1495–1503.

Drummond, R. J., and Phillips, A. T. (1974). L-Glutamic acid decarboxylase in non-neural tissues of the mouse. *J. Neurochem.*, **23**:1207–1213.

Efe, C., Straathof, A. J. J., and van der Wielen, L. A. M. (2008). Options for biochemical production of 4-hydroxybutyrate and its lactone as a substitute for petrochemical production. *Biotechnol. Bioeng.*, **99**:1392–1406.

El Mahjoub, A., and Staub, C. (2000). Stability of benzodiazepines in whole blood samples stored at varying temperatures. *J. Pharm. Biomed. Anal.*, **23**:1057–1063.

El-Kabbani, O., Green, N. C., Lin, G., Carson, M., Narayana, S. V., Moore, K. M., Flynn, T. G., and DeLucas, L. J. (1994). Structures of human and porcine aldehyde reductase: an enzyme implicated in diabetic complications. *Acta Crystallogr. D*, **50**:859–868.

Elliott, S. P. (2003). Gamma-hydroxybutyric acid (GHB) concentrations in humans and factors affecting endogenous production. *Forensic Sci. Int.*, **133**:9–16.

Elliott, S. P. (2004). Nonfatal instances of intoxication with γ-hydroxybutyrate in the United Kingdom. *Ther. Drug Monit.*, **26**:432–440.

Elliott, S. M. (2008). Drug-facilitated sexual assault: educating women about the risks. *Nurs. Womens Health*, **12**:30–37.

Elliott, S., Lowe, P., and Symonds, A. (2004). The possible influence of micro-organisms and putrefaction in the production of GHB in post-mortem biological fluid. *Forensic Sci. Int.*, **139**:183–190.

Emi, Y., Sumiyoshi, Y., Oki, E., Kakeji, Y., Fukui, Y., and Maehara, Y. (2007). Pharmacokinetics of gamma-hydroxybutyric acid (GHB) and gamma-butyrolactone (GBL), the anti-angiogenic metabolites of oral fluoropyrimidine UFT, in patients with gastric cancer. *Fukuoka Igaku Zasshi*, **98**:418–424.

Enerson, B. E., and Drewes, L. R. (2003). Molecular features, regulation, and function of monocarboxylate transporters: implications for drug delivery. *J. Pharm. Sci.*, **92**:1531–1544.

Escobedo, I., O'Shea, E., Orio, L., Sanchez, V., Segura, M., de la Torre, R., Farre, M., Green, A. R., and Colado, M. I. (2005). A comparative study on the acute and long-term effects of MDMA and 3,4-dihydroxymethamphetamine (HHMA) on brain monoamine levels after i.p. or striatal administration in mice. *Br. J. Pharmacol.*, **144**:231–241.

Feist, A. M., Henry, C. S., Reed, J. L., Krummenacker, M., Joyce, A. R., Karp, P. D., Broadbelt, L. J., Hatzimanikatis, V., and Palsson, B. O. (2007). A genome-scale metabolic reconstruction for *Escherichia coli* K-12MG1655 that accounts for 1260 ORFs and thermodynamic information. *Mol. Syst. Biol.*, **3**:121.

Feng, J., Wang, L., Dai, I., Harmon, T., and Bernert, J. T. (2007). Simultaneous determination of multiple drugs of abuse and relevant metabolites in urine by LC-MS-MS. *J. Anal. Toxicol.*, **31**:359–368.

Fickert, H., Fransson, I. G., and Hahn, U. (2006). Aptamers to small molecules. In: Klussmann, S., (Ed.), *Aptamer Handbook*. Weinheim, Germany: Wiley-VCH, pp. 95–115.

Fishbein, W. N., and Bessman, S. P. (1966). Purification and properties of an enzyme in human blood and rat liver microsomes catalyzing the formation and hydrolysis of γ-lactones: II. Metal ion effects, kinetics, and equilibria. *J. Biol. Chem.*, **241**:4842–4847.

Funada, M., Sato, M., Aoo, N., and Wada, K. (2005). Modifications of gene expression by the abused drug. *Seitai Kagaku*, **56**:323–327.

Fung, H.-L., Haas, E., Raybon, J., Xu, J., and Fung, S.-M. (2004). Liquid chromatographic–mass spectrometric determination of endogenous γ-hydroxybutyrate concentrations in rat brain regions and plasma. *J. Chromatogr. B*, **807**:287–291.

Fung, H.-L., Tsou, P.-S., Bulitta, J. B., Tran, D. C., Page, N. A., Soda, D., and Fung, S. M. (2008). Pharmacokinetics of 1,4-butanediol in rats: bioactivation to γ-hydroxybutyric acid, interaction with ethanol, and oral bioavailability. *AAPS J.*, **10**:56–69.

Gable, R. S. (2004). Comparison of acute lethal toxicity of commonly abused psychoactive substances. *Addiction*, **99**:686–696.

Galperin, M. Y., Walker, D. R., and Koonin, E. V. (1998). Analogous enzymes: independent inventions in enzyme evolution. *Genome Res.*, **8**:779–790.

Ganapathy, V., Thangaraju, M., Gopal, E., Martin, P. M., Itagaki, S., Miyauchi, S., and Prasad, P. D. (2008). Sodium-coupled monocarboxylate transporters in normal tissues and in cancer. *AAPS J.*, **10**:193–199.

Gao, B., Kilic, E., Baumann, C. R., Hermann, D. M., and Bassetti, C. L. (2008). Gamma-hydroxybutyrate accelerates functional recovery after focal cerebral ischemia. *Cerebrovasc. Dis.*, **26**:413–419.

Gassmann, M., and Bettler, B. (2007). Metabotropic GABA receptors. In: Sibley, D. R., Hanin, I., Kuhar, M., and Skolnick, P. (Eds.) *Handbook of Contemporary Neuropharmacology*, Vol. 1. Hoboken, NJ: Wiley, pp. 569–615.

Gerak, L. R., Hicks, A. R., Winsauer, P. J., and Varner, K. J. (2004). Interaction between 1,4-butanediol and ethanol on operant responding and the cardiovascular system. *Eur. J. Pharmacol.*, **506**:75–82.

Gessa, G. L., Vargiu, L., Crabai, F., Adamo, F., Boero, G. C., and Camba, R. (1967). Effects of the injection of sodium γ-hydroxybutyrate and of γ-butyrolactone into various regions of the brain and cerebrospinal system: I. Behavioral and electroencephalographic effects. *Boll. Soc. Ital. Biol. Sper.*, **43**:283–286.

Gibson, K. M., and Nyhan, W. L. (1989). Metabolism of [U-^{14}C]-4-hydroxybutyric acid to intermediates of the tricarboxylic acid cycle in extracts of rat liver and kidney mitochondria. *Eur. J. Drug Metab. Pharmacokinet.*, **14**:61–70.

Gibson, K. M., Christensen, E., Jakobs, C., Fowler, B., Clarke, M. A., Hammersen, G., Raab, K., Kobori, J., Moosa, A., Vollmer, B., et al. (1997). The clinical phenotype of succinic semialdehyde dehydrogenase deficiency (4-hydroxybutyric aciduria): case reports of 23 new patients. *Pediatrics*, **99**:567–574.

Gibson, K. M., Sweetman, L., Kozich, V., Pijackova, A., Tscharre, A., Cortez, A., Eyskens, F., Jakobs, C., Duran, M., and Poll-The, B. T. (1998). Unusual enzyme findings in five patients with metabolic profiles suggestive of succinic semialdehyde dehydrogenase deficiency (4-hydroxybutyric aciduria). *J. Inherit. Metab. Dis.*, **21**:255–261.

Gibson, K. M., Schor, D. S. M., Gupta, M., Guerand, W. S., Senephansiri, H., Burlingame, T. G., Bartels, H., Hogema, B. M., Bottiglieri, T., Froestl, W., Snead, O. C., Grompe, M., and Jakobs, C. (2002). Focal neurometabolic alterations in mice deficient for succinate semialdehyde dehydrogenase. *J. Neurochem.*, **81**:71–79.

Gladden, L. B. (2007). Is there an intracellular lactate shuttle in skeletal muscle? *J. Physiol.*, **582**:899–899.

Gobaille, S., Hechler, V., Andriamampandry, C., Kemmel, V., and Maitre, M. (1999). Gamma-hydroxybutyrate modulates synthesis and extracellular concentration of gamma-aminobutyric acid in discrete rat brain regions in vivo. *J. Pharmacol. Exp. Ther.*, **290**:303–309.

Goddard, A. F. (1998). Factors influencing antibiotic transfer across the gastric mucosa. *Aliment. Pharmacol. Ther.*, **12**:1175–1184.

Gold, B. I., and Roth, R. H. (1977). Kinetics of in vivo conversion of γ-[^{3}H]aminobutyric acid to γ-[^{3}H]hydroxybutyric acid by rat brain. *J. Neurochem.*, **28**:1069–1073.

Goldberg, R. N., and Alberty, R. A. (2008). In: Lide, D. R. (Ed.), *CRC Handbook of Chemistry and Physics*. Boca Raton, FL: Taylor & Francis, pp. 7–10.

Goldberg, R. N., Tewari, Y. B., Bell, D., Fasio, K., and Anderson, E. (1993). Thermodynamics of enzyme catalyzed reactions: 1. Oxidoreductases. *J. Phys. Chem. Ref. Data*, **22**:515–582.

Gong, X. Y., Kuban, P., Scholer, A., and Hauser, P. C. (2008). Determination of gamma-hydroxybutyric acid in clinical samples using capillary electrophoresis with contactless conductivity detection. *J. Chromatogr. A*, **1213**:100–104.

Gonzalez, A., and Nutt, D. J. (2005). Gamma-hydroxybutyrate abuse and dependency. *J. Psychopharmacol.*, **19**:195–204.

Goodwin, A. K., Brown, P. R., Jansen, E. E. W., Jakobs, C., Gibson, K. M., and Weerts, E. M. (2009). Behavioral effects and pharmacokinetics of gamma-hydroxybutyrate (GHB) precursors gamma-butyrolactone (GBL) and 1,4-butanediol (1,4-BD) in baboons. *Psychopharmacology*, **204**:465–476.

Grant, A., Staffas, L., Mankowiz, L., Kelly, V., Manson, M. M., DePierre, J. W., Hayes, J. D., and Ellis, E. M. (2001). Expression of rat aldehyde reductase AKR7A1: influence of age and sex and tissue-specific inducibility. *Biochem. Pharmacol.*, **62**:1511–1519.

Greer, D. (2006). Plastic from plants, not petroleum. *Biocycle*, **47**:43–45.

Griffiths, R. R., and Johnson, M. W. (2005). Relative abuse liability of hypnotic drugs: a conceptual framework and algorithm for differentiating among compounds. *J. Clin. Psychiatry*, **66**(Suppl. 9):31–41.

Gross, R. A., and Kalra, B. (2002). Biodegradable polymers for the environment. *Science*, **297**:803–807.

Grossman, S. I., Campbell, J. G., Loane, C. J., inventors; Bloomsbury Innovations Ltd., assignee. (2005). Apparatus for detection of drugs in a beverage. UK patent GB 2411958, Sept. 14.

Gupta, M., Polinsky, M., Senephansiri, H., Snead, O. C., Jansen, E. E. W., Jakobs, C., and Gibson, K. M. (2004). Seizure evolution and amino acid imbalances in murine succinate semialdehyde dehydrogenase (SSADH) deficiency. *Neurobiol. Dis.*, **16**:556–562.

Halestrap, A. P., and Price, N. T. (1999). The proton-linked monocarboxylate transporter (MCT) family: structure, function and regulation. *Biochem. J.*, **343**:281–299.

Haller, C., Thai, D., Jacob, P., 3rd, and Dyer, J. E. (2006). GHB urine concentrations after single-dose administration in humans. *J Anal Toxicol* **30**:360–364.

Harper, R. D., Law, W., Parsons, S. M., and Romero-Perez, L. M. (2008). Development of a solid phase test strip for determination of gamma-hydroxybutyrate in urine. *Am. Assoc. Clin. Chem. Abstr. Titles Guide*, **E-16**:45–45.

Hartl, D. L. (1989). The physiology of weak selection. *Genome*, **31**:183–189.

Hartl, D. L., Dykhuizen, D. E., and Dean, A. M. (1985). Limits of adaptation: the evolution of selective neutrality. *Genetics*, **111**:655–674.

Hashimoto, T., and Brooks, G. A. (2008). Mitochondrial lactate oxidation complex and an adaptive role for lactate production. *Med. Sci. Sports Exerc.*, **40**:486–494.

Hashimoto, T., Hussien, R., Cho, H.-S., Kaufer, D., and Brooks, G. A. (2008). Evidence for the mitochondrial lactate oxidation complex in rat neurons: demonstration of an essential component of brain lactate shuttles. *PLoS ONE*, **3**:e2915.

Hauptmann, S., and Gabler, W. (1968). Reaction of tris(hydroxymethyl)aminomethane with some aldehydes. *Z. Naturforsch.*, **23**:111–112.

Hearl, W. G., and Churchich, J. E. (1984). Interactions between 4-aminobutyrate aminotransferase and succinic semialdehyde dehydrogenase, two mitochondrial enzymes. *J. Biol. Chem.*, **259**:11459–11463.

Hearl, W. G., and Churchich, J. E. (1985). A mitochondrial $NADP^+$-dependent reductase related to the 4-aminobutyrate shunt: purification, characterization, and mechanism. *J. Biol. Chem.*, **260**:16361–16366.

Hechler, V., Ratomponirina, C., and Maitre, M. (1997). γ-Hydroxybutyrate conversion into GABA induces displacement of $GABA_B$ binding that is blocked by valproate and ethosuximide. *J. Pharmacol. Exp. Ther.*, **281**:753–760.

Hedberg, J. J., Grafstrom, R. C., Vondracek, M., Sarang, Z., Warngard, L. and Hoog, J.-O. (2001). Micro-array chip analysis of carbonyl-metabolizing enzymes in normal, immortalized and malignant human oral keratinocytes. *Cell. Mol. Life Sci.*, **58**:1719–1726.

Helms, C. M., Rogers, L. S. M., Waters, C. A., and Grant, K. A. (2008). Zolpidem generalization and antagonism in male and female cynomolgus monkeys trained to discriminate 1.0 or 2.0g/kg ethanol. *Alcohol Clin. Exp. Res.*, **32**:1197–1206.

Henry, C. S., Broadbelt, L. J., and Hatzimanikatis, V. (2007). Thermodynamics-based metabolic flux analysis. *Biophys. J.*, **92**:1792–1805.

Herd, M. B., Belelli, D., Lambert, J. J. (2007). Neurosteroid modulation of synaptic and extrasynaptic $GABA_A$ receptors. *Pharmacol. Ther.*, **116**:20–34.

Hicks, A. R., and Varner, K. J. (2008). Cardiovascular responses elicited by intragastric administration of BDL and GHB. *J. Recept. Signal Transduct.*, **28**:429–436.

Hijazi, Y., Bolon, M., and Boulieu, R. (2001). Stability of ketamine and its metabolites norketamine and dehydronorketamine in human biological samples. *Clin. Chem.*, **47**:1713–1715.

Hinshelwood, A., McGarvie, G., and Ellis, E. (2002). Characterisation of a novel mouse liver aldo-keto reductase AKR7A5. *FEBS Lett.*, **523**:213–218.

Hinshelwood, A., McGarvie, G., and Ellis, E. M. (2003). Substrate specificity of mouse aldo-keto reductase AKR7A5. *Chem. Biol. Interact.*, **143–144**, 263–269.

Hipolito, L., Sanchez, M. J., Polache, A., and Granero, L. (2007). Brain metabolism of ethanol and alcoholism: an update. *Curr. Drug Metab.*, **8**:716–727.

Hoffmann, R., and Valencia, A. (2004). A gene network for navigating the literature. *Nat. Genet.*, **36**:664.

Hogema, B. M., Gupta, M., Senephansiri, H., Burlingame, T. G., Taylor, M., Jakobs, C., Schutgens, R. B. H., Froestl, W., Snead, O. C., Diaz-Arrastia, R., Bottiglieri, T., Grompe, M., and Gibson, K. M. (2001). Pharmacologic rescue of lethal seizures in mice deficient in succinate semialdehyde dehydrogenase. *Nat. Genet.*, **29**:212–216.

Holmgren, P., Druid, H., Holmgren, A., and Ahlner, J. (2004). Stability of drugs in stored postmortem femoral blood and vitreous humor. *J. Forensic Sci.*, **49**:820–825.

Hoover, G. J., Van Cauwenberghe, O. R., Breitkreuz, K. E., Clark, S. M., Merrill, A. R., and Shelp, B. J. (2007). Characteristics of an *Arabidopsis* glyoxylate reductase: general biochemical properties and substrate specificity for the recombinant protein, and developmental expression and implications for glyoxylate and succinic semialdehyde metabolism in planta. *Can. J. Bot.*, **85**:883–895.

Horowitz, A., Galanter, M., Dermatis, H., and Franklin, J. (2008). Use of and attitudes toward club drugs by medical students. *J. Addict. Dis.*, **27**:35–42.

Hovda, K. E., Bjornaas, M. A., Skog, K., Opdahl, A., Drottning, P., Ekeberg, O., and Jacobsen, D. (2008). Acute poisonings treated in hospitals in Oslo: a one-year prospective study: I. Pattern of poisoning. *Clin. Toxicol.*, **46**:35–41.

http://bioinfo2.weizmann.ac.il/cgi-bin/genenote/home_page.pl. GeneNote. Gene Normal Tissue Expression. Accessed July 10, 2009.

http://genome.ucsc.edu. UCSC Genome Bioinformatics. Accessed July 11, 2009.

http://openregs.com/regulations/view/103806/exempt_chemical_mixtures_containing_gamma-butyrolactone.

http://pfam.sanger.ac.uk/family?acc=PF00067. Family: P450 (PF00067). Accessed Feb. 18, 2010.

http://pfam.sanger.ac.uk/family?acc=PF00465. Family: Fe-ADH (PF00465). Accessed July 15, 2008.

http://toxnet.nlm.nih.gov/. TOXNET: Toxicology Data Network. Accessed Apr. 22, 2008.

http://www.astm.org/Standard/index.shtml. ASTM: Standards. Accessed June 30, 2008.

http://www.biocompare.com/jump/2831/Immunoassays.html. Biocompare: Assay Kit Search. Accessed June 18, 2008.

http://www.bioplastics24.com/. Bioplastics24.com. Accessed June 30, 2008.

http://www.brenda-enzymes.info/. BRENDA: The Comprehensive Enzyme Information System. Accessed July 15, 2008.

http://www.cas.org/products/sfacad/index.html. Accessed Feb. 18, 2010.

http://www.drugabuse.gov/Infofacts/clubdrugs.html. National Institute on Drug Abuse: NIDA InfoFacts: Club Drugs. Accessed Apr. 22, 2008.

http://www.genecards.org/cgi-bin/carddisp.pl?gene=ADHFE1&search=adhfe1. Accessed July 13, 2009.

http://www.genome.ad.jp/kegg/pathway.html. KEGG Pathway Database: Wiring diagrams of molecular interactions, reactions, and relations. Accessed July 15, 2008.

http://www.ihop-net.org/. Hoffmann R, Valencia A. Information Hyperlinked Over Proteins. Accessed May 2, 2008.

http://www.med.upenn.edu/akr. Hyndman, D., and Penning, T. M. AKR Superfamily. Accessed May 2, 2008.

http://www.metabolibrary.ca/. Human Metabolome Library. Accessed Apr. 3, 2009.

http://www.ncbi.nlm.nih.gov/IEB/Research/Acembly/. National Center for Biotechnology Information: AceView. Accessed July 29, 2008.

http://www.ncbi.nlm.nih.gov/. National Center for Biotechnology Information: Protein Database. Accessed May 15, 2008.

http://www.rcsb.org/pdb/home/home.do. Protein Data Bank: An Information Portal to Biological Macromolecular Structures. Accessed May 15, 2008.

http://www.uniprot.org/uniprot/Q8IWW8#Q8IWW8-2. Accessed July 13, 2009.

http://www.uniprot.org/. UniProt Consortium: Protein Knowledge Base. Accessed July 15, 2008.

http://www.varianinc.com/cgi-bin/nav?products/dat/index&cid=KJOQJOKKFJ. Varian: diagnostic products. Accessed June 18, 2008.

Huber, H., Gallenberger, M., Jahn, U., Eylert, E., Berg, I. A., Kockelkorn, D., Eisenreich, W., and Fuchs, G. (2008). A dicarboxylate/4-hydroxybutyrate autotrophic carbon assimilation cycle in the hyperthermophilic *Archaeum ignicoccus hospitalis*. *Proc. Natl. Acad. Sci. USA*, **105**:7851–7856.

Hulo, N., Bairoch, A., Bulliard, V., Cerutti, L., De Castro, E., Langendijk-Genevaux, P. S., Pagni, M., and Sigrist, C. J. (2006). The PROSITE database. *Nucleic Acids Res.*, **34**:D227–D230.

Humphrey, A. J., and O'Hagan, D. (2001). Tropane alkaloid biosynthesis: a century old problem unresolved. *Nat. Prod. Rep.*, **18**:494–502.

Hurd, C. D., and Saunders, W. H., Jr. (1952). Ring-chain tautomerism of hydroxy aldehydes. *J. Am. Chem. Soc.*, **74**:5324–5329.

Hyndman, D., Bauman, D. R., Heredia, V. V., and Penning, T. M. (2003). The aldo-keto reductase superfamily homepage. *Chem. Biol. Interact.*, **143–144**:621–631.

Ikonomidou, C., Bittigau, P., Ishimaru, M. J., Wozniak, D. F., Koch, C., Genz, K., Price, M. T., Stefovska, V., Hörster, F., Tenkova, T., Dikranian, K., and Olney, J. W. (2000). Ethanol-induced apoptotic neurodegeneration and fetal alcohol syndrome. *Science*, **287**:1056–1060.

Ingels, M., Rangan, C., Bellezzo, J., and Clark, R. F. (2000). Coma and respiratory depression following the ingestion of GHB and its precursors: three cases. *J. Emerg. Med.*, **19**:47–50.

Ireland, L. S., Harrison, D. J., Neal, G. E., and Hayes, J. D. (1998). Molecular cloning, expression and catalytic activity of a human AKR7 member of the aldo-keto reductase superfamily: evidence that the major 2-carboxybenzaldehyde reductase from human liver is a homolog of rat aflatoxin B1-aldehyde reductase. *Biochem. J.*, **332**:21–34.

Irwin, R. D. (2006). A review of evidence leading to the prediction that 1,4-butanediol is not a carcinogen. *J. Appl. Toxicol.*, **26**:72–80.

Iwanaga, T., Takebe, K., Kato, I., Karaki, S.-I., and Kuwahara, A. (2006). Cellular expression of monocarboxylate transporters (MCT) in the digestive tract of the mouse, rat, and humans, with special reference to slc5a8. *Biomed. Res.*, **27**:243–254.

Jaeger, M., Rothacker, B., and Ilg, T. (2008). Saturation transfer difference NMR studies on substrates and inhibitors of succinic semialdehyde dehydrogenases. *Biochem. Biophys. Res. Commun.*, **372**:400–406.

Jakobs, C., Bojasch, M., Manch, E., Rating, D., Siemes, H., and Hanefeld, F. (1981). Urinary excretion of gamma-hydroxybutyric acid in a patient with neurological abnormalities: the probability of a new inborn error of metabolism. *Clin. Chim. Acta*, **111**:169–178.

Jin, Y., and Penning, T. M. (2007). Aldo-keto reductases and bioactivation/detoxication. *Annu. Rev. Pharmacol. Toxicol.*, **47**:263–292.

Johnston, G. A. R., Chebib, M., Hanrahan, J. R., and Mewett, K. N. (2003). GABA$_C$ receptors as drug targets. *Curr. Drug. Targets CNS Neurol. Disord.*, **2**:260–268.

Jonasson, B., Jonasson, U., and Saldeen, T. (2000). Among fatal poisonings dextropropoxyphene predominates in younger people, antidepressants in the middle aged and sedatives in the elderly. *J. Forensic Sci.*, **45**:7–10.

Jones, A. W., Holmgren, A., and Kugelberg, F. C. (2007). Gamma-hydroxybutyrate concentrations in the blood of impaired drivers, users of illicit drugs, and medical examiner cases. *J. Anal. Toxicol.*, **31**:566–572.

Jones, A. W., Eklund, A., and Kronstrand, R. (2009). Case report: Concentration–time profiles of gamma-hydroxybutyrate in blood after recreational doses are best described by zero-order rather than first-order kinetics. *J. Anal. Toxicol.*, **33**:332–335.

Juhascik, M. P., Negrusz, A., Faugno, D., Ledray, L., Greene, P., Lindner, A., Haner, B., and Gaensslen, R. E. (2007). An estimate of the proportion of drug-facilitation of sexual assault in four U.S. localities. *J. Forensic Sci.*, **52**:1396–1400.

Kale, G., Kijchavengkul, T., Auras, R., Rubino, M., Selke, S. E., and Singh, S. P. (2007). Compostability of bioplastic packaging materials: an overview. *Macromol. Biosci.*, **7**:255–277.

Kang, J. H., Park, Y. B., Huh, T.-L., Lee, W.-H., Choi, M.-S., and Kwon, O.-S. (2005). High-level expression and characterization of the recombinant enzyme, and tissue distribution of human succinic semialdehyde dehydrogenase. *Protein Expr. Purif.*, **44**:16–22.

Kapadia, R., Böhlke, M., and Maher, T. J. (2007). Detection of γ-hydroxybutyrate in striatal microdialysates following peripheral 1,4-butanediol administration in rats. *Life Sci.*, **80**:1046–1050.

Karch, S. B., Stephens, B. G., and Nazareno, G. V. (2001). GHB. Club drug or confusing artifact? *Am. J. Forensic Med. Pathol.*, **22**:266–269.

Kardon, T., Noël, G., Vertommen, D., and Schaftingen, E. V. (2006). Identification of the gene encoding hydroxyacid–oxoacid transhydrogenase, an enzyme that metabolizes 4-hydroxybutyrate. *FEBS Lett.*, **580**:2347–2350.

Kasprzak, K., Adamowicz, P., and Kala, M. (2006). Determination of gamma-hydroxybutyrate (GHB) in urine by gas chromatography–mass spectrometry with positive chemical ionisation (PCI-GC-MS). *Z. Zagadnien. Nauk. Sadowych.*, **67**:289–300.

Kastrissios, H., Gaensslen, R. E., and Negrusz, A. (2005). Analytical issues in forensic toxicokinetics. *Z. Zagadnien. Nauk. Sadowych.*, **64**:382–394.

Kaufman, E. E., and Nelson, T. (1991). An overview of γ-hydroxybutyrate catabolism: the role of the cytosolic NADP$^+$-dependent oxidoreductase EC 1.1.1.19 and of a mitochondrial hydroxyacid-oxoacid transhydrogenase in the initial, rate-limiting step in this pathway. *Neurochem. Res.*, **16**:965–974.

Kaufman, E. E., Nelson, T., Goochee, C. F., and Sokoloff, L. (1979). Purification and characterization of an NADP$^+$ linked alcohol oxido-reductase which catalyzes the interconversion of γ-hydroxybutyrate and succinic semialdehyde. *J. Neurochem.*, **32**:699–712.

Kaufman, E. E., Nelson, T., Fales, H. M., and Levin, D. M. (1988). Isolation and characterization of a hydroxyacid–oxoacid transhydrogenase from rat kidney mitochondria. *J. Biol. Chem.*, **263**:16872–16879.

Kavanagh, P. V., Kenny, P., and Feely, J. (2001). The urinary excretion of γ-hydroxybutyric acid in man. *J. Pharm. Pharmacol.*, **53**:399–402.

Keenan, C., Ghaffar, S., Grant, A. W., Hinshelwood, A., Li, D., McGarvie, G., and Ellis, E. M. (2006). Succinic semialdehyde reductases: contribution to γ-hydroxybutyrate catabolism and subcellular localization. In: Weiner, H. (Ed.), *Enzymology and Molecular Biology of Carbonyl Metabolism*, Vol. 12. West Lafayette, IN: Purdue University Press, pp. 388–395.

Kelly, V., Ireland, L., Ellis, E., and Hayes, J. (2000). Purification from rat liver of a novel constitutively expressed member of the aldo-keto reductase 7 family that is widely distributed in extrahepatic tissues. *Biochem. J.*, **348**:389–400.

Kelly, V. P., Sherratt, P. J., Crouch, D. H., and Hayes, J. D. (2002). Novel homodimeric and heterodimeric rat γ-hydroxybutyrate synthases that associate with the Golgi apparatus define a distinct subclass of aldo-keto reductase 7 family proteins. *Biochem. J.*, **366**:847–861.

Kelly, V. P., O'Connor, T., Ellis, E. M., Ireland, L. S., Slattery, C. M., Sherratt, P. J., Crouch, D. H., Cavin, C., Schilter, B., Gallina, A., and Hayes, J. D. (2004). Aflatoxin aldehyde reductases. *ACS Symp. Ser.*, **865**:155–170.

Kemmel, V., Taleb, O., Perard, A., Andriamampandry, C., Siffert, J. C., Mark, J., and Maitre, M. (1998). Neurochemical and electrophysiological evidence for the existence of a functional γ-hydroxybutyrate system in NCB-20 neurons. *Neuroscience*, **86**:989–1000.

Kemmel, V., Miehe, M., Roussel, G., Taleb, O., Nail-Boucherie, K., Marchand, C., Stutz, C., Andriamampandry, C., Aunis, D., and Maitre, M. (2006). Immunohistochemical localization of a GHB receptor-like protein isolated from rat brain. *J. Comp. Neurol.*, **498**:508–524.

Khizhnyak, A. A. (1976). Distribution of sodium hydroxybutyrate in the body during experimental hemorrhagic shock. *Sb. Nauchn. Rab. Reanimatol. (Mater. Nauchn. Konf.)*, pp. 156–158.

Kim, J. Y., Tillison, K. S., Zhou, S., Lee, J. H., and Smas, C. M. (2007a). Differentiation-dependent expression of Adhfe1 in adipogenesis. *Arch. Biochem. Biophys.*, **464**:100–111.

Kim, S., Anderson, I. B., Dyer, J. E., Barker, J. C., and Blanc, P. D. (2007b). High-risk behaviors and hospitalizations among gamma-hydroxybutyrate (GHB) users. *Am. J. Drug Alcohol Abuse*, **33**:429–438.

Kim, Y.-G., Lee, S., Kwon, O.-S., Park, S.-Y., Lee, S.-J., Park, B.-J., and Kim, K.-J. (2009). Redox-switch modulation of human SSADH by dynamic catalytic loop. *EMBO J.*, **28**:959–968.

Kintz, P. (2007). Bioanalytical procedures for detection of chemical agents in hair in the case of drug-facilitated crimes. *Anal. Bioanal. Chem.*, **388**:1467–1474.

Klein, C., Kemmel, V., Taleb, O., Aunis, D., and Maitre, M. (2009). Pharmacological doses of gamma-hydroxybutyrate (GHB) potentiate histone acetylation in the rat brain by histone deacetylase inhibition. *Neuropharmacology*, **57**:137–147.

Kleinzeller, A. (1944). The metabolism of butyric and related acids in animal tissue. *Biochem. J.*, **37**:678–682.

Knerr, I., Pearl, P. L., Bottiglieri, T., Snead, O. C., Jakobs, C., and Gibson, K. M. (2007). Therapeutic concepts in succinate semialdehyde dehydrogenase (SSADH; ALDH5a1) deficiency (gamma-hydroxybutyric aciduria): hypotheses evolved from 25 years of patient evaluation, studies in Aldh5a1-/- mice and characterization of gamma-hydroxybutyric acid pharmacology. *J. Inherit. Metab. Dis.*, **30**:279–294.

Knudsen, K., Greter, J., and Verdicchio, M. (2008). High mortality rates among GHB abusers in western Sweden. *Clin. Toxicol.*, **46**:187–192.

Koelker, S., Sauer, S. W., Hoffmann, G. F., Mueller, I., Morath, M. A., and Okun, J. G. (2008). Pathogenesis of CNS involvement in disorders of amino and organic acid metabolism. *J. Inherit. Metab. Dis.*, **31**:194–204.

Kohlweyer, U., Thiemer, B., Schrader, T., and Andreesen, J. R. (2000). Tetrahydrofuran degradation by a newly isolated culture of *Pseudonocardia* sp. strain K1. *FEMS Microbiol. Lett.*, **186**:301–306.

Kraemer, T., and Paul, L. D. (2007). Bioanalytical procedures for determination of drugs of abuse in blood. *Anal. Bioanal. Chem.*, **388**:1415–1435.

Krajewski, D., Neugebauer, W., Amajoyi, I. K., Schreier, P., and Bicchi, C. (1995). Enantiodifferentiation of solerol isomers in dried figs and dates. *Z. Lebensm. Unters. Forsch.*, **201**:378–380.

Kueh, D., Iwamoto, K., Poling, A., and Baker, L. E. (2008). Effects of γ-hydroxybutyrate (GHB) and its metabolic precursors on delayed-matching-to-position performance in rats. *Pharmacol. Biochem. Behav.*, **89**:179–187.

Kugler, P. (1993). In situ measurements of enzyme activities in the brain. *Histochem. J.*, **25**:329–338.

Kunsman, G. W., Levine, B., Kuhlman, J. J., Jones, R. L., Hughes, R. O., Fujiyama, C. I., and Smith, M. L. (1996). MDA–MDMA concentrations in urine specimens. *J. Anal. Toxicol.*, **20**:517–521.

Kurdikar, D., Fournet, L., Slater, S. C., Paster, M., Gerngross, T. U., and Coulon, R. (2000). Greenhouse gas profile of a plastic material derived from a genetically modified plant. *J. Ind. Ecol.*, **4**:107–122.

Ladona, M. G., Gonzalez, M. L., Rane, A., Peter, R. M., and De la Torre, R. (2000). Cocaine metabolism in human fetal and adult liver microsomes is related to cytochrome P450 3A expression. *Life Sci.*, **68**:431–443.

Lalonde, B. R., and Wallage, H. R. (2004). Postmortem blood ketamine distribution in two fatalities. *J. Anal. Toxicol.*, **28**:71–74.

Lane, S. D., Cherek, D. R., and Nouvion, S. O. (2008). Modulation of human risky decision making by flunitrazepam. *Psychopharmacology (Berl.)*, **196**:177–188.

Langman, L. J. (2007). The use of oral fluid for therapeutic drug management: clinical and forensic toxicology. *Ann. NY Acad. Sci.*, **1098**:145–166.

Larson, S. J., Putnam, E. A., Schwanke, C. M., and Pershouse, M. A. (2007). Potential surrogate markers for gamma-hydroxybutyrate administration may extend the detection window from 12 to 48 hours. *J. Anal. Toxicol.*, **31**:15–22.

Lawyer, S., Resnick, H., Bakanic, V., Burkett, T., and Kilpatrick, D. (2010). Forcible, drug-facilitated, and incapacitated rape and sexual assault among undergraduate women. *J. Am. Coll. Health*, **58**:453–460.

Le Henaff, P. (1967). Equilibria and rates of hydration and hemiacetalization of aliphatic aldehydes. *C. R. Seances Acad. Sci. C*, **265**:175–178.

LeBeau, M. A., Montgomery, M. A., Morris-Kukoski, C., Schaff, J. E., and Deakin, A. (2006). A comprehensive study on the variations in urinary concentrations of endogenous gamma-hydroxybutyrate (GHB). *J. Anal. Toxicol.*, **30**:98–105.

LeBeau, M. A., Montgomery, M. A., Morris-Kukoski, C., Schaff, J. E., and Deakin, A. (2007). Further evidence of in vitro production of gamma-hydroxybutyrate (GHB) in urine samples. *Forensic Sci. Int.*, **169**:152–156.

Lee, C. R. (1977). Evidence for the β-oxidation of orally administered 4-hydroxybutyrate in humans. *Biochem. Med.*, **17**:284–291.

Lee, S. J., and Levounis, P. (2008). Gamma-hydroxybutyrate: an ethnographic study of recreational use and abuse. *J. Psychoactive Drugs*, **40**:245–253.

Lee, S.-L., Chau, G.-Y., Yao, C.-T., Wu, C.-W., and Yin, S.-J. (2006). Functional assessment of human alcohol dehydrogenase family in ethanol metabolism: significance of first-pass metabolism. *Alcohol Clin. Exp. Res.*, **30**:1132–1142.

Lenz, D., Rothschild, M. A., and Kroner, L. (2008). Intoxications due to ingestion of gamma-butyrolactone: organ distribution of gamma-hydroxybutyric acid and gamma-butyrolactone. *Ther. Drug Monit.*, **30**:755–761.

LeTourneau, J. L., Hagg, D. S., and Smith, S. M. (2008). Baclofen and gamma-hydroxybutyrate withdrawal. *Neurocrit. Care*, **8**:430–433.

Lettieri, J., and Fung, H.-L. (1976). Absorption and first-pass metabolism of ^{14}C-gamma-hydroxybutyric acid. *Res. Commun. Chem. Pathol. Pharmacol.*, **13**:425–437.

Levitt, M. D., Li, R., DeMaster, E. G., Elson, M., Furne, J., and Levitt, D. G. (1997). Use of measurements of ethanol absorption from stomach and intestine to assess human ethanol metabolism. *Am. J. Physiol.*, **273**(4, Pt. 1):G951–G957.

Liechti, M. E., Kunz, I., Greminger, P., Speich, R., and Kupferschmidt, H. (2006). Clinical features of gamma-hydroxybutyrate and gamma-butyrolactone toxicity and concomitant drug and alcohol use. *Drug Alcohol Depend.*, **81**:323–326.

Lin, S.-J., and Guarente, L. (2003). Nicotinamide adenine dinucleotide, a metabolic regulator of transcription, longevity and disease. *Curr. Opin. Cell Biol.*, **15**:241–246.

Lingenhoehl, K., Brom, R., Heid, J., Beck, P., Froestl, W., Kaupmann, K., Bettler, B., and Mosbacher, J. (1999). Gamma-hydroxybutyrate is a weak agonist at recombinant GABA(B) receptors. *Neuropharmacology*, **38**:1667–1673.

Liu, W., Shen, M., Liu, X.-Q., Shen, B.-H., and Xiang, P. (2006). Study on determination and distribution of γ-hydroxybutyric acid in biological fluids and tissues of acute poisoned rats. *Fayixue Zazhi*, **22**:55–57.

Lora-Tamayo, C., Tena, T., Rodriguez, A., Sancho, J. R., and Molina, E. (2003). Intoxication due to 1,4-butanediol. *Forensic Sci. Int.*, **133**:256–259.

Louagie, H. K., Verstraete, A. G., De Soete, C. J., Baetens, D. G., and Calle, P. A. (1997). A sudden awakening from a near coma after combined intake of γ-hydroxybutyric acid (GHB) and ethanol. *J. Toxicol. Clin. Toxicol.*, **35**:591–594.

Lowe, X. R., Lu, X., Marchetti, F., and Wyrobek, A. J. (2007). The expression of troponin T1 gene is induced by ketamine in adult mouse brain. *Brain Res.*, **1174**:7–17.

Lyon, R. C., Johnston, S. M., Watson, D. G., McGarvie, G., and Ellis, E. M. (2007). Synthesis and catabolism of gamma-hydroxybutyrate in SH-SY5Y human neuroblastoma cells: role of the aldo-keto reductase AKR7A2. *J. Biol. Chem.*, **282**:25986–25992.

Lyon, R. C., Johnston, S. M., Panopoulos, A., Alzeer, S., McGarvie, G., and Ellis, E. M. (2009). Enzymes involved in the metabolism of γ-hydroxybutyrate in SH-SY5Y cells: identification of an iron-dependent alcohol dehydrogenase ADHFe1. *Chem. Biol. Interact.*, **178**:283–287.

MacMillan, V. (1979). A comparison of the effects of γ-hydroxybutyrate and γ-butyrolactone on cerebral carbohydrate metabolism. *Can. J. Physiol. Pharmacol.*, **57**:787–797.

Magner, E., and Klibanov, A. M. (1995). The oxidation of chiral alcohols catalyzed by catalase in organic solvents. *Biotechnol. Bioeng.*, **46**:175–179.

Maitre, M. (1997). The γ-hydroxybutyrate signaling system in brain: organization and functional implications. *Prog. Neurobiol.*, **51**:337–361.

Maitre, M., Andriamampandry, C., Kemmel, V., Schmidt, C., Hodé, Y., Hechler, V., and Gobaille, S., (2000). Gamma-hydroxybutyric acid as a signaling molecule in brain. *Alcohol*, **20**:277–283.

Malaspina, P., Picklo, M. J., Jakobs, C., Snead, O. C., and Gibson, K. M. (2009). Comparative genomics of aldehyde dehydrogenase 5a1 (succinate semialdehyde dehydrogenase) and accumulation of gamma-hydroxybutyrate associated with its deficiency. *Hum. Genom.*, **3**:106–120.

Mamelak, M. (2007). Alzheimer's disease, oxidative stress and gamma-hydroxybutyrate. *Neurobiol. Aging*, **28**:1340–1360.

Marchitti, S. A., Brocker, C., Stagos, D., and Vasiliou, V. (2008). Non-P450 aldehyde oxidizing enzymes: the aldehyde dehydrogenase superfamily. *Expert Opin. Drug Metab. Toxicol.*, **4**:697–720.

Matsuo, N., and Morita, T. (2007). Efficacy, safety, and cost effectiveness of intravenous midazolam and flunitrazepam for primary insomnia in terminally ill patients with cancer: a retrospective multicenter audit study. *J. Palliat. Med.*, **10**:1054–1062.

Maurer, H. H. (2007). Current role of liquid chromatography–mass spectrometry in clinical and forensic toxicology. *Anal. Bioanal. Chem.*, **388**:1315–1325.

Maxwell, R., and Roth, R. H. (1972). Conversion of 1,4-butanediol to γ-hydroxybutyric acid in rat brain and in peripheral tissue. *Biochem. Pharmacol.*, **21**:1521–1533.

Mayevsky, A., and Chance, B. (2007). Oxidation–reduction states of NADH in vivo: from animals to clinical use. *Mitochondrion*, **7**:330–339.

Mazarr-Proo, S., and Kerrigan, S. (2005). Distribution of GHB in tissues and fluids following a fatal overdose. *J. Anal. Toxicol.*, **29**:398–400.

McCusker, R. R., Paget-Wilkes, H., Chronister, C. W., Goldberger, B. A., and ElSohly, M. A. (1999). Analysis of gamma-hydroxybutyrate (GHB) in urine by gas chromatography–mass spectrometry. *J. Anal. Toxicol.*, **23**:301–305.

Meldrum, B. S., and Rogawski, M. A. (2007). Molecular targets for antiepileptic drug development. *Neurotherapeutics*, **4**:18–61.

Meyers, J. E., and Almirall, J. R. (2004). A study of the effectiveness of commercially available drink test coasters for the detection of "date rape" drugs in beverages. *J. Anal. Toxicol.*, **28**:685–688.

Miller, L. M. (1979). Investigation of selected potential environmental contaminants: ethylene glycol, propylene glycols and butylene glycols. *US Environmental Protection Agency Report 560/11-79-006*, pp. 195–195.

Miller, A. L., and Pitts, F. N., Jr. (1967) Brain succinate semialdehyde dehydrogenase: III. Activities in twenty-four regions of the human brain. *J. Neurochem.*, **14**:579–584.

Mody, I. (2008). Extrasynaptic $GABA_A$ receptors in the crosshairs of hormones and ethanol. *Neurochem. Int.*, **52**:60–64.

Mody, I., Glykys, J., and Wei, W. (2007). A new meaning for "gin & tonic": tonic inhibition as the target for ethanol action in the brain. *Alcohol*, **41**:145–153.

Mohler, H., Patel, A. J., and Balazs, R. (1976). Gamma-hydroxybutyrate degradation in the brain in vivo: negligible direct conversion to GABA. *J. Neurochem.*, **27**:253–258.

Monroy-Noyola, A., Rojas, P., Vilanova, E., and Sogorb, M. A. (2007). Comparative hydrolysis of *O*-hexyl *O*-2,5-dichlorophenyl phosphoramidate and paraoxon in different tissues of vertebrates. *Arch. Toxicol.*, **81**:689–695.

Moore, T., Adhikari, R., and Gunatillake, P. (2005). Chemosynthesis of bioresorbable poly(γ-butyrolactone) by ring-opening polymerisation: a review. *Biomaterials*, **26**:3771–3782.

Moriya, F., and Hashimoto, Y. (2005). Site-dependent production of gamma-hydroxybutyric acid in the early postmortem period. *Forensic Sci. Int.*, **148**:139–142.

Morris, M. E., and Felmlee, M. A. (2008). Overview of the proton-coupled MCT (SLC16A) family of transporters: characterization, function and role in the transport of the drug of abuse gamma-hydroxybutyric acid. *AAPS J.*, **10**:311–321.

Morton, N. S. (2008). Ketamine for procedural sedation and analgesia in pediatric emergency medicine: a UK perspective. *Paediatr. Anaesth.*, **18**:25–29.

Muller, C., Viry, S., Miehe, M., Andriamampandry, C., Aunis, D., and Maitre, M. (2002). Evidence for a γ-hydroxybutyrate (GHB) uptake by rat brain synaptic vesicles. *J. Neurochem.*, **80**:899–904.

Munir, V. L., Hutton, J. E., Harney, J. P., Buykx, P., Weiland, T. J., Dent, A. W. (2008). Gamma-hydroxybutyrate: a 30 month emergency department review. *Emerg. Med. Australasia*, **20**:521–530.

Musshoff, F., and Madea, B. (2007a). New trends in hair analysis and scientific demands on validation and technical notes. *Forensic Sci. Int.*, **165**:204–215.

Musshoff, F., and Madea, B. (2007b). Analytical pitfalls in hair testing. *Anal. Bioanal. Chem.*, **388**:1475–1494.

Negrusz, A., Adamowicz, P., Saini, B. K., Webster, D. E., Juhascik, M. P., Moore, C. M., and Schlemmer, R. F. (2005). Detection of ketamine and norketamine in urine of nonhuman primates after a single dose of ketamine using microplate enzyme-linked immunosorbent assay (ELISA) and NCI-GC-MS. *J. Anal. Toxicol.*, **29**:163–168.

Nelson, T., Kaufman, E., Kline, J., and Sokoloff, L. (1981). The extraneural distribution of γ-hydroxybutyrate. *J. Neurochem.*, **37**:1345–1348.

Nirenberg, M. W., and Jakoby, W. B. (1960). Enzymatic utilization of γ-hydroxybutyric acid. *J. Biol. Chem.*, **236**:954–960.

Nishimura, M., and Naito, S. (2005). Tissue-specific mRNA expression profiles of human ATP-binding cassette and solute carrier transporter superfamilies. *Drug Metab. Pharmacokinet.*, **20**:452–477.

Niwa, T., Maeda, K., Asada, H., Shibata, M., Ohki, T., Saito, A., and Furukawa, H. (1982). Gas chromatographic–mass spectrometric analysis of organic acids in renal tissue biopsy. Identification of 4-hydroxybutyric acid and 4-hydroxy-2-butenoic acid. *J. Chromatogr. Biomed.*, **230**:1–6.

Niwa, T., Yamamoto, N., Kawanishi, A., Saiki, J., Maeda, K., Yamada, K., Ohki, T., and Saito, A. (1983). Gas chromatographic–mass spectrometric investigation of renal tissue metabolism. *Iyo Masu Kenkyukai Koenshu*, **8**:275–278.

O'Connor, T., Ireland, L. S., Harrison, D. J., and Hayes, J. D. (1999). Major differences exist in the function and tissue-specific expression of human aflatoxin B1 aldehyde reductase and the principal human aldo-keto reductase AKR1 family members. *Biochem. J.*, **343**:487–504.

Odujebe, O., Landman, A., and Hoffman, R. S. (2007). GHB urine concentrations after single-dose administration in humans. *J. Anal. Toxicol.*, **31**:179–180.

Ogilvie, J. W., and Whitaker, S. C. (1976). Reaction of tris with aldehydes: effect of Tris on reactions catalyzed by homoserine dehydrogenase and glyceraldehyde-3-phosphate dehydrogenase. *Biochim. Biophys. Acta*, **445**:525–536.

Okino, Y. (2005a). Poly(butylene succinate) (PBS)-type biodegradable plastics: the market trend of Bionolle. *Sangyo to Kankyo*, **34**:40–42.

Okino, Y. (2005b). Distinctive feature and latest information on poly(butylene succinate) (PBS)-type green plastics. *Kankyo Gijutsu*, **34**:406–410.

Okon, T. (2007). Ketamine: an introduction for the pain and palliative medicine physician. *Pain Physician*, **10**:493–500.

Orrenius, S. (1969). Cytochrome P-450 in the ω-oxidation of fatty acids. *Biochem. J.*, **115**:25P–26P.

Palatini, P., Tedeschi, L., Frison, G., Padrini, R., Zordan, R., Gallimberti, L., Gessa, G. L., and Ferrara, S. D. (1993). Dose-dependent absorption and elimination of gamma-hydroxybutyric acid in healthy volunteers. *Eur. J. Clin. Pharmacol.*, **45**:353–356.

Palmer, R. B. (2004). γ-Butyrolactone and 1,4-butanediol: abused analogues of γ-hydroxybutyrate. *Toxicol. Rev.*, **23**:21–31.

Pan, Z. H., Zhang, X., and Lipton, S. A. (2000). Redox modulation of recombinant human $GABA_A$ receptors. *Neuroscience*, **98**:333–338.

Pardi, D., and Black, J. (2006). γ-Hydroxybutyrate/sodium oxybate: neurobiology, and impact on sleep and wakefulness. *CNS Drugs*, **20**:993–1018.

Parrott, A. C. (2007). The psychotherapeutic potential of MDMA (3,4-methylenedioxymethamphetamine): an evidence-based review. *Psychopharmacology*, **191**:181–193.

Pearl, P. L., Gibson, K. M., Cortez, M. A., Wu, Y., Snead, O. C., III., Knerr, I., Forester, K., Pettiford, J. M., Jakobs, C., and Theodore, W. H. (2009). Succinic semialdehyde dehydrogenase deficiency: lessons from mice and men. *J. Inherit. Metab. Dis.*, **32**(3):343–352.

Perez-Prior, M. T., Manso, J. A., Garcia-Santos, M. del P., Calle, E., and Casado, J. (2005). Reactivity of lactones and GHB formation. *J. Org. Chem.*, **70**:420–426.

Peters, F. T. (2007). Stability of analytes in biosamples: an important issue in clinical and forensic toxicology? *Anal. Bioanal. Chem.*, **388**:1505–1519.

Picklo, M. J., Olson, S. J., Hayes, J. D., Markesbery, W. R., and Montine, T. J. (2001a). Elevation of AKR7A2 (succinic semialdehyde reductase) in neurodegenerative disease. *Brain Res.*, **916**:229–238.

Picklo, M. J., Sr., Olson, S. J., Markesbery, W. R., and Montine, T. J. (2001b). Expression and activities of aldo-keto oxidoreductases in Alzheimer disease. *J. Neuropathol. Exp. Neurol.*, **60**:686–695.

Pierre, K., and Pellerin, L. (2005). Monocarboxylate transporters in the central nervous system: distribution, regulation and function. *J. Neurochem.*, **94**:1–14.

Pietruszko, R., Voigtlander, K., and Lester, D. (1978) Alcohol dehydrogenase from human and horse liver: substrate specificity with diols. *Biochem. Pharmacol.*, **27**:1296–1297.

Pizzolato, T. M., Lopez de Alda, M. J., and Barcelo, D. (2007). LC-based analysis of drugs of abuse and their metabolites in urine. *Trends Anal. Chem.*, **26**:609–624.

Pocker, Y., and Dickerson, D. G. (1969). Hydration of propionaldehyde, isobutyraldehyde, and pivalaldehyde: thermodynamic parameters, buffer catalysis and transition state characterization. *J. Phys. Chem.*, **73**:4005–4012.

Poggioli, R., Vitale, G., Colombo, G., Ottani, A., and Bertolini, A. (1999). Gamma-hydroxybutyrate increases gastric emptying in rats. *Life Sci.*, **64**:2149–2154.

Poirier, Y. (2001). Production of polyesters in transgenic plants. *Adv. Biochem. Eng. Biotechnol.*, **71**:209–240.

Poldrugo, F., and Snead, O. C., III (1984). 1,4-Butanediol, γ-hydroxybutyric acid and ethanol: relationships and interactions. *Neuropharmacology*, **23**:109–113.

Poldrugo, F., and Snead, O. C., III (1986). 1,4-Butanediol and ethanol compete for degradation in rat brain and liver in vitro. *Alcohol*, **3**:367–370.

Poldrugo, F., Snead, O. C., and Barker, S. (1985). Chronic alcohol administration produces an increase in liver 1,4-butanediol concentration. *Alcohol*, **20**:251–253.

Poldrugo, F., Monti, J. A., Snead, O. C., and Ostrovskii, S. Y. (1989). Characterization of gamma-hydroxybutyric acid and succinic semialdehyde metabolism in rat liver. *Dokl. Akad. Nauk. BSSR*, **33**:750–753.

Poulos, A., Sharp, P., Singh, H., Johnson, D. W., Carey, W. F., and Easton, C. (1993). Formic acid is a product of the α-oxidation of fatty acids by human skin fibroblasts: deficiency of formic acid production in peroxisome-deficient fibroblasts. *Biochem. J.*, **292**:457–461.

Poyot, T., Nachon, F., Rochu, D., Fournier, D., and Masson, P. (2007). Enzyme optimization, mutagenesis and directed evolution. *Ann. Pharm. Fr.*, **65**:119–125.

Pozharski, E., Moulin, A., Hewagama, A., Shanafelt, A. B., Petsko, G. A., and Ringe, D. (2005). Diversity in hapten recognition: structural study of an anti-cocaine antibody M82G2. *J. Mol. Biol.*, **349**:570–582.

Pragst, F. (2007). Application of solid-phase microextraction in analytical toxicology. *Anal. Bioanal. Chem.*, **388**:1393–1414.

Quang, L. S., Shannon, M. W., Woolf, A. D., Desai, M. C., and Maher, T. J. (2002). Pretreatment of CD-1 mice with 4-methylpyrazole blocks toxicity from the gamma-hydroxybutyrate precursor, 1,4-butanediol. *Life Sci.*, **71**:771–778.

Quang, L. S., Desai, M. C., Shannon, M. W., Woolf, A. D., and Maher, T. J. (2004). 4-Methylpyrazole decreases 1,4-butanediol toxicity by blocking its in vivo biotransformation to gamma-hydroxybutyric acid. *Ann. NY Acad. Sci.*, **1025**:528–537.

Quest, D. W., and Horsley, J. (2007). Field-test of a date-rape drug detection device. *J. Anal. Toxicol.*, **31**:354–357.

Rabetafika, H. N., Paquot, M., and Dubois, P. (2006). Features of plant-based polymers with special applications in plastic field. *Biotechnol. Agron. Soc. Environ.*, **10**:185–196.

Raeder, J. (2003). Ketamine, revival of a versatile intravenous anaesthetic. *Adv. Exp. Med. Biol.*, **523**:269–277.

Rao, V. L. R., and Murthy, C. R. K. (1993). Uptake, release and metabolism of glutamate and aspartate by rat cerebellar subcellular preparations. *Biochem. Mol. Biol. Int.*, **29**:711–717.

Ren, X., and Mody, I. (2006). γ-Hydroxybutyrate induces cyclic AMP-responsive element-binding protein phosphorylation in mouse hippocampus: an involvement of $GABA_B$ receptors and cAMP-dependent protein kinase activation. *Neuroscience*, **141**:269–275.

Rendle, D. F. (2005). Advances in chemistry applied to forensic science. *Chem. Soc. Rev.*, **34**:1021–1030.

Ricaurte, G. A., and McCann, U. D. (2005). Recognition and management of complications of new recreational drug use. *Lancet*, **365**:2137–2145.

Richard, D., Ling, B., Authier, N., Faict, T. W., Eschalier, A., and Coudore, F. (2005). GC/MS profiling of γ-hydroxybutyrate and precursors in various animal tissues using automatic solid-phase extraction: preliminary investigations of its potential interest in postmortem interval determination. *Anal. Chem.*, **77**:1354–1360.

Robinson, D. M., and Keating, G. M. (2007). Sodium oxybate: a review of its use in the management of narcolepsy. *CNS Drugs*, **21**:337–354.

Rocha, S. M., Coutinho, P., Delgadillo, I., Cardoso, A. D., and Coimbra, M. A. (2005). Effect of enzymatic aroma release on the volatile compounds of white wines presenting different aroma potentials. *J. Sci. Food Agric.*, **85**:199–205.

Rogers, S., Girolami, M., Kolch, W., Waters, K. M., Liu, T., Thrall, B., and Wiley, H. S. (2008). Investigating the correspondence between transcriptomic and proteomic expression profiles using coupled cluster models. *Bioinformatics*, **24**:2894–2900.

Rosenkranz, H. S. (2001). Computational toxicology and the generation of mechanistic hypotheses: γ-butyrolactone. *SAR QSAR Environ. Res.*, **12**:435–444.

Roth, R. H., and Giarman, N. J. (1966). γ-Butyrolactone and γ-hydroxybutyric acid: I. Distribution and metabolism. *Biochem. Pharmacol.*, **15**:1333–1348.

Rowland, L. M. (2005). Subanesthetic ketamine: how it alters physiology and behavior in humans. *Aviat. Space Environ. Med.*, **76**:C52–C58.

Rumigny, J. F., Cash, C., Mandel, P., Vincendon, G., and Maitre, M. (1981). Evidence that a specific succinic semialdehyde reductase is responsible for γ-hydroxybutyrate synthesis in brain tissue slices. *FEBS Lett.*, **134**:96–98.

Ryzlak, M. T., and Pietruszko, R. (1988). Human brain "high K_m" aldehyde dehydrogenase: purification, characterization, and identification as NAD^+-dependent succinic semialdehyde dehydrogenase. *Arch. Biochem. Biophys.*, **266**:386–396.

Rzem, R., Veiga-da-Cunha, M., Noël, G., Sophie, S., Nassogne, M.-C., Tabarki, B., Schöller, C., Marquardt, T., Vikkula, M., and Van Schaftingen, E. (2004). A gene encoding a putative FAD-dependent L-2-hydroxyglutarate dehydrogenase is mutated in L-2-hydroxyglutaric aciduria. *Proc. Natl. Acad. Sci. USA*, **101**:16849–16854.

Sadasivan, S., Maher, T. J., and Quang, L. S. (2006). γ-hydroxybutyrate (GHB), γ-butyrolactone (GBL), and 1,4-butanediol (1,4-BD) reduce the volume of cerebral infarction in rodent transient middle cerebral artery occlusion. *Ann. NY Acad. Sci.*, **1074**:537–544.

Saint-Martin, P., Furet, Y., O'Byrne, P., Bouyssy, M., Paintaud, G., and Autret-Leca, E. (2006). Chemical submission: a literature review. *Therapie*, **61**:145–150.

Sakata, T. (1990). Structural and stereoisomeric specificity of serum-borne sugar acids related to feeding control by rats. *Brain Res. Bull.*, **25**:969–914.

Sakurada, K., Kobayashi, M., Iwase, H., Yoshino, M., Mukoyama, H., Takatori, T., and Yoshida, K. (2002). Production of gamma-hydroxybutyric acid in postmortem liver increases with time after death. *Toxicol. Lett.*, **129**:207–217.

Salamone, S. J., and Tsai, J. S.-C. (2002). OnTrak TesTstik device. In: Jenkins, A. J., and Goldberger, B. A. (Eds.) *On-Site Drug Testing*. Totowa, NJ: Humana Press, pp. 185–198.

Samyn, N., Laloup, M., and De Boeck, G. (2007). Bioanalytical procedures for determination of drugs of abuse in oral fluid. *Anal. Bioanal. Chem.*, **388**:1437–1453.

Santaniello, E., Kienle, M. G., Bosisio, E., and Manzocchi, A. (1978). The stereochemistry of 4-aminobutyric acid transamination in mouse brain. *Symposium Papers: IUPAC 11th International Symposium on the Chemistry of Natural Products*, Vol. 1, pp. 197–200.

Santos, N. (2007). Scientific Working Group for the Analysis of Seized Drugs. http://www.swgdrug.org/Documents/SWGDRUG%20Recommendations.pdf. Accessed June 6, 2008.

Saude, E. J., and Sykes, B. D. (2007). Urine stability for metabolomic studies: effects of preparation and storage. *Metabolomics*, **3**:19–27.

Sauer, S. W., Koelker, S., Hoffmann, G. F., Brink, H. J., Jakobs, C., Gibson, K. M., and Okun, J. G. (2007). Enzymatic and metabolic evidence for a region specific mitochondrial dysfunction in brains of murine succinic semialdehyde dehydrogenase deficiency (Aldh5a1 -/- mice). *Neurochem. Int.*, **50**:653–659.

Schaller, M., Schaffhauser, M., Sans, N., and Wermuth, B. (1999). Cloning and expression of succinic semialdehyde reductase from human brain: identity with aflatoxin B1 aldehyde reductase. *Eur. J. Biochem.*, **265**:1056–1060.

Scharf, M. B., Lai, A. A., Branigan, B., Stover, R., and Berkowitz, D. B. (1998). Pharmacokinetics of gamma-hydroxybutyrate (GHB) in narcoleptic patients. *Sleep*, **21**:507–514.

Schmidt, M. (2008). $GABA_C$ receptors in retina and brain. *Results Probl. Cell Diff.*, **44**:49–67.

Schousboe, A., Saito, K., and Wu, J.-Y. (1980). Characterization and cellular and subcellular localization of GABA-transaminase. *Brain Res. Bull.*, **5**(Suppl. 2):71–76.

Scott-Ham, M., and Burton, F. C. (2005). Toxicological findings in cases of alleged drug-facilitated sexual assault in the United Kingdom over a 3-year period. *J. Clin. Forensic Med.*, **12**:175–186.

Seamans, J. (2008). Losing inhibition with ketamine. *Nat. Chem. Biol.*, **4**:91–93.

Sgaragli, G., and Zilletti, L. (1970). Passage of 4-hydroxybutyrate and gamma-butyrolactone into human erythrocyte. *Pharmacol. Res. Commun.*, **2**:31–38.

Sgaravatti, A. M., Sgarbi, M. B., Testa, C. G., Durigon, K., Pederzolli, C. D., Prestes, C. C., Wyse, A. T. S., Wannmacher, C. M. D., Wajner, M., and Dutra-Filho, C. S. (2007). γ-Hydroxybutyric acid induces oxidative stress in cerebral cortex of young rats. *Neurochem. Int.*, **50**:564–570.

Sgaravatti, A. M., Magnusson, A. S., Oliveira, A. S., Mescka, C. P., Zanin, F., Sgarbi, M. B., Pederzolli, C. D., Wyse, A. T. S., Wannmacher, C. M. D., Wajner, M., and Dutra-Filho, C. S. (2009). Effects of 1,4-butanediol administration on oxidative stress in rat brain: study of the neurotoxicity of γ-hydroxybutyric acid in vivo. *Metab. Brain Dis.*, **24**:271–282.

Shaw, L. M. J., and Westerfeld, W. W. (1968). Enzymic reactions involved in the formation of 5-hydroxy-4-ketohexanoic acid and its isomer, 5-keto-4-hydroxy-hexanoic acid. *Biochemistry*, **7**:1333–1338.

Shi, S.-R., Gu, J., Kalra, K. L., Chen, T., Cote, R. J., and Taylor, C. R. (1995). Antigen retrieval technique: a novel approach to immunohistochemistry on routinely processed tissue sections. *Cell Vision*, **2**:6–22.

Shima, N., Miki, A., Kamata, T., Katagi, M., and Tsuchihashi, H. (2005). Urinary endogenous concentrations of GHB and its isomers in healthy humans and diabetics. *Forensic Sci. Int.*, **149**:171–179.

Shinka, T., Inoue, Y., Ohse, M., Ito, A., Ohfu, M., Hirose, S., and Kuhara, T. (2002). Rapid and sensitive detection of urinary 4-hydroxybutyric acid and its related compounds by gas chromatography–mass spectrometry in a patient with succinic semialdehyde dehydrogenase deficiency. *J. Chromatogr. B*, **776**:57–63.

Shinka, T., Ohfu, M., Hirose, S., and Kuhara, T. (2003). Effect of valproic acid on the urinary metabolic profile of a patient with succinic semialdehyde dehydrogenase deficiency. *J. Chromatogr. B*, **792**:99–106.

Shumate, J. S., and Snead, O. C., III (1979). Plasma and central nervous system kinetics of gamma-hydroxybutyrate. *Res. Commun. Chem. Pathol. Pharmacol.*, **25**:241–256.

Simpson, I. A., Carruthers, A., and Vannucci, S. J. (2007). Supply and demand in cerebral energy metabolism: the role of nutrient transporters. *J. Cereb. Blood Flow Metab.*, **27**:1766–1791.

Singh, M. K., Srivastava, S., Raghava, G. P. S., and Varshney, G. C. (2006). HaptenDB: a comprehensive database of haptens, carrier proteins and anti-hapten antibodies. *Bioinformatics*, **22**:253–255.

Sinner, B., and Graf, B. M. (2008). Ketamine. *Handb. Exp. Pharmacol.*, **182**:313–333.

Smith, M. L., Vorce, S. P., Holler, J. M., Shimomura, E., Magluilo, J., Jacobs, A. J., and Huestis, M. A. (2007). Review: modern instrumental methods in forensic toxicology. *J. Anal. Toxicol.*, **31**:237–253.

Snead, O. C. γ-Hydroxybutyric acid in subcellular fractions of rat brain. (1987). *J. Neurochem.*, **48**:196–201.

Snead, O. C., III, and Gibson, K. M. (2006). γ-Hydroxybutyric acid. *N. Engl. J. Med.*, **352**:2721–2732. Erratum in *N. Engl. J. Med.*, 2006;**354**:537–537.

Snead, O. C., III, Liu, C. C., and Bearden, L. J. (1982). Studies on the relation of γ-hydroxybutyric acid (GHB) to γ-aminobutyric acid (GABA): evidence that GABA is not the sole source for GHB in rat brain. *Biochem. Pharmacol.*, **31**:3917–3923.

Snead, O. C., III, Furner, R., and Liu, C. C. (1989). In vivo conversion of γ-aminobutyric acid and 1,4-butanediol to γ-hydroxybutyric acid in rat brain: studies using stable isotopes. *Biochem. Pharmacol.*, **38**:4375–4380.

Snow, N. H. (2007). Biological, clinical, and forensic analysis using comprehensive two-dimensional gas chromatography. *Adv. Chromatogr.*, **45**:215–243.

Song, S. S., Ma, H., Gao, Z. X., Jia, Z., and Zhang, X. (2005). Construction of recombinant *Escherichia coli* strains producing poly(4-hydroxybutyric acid) homopolyester from glucose. *Wei Sheng Wu Xue Bao*, **45**:382–386.

Sporer, K. A., Chin, R. L., Dyer, J. E., and Lamb, R. (2003). Gamma-hydroxybutyrate serum levels and clinical syndrome after severe overdose. *Ann. Emerg. Med.*, **42**:3–8.

Struys, E. A. (2006). D-2-Hydroxyglutaric aciduria: unravelling the biochemical pathway and the genetic defect. *J. Inherit. Metab. Dis.*, **29**:21–29.

Struys, E. A., Verhoeven, N. M., Brunengraber, H., and Jakobs, C. (2004). Investigations by mass isotopomer analysis of the formation of D-2-hydroxyglutarate by cultured lymphoblasts from two patients with D-2-hydroxyglutaric aciduria. *FEBS Lett.*, **557**:115–120.

Struys, E. A., Jansen, E. E. W., Gibson, K. M., and Jakobs, C. (2005a). Determination of the GABA analogue succinic semialdehyde in urine and cerebrospinal fluid by dinitrophenylhydrazine derivatization and liquid chromatography–tandem mass spectrometry: application to SSADH deficiency. *J. Inherit. Metab. Dis.*, **28**:913–920.

Struys, E. A., Verhoeven, N. M., Brink, H. J., Wickenhagen, W. V., Gibson, K. M., and Jakobs, C. (2005b). Kinetic characterization of human hydroxyacid–oxoacid transhydrogenase: relevance to D-2-hydroxyglutaric and γ-hydroxybutyric acidurias. *J. Inherit. Metab. Dis.*, **28**:921–930.

Struys, E. A., Verhoeven, N. M., Salomons, G. S., Berthelot, J., Vianay-Saban, C., Chabrier, S., Thomas, J. A., Tsai, A. C.-H., Gibson, K. M., and Jakobs, C. (2006a). D-2-Hydroxyglutaric aciduria in three patients with proven SSADH deficiency: genetic coincidence or a related biochemical epiphenomenon? *Mol. Genet. Metab.*, **88**:53–57.

Struys, E. A., Verhoeven, N. M., Jansen, E. E. W., Brink, H. J., Gupta, M., Burlingame, T. G., Quang, L. S., Maher, T., Rinaldo, P., Snead, O. C., Goodwin, A. K., Weerts, E. M., Brown, P. R., Murphy, T. C., Picklo, M. J., Jakobs, C., and Gibson, K. M. (2006b). Metabolism of γ-hydroxybutyrate to D-2-hydroxyglutarate in mammals: further evidence for D-2-hydroxyglutarate transhydrogenase. *Metabolism*, **55**:353–358.

Tagliaro, F., Bortolotti, F., and Pascali, J. P. (2007). Current role of capillary electrophoretic/electrokinetic techniques in forensic toxicology. *Anal. Bioanal. Chem.*, **388**:1359–1364.

Tanaka, C. (1985). γ-Aminobutyric acid in peripheral tissues. *Life Sci.*, **37**:2221–2235.

Tanaka, E., Terada, M., Shinozuka, T., and Honda, K. (2003). Gamma-hydroxybutyric acid (GHB): its pharmacology and toxicology. *Forensic Toxicol.*, **21**:210–217.

Tao, L., Han, J., and Tao, F.-M. (2008). Correlations and predictions of carboxylic acid pK_a values using intermolecular structure and properties of hydrogen-bonded complexes. *J. Phys. Chem. A*, **112**:775–782.

Teiber, J. F., Draganov, D. I., and La Du, B. N. (2003). Lactonase and lactonizing activities of human serum paraoxonase (PON1) and rabbit serum PON3. *Biochem. Pharmacol.*, **66**:887–896.

Thai, D., Dyer, J. E., Jacob, P., and Haller, C. A. (2007). Clinical pharmacology of 1,4-butanediol and gamma-hydroxybutyrate after oral 1,4-butanediol administration to healthy volunteers. *Clin. Pharmacol. Ther.*, **81**:178–184.

Thiriet, N., Ladenheim, B., McCoy, M. T., and Cadet, J. L. (2002). Analysis of ecstasy (MDMA)-induced transcriptional responses in the rat cortex. *FASEB J.*, **16**:1887–1894.

Ticku, M. K., and Mehta, A. K. (2008). Characterization and pharmacology of the GHB receptor. *Ann. NY Acad. Sci.*, **1139**:374–385.

Tillakaratne, N. J. K., Medina-Kauwe, L., and Gibson, K. M. (1995). Gamma-aminobutyric acid (GABA) metabolism in mammalian neural and non-neural tissues. *Comp. Biochem. Phys. A*, **112A**:247–263.

Tokiwa, Y., and Pranamuda, H. (2001). Microbial degradation of aliphatic polyesters. *Biopolymers*, **3b**:85–103.

Tsunemasa, N. (1998). Production of biopolyester by activated sludge. *Yosui to Haisui*, **40**:981–986.

Tunnicliff, G. (2002). Membrane transport of γ-hydroxybutyrate. In: Tunnicliff, G., and Cash, C. D., (Eds.), *Gamma-hydroxybutyrate*. London: Taylor & Francis, pp. 64–74.

Ulrich, D., and Bettler, B. (2007). $GABA_B$ receptors: synaptic functions and mechanisms of diversity. *Curr. Opin. Neurobiol.*, **17**:298–303.

Valentin, H. E., Zwingmann, G., Schönebaum, A., and Steinbüchel, A. (1995). Metabolic pathway for biosynthesis of poly(3-hydroxybutyrate-*co*-4-hydroxybutyrate) from 4-hydroxybutyrate by *Alcaligenes eutrophus*. *Eur. J. Biochem.*, **227**:43–60.

Valentin, H. E., Reiser, S., and Gruys, K. J. (2000). Poly(3-hydroxybutyrate-*co*-4-hydroxybutyrate) formation from gamma-aminobutyrate and glutamate. *Biotechnol. Bioeng.*, **67**:291–299.

Vamecq, J., and Poupaert, J. H. (1990). Studies on the metabolism of glycolyl-CoA. *Biochem. Cell Biol.*, **68**:846–851.

Van Bemmelen, F. J., Schouten, M. J., Fekkes, D., and Bruinvels, J. (1985). Succinic semialdehyde as a substrate for the formation of γ-aminobutyric acid. *J. Neurochem.*, **45**:1471–1474.

Van der Walle, G. A. M., De Koning, G. J. M., Weusthuis, R. A., and Eggink, G. (2001). Properties, modifications and applications of biopolyesters. *Adv. Biochem. Eng. Biotechnol.*, **71**:263–291.

van Herwaarden, M. A., Samsom, M., and Smout, A. J. (1999). 24-h recording of intragastric pH: technical aspects and clinical relevance. *Scand. J. Gastroenterol. Suppl.*, **230**:9–16.

van Marken Lichtenbelt, W. D., Vanhommerig, J. W., Smulders, N. M., Drossaerts, J. M. A. F. L., Kemerink, G. J., Bouvy, N. D., Schrauwen, P., and Teule, G. J. J. (2009). Cold-activated brown adipose tissue in healthy men. *N. Engl. J. Med.*, **360**:1500–1508.

van Nieuwenhuijzen, P. S., McGregor, I. S., and Hunt, G. E. (2009). The distribution of γ-hydroxybutyrate–induced Fos expression in rat brain: comparison with baclofen. *Neuroscience*, **158**:441–455.

van Nieuwenhuijzen, P. S., Kashem, M. A., Matsumoto, I., Hunt, G. E., and McGregor, I. S. (2010). A long hangover from party drugs: Residual proteomic changes in the hippocampus of rats 8 weeks after γ-hydroxybutyrate (GHB), 3,4-methylenedioxymethamphetamine (MDMA) or their combination. *Neurochem. Int.*, **56**:871–877.

Van Sassenbroeck, D. K., De Neve, N., De Paepe, P., Belpaire, F. M., Verstraete, A. G., Calle, P. A., and Buylaert, W. A. (2007). Abrupt awakening phenomenon associated with gamma-hydroxybutyrate use: a case series. *Clin. Toxicol.*, **45**:533–538.

Vasiliou, V., and Nebert, D. W. (2005). Analysis and update of the human aldehyde dehydrogenase (ALDH) gene family. *Hum. Genom.*, **2**:138–143.

Vastag, B. (2010). Can the peace drug help clean up the war mess? http://www.scientific american.com/article.cfm?id=mdma-drug-ptsd-trauma-psychedelic&sc=DD_20100421. Accessed June 15.

Vayer, P., Mandel, P., and Maitre, M. (1985a). Conversion of γ-hydroxybutyrate to γ-aminobutyrate in vitro. *J. Neurochem.*, **45**:810–814.

Vayer, P., Dessort, D., Bourguignon, J. J., Wermuth, C. G., Mandel, P., and Maitre, M. (1985b). Natural occurrence of trans-gamma-hydroxycrotonic acid in rat brain. *Biochem. Pharmacol.*, **34**:2401–2404.

Vayer, P., Ehrhardt, J. D., Gobaille, S., Mandel, P., and Maitre, M. (1988). Gamma-hydroxybutyrate distribution and turnover rates in discrete brain regions of the rat. *Neurochem. Int.*, **12**:53–59.

Veech, R. L., Eggleston, L. V., and Krebs, H. A. (1969). Redox state of free nicotinamide adenine dinucleotide phosphate in the cytoplasm of rat liver. *Biochem. J.*, **115**:609–619.

Vengeliene, V., Bilbao, A., Molander, A., and Spanagel, R. (2008). Neuropharmacology of alcohol addiction. *Br. J. Pharmacol.*, **154**:299–315.

Verschraagen, M., Maes, A., Ruiter, B., Bosman, I. J., Smink, B. E., and Lusthof, K. J. (2007). Post-mortem cases involving amphetamine-based drugs in the Netherlands: comparison with driving under the influence cases. *Forensic Sci. Int.*, **170**:163–170.

Verstraete, A. G. (2004). Detection times of drugs of abuse in blood, urine, and oral fluid. *Ther. Drug Monit.*, **26**:200–205.

Villain, M., Cirimele, V., Ludes, B., and Kintz, P. (2003). Ultra-rapid procedure to test for gamma-hydroxybutyric acid in blood and urine by gas chromatography-mass spectrometry. *J. Chromatogr. B*, **792**:83–87.

Vinckier, N. K., Dekker, L., McSpadden, E. D., Ostrand, J. T., and Parsons, S. M. (2011). Substrate selectivity and kinetics for cloned Fe(II)-dependent gamma-hydroxybutyrate dehydrogenase from *Ralstonia eutropha*, unpublished observations.

Visser, W. F., van Roermund, C. W. T., Ijlst, L., Waterham, H. R., and Wanders, R. J. A. (2007). Metabolite transport across the peroxisomal membrane. *Biochem. J.*, **401**:365–375.

Volpi, R., Chiodera, P., Caffarra, P., Scaglioni, A., Malvezzi, L., Saginario, A., and Coiro, V. (2000). Muscarinic cholinergic mediation of the GH response to gamma-hydroxybutyric acid: neuroendocrine evidence in normal and parkinsonian subjects. *Psychoneuroendocrinology*, **25**:179–185.

Vree, T. B., Van Dalen, R., Van der Kleijn, E., and Gimbrere, J. S. F. (1978). Pharmacokinetics of 1,4-butanediol and 4-hydroxybutyric ethylester in man, rhesus monkey and dog. *Anaesthesiol. Intensivmed.*, **110**:66–73.

Walkenstein, S. S., Wiser, R., Gudmundsen, C., and Kimmel, H. (1964). Metabolism of γ-hydroxybutyric acid. *Biochim. Biophys. Acta*, **86**:640–642.

Wang, G., and Bai, N. (1998). Structure–activity relationships for rat and mouse LD_{50} of miscellaneous alcohols. *Chemosphere*, **36**:1475–1483.

Wang, Q., and Morris, M. E. (2007). The role of monocarboxylate transporter 2 and 4 in the transport of γ-hydroxybutyric acid in mammalian cells. *Drug Metab. Dispos.*, **35**:1393–1399.

Wang, M.-H., Wade, D., Chen, L., White, S., and Yang, C. S. (1995). Probing the active sites of rat and human cytochrome P450 2E1 with alcohols and carboxylic acids. *Arch. Biochem. Biophys.*, **317**:299–304.

Wang, X., Wang, Q., and Morris, M. E. (2008). Pharmacokinetic interaction between the flavonoid luteolin and gamma-hydroxybutyrate in rats: potential involvement of monocarboxylate transporters. *AAPS J.*, **10**:47–55.

Wedin, G. P., Hornfeldt, C. S., and Ylitalo, L. M. (2006). The clinical development of γ-hydroxybutyrate (GHB). *Curr. Drug Saf.*, **1**:99–106.

Wickenhagen, W. V., Salomons, G. S., Gibson, K. M., Jakobs, C., and Struys, E. A. (2009). Measurement of *d*-2-hydroxyglutarate dehydrogenase activity in cell homogenates derived from *d*-2-hydroxyglutaric aciduria patients. *J. Inherit. Metab. Dis.*, **32**:264–268.

Wierenga, R. K., De Maeyer, M. C. H., and Hol, W. G. J. (1985). Interaction of pyrophosphate moieties with α-helices in dinucleotide-binding proteins. *Biochemistry*, **24**:1346–1357.

Wiesner, R. J., Kreutzer, U., Roesen, P., and Grieshaber, M. K. (1988). Subcellular distribution of malate–aspartate cycle intermediates during normoxia and anoxia in the heart. *Bioenergetics*, **936**:114–123.

Witzmann, F. A., Fultz, C. D., Grant, R. A., Wright, L. S., Kornguth, S. E., and Siegel, F. L. (1998). Differential expression of cytosolic proteins in the rat kidney cortex and medulla: preliminary proteomics. *Electrophoresis*, **19**:2491–2497.

Wojtowicz, J. M., Yarema, M. C., and Wax, P. M. (2008). Withdrawal from gamma-hydroxybutyrate, 1,4-butanediol and gamma-butyrolactone: a case report and systematic review. *Can. J. Emerg. Med.*, **10**:69–74.

Wolff, R. A., and Kenealy, W. R. (1995). Purification and characterization of the oxygen-sensitive 4-hydroxybutanoate dehydrogenase from *Clostridium kluyveri*. *Protein Expr. Purif.*, **6**:206–212.

Wolff, K., and Winstock, A. R. (2006). Ketamine: from medicine to misuse. *CNS Drugs*, **20**:199–218.

Wong, C. G. T., Chan, K. F. Y., Gibson, K. M., and Snead, O. C., III (2004a). γ-Hydroxybutyric acid: neurobiology and toxicology of a recreational drug. *Toxicol. Rev.*, **23**:3–20.

Wong, C. G. T., Gibson, K. M., and Snead, O. C., III (2004b). From the street to the brain: neurobiology of the recreational drug γ-hydroxybutyric acid. *Trends Pharmacol. Sci.*, **25**:29–34.

Wood, M., Laloup, M., Samyn, N., Morris, M. R., de Bruijn, E. A., Maes, R. A., Young, M. S., Maes, V., and De Boeck, G. (2004). Simultaneous analysis of gamma-hydroxybutyric acid and its precursors in urine using liquid chromatography-tandem mass spectrometry. *J. Chromatogr. A*, **1056**:83–90.

Wood, D. M., Warren-Gash, C., Ashraf, T., Greene, S. L., Shather, Z., Trivedy, C., Clarke, S., Ramsey, J., Holt, D. W., and Dargan, P. I. (2008). Medical and legal confusion surrounding gamma-hydroxybutyrate (GHB) and its precursors gamma-butyrolactone (GBL) and 1,4-butanediol (1,4BD). *Q. J. Med.*, **101**:23–29.

Wright, E. M., and Pietras, R. J. (1974). Routes of nonelectrolyte permeation across epithelial membranes. *J. Membr. Biol.*, **17**:293–312.

Wu, L.-T., Schlenger, W. E., and Galvin, D. M. (2006). Concurrent use of methamphetamine, MDMA, LSD, ketamine, GHB, and flunitrazepam among American youths. *Drug Alcohol Depend.*, **84**:102–113.

Yeatman, D. T., and Reid, K. (2003). A study of urinary endogenous gamma-hydroxybutyrate (GHB) levels. *J. Anal. Toxicol.*, **27**:40–42.

Yoshida, Y., Holloway, G. P., Ljubicic, V., Hatta, H., Spriet, L. L., Hood, D. A., and Bonen, A. (2007). Negligible direct lactate oxidation in subsarcolemmal and intermyofibrillar mitochondria obtained from red and white rat skeletal muscle. *J. Physiol.*, **582**:1317–1335.

Zacharis, C. K., Raikos, N., Giouvalakis, N., Tsoukali-Papadopoulou, H., and Theodoridis, G. A. (2008). A new method for the HPLC determination of gamma-hydroxybutyric acid (GHB) following derivatization with a coumarin analogue and fluorescence detection. *Talanta*, **75**:356–361.

Zambonin, P. G., Guerrieri, A., Rotunno, T., and Palmisano, F. (1991). Simultaneous determination of γ-aminobutyric acid and polyamines by *o*-phthalaldehyde-β-mercaptoethanol precolumn derivatization and gradient elution liquid chromatography with electrochemical detection. *Anal. Chim. Acta*, **251**:101–107.

Zhang, Q., and Wang, C. (1994). Polyhydroxybutyrate produced from cheap resources: I. Crystallization and melting behavior. *J. Appl. Polym. Sci.*, **54**:515–518.

Zhang, Q., Piston, D. W., and Goodman, R. H. (2002). Regulation of corepressor function by nuclear NADH. *Science*, **295**:1895–1897.

Zhu, X., Lapthorn, A. J., and Ellis, E. M. (2006). Crystal structure of mouse succinic semi-aldehyde reductase AKR7A5: structural basis for substrate specificity. *Biochemistry*, **45**:1562–1570.

Zimatkin, S. M., Pronko, S. P., Vasiliou, V., Gonzalez, F. J., and Deitrich, R. A. (2006). Enzymatic mechanisms of ethanol oxidation in the brain. *Alcohol Clin. Exp. Res.*, **30**:1500–1505.

Zvosec, D. L., Smith, S. W., Litonjua, R., and Westfal, R. E. J. (2007). Physostigmine for gamma-hydroxybutyrate coma: inefficacy, adverse events, and review. *Clin. Toxicol.*, **45**:261–265.

CHAPTER 12

Forensic and Clinical Issues in Alcohol Analysis

RICHARD STRIPP

Department of Sciences, John Jay College of Criminal Justice, The City University of New York, New York

Summary Ethanol is a clear volatile liquid that is soluble in water and has a characteristic taste. Ethanol is a central nervous system (CNS) depressant and causes most of its effects on the body by depressing brain function. CNS depression correlates directly to the concentration of alcohol in the blood. The estimation of blood alcohol concentration for a person is based on important parameters such as body weight, ethanol concentration of the beverage consumed and number of such beverages consumed, and length of time and pattern of the drinking. Because men and women have different body water amounts (men average 68% and women 55%), there are differences between the ethanol concentration achieved in men and women of similar weight for the same amount of alcohol. Various methods are described that can help to determine blood alcohol concentration in the field as well as in the laboratory.

12.1	Introduction	436
12.2	Blood alcohol concentration	437
12.3	Alcohol impairment and driving skills	441
12.4	Field sobriety tests	443
12.5	Blood alcohol measurements	444
	12.5.1　Enzymatic methods	444
	12.5.2　Headspace gas chromatography	445
	12.5.3　Breath alcohol testing	446
	12.5.4　Breath alcohol instrumentation	447
	12.5.5　Conversion from BrAC to BAC	449
	12.5.6　Urine and saliva	450
	12.5.7　Ethyl glucuronide	450

Forensic Chemistry Handbook, First Edition. Edited by Lawrence Kobilinsky.
© 2012 John Wiley & Sons, Inc. Published 2012 by John Wiley & Sons, Inc.

12.5.8	Postmortem determination of alcohol	451
12.5.9	Quality assurance of alcohol testing	452
References		453

12.1 INTRODUCTION

Alcohol (Ethanol: CH_3CH_2OH) Disposition

Half-life: plasma half-life, dose dependent

Volume of distribution: about 0.6 L/kg

Distribution in blood: plasma/whole blood ratio = 1.2

Saliva: plasma/saliva ratio is about 0.93

Protein binding: in plasma, not significantly bound

Ethyl alcohol has been consumed by humans for millennia and is probably one of the oldest and most commonly consumed drugs in most societies. A very high percentage of the adult population regularly consumes some form of alcohol to varying degrees. Approximately 10% use alcohol to an extent that may result in the manifestation of adverse heath effects. People who regularly consume six or more drinks per day are known to have higher mortality rates, and alcohol contributes to deaths in a variety of ways, including violence, accidents, and other drug-related deaths. Acute ethanol intoxication and its effect is the single most commonly encountered subject in forensic toxicology today. The blood alcohol concentration is commonly determined in medicolegal and forensic cases, and it is the ethanol in blood that is correlated with impairment of human performance. Ethanol is a product of fermentation by the action that yeasts on sugars found in fruit and grains. The content of alcoholic beverages varies for different types of drinks (see Table 12.1).

Alcohols are a class of organic chemicals that are characterized by the presence of a hydroxyl group (OH) on the hydrocarbon chain of the molecule (Figure 12.1). They all share common properties and ethanol is one of many alcohols. Other commonly encountered alcohols are methanol and isopropanol. Ethanol is a clear volatile liquid that is soluble in water and has a characteristic taste. As a drug, ethanol falls into the category of a *central nervous system* (CNS) *depressant* (Garriott, 2003). This means that the predominant effects of alcohol are to depress brain function. CNS depression is directly correlated to the concentration of alcohol in the blood (BAC). In a progressive manner, as the BAC rises, so does the degree of CNS depression. The most common reported units for BAC (%) are expressed as grams of ethanol per 100 mL of blood. Therefore, a BAC of 0.05 g/dL is equivalent to 0.05%. The American Medical Association has defined the BAC where impairment begins to be 0.04%, but for some functions it may be much lower. In most states in the United States a person is legally intoxicated at a BAC of 0.08% and is considered to be impaired at 0.05%. The consequences of impairment are

TABLE 12.1 Alcohol Content of Various Beverages

Beverage	Alcohol Content (%)
Beers (lager)	3.2–4.0
Ales	4.5
Porter	6.0
Stout	6.0–8.0
Malt liquor	3.2–7.0
Sake	14.0–16.0
Table wines	7.1–14.0
Sparkling wines	8.0–14.0
Fortified wines	14.0–24.0
Brandies	40.0–43.0
Whiskies	40.0–75.0
Vodkas	40.0–50.0
Gin	40.0–48.5
Rum	40.0–95.0
Tequila	45.0–50.5

CH_3OH	CH_3CH_2OH	$CH_3CH_2CH_2OH$	$CH_3CH(OH)CH_3$
Methanol	Ethanol	Propanol	Isopropanol
b.p. 64.7°C	b.p. 78.4°C	b.p. 97–98°C	b.p. 82.5°C

Figure 12.1 Structure of common alcohols.

the major reasons for the intense forensic interest in alcohol consumption. The individual assessment of impairment due to acute alcohol intoxication is of great medical–legal importance and is a complex matter that is influenced by many factors. There is significant variation between individual persons and their alcohol-related effects, but for the most part, predictable effects at various levels of BAC are encountered for a large percentage of the general population (Table 12.2).

It must be emphasized that a specific individual would show signs that can be markedly different for the ranges that are indicated. As already noted, many tolerant individuals may show little sign of physical impairment at the higher end of the table, although significant cognitive, judgmental, and psychological impairment may reasonably be presumed to be present.

12.2 BLOOD ALCOHOL CONCENTRATION

The estimation of blood alcohol concentration must be based on important parameters such as body weight, ETOH concentration of the beverage consumed

TABLE 12.2 Stages of Alcohol Intoxication

BAC (g/100 mL of blood or g/210 L of breath)	Stage	Clinical Symptoms
0.01–0.05	Subclinical	Behavior nearly normal by ordinary observation
0.03–0.12	Euphoria	Mild euphoria, sociability, talkativeness Increased self-confidence; decreased inhibitions Diminution of attention, judgment and control Beginning of sensory-motor impairment Loss of efficiency in finer performance tests
0.09–0.25	Excitement	Emotional instability; loss of critical judgment Impairment of perception, memory and comprehension Decreased sensatory response; increased reaction time Reduced visual acuity; peripheral vision and glare recovery Altered sensory-motor coordination; impaired balance Drowsiness
0.18–0.30	Confusion	Disorientation, mental confusion; dizziness Exaggerated emotional states Disturbances of vision and of perception of color, form, motion and dimensions Increased pain threshold Increased muscular loss in coordination; staggering gait; slurred speech Apathy, lethargy
0.25–0.40	Stupor	General inertia; approaching loss of motor functions Markedly decreased response to stimuli Marked muscular coordination impairment; inability to stand or walk Vomiting; incontinence Impaired consciousness; sleep or stupor
0.35–0.50	Coma	Complete unconsciousness Depressed or abolished reflexes Subnormal body temperature Incontinence Impairment of circulation and respiration Possible death
0.45+	Death	Death from respiratory arrest

(Table 12.3), number of alcoholic beverages consumed, and length and pattern of the drinking.

Other factors that will affect the rate of absorption from the gastrointestinal tract will further affect the final BAC curve (Morgan Jones and Vega, 1972; Karch, 1997; Drummer, 1999; Garriott, 2003). Approximation of the time frame of when the alcohol was consumed is usually of crucial importance (Morgan Jones and Vega, 1972; Karch, 1997). It is also essential to determine if the subject's BAC was on the ascending or descending limb of the BAC curve. The blood alcohol concentration reached after drinking depends not only on the volume of ethanol consumed but

TABLE 12.3 Estimated BAC by Number of Drinks in Relation to Body Weight[a]

Body Weight (lt)	Number of Drinks											
	1	2	3	4	5	6	7	8	9	10	11	12
100	.038	.075	.113	.150	.188	.225	.263	.300	.338	.375	.413	.450
110	.034	.066	.103	.137	.172	.207	.241	.275	.309	.344	.379	.412
120	.031	.063	.094	.125	.156	.188	.219	.250	.281	.313	.344	.375
130	.029	.058	.087	.116	.145	.174	.203	.232	.261	.290	.320	.348
140	.027	.054	.080	.107	.134	.161	.188	.214	.241	.268	.295	.321
150	.025	.050	.075	.100	.125	.151	.176	.201	.226	.251	.276	.301
160	.023	.047	.070	.094	.117	.141	.164	.188	.211	.234	.258	.281
170	.022	.045	.066	.088	.110	.132	.155	.178	.200	.221	.244	.265
180	.021	.042	.063	.083	.104	.125	.146	.167	.188	.208	.229	.250
190	.020	.040	.059	.079	.099	.119	.138	.158	.179	.198	.217	.237
200	.019	.038	.056	.075	.094	.113	.131	.150	.169	.188	.206	.225
210	.018	.036	.053	.071	.090	.107	.125	.143	.161	.179	.197	.215
220	.017	.034	.051	.068	.085	.102	.119	.136	.153	.170	.188	.205
230	.016	.032	.049	.065	.081	.098	.115	.130	.147	.163	.180	.196
240	.016	.031	.047	.063	.078	.094	.109	.125	.141	.156	.172	.188

[a] One standard drink equals: 12 oz regular beer, 5 oz table wine, 1 oz 80 proof liquor (National Highway Traffic Safety Administration (NHTSA)).

also on the speed and pattern of drinking and the rates of absorption, distribution, metabolism, and excretion (ADME) of ethanol, all of which in turn may be affected by a multitude of other factors (Drummer, 1999). Age, gender, race, body mass index, presence of other drugs, beverage type, food, genetics, and health status are a few examples of factors that can alter the ADME of ethanol and hence are important considerations when interpreting BAC levels in a person (Millar et al., 1992; Jones and Jonsson, 1994; Ammon et al., 1996; Karch, 1997; Drummer, 1999; Garriott, 2003). Once ethanol is absorbed it is distributed throughout the body via the blood to other tissues. Ethanol is distributed to the body water, and the greater the water contents of a tissue, the greater the ethanol concentration. The individual factors that determine BAC are body weight and adiposity, and distribution ratio (based on the total body water content) of ethanol. It follows that the concentration of alcohol within a person depends primarily on the amount of alcohol he or she drinks, the person's drinking pattern and total body water, and various other life and environmental issues.

Because men and women have different body water amounts (men average 68% and women 55%), there are differences between the ethanol concentration achieved in men and women of similar weight for the same amount of alcohol consumed. In other words, if a man and a woman of the same weight drink equivalent amounts of ethanol, the female will have a higher BAC on average. Similarly, if two men of different body weights drink the same amount of ethanol, the lighter man will have a higher BAC compared to the heavier man, and so on. In general, the less you weigh, the more you will be affected by a given amount of alcohol, all other factors assumed to be equal for this purpose.

The volume of distribution for ethanol was first expressed by Widmark using a formula that still bears his name (Garriott, 2003):

$$A = \frac{WRC}{0.8}$$

where A is the volume of pure ethanol consumed, W the body weight, R the distribution ratio, C the blood ethanol concentration, and 0.8 the specific gravity of ethanol.

There is a time lag from the period of intake of alcohol until the peak blood alcohol concentration is reached. This is regulated by the rate of alcohol absorption after oral intake. As described above, alcohol absorption is affected by numerous factors. Because of these factors, the time to reach peak blood alcohol concentrations varies greatly and can range between 15 min and 2 h. Hence, if alcohol absorption is not complete, peak alcohol levels will not yet be reached. The BAC is a balance between absorption and elimination of ethanol (Winek and Esposito, 1985).

There can be a threefold variation in ethanol elimination rates among individuals. Gender, age, body weight, and even the time of day can influence the rate of ethanol metabolism. To back-extrapolate from a blood alcohol concentration at one time to a value at an earlier time, one needs to know an accurate value for the ethanol elimination rate, which is difficult in view of the large interindividual variability in ethanol elimination kinetics. Also, to back-extrapolate, one uses linear kinetics, which may not be accurate, especially for the concentration range under consideration. The variability in individual rates of ethanol elimination, the difficulty in knowing exactly when absorption of ethanol was completed and how rates of ethanol elimination can change at different blood ethanol concentrations, cast some uncertainty on such retrograde calculations to an earlier blood ethanol concentration.

Since BAC is a function of the balance between how much is entering the blood versus how much is cleared or removed, the liver, via metabolism, plays a very important role. The liver removes most of the ethanol via oxidation, with lesser amounts excreted in the breath, urine, feces, and saliva or converted to ethyl glucuronide (Figure 12.2). The cytochrome P450 enzymes in liver microsomes catalyze the conversion of ethanol to acetaldehyde and water in the presence of molecular oxygen and NADPH (Figure 12.3). The toxic acetaldehyde product is quickly converted to acetic acid via acetaldehyde dehydrogenase, as shown in Fig. 2. The rate at which ethanol is cleared from the blood varies from one person to another. Heavy drinkers and alcoholics tend to eliminate alcohol much faster than does the inexperienced drinker. In most people the rate of clearance from blood ranges from 0.01 to 0.025% per hour, and is even higher in some very tolerant drinkers, with a mean of about 0.017%. This is very important for the forensic toxicologist.

Assume, for example, that an accident occurs at midnight and the alcohol content of a blood sample taken from the driver at a hospital at 2:00 A.M. (two hours later) was found to be 0.07%. Does this mean that the level was below 0.08% at the

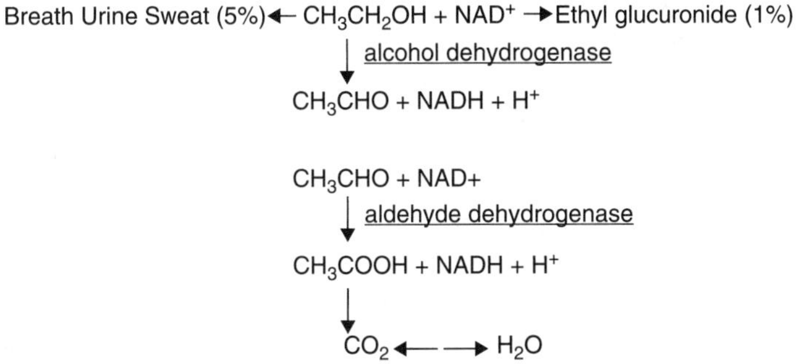

Breath Urine Sweat (5%) ← $CH_3CH_2OH + NAD^+$ → Ethyl glucuronide (1%)
$$\downarrow \text{alcohol dehydrogenase}$$
$$CH_3CHO + NADH + H^+$$

$$CH_3CHO + NAD+$$
$$\downarrow \text{aldehyde dehydrogenase}$$
$$CH_3COOH + NADH + H^+$$
$$\downarrow$$
$$CO_2 \leftarrow\!\longrightarrow H_2O$$

Figure 12.2 General scheme for ethanol oxidation. (a) <10% ethanol excreted in breath, sweat, and urine; (b) ~90% ethanol removed by oxidation; (c) most of this ethanol oxidation occurs in the liver; (d) ethanol cannot be stored in the liver; (e) no major feedback mechanisms to pace the rate of ethanol metabolism to the physiological conditions of the liver cell.

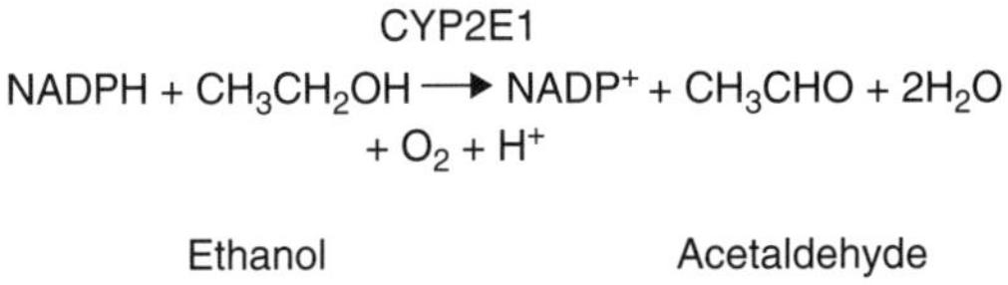

$$\text{CYP2E1}$$
$$NADPH + CH_3CH_2OH \longrightarrow NADP^+ + CH_3CHO + 2H_2O$$
$$+ O_2 + H^+$$

Ethanol   Acetaldehyde

Figure 12.3 Microsomal (cytochrome P450) oxidation of ethanol in the liver.

time of the accident? Well... it depends. Let's assume that the person stopped drinking at a time prior to the accident, so that all the ethanol was absorbed from the person's gastrointestinal tract at midnight (time lag). We can extrapolate the BAC to the time of the accident as follows: Using the average rate of 0.017 g/dL per hour for two hours, the BAC two hours earlier (12:00 midnight) would yield a BAC of 0.104 g/dL, using the average "burn-off" rate of 0.017 g/dL per hour. Under these circumstances this person would be considered legally intoxicated at the time of the accident. On the other hand, if the person stopped drinking very shortly before the time of the accident, the BAC would still be rising at midnight and would continue to rise and peak sometime between midnight and 2:00 A.M. This clearly complicates the estimation of what the level would have been at midnight, and caution needs to be applied to such assessments. Therefore, a considerable range of BACs may be possible, depending on the drinking pattern, and this must be taken into consideration when interpreting the data.

12.3 ALCOHOL IMPAIRMENT AND DRIVING SKILLS

Driving or operating automobiles or any other vehicles while under the influence of alcohol is dangerous and perhaps the most commonly encountered forensic issues associated with this drug. Legislation around the country has enacted various rules mandating fines and penalties for driving under the influence of alcohol. Alcohol is a central nervous system depressant. The precise mechanism by which alcohol affects cognition and psychomotor control is not known completely, but the fact that it does so is very well documented and is well understood by most members of the public. To appreciate the effects of alcohol on driving skill, the complexity of driving itself must be recognized. Driving is a multifaceted process that requires a combination of hand–eye coordination, psychomotor function, muscle control, and cognitive function. Furthermore, driving is a divided-attention task, and the driver does not have the comfort of focusing exclusively on any one of these components, but rather, must concentrate on all simultaneously (i.e., monitoring a number of sensory inputs and responding appropriately with all the changing circumstances). Alcohol use can affect the implementation of individual tasks, and its effects are therefore more pronounced on more complex tasks, of which driving is a perfect example.

Alcohol causes a slowing of nerve conduction, which results in slower reaction times, difficulty in processing information, and thus impaired performance in divided-attention tasks. People are clearly affected differently at different BAC values for various reasons discussed previously. Sometimes the effects of alcohol are not apparent without specialized testing and must be elicited by tasks that challenge those abilities (see field sobriety tests for impairment). A person can be affected by alcohol to the extent that it can impair driving skills before overt signs of intoxication (e.g., staggering and slurred speech) materialize. The BAC value at which driving ability is affected by alcohol use is a recurrent issue confronting forensic toxicologists, although most toxicologists agree that any person with a BAC above 0.1 g/dL would have impaired driving skills, and those above 0.2 g/dL would show signs of overt intoxication on close inspection, even by a nonexpert.

The pharmacology and toxicology of alcohol differ from person to person. Various behaviors and physiological functions affected by ethanol result in impairment of driving skills, including reaction time, tracking, vigilance or concentrated attention, divided attention, information processing, visual functioning, perception, and psychomotor functions. Detrimental effects on reaction time are known to occur at BAC values in excess of 0.07 g/dL. Bloods levels of ethanol ranging from 0.012 to 0.12 g/dL increase the reaction time more than threefold and BAC values over 0.12 g/dL impair reaction times even further. With more complicated tasks that require more than one attentive task occurring simultaneously (e.g., driving), increased reaction time is even more evident. This situation is known to occur with the impairing effects of ethanol on driving skills. Driving requires that a person control speed, track position, and maintain lane control, and monitor various outside signals visually and react to them, all of which are impaired by increasing BAC. Since alcohol is a central nervous system depressant, it affects a wide range of these functions. (Connors and Maisto, 1980; Linnoila et al., 1980; Moskowitz et al., 1985; Steele and Josephs, 1988; Maylor et al., 1989 Lipscomb and Nathan, 1980). Statistics consistently show that the crash risk is increased significantly when BAC values are above 0.04 g/dL. Several studies with automobile simulators have shown that risk taking is affected at higher levels of alcohol, described as 0.106 g/dL (Mongrain, 1974).

The area most often showing the greatest impairment with alcohol consumption is divided-attention performance. Driving is a task that requires considerable divided attention, and decrements have been shown to occur with BAC as low as 0.02 g/dL and worsens with increasing BAC. The second-fastest rise in impairment occurs with tracking performance, which can occur at or below 0.05 g/dL. Driving is a time-sharing task that requires two activities: compensatory tracking and visual search. Divided-attention tasks are particularly sensitive to the effects of rising BAC, and such impairment may begin to occur at very low levels of BAC (<0.02 g/dL) for some people. There is sufficient evidence to suggest that BAC values of 0.05 g/dL and above are capable of producing impairment of the major components of driving skills, including reaction time, tracking, divided-attention performance, information processing, visual functions, and psychomotor impairment (Nito et al., 1964; Money et al., 1965; Aschan and Gergstedt, 1975; Baloh et al., 1979; Fagan et al., 1987). A major factor that determines the magnitude of such impairing effects in a person is tolerance (see the discussion below). Tolerance to some of the impairment parameters of ethanol can occur, but the extent to which it develops is very difficult to predict (Levine et al., 1975; Lipscomb et al., 1980; Wilson et al., 1984; Niaura et al., 1988). However, it is known that the impairing effects of alcohol on experienced versus naive drinkers is attenuated for some of these tasks in heavy drinkers. Even so, serious effects of alcohol on driving skills occur in all subjects at BAC levels of 0.10 g/dL or higher, and cognitive function will also be affected in these subjects.

12.4 FIELD SOBRIETY TESTS

Field sobriety tests, including the walk and turn, one-leg stand, and horizontal gaze nystagmus, are commonly used in field testing to detect impairment (Aschan, 1958). Fundamentally, these are only screening tests that presumptively detect impairment and must be followed by some other analytical measurement of blood alcohol content directly or via extrapolation. Horizontal gaze nystagmus is more reliable at predicting blood alcohol content than the walk and turn and the one-leg tests. Field sobriety testing is generally applicable to blood alcohol levels of 0.08 g/dL and above. At lower levels more studies are needed. Consequently, at levels lower than 0.08 g/dL, field sobriety tests will not be helpful. Furthermore, the field sobriety test must take into account factors such as prescription medications or altered health status, such as neurological or orthopedic problems.

Pharmacologically what happens when a person starts to drink alcohol? At first, as the BAC begins to rise, there is an apparent stimulatory effect. The person may become more talkative and they may seem more social and have increased self-confidence. This effect is not a stimulatory effect of ethanol but a depression of the inhibitory centers of the brain. As the BAC increases, impairment of judgment, decision making, perception, motor function, and reaction time occurs. The impairment of motor function and muscular coordination starts to affect the ability to walk, speak, and maintain balance. Significant and progressive reduction in mental and physical abilities continues as the BAC rises and the person is visibly intoxicated. In relatively low-tolerant or inexperienced drinkers, a BAC value over 0.30% can result in loss of consciousness and possible death, and with concentrations rising over 0.40%, death becomes increasingly likely. Chronic heavy use leads to a high incidence of brain damage, liver disease, heart disease, stroke, metabolic disorders, and oral, lung, and liver cancers, and abuse during pregnancy increases the risk of fetal malformations and development of fetal alcohol syndrome.

Tolerance is the lessening of the effectiveness of a drug after a period of prolonged or heavy use (Levine et al., 1975; Lipscomb et al., 1980; Wilson et al., 1984; Niaura et al., 1988). Studies have shown that chronic alcohol users can have as much as twice the tolerance for alcohol as an average person. Tolerance develops from two separate mechanisms: dispositional (or metabolic) and functional. *Metabolic tolerance* occurs because the drug is metabolized or inactivated at a faster rate after prolonged use (Lipscomb and Nathan, 1980). Thus, a given dose produces a lower BAC value. *Functional tolerance* is an actual change in a person's sensitivity to the drug at an organ or biochemical system level. With functional tolerance, greater amounts of the drug are required to elicit similar effects with increasing use of the drug. It is important to note that even in heavy users of alcohol, impairment is clearly measurable at the BAC levels that are currently used for traffic law enforcement. Tolerance is not complete for all systems and may be marked in some but not others. However, it is significant to note that even though cognitive impairment is present, many highly tolerant persons are capable of engaging in physical activities and actions (including fighting) at levels that would result in unconsciousness or even death in nontolerant persons.

12.5 BLOOD ALCOHOL MEASUREMENTS

Early blood alcohol testing involved distillation methods based on alcohol from the blood, followed by oxidation and titration. One of the carryovers of these techniques is the use of whole blood as the sample of choice for legal blood alcohol determinations in body fluids. Although popular up until the 1970s, such methods have largely been replaced by enzymatic and gas chromatographic methods (Jones and Pounder, 1997).

12.5.1 Enzymatic Methods

Enzymatic methods are attractive because they can be automated readily and thus allow for high-throughput analysis. Typically, enzymatic ethanol determinations are often done as part of a test panel on plasma or serum for other clinical chemistry parameters. This is often done in a medical laboratory setting and is not considered "forensic" testing. The use of plasma is an important consideration, since the reports issued by hospital laboratories often do not specify the specimen type.

Plasma contains more water than whole blood, and correspondingly more alcohol per milliliter. The accepted practice for converting a plasma or serum alcohol into a whole blood alcohol is to divide by 1.18, although a range of 1.10 to 1.30 gives a 95% confidence interval for the corresponding whole BAC. The fact that there is a range associated with this conversion makes measurement directly in whole blood the method of choice.

Enzymatic methods rely on the enzyme-specific oxidation of ethanol to acetaldehyde with alcohol dehydrogenase. This oxidation requires reduction of the cofactor oxidized nicotinamide–adenine dinucleotide (NAD^+) to NADH, which is accompanied by a change in absorbance that can be monitored spectrophotometrically. Enzymatic methods can be susceptible to interference from other alcohols oxidized by the same enzyme, but the differential kinetics of these reactions lends some specificity to the method. When calibrated using the appropriate standards, conducted in duplicate, and analyzed alongside matrix-matched controls with acceptable results, enzymatic oxidation can be an acceptable forensic method of determining BAC. These precautions are rarely followed in a clinical setting, however, and the use of hospital alcohol results for forensic purposes should be interpreted with this limitation in mind.

12.5.2 Headspace Gas Chromatography

By far the most popular, precise, and accurate method of determining blood alcohol is by gas chromatography (GC). This technique has the advantages of being semiautomatable, extremely reproducible, and compatible with multiple sample types. Currently, both packed column and capillary column procedures are popular and acceptable for forensic blood alcohol analysis. GC assays provide the greatest amount of flexibility and specificity in analyzing for these volatile compounds. Direct injection of biological samples into GC columns has been used in the past as the method of sample introduction. This typically leads to column contamination and decreased performance.

The incorporation of headspace sampling into the method prevents the buildup of nonvolatile contamination at the head of the column and helps to maintain consistent performance and extend column lifetime. Analysis time and resolution are two critical factors when developing a GC assay for ethanol. Analysis time for each sample should be as short as possible, while maintaining baseline resolution for all analytes.

An aliquot of blood (200 μL) in sodium fluoride and potassium oxalate is removed in duplicate to test for ethanol and other volatiles of low relative molecular mass. This is performed by mixing the blood using an automated pipettor or diluter with 2 mL of a 3 M NaCl solution that contains 1.0 g/L of 1-propanol as the internal standard. The mixed sample is dispensed into a 10-mL septum-topped vial and sealed. The vials are loaded into an automated headspace autosampler attached to a GC equipped with flame ionization detection (FID) in duplicate. Each sample is run in combination with a set of contemporaneous calibrators (0.4, 0.8, 1.6, and 3.2 g/L), a blank following the highest standard, and an externally purchased reference control after every tenth sample. The preservatives are important in this analysis. The sodium fluoride is an enzyme poison that will retard the formation or degradation of alcohol by bacteria. This is especially important in postmortem casework in which there is significant opportunity for bacterial contamination after death or at autopsy. It is less significant in blood drawn under sterile conditions from a living person. The potassium oxalate is an anticoagulant, which ensures that the blood remains homogeneous and does not separate into red cells and serum, since the water-enriched serum sample would contain more ethanol than would the whole blood. Since GC–mass spectrometry (GC–MS) is typically not used to test volatiles of low relative molecular mass, the use of complementary GC columns for duplicate testing is highly desirable, since (as in this case) the retention times obtained as a result of the different selectivity of the two columns provide an added dimension in the identification of a compound. Frequent calibration is desirable, but not essential. However, if a laboratory performs quantifications using a historical calibration curve, the reliability of the curve should be checked frequently with appropriate and comprehensive controls. Control samples obtained from and certified by an external vendor allow the laboratory to demonstrate the accuracy of its procedure by comparison with other laboratories.

When blood samples are analyzed for forensic purposes, civil or criminal, a standard procedure is to make duplicate determinations (Gullberg and Jones, 1994; Andreasson and Jones, 1996). The concentration of blood alcohol should be reported with the confidence limits such as 95% or 99% (Goldstein, 1983; Jones and Schubertz, 1989). Therefore, it is crucial that in hospital laboratories, the calibration of standards test by biological specimens along with the unknowns be done and kept on record for forensic evidence (Fraser, 1986). Cases have been thrown out of court due to the absence of the raw standardization data. Analyses done by a clinical toxicology laboratory and a certified forensic lab are dissimilar. Forensic laboratories perform analyses on samples of whole blood, whereas clinical laboratories usually measure plasma or serum ethanol concentration. The water content of plasma and serum is higher than that of whole blood and therefore requires an appropriate conversion (Winek and Carfagna, 1987; Frajola, 1993). Typically, the range is between 15 and 20% for plasma as compared to whole blood, and this conversion factor should be used in forensic work (Rainey, 1993). This difference may be significant in some cases, and should not be missed.

12.5.3 Breath Alcohol Testing

Driving under the influence (DUI or DWI) of alcohol and drugs is one of the most significant issues associated with abuse of these materials. Several testing strategies have been developed and introduced to assess the role of alcohol intoxication in impaired driving and to detect and prosecute offenders. Alcohol is still the drug that contributes most often to impaired driving. The smell of alcohol beverages on the breath has long been recognized as an indication that a person has consumed ethanol. Alcohol is eliminated unchanged in the breath and hence provides a means by which alcohol can be measured in the body. Breath alcohol instruments were developed to provide a fast and easy way of monitoring BAC indirectly. Analysis of a person's expired air provides an indirect way of monitoring volatile substances in the pulmonary blood. Breath has become the biological specimen of choice for measuring alcohol concentration in association with the enforcement and prosecution of drivers suspected of being impaired. Some of the reasons for this preference include:

- Breath collection is generally considered to be noninvasive.
- Rapid and inexpensive analysis allows for timely results, often as a field test.
- Analysis is straightforward and requires minimal operator training.
- Analysis does not require sophistcated laboratory equipment to obtain results.

These factors have combined to establish breath alcohol analysis as the most common forensic analysis employed in drunk-driving enforcement and workplace alcohol testing. Typically applied as a field test by a trained user, the interpretation of results begins with a clear understanding of the purpose of such measurement. Two classes of instrument for breath alcohol analysis can be distinguished, based on whether the results are intended for qualitative screening or as evidentiary testing for prosecution purposes. For example, prearrest screening breath tests clearly have a different purpose from that of the subsequent evidential breath test, which may be used as binding evidence in court. The serious implications associated with a drunk-driving conviction necessitate, at a very minimum, a sound forensic breath test protocol, which includes:

1. Trained and qualified instrument operators
2. At least 15 min of pre-exhalation observation
3. Internal instrument standards to verify calibration
4. Collection of duplicate breath samples
5. Appropriate agreement (i.e., $\pm 10\%$ of the mean) between duplicates
6. External control standards
7. Blank tests performed between all measurments
8. Appropriate printout of all data
9. Error detection systems
10. Instruments evaluated periodically according to an established quality assurance procedure to ensure optimal analytical performance, including accuracy, precision, linearity, and critical-system evaluations (e.g., acetone detection, radio-frequency interference detection, mouth alcohol detection)

12.5.4 Breath Alcohol Instrumentation

Ethanol is detectable in the breath and hence provides a means by which evidentiary testing can be accomplished to provide direct onsite collection and analysis of blood alcohol concentration for forensic purposes, principally for traffic law enforcement. The important characteristic of ethanol that allows for its measurement in the field is that ethanol is volatile. The theory that provides the foundation for such testing is based scientifically on the fact that the distribution of ethanol between the blood and the alveolar air obeys Henry's law. Therefore, at a given temperature, a direct relationship exists between the amount of alcohol (volatile substance) dissolved in blood (liquid) and the amount of alcohol in the vapor above the solution (alveolar air). At the temperature of expired air (34°C), the amount of ethanol in 2100 mL of alveolar air is assumed to be equivalent to the amount of ethanol in 1 mL of venous blood. Various instruments have been developed to take advantage of this ratio (2100:1). Various handheld devices are available for roadside testing and prearrest screening. For more controlled quantitative analysis of BrAC for evidential purposes, the instruments are much larger and more sophisticated, and include ways to check calibration, analyze alcohol free air, and produce hardcopies of the results. The best-known and most robust chemical handheld roadside breath alcohol screening tests generally utilize a fuel cell–based technology. These sensors oxidize ethanol to acetaldehyde and in the process produce free electrons, with the electric current generated directly proportional to the amount of alcohol present in the sample. Other alcohols present in the breath, such as methanol and isopropanol, also undergo chemical oxidation, although at different rates. Acetone (e.g., untreated diabetes) exhaled in the breath is not oxidized and will not cause a false positive. However, if acetone levels are exceedingly high, isopropanol can be formed in the body by reduction of acetone. Also for alcohol, enzymatic tests with a color endpoint have been developed for the field measurement of saliva alcohol. Such tests may have some potential value as screening tests in instances where the subject cannot provide breath, such as if injured and immobile.

In 1938, Rolla Harger developed the first commercially available forensic breath alcohol instrument intended for law enforcement applications, called the Drunk-o-meter. To use the Drunk-o-meter, the person being tested blew into a balloon and the air in the balloon was then released into a chemical solution containing potassium permanganate. If there was alcohol in the breath, the chemical solution changed color. The greater the color change, the more alcohol in the breath. The level of alcohol in a person's blood could then be estimated by a simple equation. This was a wet chemistry method based on optical measurement of the color change that occurred in potassium permanganate resulting from the oxidation of alcohol in the breath sample. This was largely replaced by the much more portable Breathalyzer developed in the 1950s. The Breathalyzer also utilized a colorimetric technique following the oxidation of alcohol in a potassium dichromate solution. As concerns began to develop that there was a lack of specificity and accuracy of wet chemical methods, instruments that employed the optical absorption of infrared energy according to Beer's law began to replace these methods for most forensic applications during the 1970s. Several variations of the infrared technology have been applied, including the use of multiple wavelengths to increase the specificity for ethanol. Most evidential breath-testing instruments used today identify and measure the concentration of alcohol by its absorption of infrared energy at wavelengths of 3.4 and/or 9.5 μm, which correspond, respectively, to the C—H and C—O vibration stretching in the ethanol molecules.

During the 1980s, fuel cell technology emerged as a useful analytical method in forensic breath alcohol measurement. Although fuel cells have been employed typically in screening devices, their technology has improved dramatically in recent years. Selectivity has been enhanced in one modern evidential instrument, the Alcotest 7110 (Draeger Instruments, Durango, CO), which now employs both infrared and fuel cell methodologies to quantify ethanol in breath and represents the first dual technology breath alcohol instrument. Employing two different and independent measuring systems, infrared spectroscopy and electrochemical cell technology, it provides the highest level of forensic and legal veracity. Such true dual-sensor instruments are now readily available on the market. By means of these instruments, both the electrochemical sensor and the infrared-optical sensor operating at 9.5-μm-wavelength quantify BrAC.

Modern breath alcohol instruments offer several advantages over their predecessors. The modern instruments are computerized, which greatly reduces the potential for operator error and ensures compliance with specific protocol requirements. In addition, automated instruments enable rapid analysis, error detection, ensured compliance with breath sampling parameters, automated purging of sample chamber and signal processing, regulation of test protocol and sequence, data collection and hardcopy printout of results. Measurement quality control is enhanced, therefore, by allowing only those test results that fully conform with predetermined standards to be printed out and to qualify as evidence in court. Contemporary instruments also allow features of "intelligent measurement" to be incorporated. These features permit the instrument to "make decisions" regarding test performance, run, or analytical properties. Data collection features also allow for the development of control charts and other methods to evaluate instrument and programmed performance over time. Various forms of data analysis can provide relevant information to the courts and enhance confidence in the results. Many features of the emergent analytical technology have been integrated advantageously into forensic breath test instruments.

12.5.5 Conversion from BrAC to BAC

The extrapolation from breath alcohol content to blood alcohol content level at the time of accident or arrest is one of the most important steps in the forensic process. While the conversion process takes into account simple mathematical conversion, it is imperative to understand that in most cases the extrapolation of the amount of blood alcohol concentration to a time other than the time when the specimen of body fluid or breath was taken from the subject is applicable only to the average person in a normal physical condition and may not be relevant otherwise. Current research has shown that there are multiple variables, such as blood breath ratio, absorption rates of alcohol, elimination rates of alcohol, and difference in total body water (Winek and Esposito, 1985; Karch, 1997). The analyst should be aware that the BAC/BrAC ratio of alcohol changes as a function of time after drinking, depending on whether the test was done on the absorptive or postabsorptive phase of the alcohol curve, which my have an important bearing on the interpretation of these results.

12.5.6 Urine and Saliva

Urine has been used in some jurisdictions as a specimen for forensic alcohol testing. It is amenable to analysis by both the enzymatic and chromatographic methods described above. Its major limitations are collection and interpretation. Urine collection, if observed, raises privacy issues, and if not observed, raises issues of chain of custody and the potential for adulteration, substitution or dilution. With respect to interpretation, urinary alcohol concentration is an indirect measure of BAC. Urine is formed in the kidneys as an ultrafiltrate of blood. As with plasma, therefore, a factor must be applied to account for the higher water content. Of more importance, however, is the fact that since urine is formed over time, the urine alcohol concentration is an average corresponding to the blood concentration over the time during which the urine was formed and collected in the bladder. Since this can cover an appreciable amount of time (up to several hours), even a urine sample corrected for water content may not reflect the true BAC at the time of urine collection. To correct for this, for forensic purposes urine is generally collected for analysis about an hour after an initial void. The subject is directed to empty his or her bladder, after which he or she is allowed a period of time (approximately an hour) during which new urine corresponding to their current BAC is formed and collected for testing. Such complications and the associated problems of interpretation mean that urine alcohol determination is not a preferred forensic approach, although it is acceptable in many jurisdictions. Problems with interpretation are generally legislated away with a statutory per se urine alcohol offence.

Saliva or oral fluid is a plasma ultrafiltrate produced through the parotid and other glands in the oral cavity. Drugs, including alcohol, that are present in the blood will also be found in the saliva (Jones, 1981, 1993). Care must be taken when examining oral fluid for alcohol because it can be subject to contamination from drugs and alcohol simply present in the oral cavity. The length of time of the contamination depends on the drug's lipophilicity and water solubility. Mouth alcohol is generally dissipated by absorption and exhalation within 15 min. This is why at least a 15- to 20-min observation and waiting period is stipulated prior to the subject being offered an evidential breath test, and a similar waiting period is essential for a test of salivary alcohol. A number of companies have marketed oral fluid alcohol test kits, which are based on the principle of enzymatic oxidation. These are to be used as screening devices because their quantitative accuracy is variable.

Generally, from an analytical standpoint there is a good correlation between circulating blood alcohol and oral fluid alcohol concentrations, and a properly collected and preserved oral fluid sample can be subjected to an analysis, usually by headspace GC, with good-quality results. Samples are usually collected in a tube with minimal headspace to prevent loss, together with appropriate preservative (enzyme inhibitor) to prevent postcollection alcohol production.

12.5.7 Ethyl Glucuronide

This is a compound formed from a minor metabolic pathway of ethanol (Yegles et al., 2004). Less than 1% of ingested ethanol is conjugated in the liver with glucuronic acid to form a water-soluble compound that is present in blood and excreted in urine. It has also been detected in hair, and this is currently being used in some alcohol and drug monitoring plans clinically (Jurado et al., 2004; Skipper et al., 2004). Ethyl glucuronide is a stable compound and a specific indicator for ethanol ingestion and has been applied to some forensic applications; however, measurement of EG levels in both blood and urine do not bear a constant relationship to BAC (Schmitt et al., 1997; Bergstrom et al., 2003; Jurado et al., 2004; Skipper et al., 2004; Yegles et al., 2004). The demonstration of EG in a specimen only indicates recent (within the last 40 h or even longer) ingestion of ethanol, which limits its use. It is capable of determining whether or not the person had been drinking in the recent past. A serum ethyl glucuronide concentration of greater than 5 mg/L when the BAC is less than 0.10 g/dL has been suggested as representing alcohol misuse (Drummer, 1999).

12.5.8 Postmortem Determination of Alcohol

One of the primary roles of the forensic toxicologist is to determine the role of drugs and alcohol in establishing a cause of death (Pounder and Jones, 2006). Ethanol is a major analyte in postmortem forensic determinations (Turkel and Clifton, 1957; Zumwalt et al., 1982). Commonly, the forensic toxicologist must opine on the role that alcohol may have played in a forensic case in both criminal and civil cases. This includes traumatic injuries such as motor vehicle accidents, alcohol poisoning, criminal behavior and human performance toxicology, and occupational incidents, to name but a few. There are major issues associated with the interpretation of the analytical alcohol results on samples obtained at autopsy. These are mostly due to the lack of homogeneity of blood samples, postmortem alcohol production or loss, and disruption of the gastrointestinal tract following traumatic injury, resulting in the release of alcohol into the surrounding tissues. However, with proper care, ethanol analysis can be accomplished with high confidence on postmortem samples. Postmortem forensic analysis is discussed in detail in Chapter 13, so the topic is only summarized here.

Blood is the usual specimen provided for postmortem ethanol analysis. The site from which postmortem blood is collected is important (Plueckhahn, 1968; Prouty and Anderson, 1987). The preferred site is the femoral vein. Blood from the pericardial sac or chest cavity can have falsely elevated alcohol levels due to postmortem diffusion of alcohol from the stomach (Plueckhahn and Ballard, 1967; Pounder and Smith, 1995). As with antemortem blood, samples should be collected into containers with appropriate preservative (sodium fluoride) and anticoagulant (potassium oxalate), depending on how the analysis is to be performed. Because of ethanol's uniform distribution through the body, other fluids and tissues can be used for analysis as well. Vitreous fluid from within the eye consists of 99% water with small amounts of salts and mucoproteins. It is in an enclosed space and is not normally subject to contamination by microorganisms, and these specimens are easily collected at autopsy. Vitreous is important because due to its anatomical location, it is usually well preserved after death and is generally unaffected by postmortem changes. The high water content of vitreous means that measured ethanol levels should be comparable to blood levels. The ratio of blood to vitreous concentration is between 0.57 and 0.96, with a value of 0.85 the most commonly used best estimate. Blood has a lower water content than vitreous, so the expectation is that the equilibrium blood vitreous alcohol ratio will be less than 1. In cases where the blood/vitreous ratio exceeds 1 it is likely that death occurred while the person was in the absorptive phase and before diffusion equilibrium had been obtained, and this observation may be of forensic significance (Sturner and Coumbis, 1966; Felby and Olson, 1969; Kraut, 1995; Yip, 1995).

Urine collected from the bladder like vitreous is resistant to postmortem contamination by microorganisms. Postmortem alcohol production in urine occurs only when the urine contains large amounts of glucose (diabetes) or there was a urinary tract infection during life (Drummer, 1999). Urine is easily sampled at autopsy provided that the bladder is intact. A urine alcohol measurement does not reflect the alcohol content of urine at any point in time (i.e., time of death). It may be a useful specimen to use to assess possible prior consumption of alcohol. Urine concentrations are generally similar to those of blood but may vary considerably and should not be used to predict BAC without incurring a large error. As long as this is recognized, urinary alcohol estimation can be a useful measurement.

12.5.9 Quality Assurance of Alcohol Testing

Monitoring of the day-to-day performance of the collection, storage, and analytical methodology is a vital component of a proper forensic alcohol analysis (Moffat et al., 2004). The laboratory must have the highest confidence that the quality of the data produced is of utmost precision and accuracy because the results are often used for criminal prosecution or civil litigation. Use of a range of replicate determinations is an important means of controlling the quality assurance of test results. All specimens are carefully inspected when they arrive and the condition is noted. When needed, preservatives and anticoagulant are used to inhibit enzyme activity to prevent post-collection microbiological production of alcohol and blood clotting in the samples, respectively. Proper labeling is essential, with all the required identifying information to be included in a suitable label. The collection container must be properly sealed to prevent unauthorized tampering or loss of sample during transport and storage. For specimens that are stored, appropriate refrigeration ($-4°C$) of the samples is needed to prevent loss of the volatile alcohol, and the rate of lost alcohol during this time needs to be established. Calibration and measurements of controls and standards must be completed and documented by the measuring laboratory, and copies of the analytical data, including chromatograms should be maintained in a secure storage facility so they are available for scrutiny or produced as evidence. The laboratory performing the alcohol analysis must be properly accredited and must participate in external proficiency testing to establish that all results produced are precise and accurate. Failure to do so greatly reduces the confidence in the data produced and may result in the data being inadmissible in court.

REFERENCES

Ammon, E., Schafer, C., Hoffman, U., and Klotz, U. (1996). Disposition and first-pass metabolism of ethanol in humans: Is it gastric or hepatic and does it depend on gender. *Clin. Pharmacol. Ther.,* **59**:503.

Andreasson, R., and Jones, A. W. (1996). The life and work of Erik MP Widmark. *Am. J. Forensic Med. Pathol.,* **17**:177.

Aschan, G. (1958) Different types of alcohol nystagmus. *Acta Otolaryngol.,* Suppl. **140**:69.

Aschan, G., and Gergstedt, M. (1975). Positional alcoholic nystagmus (PAN) in man following repeated alcohol doses. *Acta Otolaryngol.,* Suppl. **330**:15.

Baloh, R. W., Sharma, S., Moskowitz, H., and Griffith, R. (1979). Effect of alcohol and marijuana on eye movements. *Aviat., Space Environ. Med.,* **50**:18.

Bergstrom, J., Helander, A., and Jones, A. W. (2003). Ethyl glucuronide concentrations in two successive urinary voids from drinking drivers: relationship to creatinine content and blood and urine ethanol concentrations. *Forensic Sci. Int.,* **133(1–2)**:86–94.

Connors, G. J., and Maisto, S. A. (1980). Effects of alcohol instructions and consumption rate on motor performance. *J. Stud. Alcohol*, **41**:509.

Drummer, O. H. (1999). *The Forensic Pharmacology of Drug Abuse*. London: Edward Arnold, p. 300.

Fagan, D., Tiplady, B., and Scott, D. B. (1987). Effects of ethanol on psychomotor performance. *Br. J. Anaesth.,* **59**:961.

Felby, S., and Olson, J. (1969). Comparative studies of postmortem ethyl alcohol in vitreous humor, blood, and muscle. *J. Forensic Sci.,* **15**:185.

Frajola, W. J. (1993). Blood alcohol testing in the clinical laboratory: Problems and suggested remedies. *Clin. Chem.,* **38**:377.

Fraser, C. G. (1986). *Interpretation of Clinical Chemistry Laboratory Data*. Oxford, UK: Blackwell Scientific.

Garriott, J. (2003). *Medical–Legal Aspects of Alcohol*, 4th ed. Tucson, AZ: Lawyers & Judges Publishing Company.

Goldstein, D. B. (1983). *Pharmacology of Alcohol*. New York: Oxford University Press.

Gullberg, R. G., and Jones, A. W. (1994). Guidelines for estimating the amount of alcohol consumed from a single measurement of blood alcohol concentration; re-evaluation of Widmark's equation. *Forensic Sci. Int.,* **69**:119.

Jones, A. W. (1981). Quantitative relationships among ethanol concentrations in blood, breath, saliva and urine during ethanol metabolism in man. In: Goldberg, L. (Ed.), *Proceedings of the 8th International Conference on Alcohol, Drugs and Traffic Safety*. Stockholm: Almqvist and Wiksell, p. 550.

Jones, A. W. (1993). Pharmacokinetics of ethanol in saliva; Comparison with blood and breath alcohol profiles, subjective feelings of intoxication and diminished performance. *Clin. Chem.,* **39**:1837.

Jones, A. W., and Jonsson, K. A. (1994). Food-induced lowering of blood ethanol profiles and increased rate of elimination immediately after a meal. *J. Forensic Sci.,* **39**:1084.

Jones, A. W., and Pounder, D. J. (1997). Measuring blood-alcohol concentration for clinical and forensic purposes. In: Karch, S. B. (Ed.), *Drug Abuse Handbook*. Boca Raton, FL: CRC Press, Chap. 5.2.

Jones, A. W., and Schuberth, J. O. (1989). Computer-aided headspace gas chromatography applied to blood-alcohol analysis: importance of on-line process control. *J. Forensic Sci.*, **34**:1116.

Jurado, C., Soriano, T., Gimenez, M. P., and Menendez, M. (2004). Diagnosis of chronic alcohol consumption. Hair analysis for ethyl-glucuronide. *Forensic Sci. Int.*, **145(2–3)**:161–166.

Karch, S. B. (Ed.) (1997). *Drug Abuse Handbook*. Boca Raton, FL: CRC Press, pp. 1152.

Kraut, A. (1995). Vitreous alcohol. *Forensic Sci. Int.*, **73**:157.

Levine, J. M., Kramer, J., and Levine, E. (1975). Effects of alcohol on human performance. *J. Appl. Psychol.*, **60**:508.

Linnoila, M., Erwin, C. W., Ramm, D., and Cleveland, W. P. (1980). Effects of age and alcohol on psychomotor performance of men. *J. Stud. Alcohol*, **41**:488.

Lipscomb, T. R., and Nathan, P. E. (1980). Effect of family history of alcoholism, drinking pattern, and tolerance on blood alcohol level discrimination. *Arch. Gen. Psychiatry*, **37**:576.

Lipscomb, T. R., Nathan, P. E., Wilson, G. T., and Abrams, D. B. (1980). Effects of tolerance on the anxiety-reducing functions of alcohol. *Arch. Gen. Psychiatry*, **37**:577.

Maylor, E. A., Rabbitt, P. M., and Connolly, S. A. (1989). Rate of processing and judgment of response speed: comparing the effects of alcohol and practice. *Percep. Psychophys.*, **45**:431.

Millar, K., Hammersley, R. H., and Finnigan, F. (1992). Reduction of alcohol-induced performance by prior ingestion of food. *Br. J. Psychol.*, **83**:261.

Moffat, A. (2004). (Eds.), In: Moffat, A., Ossselton, M. D., and Widdop, B. *Clarke's Analysis of Drugs and Poisons*, NY: Pharmaceutical Press.

Money, K. E., Johnson, W. H., and Corlett, B. M. (1965). Role of semicircular canals in positional alcohol nystagmus. *Am. J. Physiol.*, **208**:1065.

Mongrain, S. (1974). Standing, L Impairment of cognition, risk-taking, and self-perception by alcohol. *Percept. Motor Skills*, **69(1)**:19.

Morgan Jones, B., and Vega, A. (1972). Original investigations: cognitive performance measured on the ascending and descending limb of the blood alcohol curve. *Psychopharmacologia (Berlin)*, **23**:99–114.

Moskowitz, H., Burns, M. M., and Williams, A. F. (1985). Skilled performance at low blood alcohol levels. *J. Stud. Alcohol*, **46**:482.

Niaura, R. S., Wilson, G. T., and Westrick, E. (1988). Self-awareness, alcohol consumption, and reduced cardiovascular reactivity. *Psychosomatic Med.*, **50**:360.

Nito, Y., Johnson, W. H., and Money, K. E. (1964). The non-auditory labyrinth and positional alcohol nystagmus. *Acta Otolaryngol.*, 58:65.

Plueckhahn, V. D. (1968). Alcohol levels in autopsy heart blood. *J. Forensic Med.*, **15**:1221.

Plueckhahn, V. D., and Ballard, B. (1967). Diffusion of stomach alcohol and heart blood alcohol concentration at autopsy. *J. Forensic Sci.*, **12**:463–470.

Pounder, D. J., and Jones, A. W. (2006). Measuring alcohol postmortem. In: Karch, S. B. (Ed.), *Drug Abuse Handbook*, 2nd ed. Boca Raton, FL: CRC Press, Chap. 5.3.42.

Pounder, D. J., and Smith, D. R. W. (1995). Postmortem diffusion of alcohol from the stomach. *Am. J. Forensic Med. Pathol.*, **16(2)**:89–96.

Prouty, R., and Anderson, W. (1987). A comparison of post-mortem heart blood and femoral blood ethyl alcohol concentration. *J. Anal. Toxicol.,* **11(5)**:191–197.

Rainey, P. M. (1993). Relation between serum and whole blood ethanol concentrations. *Clin. Chem.,* **39**:2288.

Schmitt, G., Droenner, P., Skopp, G., and Aderjan, R. (1997). Ethyl glucuronide concentration in serum of human volunteers, teetotalers and suspected drinking drivers. *J. Forensic Sci.,* **42(6)**:1099–1102.

Skipper, G. E., Weinmann, W., Thierauf, A., Schaefer, P., Wiesbeck, G., Allen, J. P., Miller, M., and Wurst, F. M. (2004). Ethyl glucuronide: a biomarker to identify alcohol use by health professionals recovering from substance use disorders. *Alcohol,* **5**:445–449.

Steele, C. M., and Josephs, R. A. (1988). Drinking your troubles away: II. An attention-allocation model of alcohol's effects on stress. *J. Abnormal Psychol.,* **97**:196.

Sturner, W. Q., and Coumbis, R. J. (1966). The quantitation of ethyl alcohol in vitreous humor and blood by gas chromatography. *Am. J. Clin. Pathol.,* **46**:349–351.

Turkel, H. W., and Clifton, H. (1957). Erroneous blood alcohol findings at autopsy: avoidance by proper sampling technique. *J. Am. Med. Assoc.,* **165**:1077.

Wilson, J., Erwin, G., and McClearn, G. (1984). Effects of ethanol: Behavioral sensitivity and acute behavioral tolerance. *Alcoholism: Clin. Exp. Res.,* **8**:366.

Winek, C. E., and Carfagna, M. (1987). Comparison of plasma, serum and whole blood ethanol concentrations. *J. Anal. Toxicol.,* **11**:267.

Winek, C. L., and Esposito, F. M. (1985). Blood alcohol concentrations: factors affecting predictions. *J. Legal Med.,* 34–61.

Yegles, M., Labarthe, A., Autwarter, V., Hartwig, S., Vater, H., and Wennig, R. (2004). Comparison of ethyl glucuronide and fatty acid ethyl ester concentration in the hair of alcoholics, social drinkers and teetotallers. *Forensic Sci. Int.,* **145(2–3)**:167–173.

Yip, D. C. P. (1995). Vitreous humor alcohol. *Forensic Sci. Int.,* **73**:15.

Zumwalt, R. E., Bost, R. O., and Sunshine, I. (1982). Evaluation of ethanol concentrations in decomposed bodies. *J. Forensic Sci.,* 27:549–554.

CHAPTER 13

Fundamental Issues of Postmortem Toxicology

DONALD B. HOFFMAN

Department of Sciences, John Jay College of Criminal Justice, The City University of New York, New York

BETH E. ZEDECK

Pediatric Nurse Practitioner, New York

MORRIS S. ZEDECK

(Retired) Department of Sciences, John Jay College of Criminal Justice, New York

Summary The basic principles of forensic postmortem toxicology are presented. In this chapter we discuss the acquisition and usefulness of various specimens, current analytical techniques, and interpretation of findings. Special problems associated with the interpretation of drug levels include the conditions of the specimens and the effects of postmortem redistribution, postmortem drug changes, pharmacogenomics, drug interactions, and embalming fluid.

13.1	Introduction		458
13.2	Tissue and fluid specimens		460
	13.2.1	Blood	460
	13.2.2	Urine	461
	13.2.3	Vitreous humor and cerebrospinal fluid	461
	13.2.4	Gastric contents	462
	13.2.5	Meconium	463
	13.2.6	Brain	464
	13.2.7	Liver and bile	464
	13.2.8	Lung, spleen, kidney, and skin	465
	13.2.9	Muscle	465
	13.2.10	Bone, teeth, nails, and hair	465
	13.2.11	Other materials for analysis	466
13.3	Specimen collection and storage		466

Forensic Chemistry Handbook, First Edition. Edited by Lawrence Kobilinsky.
© 2012 John Wiley & Sons, Inc. Published 2012 by John Wiley & Sons, Inc.

13.4	Extraction procedures	467
13.5	Analytical techniques	467
13.6	Interpretation	470
	13.6.1 Postmortem redistribution	470
	13.6.2 Pharmacogenomics	471
	13.6.3 Drug interactions	472
	13.6.4 Drug stability and decomposed tissue	473
	13.6.5 Effects of embalming fluid	474
	Conclusion	475
	References	476

13.1 INTRODUCTION

One aspect of forensic chemistry is forensic toxicology, the discipline that analyzes biological fluids and tissues for the presence of drugs and poisons, determines their concentration, and interprets the findings. The results may be helpful in determining whether one or more drugs played a role in a person's injury, illness, behavior, or death. Thus, forensic toxicology can be divided into clinical toxicology, human performance toxicology, employment drug testing, and postmortem toxicology. This fascinating and challenging profession draws on the disciplines of pharmacology and of biochemical, analytic, and organic chemistry. It is not our purpose here to present an inclusive detailed treatise on forensic toxicology, as excellent source materials of this nature are available. Methods for the extraction of drugs from various tissues and the analytical procedures used for the detection and measurement of drug concentration have been described in many textbooks and reference books (Chamberlain, 1995; Wong and Sunshine, 1997; Levine, 2006; Smith et al., 2007), as have the effects of drugs (Marquardt et al., 1999; Brunton, et al., 2005; Klaassen, 2008). The reader interested in pursuing these topics in more detail is encouraged to utilize these and all other relevant sources of information.

In this chapter we focus on a special area of forensic toxicology: postmortem toxicology. We wish to present to the forensic scientist an overview of the current key issues involved in the practice of postmortem toxicology. Several earlier articles discussing this particular area of forensic science are cited (Prouty and Anderson, 1990; Hilberg et al., 1999b; Drummer, 2004, 2007; Skopp, 2004; Dolinak, 2005; Levine, 2006; Rodda and Drummer, 2006; Spitz and Spitz, 2006; Watterson et al., 2006; Karch, 2007; Garriott, 2008).

Postmortem forensic toxicology is an essential and integral component of any comprehensive medicolegal investigation into the cause and contributing circumstances of death. As such, it is essential that it be carried out appropriately and efficiently in order to provide the necessary information. An inclusive toxicological investigation encompasses knowledge of case history, specimen selection, choice of extraction procedures, application of multiple analytic techniques, and interpretation of findings.

INTRODUCTION

Postmortem toxicology involves the analysis of tissues and fluids taken from cadavers, either shortly after death or from decomposed or exhumed bodies, and also from formalin-fixed specimens. Each presents special situations that must be taken into consideration to ensure proper interpretation of the data. Drug samples, drug paraphernalia, and prescription items may also be found at the site of death and usually become part of the postmortem toxicological examination.

Postmortem tissues taken from a person who had been confined to a hospital for an extended period may have limited value because of drug clearance occurring over a period of time prior to death. However, the analysis of antemortem blood and urine, when available and especially if taken around the time of hospital admission, may become essential in postmortem protocols when there has been a prolonged period of survival, or when extensive administration of transfusions occurred prior to death.

Knowledge of the case history is very important. The circumstances surrounding the death or the indicated use of specific agents may require the use of non-routine procedures to check for the possible presence of certain classes of drugs or poisons that are not necessarily included in the routine protocols. Examples of such substances are digoxin and other cardiac glycosides, insulin, quaternary ammonium muscle relaxants, water-soluble or very highly polar compounds [e.g., γ-hydroxybutyrate (GHB)], and salts of heavy metals. Such substances have been encountered in homicidal or other criminal poisonings, and some following therapeutic treatments. There must be knowledge as to appropriate specimen selection and preservation. Certain specimens may be particularly appropriate for a particular drug or where the condition of the body is a factor. Issues such as postmortem redistribution of drug from its site at the time of death (discussed below) and postmortem drug stability or transformation play key roles both in the choice of specimen as well as in the subsequent interpretation of findings (Drummer, 2004; Skopp, 2004). Cases involving decomposed, dismembered, embalmed and exhumed bodies require special attention in order to maximize the utilization of such specimens as are available at autopsy.

Once an analyte has been detected in bodily fluids and tissues, questions arise. Is the concentration of analyte reflective of its concentration at the time of death, or is it higher or lower than at the time of death? Conditions that can alter drug concentration include changes resulting from human enzymes, bacterial enzymes, pH, temperature, and hydrolysis. Do the levels of drug found represent an overdose, intentional or otherwise, or a therapeutic level? Are tolerance and/or pharmacogenomics issues? Information as to the interval of time between death and the finding of the body, or information concerning therapeutic treatments of the person prior to death, is often not available. An additional factor to be considered when interpreting drug levels is the situation when a decedent had received significant amounts of replacement fluids prior to death. This is not, however, a simple "dilution" effect since the infusion of fluid results in efforts to reestablish new equilibrium between tissues and blood. Thus, the forensic toxicologist must consider a host of factors in reaching a decision as to the manner and cause of death. A primary objective of this chapter is to discuss those factors that can make it difficult to interpret postmortem toxicological findings in deciding whether or not a drug played a role in causing a person's death.

13.2 TISSUE AND FLUID SPECIMENS

Postmortem specimens taken at autopsy may include blood, urine, bile, cerebrospinal fluid, vitreous humor, gastric contents, meconium, various organs, bone, teeth, hair, and nails. If an autopsy is not performed, the postmortem specimens collected usually include peripheral blood, urine, and vitreous humor.

13.2.1 Blood

It needs to be emphasized that postmortem blood, even if obtained quickly after death, has already been affected by processes that occur at the time of death, such as loss of cell membrane potential and the initiation of hemolysis. Other postmortem processes begin early on and generally increase with increasing postmortem interval. Hence, any sample of postmortem blood is not physiologically or toxicologically equivalent to a clinical sample. Proper selection of the sites for blood sampling is crucial because of the phenomenon of postmortem redistribution. This is defined as the postmortem movement of drug away from its location at the time of death to another site in the body. Several basic mechanisms are believed to be involved (Pélissier-Alicot et al., 2003).

One mechanism is diffusion of drug from inside cells into the surrounding blood. This starts with loss of integrity of the cell membrane, which occurs at the moment of death and generally results in unregulated flow of substances, both endogenous and exogenous, out of cells. Because of this mechanism, postmortem blood is not used for the determination of electrolytes and glucose.

A second mechanism is release of drug from certain tissue depots and the subsequent formation of drug concentration gradients directed toward blood in the heart chambers. Of particular importance is antemortem accumulation of drug in lung tissue, which can act as a depot for many drugs, especially those that are lipid soluble. The postmortem release of drugs from lung depots results in the movement of drug via the pulmonary veins into the heart chambers. Such postmortem diffusion generally occurs with those drugs that have high volumes of distribution (≥ 3 L/kg) and attain high lung levels; however, there is no hard-and-fast rule for this (Hilberg et al., 1999a). The release of drug from tissue depots is believed to involve the disruption of nonspecific drug–macromolecular binding as a result of pH changes and denaturation. Those drugs that concentrate in heart tissue will be released directly into heart blood, and sampling from the chambers would give an incorrect concentration of drug in blood at the time of death. It is known, however, that highly lipophilic drugs may move from blood into adipose tissue.

A third mechanism is the diffusion of drug away from the gastrointestinal and hepatic systems, which ultimately enters heart blood. This mechanism is particularly relevant when death occurs in the nonequilibrium phase; that is, significant amounts of drug are still present in the stomach, intestines, and in the portal vein–hepatic system. Thus, for example, ethanol from the gastrointestinal tract can redistribute to the pericardial and also to the pleural fluids (Plueckhahn, 1967). Thus, it is essential that when blood is taken for analysis, a sample be drawn from peripheral sites such as the femoral veins (or subclavian veins), as quantitation on femoral blood is considered to be the most reliable measure of the antemortem concentration just prior to death (Dalpe-Scott et al., 1995; Hilberg et al., 1999b). However, whenever possible, the results derived from femoral blood should be compared with results from other tissues and fluids.

The production of gases during decomposition and the resulting pressures can result in the mixing of blood from different sites (Prouty and Anderson, 1990). A sample of femoral blood may contain blood from the inferior vena cava, and cardiac blood may contain blood from the vena cava, subclavian, pulmonary, and aortic vessels. Other factors that make interpretation of postmortem blood data difficult include the degree of hemolysis, or any other changes in the blood matrix. Therefore, unlike clinical toxicology, where blood is separated into plasma or serum, whole blood is almost always analyzed in postmortem toxicology. Analysis of blood clots, especially after trauma such as head injury, may be useful for determining drug exposure prior to death, especially in cases of prolonged survival, during which time the circulating blood may have been cleared of drug. The concentration of drug in a blood clot may reflect the concentration in blood at the time of clotting, assuming that formation occurred at the time of injury. But care in quantitative interpretation is important since the consistency of the clot differs from that of circulating blood, because formation of the clot cannot always be assumed to be instantaneous (where there is a survival time before death) and because clots themselves may partially break down and release drug.

13.2.2 Urine

Unless death is so rapid as to preclude the drug from being excreted into the bladder, urine is useful for initial screening for the different classes of drugs and for their metabolites. Finding drugs in urine indicates prior usage but cannot itself be used to indicate any level of impairment or the concentration of drug in blood prior to death. Urine analysis presents fewer problems than blood analysis; for example, urine does not readily support bacterial growth.

13.2.3 Vitreous Humor and Cerebrospinal Fluid

Analysis of vitreous humor serves several useful roles. It must be emphasized, however, that vitreous humor is useful only when bodily decomposition has not set in. There is some evidence that determination of potassium levels in vitreous humor may be useful in some cases for approximating the postmortem interval (PMI). Potassium levels rise as the interval between death and collection of a sample increases; the correlation, however, is not always accurate (Stephens and Richards, 1987). More recent studies (Muñoz et al., 2001, 2002; Madea and Rödig, 2006) have demonstrated that certain statistical treatments of the data observed provide a somewhat more accurate estimation of the PMI. Nonetheless, the relationship between the concentration of vitreous potassium and PMI remains a controversial issue.

Vitreous humor has been used to analyze for the presence of many drugs and alcohol (Hepler and Isenschmid, 1998; Jones, 1998). Vitreous humor contains very little protein compared to blood and tissues. Only free drug in blood enters the vitreous humor, and since most drugs exhibit at least some significant degree of binding to proteins and other macromolecules in blood, vitreous drug levels are lower than those in femoral blood. Vitreous humor is a protected site; usually, there is little bacterial contamination and therefore little fermentation. Thus, it may be useful for determining whether the presence of alcohol in tissues was due to consumption prior to death or formed as a result of postmortem fermentation. Vitreous humor is a good backup sample for blood ethanol. The levels of alcohol in vitreous humor pre- and post-embalming are in fairly good agreement (Scott et al., 1974; Coe, 1976). Since the pre- and post-embalming samples in these studies were taken within a relatively short interval of each other, one cannot draw conclusions about the effects of embalming fluid over long periods of time. However, caution must be exercised whenever inferring a blood alcohol level from vitreous humor. The vitreous/blood (femoral) ratio of alcohol can be a very useful indicator of whether death occurred in the absorptive or post-absorptive state. At equilibrium (post-absorptive state), the ratio of vitreous alcohol to blood alcohol averages about 1.10 to 1.20 (Garriott, 2008). But if death occurs during the absorptive state (rising blood alcohol level), this ratio will be lower, possibly even less than 1.00. Postmortem redistribution from ocular tissue into vitreous humor may also occur.

Vitreous humor can be very useful for the determination of such antemortem conditions as hypo- or hypernatremia, dehydration or uremic conditions, and hyperglycemia. Vitreous glucose is known to decline, often erratically, with increasing postmortem interval, thus making determination of hypoglycemia generally impractical, if not impossible. However, under certain very limited conditions, a precipitously low level found very shortly after death may be indicative of insulin overdose (Coe, 1972, 1974).

Many drugs are known to enter the cerebrospinal fluid (CSF). Absent blood, CSF has been found useful for the qualitative determination of drugs. Analysis of CSF may reveal recent use of heroin, as 6-monoacetylmorphine has a longer half-life in this specimen than that of other fluids (Jenkins and Lavins, 1998; Engelhart and Jenkins, 2007).

13.2.4 Gastric Contents

The entire gastric contents should be taken for analysis and the total weight should be recorded. Any undigested tablets or other medicinals should be analyzed separately. The total amount of drug present in gastric contents, including any undigested material, is of forensic importance because it may be useful in interpreting the manner of death (i.e., it may be indicative of suicidal overdose). Since gastric content is normally acidic, detecting trace amounts of basic drugs in stomach contents may be due to back diffusion into the stomach as well as representing residual drug following ingestion.

13.2.5 Meconium

Exposure to maternal substance abuse during gestation is associated with many adverse social and significant perinatal complications, including a high incidence of stillbirths, premature rupture of the embryonic sac membranes, maternal hemorrhage, spontaneous abortion, fetal distress, and increased rate of mortality and morbidity (developmental and cognitive impairment) (Ostrea et al. 1989; Moriya et al., 1994; Moore and Negrusz, 1995; Ostrea, 2001). Previously, the history of in utero exposure to drugs was based on urinalysis and on maternal self-reporting, which may be unreliable, due to patient denial about addiction or fear of the consequences (Ostrea et al., 1989; Coles et al. 2005). In stillbirth infants and in infants that die after birth, meconium may be the specimen of choice for drugs of abuse testing because of the difficulty in obtaining urine specimens, and the use of organs may have limited value if the mother had stopped using drugs several days before delivery (Moriya et al., 1994).

Meconium, the first fecal matter of the neonate, which is normally expelled from the intestine during the first 24 to 48 h after birth, is a thick greenish-black material composed of epithelial cells, digestive tract secretions, and residue of swallowed amniotic fluid (Ostrea et al., 1992; Moore and Negrusz, 1995). Drugs accumulate in meconium either by direct deposition from bile or through swallowing of amniotic fluid (Ostrea et al., 1992; Moore and Negrusz, 1995; ElSohly et al., 1999b; Ortega García et al., 2006).

Meconium starts to accumulate in the intestines between weeks 12 and 16 of gestation and can provide a long history of in utero drug exposure as well as indicate during which period of gestation drug use took place (Callahan et al., 1992; Ostrea et al., 1994; Ortega García et al., 2006). Unlike urine, blood, and gastric contents, which allow the detection of drug exposure for 2 to 3 days, meconium extends this window to about 20 weeks (Ostrea et al., 1989; Abusada et al., 1993; Moore and Negrusz, 1995). Lewis et al. (1995) reported that meconium analysis detected threefold more cocaine-exposed babies than urinalysis did.

Meconium is readily available, and only a small sample is required for complete analysis, including confirmation analysis (Ortega García et al., 2006). Meconium testing does have some limitations. It is often an unfamiliar specimen in the clinical laboratory, being a sticky material that is more difficult to work with than urine and requires a thorough preliminary cleanup procedure prior to any analytical assay.

Solid-phase extraction (SPE) has been used to isolate drugs and metabolites from meconium, and both immunoassay and chromatographic techniques have been used for the screening of drugs in meconium. Preliminary screening results must be confirmed by gas chromatography–mass spectrometry (GC–MS). Coles et al. (2007) compared GC–MS with LC–MS–MS (liquid chromatography–tandem mass spectrometry) and found that LC–MS–MS has increased specificity and decreased interference rate, reporting fewer false-positive results. Marin et al.

(2008) developed a LC–MS–MS procedure that can detect and quantitate many benzodiazepines in meconium.

Various drugs and their metabolites that are found in neonatal or maternal urine have also been detected in meconium. Samples containing cocaine and/or cocaethylene also contained benzoylecgonine and m-hydroxybenzoylecgonine. The presence of cocaethylene in meconium suggests that both cocaine and alcohol were used during pregnancy (Callahan et al., 1992; Abusada et al., 1993; Steele et al., 1993; Moriya et al., 1994; ElSohly et al., 1999a). One study of postmortem analysis of three fetuses that had been exposed to cocaine showed the presence of cocaine in the meconium of a 17-week-old fetus, suggesting that fetal exposure can be determined early in gestation and in the very premature fetus. At this time, meconium contained a higher concentration of drugs than did neonatal urine (Ostrea et al., 1994).

Moore and Negrusz (1995) and Coles et al. (2005) found that some samples of meconium contained both 11-nor-Δ^9-tetrahydrocannabinol-9-carboxylic acid (THC-COOH) and 11-hydroxy-Δ^9-THC (11-OH-THC). Using trimethylsilyl derivatization and GC–MS, hydrocodone and hydromorphone were detected in meconium along with codeine and morphine (Moore et al., 1995). The level of codeine, hydrocodone, and hydromorphone increased dramatically following acid hydrolysis, while the morphine level also increased but not as significantly.

13.2.6 Brain

The brain is a good repository for lipophilic drugs such as anesthetics and hydrocarbons, whereas many polar compounds are not able to pass the blood–brain barrier in significant amounts. Clearly, psychotropic drugs enter the brain easily and their levels in brain are useful for determining the manner of death. Brain is also used as a backup specimen for blood alcohol. At equilibrium, the brain alcohol/blood alcohol ratio is about 0.85, but this number may be lower if death occurs during the period of a rising blood alcohol. Brain cocaine levels are useful for evaluating cocaine-related deaths (Bertol et al., 2008). From a few minutes to about 3 h after exposure, the brain/blood ratio for cocaine is approximately 4 to 10 (Spiehler and Reed, 1985). The ratio of brain to blood cocaine in cases of death within 3 to 6 h of exposure is approximately 0.4 to 0.8. These ratios may be different, however, in persons who had used cocaine chronically.

13.2.7 Liver and Bile

The liver concentrates most parent drugs and also metabolites, ethanol being the notable exception. There is an extensive body of data on the range of hepatic drug levels in therapeutic, toxic, and lethal cases. Bile is a good source for many drugs and their metabolites, in particular opioids and benzodiazepines, and because of its storage in the gallbladder, it may be the only specimen where residual drugs or metabolites may be detected when absent from other tissues. Bile may be useful in those cases where a prolonged interval elapsed between the last intake of drug and death. However, bile may contain high concentration of drug in cases of acute overdose where death occurs while large amounts of drug are still in the gastrointestinal tract and in the portal system.

13.2.8 Lung, Spleen, Kidney, and Skin

The analysis of lung tissue is useful in cases involving volatile substances such as gases (other than carbon monoxide), anesthetics, and solvents. Because spleen concentrates red blood cells, it may be useful as a last resort specimen for determining cyanide- or carbon monoxide–bound hemoglobin. Since cyanide derived from antemortem toxic exposure may be lost rapidly with advanced decomposition, spleen may be the only organ where toxicologically significant levels of cyanide may still be found. Where a routine blood sample is not available or is not considered usable, spleen may be used cautiously to estimate antemortem exposure to carbon monoxide. Analysis of kidney tissue is useful for detecting heavy metals. Analysis of unburned skin may be useful for detecting use of accelerants in arson-related cases.

13.2.9 Muscle

Muscle tissue is not a routine specimen, but is particularly useful in cases of advanced decomposition because skeletal muscle, being more slowly affected by decomposition, may still be present when little or no other soft tissue is present. Great care must be exercised when interpreting the findings quantitatively. For skeletal muscle, there is little postmortem toxicology data from which one can draw conclusions relating drug level and cause of death. The studies that have been done on fresh skeletal muscle indicate that the data are best used qualitatively. Of interest is the fact that cocaine, which is unstable in blood, has been detected in numerous cases in decomposed but dry skeletal muscle.

13.2.10 Bone, Teeth, Nails, and Hair

It is known that many drugs and heavy metals become incorporated in bone, teeth, nails, and hair. These tissues will not necessarily, however, contain drugs or toxic substances if death occurred rapidly after intake or administration, but are good indicators of chronic dosing. They are usually not analyzed postmortem except for exhumed, badly decomposed, or skeletonized bodies. Teeth are useful for analysis in bodies that have been badly burned. There is an extensive body of data on the incorporation of drugs and heavy metals into hair, which is used primarily in clinical toxicology. Hair can be cut into segments and analyzed individually to determine a time pattern of drug usage. Since hair grows at a rate of approximately 1 cm per month, one can approximate when drug usage was discontinued. Hair may reflect a single acute dose, depending on the drug and dose involved, provided that sufficient time has elapsed to allow entry of drug into the hair. In this regard the analysis of hair follicles may be particularly useful. Hair analysis requires careful and proper preliminary washing steps to remove as much as possible of any substances present as a result of external contamination.

13.2.11 Other Materials for Analysis

1. *Injection site.* If an injection site can be located, analysis may reveal useful qualitative and quantitative data. Since a drug may have come to what appears to be an injection site via the circulation, other skin samples should be taken for comparison.
2. *Insect larvae/maggots.* Larvae and maggots feed on decomposing tissue and will incorporate into their systems drugs that are in the tissues they consume. Collection of these insect specimens from decomposed or skeletonized bodies may provide much useful qualitative information.
3. *Nasal swabs.* Results from swabbing of the nasal cavities must be interpreted cautiously. Drugs are normally secreted into nasal mucosa. Finding small amounts of drug on a swab does not necessarily mean that the person snorted drugs prior to death. The amount of drug detected will help determine if snorting is a likely explanation.

13.3 SPECIMEN COLLECTION AND STORAGE

For all samples, proper collection of the sample in adequate amounts, and proper use of preservatives and storage containers (e.g., glass, plastic, nylon, foil) along with proper storage temperature are essential (Skopp, 2004). It is recommended that glass containers be used for liquids and glass or plastic containers for tissue and gastric contents. However, it must be noted that the use of certain types of plastic containers for storage can result in the loss of THC–COOH (Stout et al., 2000).

Blood should be collected in tubes containing at least 1 to 2% sodium fluoride as preservative and stored at 0 to 4°C. This will inhibit bacterial growth and enzyme activity, which can either produce alcohol or reduce the concentration of alcohol in alcohol-containing specimens. Without preservative, bacterial action at room temperature is likely, and alcohol production by bacteria can reach concentrations up to 0.10 to 0.15 g/dL. Sodium fluoride will also inhibit cholinesterase activity, which can metabolize cocaine and other ester-containing substances. Blood specimens held for long periods can be frozen. Tissues may be stored for short periods at 0 to 4°C. Most drugs are stable at this temperature range, but some drugs (e.g., cocaine) are labile. Tissues may also be frozen. However, the freezing and thawing of tissues before processing is not recommended, as this can rupture cells and cause a loss of drug-containing fluid. Where short- or long-term freezing is utilized, frozen tissues should be rapidly cut and weighed. Vitreous humor and urine can be stored at 0 to 4°C or frozen. Meconium stored at room temperature for 24 h shows a loss of drug, in particular cocaine and cannabinoids. Meconium can be stored at $-15°C$ for nine months without loss of drug concentration (Ostrea,

2001). Samples to be analyzed for volatile substances other than alcohol must be stored frozen and, when appropriate, in hermetically sealed containers. If volatile substances are suspected in a death, the pathologist should submit duplicate samples so that analysis can be performed for nonvolatile drugs as well as for volatile compounds. It should be emphasized that blood and other fluids should be collected in such a manner as to minimize the headspace area in the collection vessel. All other biological-containing evidence should be sealed and stored frozen.

13.4 EXTRACTION PROCEDURES

Choice of the extraction procedures will determine the recovery of analytes from the tissue matrix, and improper selection may result in failure to recover adequately or even to detect one or more agents. The use of liquid–liquid extraction procedures requires careful selection of solvent and pH in relation to the analytes targeted. The polarity of the extracting solvent, or solvent mixture, should be similar to the polarity range of the target analytes. The pH used for the extraction of acidic and basic compounds should be 2 units below and above, respectively, the pK_a value of the analyte. Neutral drugs are not dependent on pH. However, when performing a general unknown analysis, the selection of solvent and pH should be done so as to obtain satisfactory recovery of a wide range of possible analytes, keeping in mind such factors as pH-related drug stability, drugs such as amphetamines that require very high basic pH, and the requirements of amphoteric compounds. The use of solid-phase extraction (SPE) generally requires protein-free extracts. SPE has a high versatility, owing to the broad range of columns available: polar, nonpolar, ion-exchange, and mixed-mode. Columns typically used for general unknown procedures consist of a combination of nonpolar and cation-exchange components. SPE can be very useful for the isolation of highly polar or water-soluble drugs and metabolites. It should also be noted that the condition or state of the biological matrix being extracted (i.e., "fresh", decomposed, or embalmed) can affect the amount recovered (see below), and this needs to be taken into account when interpreting findings (Skopp, 2004). Experience has shown that many different types and classes of drugs and toxic metals can be recovered from exhumed bodies even after long periods of interment (Skopp, 2004). Great caution must be exercised, however, in the interpretation of all such findings.

13.5 ANALYTICAL TECHNIQUES

Knowledge of the basic theory, range of applicability to forensic samples, and the analytic limitations of all techniques currently utilized (i.e., immunoassay, chromatography, spectrometry, and spectroscopy) is essential. Immunoassays and chromatographic procedures are screening techniques, and all presumptive positives must be confirmed. The standard technique for confirmation is one or more of the modes of mass spectrometry, usually coupled with gas chromatography (GC) or liquid chromatography (LC).

It is beyond the scope of this chapter to discuss in detail the principles and applications of these analytic technologies. However, the following brief comments are relevant. There are ongoing advances in both GC and high-performance liquid chromatographic (HPLC) technology, resulting in the availability of more versatile and stable columns providing higher resolution and greater ranges of class retentivity.

The use of HPLC is gaining increasing popularity for several reasons. The development of mixed-mode or multimechanism columns for HPLC allows for the retention of drugs over a broad range of polarity. These columns incorporate both nonpolar and polar moieties. An analyst can choose from a multitude of normal, reversed-phase, and mixed-mode types of columns and, accordingly, tailor a separation. Compounds that do not chromatograph well on routinely used low- to intermediate-polarity GC columns because of high polarity or size or thermal instability will as a rule be effectively and conveniently separated by LC systems. In many cases this eliminates the need for derivatization required for GC. LC is essential for quaternary ammonium muscle relaxants such as tubocurarine and pancuronium. It has long been used for the acidic barbiturates (commonly analyzed along with neutral drugs as the acid/neutral fraction), and it has many advantages for the analysis of benzodiazepines (Levine, 2006; Marin et al., 2008). It is being applied increasingly to the routine analysis of basic drugs. Although still under development, the appearance of ultrafast or high speed HPLC does promise to increase the efficiency of obtaining effective and reliable separations with a very short run time. This involves the use of shorter columns, higher linear velocities, smaller porous particle size (ranging from less than 2 to 2.2 μm in size), and significantly higher operating pressures. It also generally requires cleaner sample extracts. LC is now commonly coupled to mass spectrometry by means of electrospray ionization or atmospheric pressure chemical ionization interfaces, and this is being used very effectively.

GC is, of course, still used very widely as a screening technique for basic drugs. Headspace GC is the standard method for routine analysis of the "volatile" fraction (i.e., ethanol, methanol, isopropanol, and acetone). A wide variety of columns of low, intermediate, and high polarity is available. As with LC, there is now available fast-track GC, which promises to bring effective separations in much shorter time. Fast GC typically involves the use of very rapid column heating systems, which provide much faster increases in temperature, higher carrier gas pressures, and very fast signal acquisition systems. Column length will vary, but 5-m columns are frequently used. Derivatization may be necessary for the following reasons: improvement of chromatographic characteristics, resolution of coeluting compounds, elimination of thermal instability, and in some cases, as with amphetamines, to produce a derivative with a much more characteristic mass spectral fragmentation pattern than is found with the underivatized compound. Where derivatization is necessary, this involves the replacement of "active" hydrogen with an alkyl, acyl (especially polyfluorinated acyls), or silyl group. A particular advantage to the use of polyfluorinated derivatives is that it allows for highly sensitive detection using an electron capture detector. One common application of this procedure is the gas chromatography–mass spectrometry analysis of cannabinoids.

There have been very significant developments in mass spectrometry technology which have dramatically increased its power and applicability for confirmation. The use of tandem MS, ramped fragmentation (smart fragmentation), simultaneous acquisition of both scan and select ion monitoring data, and the development of hybrid mass spectrometers have all expanded and improved the confirmatory process. Very fast mass scanning systems are available and required for use with fast chromatographic screening techniques. The use of current time-of-flight mass spectrometers can provide extremely accurate mass detection to differentiate compounds with similar masses.

Although generally no longer done routinely as part of a general unknown, the analysis of metals may be indicated based on the circumstances or suspicions associated with a particular case. Homicidal poisoning with certain metal salts (i.e., thallium or arsenic) is still encountered. The analysis of metals starts with digestion of the biological matrix. Wet ashing may involve refluxing the sample with a mixture of strong inorganic acids (i.e., a mixture of nitric, sulfuric, and perchloric acids), or for more volatile metals may involve just incubation with the acid solution. Dry ashing, which involves heating in a muffle furnace at elevated temperature to produce an ash residue, which is then treated with 5% nitric acid to dissolve the metals, cannot be used for mercury and other volatile metals and is used for selected metals only as appropriate. In general, mercury requires the use of cold vapor techniques. The state-of-the-art instrumental technique is inductively coupled plasma mass spectrometry, which provides very high sensitivity over the mass range of metals of forensic interest. However, graphite furnace atomic absorption spectroscopy (AAS) is still very widely used, and cold vapor AAS is the preferred technique for mercury.

The analysis of carboxyhemoglobin and the other hemoglobin species is today typically and conveniently achieved with the use of CO-oximeters, which involve automatic spectrophotometric determinations (Lee et al., 2002, 2003; Lewis et al., 2004). Thermocoagulated, putrified, and contaminated blood samples can pose potential problems. In such cases, sample pretreatment before oximetry has been found effective. Alternatively, the use of headspace GC on carbon monoxide released from the sample after treatment with liberating agents or the application of derivative spectroscopy to the sample itself can provide acceptable results (Perrigo and Joynt, 1989).

Cyanide may be determined conveniently by headspace GC using a nitrogen–phosphorus detector after liberation from the biological matrix under acidic conditions. Spectrophotometric and fluorometric procedures are also available (Felscher and Wulfmeyer, 1998; Gambaro et al., 2007).

It is essential that there be maintenance of all the accepted standards for quality assurance and quality control throughout the entire testing process from sample receipt and accession through all the analytic operations and continuing to sample storage, release, and disposal. This includes the generation and maintenance of all reports necessary to document every aspect of the processes. Proper maintenance logs must also be kept for all analytic instrumentation involved. If it should be necessary to convey test results verbally to an authorized party, then the results and any specific information given should be recorded appropriately to avoid any future misunderstandings.

13.6 INTERPRETATION

The interpretation of postmortem findings is becoming increasingly complex. This is due to such factors as postmortem redistribution, significant differences in individual rates of metabolism arising from genetically based polymorphic variations in metabolic enzymes, the type and condition of the specimen, drug interactions arising from the concurrent use of multiple drugs, and a particular person's tolerance for a specific drug or class of drugs and cross tolerance.

13.6.1 Postmortem Redistribution

The mechanisms and consequences of postmortem redistribution were discussed earlier. The effects of postmortem redistribution increase with increasing postmortem interval, but the mechanism involving unregulated flow of drug out of cells becomes significant at the moment at which death occurs. In general, it has been observed that cardiac blood has a higher concentration of drug than that of peripheral blood, although the magnitude of the change depends on the physiochemical properties of the drug, its volume of distribution, whether death occurred under equilibrium or nonequilibrium conditions, the organ concentrations, and the length of the postmortem interval. Variations in the cardiac/peripheral ratio can be very significant, even for a given drug (Dalpe-Scott et al., 1995; Anderson and Muto, 2000). The general importance of always including a sample of peripheral blood has been noted. If peripheral blood is not available, the interpretation of findings based only on cardiac blood is frequently perilous, and, in such situations tissue concentrations, especially liver, are necessary to minimize the possibility of an erroneous conclusion. Postmortem redistribution may also occur between organs, as high levels of drug from liver, lung, stomach, and heart are released to neighboring sites (Moriya and Hashimoto, 1999). Where there has been traumatic injury to the organs (i.e., laceration or rupture of the stomach, intestines, liver, etc.), there is a high potential for contamination from drug that may be present in the affected organ and now spills out into adjacent areas. Sampling of multiple sites and testing of different specimens may be necessary to resolve such issues.

In cases involving alcohol when blood is not available postmortem, blood alcohol levels can be estimated from correlation data derived from studies comparing alcohol levels in blood with those in other postmortem tissues. The state of alcohol absorption (pre- or post-absorptive) can be determined by analyzing the stomach contents for alcohol, and the results are then helpful in deciding which tissue to use for the correlation (Backer et al., 1980).

Because of the complex processes involved in postmortem redistribution, calculation from a single postmortem blood sample of a dose of drug taken by the deceased prior to death is unwise, and it is especially dangerous when using a single cardiac sample (Prouty and Anderson, 1990; Hilberg et al., 1999b). Back calculation of a dose from even a femoral blood concentration using a clinically derived plasma-based steady-state volume of distribution is strongly discouraged, as it can lead to an erroneous conclusion. The volume of distribution itself for a given drug can show wide interpersonal variation, and its use postmortem is contraindicated. Conversion of postmortem blood levels to time-of-death plasma or serum levels cannot be carried out accurately or reliably. Furthermore, in many cases (i.e., where an intentional or accidental overdose is involved), it cannot be assumed that a steady-state distribution existed at the time of death. The experienced toxicologist may, however, using data from different specimens, cautiously estimate what approximate minimum dose was taken or give a reasonable opinion as to whether "therapeutic" or overdose quantities were taken or administered.

13.6.2 Pharmacogenomics

The interpretation of postmortem drug levels as they relate to cause of death requires careful assessment of all the particulars of a case. Toxicologic analysis should consider genetic influences on the concentration of drug found. The field of pharmacogenomics attempts to explain a person's metabolic capabilities for drug metabolism in relation to the genotype. Poisoning, intentional or otherwise, may be the result of genetically influenced altered metabolism. A person's genetic makeup may also influence the transport of drugs across membranes and the interaction of drugs with receptors. All of these effects will play a role in attaining a concentration of drug in blood as well as the person's response to a given drug (Kupiec et al., 2006). The use of pharmacogenomics for understanding drug reactions and drug-related deaths has been termed *molecular autopsy* (Jannetto et al., 2002). A person's genetic makeup has already entered the clinical area and will be utilized increasingly in the future prior to prescribing medication to better maintain therapeutic levels.

There are many metabolic pathways for the metabolism of drugs and other xenobiotics. Basically, they are divided into phase I and phase II pathways. Phase I is responsible for oxidation, reduction, and hydrolysis reactions, preparing the metabolite for conjugation reactions by phase II enzymes. The final product is usually more water soluble than the parent drug and can easily be eliminated. Most phase I reactions are carried out by the cytochrome P450 (CYP450) monooxygenase enzyme system, located primarily in the liver but also less extensively in other tissues (Omiecinski et al. 1999). These heme-containing proteins are located in the smooth endoplasmic reticulum, and many of the isoenzymes exhibit genetic polymorphism. A list of the various isoenzymes and their substrates, along with inhibitors and inducers of the isoenzymes, is available (Flockhart, 2007). The CYP450 system may play a role in explaining drug-induced deaths by two different mechanisms.

First, based on a person's genomic makeup, one can be a "slow" or a "fast" metabolizer. Another classification of variants is ultrarapid, extensive, intermediate, or poor metabolizer. Those with the slow polymorphic form can accumulate high levels of drug that could result in death, even from therapeutic doses. About 25 to

30% of all xenobiotics are metabolized by one isoenzyme, CYP2D6 (debrisoquine hydroxylase); 5 to 10% of Caucasians have low levels of CYP2D6 or lack it completely, and doses of medications metabolized by this isoenzyme must be reduced (Kupiec et al., 2006; Roden et al., 2006). On the other hand, persons with multiple copies of the gene for CYP2D6 must be given higher doses to maintain therapeutic blood levels. The following examples will serve to illustrate the importance of such genetic variations.

Oxycodone, the opioid prescribed for relief of pain, is metabolized by CYP2D6. Three mutations of the polymorphic gene encoding for this isoenzyme have been identified and may play a role in oxycodone toxicity (Jannetto et al., 2002). Methadone, another opioid, is metabolized by CYP1A2, CYP3A4, and CYP2D6. Again, genetic variants of CYP2D6 showed a trend, although it is not statistically significant, toward poor metabolism of methadone, and higher levels were found in these people than in nonvariant controls (Wong et al., 2003). Fentanyl, another opioid used for relief of pain and as an anesthetic during surgery, is metabolized by CYP3A4 and CYP3A5. Variant alleles of the genes encoding these proteins are known and again may play a role in fentanyl toxicity (Jin et al., 2005). The response to the anticoagulant warfarin (coumadin) is dependent on cytochrome metabolism. Approximately 6% of the variation in response to warfarin is due to variation in the gene encoding CYP2C9, and 27% of the variation in response is due to variation in the gene VKORC1 encoding the warfarin target, vitamin K epoxide reductase (Berg, 2007). The metabolic rates of amphetamine and of methamphetamine are dependent on CYP2D6 activity, and genetic polymorphism for ring hydroxylation of amphetamine has been reported (Miranda-G et al., 2007).

Second, since different drugs can be metabolized by the same enzyme, taking two such drugs at the same time may lead to competition, with one drug being less metabolized than would occur had it been taken alone. This could result in an increase in drug concentration, leading to toxicity and even death. Such a result could also occur if one drug inhibits its own or another metabolic enzyme involved with other drugs. Such effects have been reported with the antidepressant selective serotonin reuptake inhibitors (SSRIs) that are both metabolized by CYP2D6 and inhibit CYP2D6 (Goeringer et al., 2000). The inhibitory effect of ethanol on the metabolism of certain other drugs, e.g. barbiturates when both are taken concurrently is well known. The inhibition of CYP3A4 by the antifungal agent ketoconazole results in terfenadine-induced cardiac toxicity. High concentrations of parent drug may therefore be due to a low rate of metabolism, to metabolic competition by other drugs, or to inhibition of the enzyme by one drug, allowing the second drug to increase in concentration. It must be remembered that frequently more than one CYP450 enzyme is involved in the metabolism of a given drug, and inhibition or induction of a nonprimary pathway may be significant.

13.6.3 Drug Interactions

Polydrug use can result in additive, synergistic, potentiating, inhibitory, or antagonistic interactions, and may involve inhibition or induction of metabolic enzymes by one agent affecting the clearance and effects of others. Some examples were noted earlier. Other common examples are the frequently encountered difficulties in interpretation of methadone and other drug abuse–related deaths (Milroy and Forrest, 2000; Corkery et al., 2004; Jönsson et al., 2007). The widespread use of all types of over-the-counter herbal products, domestic and imported, is an additional, potential complicating factor. Many of these products are known to contain components that interact with prescribed medications or may contain unlisted drugs or toxic substances (Ang-Lee et al., 2001; Saper et al., 2004; Hu et al., 2005; Dasgupta and Bernard, 2006; Woo, 2008). For example, some imported products may contain toxic licit medications which are not listed on the label. Other products have been shown to contain toxic metals. Many of these components may not be routinely detectable in standard analytical protocols. It is hoped that the more stringent regulations issued by the U.S. Food and Drug Administration in June 2007 concerning the importation of these products will reduce this danger, but toxicological vigilance is crucial (U.S. Food and Drug Administration, 2007).

13.6.4 Drug Stability and Decomposed Tissue

The postmortem chemical stability (or the lack thereof) of a drug or toxin or of the metabolites is a function of the combined effects of the chemical's structure, the length of the postmortem interval, the conditions of temperature, bacteria, oxygen, water, pH, and so on, to which the body has been subjected, and the particular matrix containing the agent. The reader is referred to the references cited for a comprehensive discussion of all these issues as they pertain to many different drugs (Grellner and Glenewinkel, 1997; Drummer, 2004, 2007; Skopp, 2004).

The interpretation of toxicological findings in decomposed, exhumed, or embalmed (discussed below) bodies requires special vigilance. Decomposed bodies reflect the processes of autolysis and putrefaction. Autolysis, digestion by natural enzymes, occurs more readily in tissues with a high enzyme content, such as the pancreas and stomach. Tissues with a low content of digestive enzymes, such as heart and liver, are less readily autolyzed. Autolysis leads to liquefaction of the tissues, which can then mix with blood and result in erroneous conclusions. Putrefaction, destruction of tissue by bacteria, depends on the bacteria that are present, the temperature and oxygen content of the environment in which the body is located, and the substrates available for bacterial action. The intestine is particularly susceptible to putrefaction and bacterial fermentation, and is not normally used for postmortem analysis.

Ethanol can be produced because of fermentation in decomposing bodies, but can also decrease from initial antemortem levels due to chemical, microbial, and evaporative processes occurring with decomposition (Kugelberg and Jones, 2007). Postmortem changes are known to cause a rapid loss of the high levels of cyanide associated with fatal poisoning, but cyanide is also subject to varying postmortem production, particularly in deteriorating bodies (Skopp, 2004). Artifactual production of cyanide has also been found to occur in some blood samples from fire death cases, and extreme caution must be exercised in the interpretation of such levels. Cyanide can be a significant component of some toxic fire gases that cause or contribute to death, but in such instances it must be reasonably shown that it derived from pyrolysis of nitrogenous material (e.g., nitrogen containing plastics) present in the fire environment.

Toxicologically significant concentrations of γ-hydroxybutyrate (GHB) may also be produced endogenously in postmortem fluids (Elliott, 2001, 2004; Moriya and Hashimoto, 2005; Beránková et al., 2006). GHB is a sedative-hypnotic agent not currently approved for routine use in the United States. It is, however, available from different sources, and is encountered both as a drug of abuse and as an agent in drug-facilitated sexual assault. Clearly, very serious misinterpretations can occur unless the potential ranges of postmortem production are taken into account.

The finding of a low or negative drug level in decomposed samples cannot be assumed to mean that no antemortem significance was involved. However, the presence in relatively intact or preserved exhumed specimens of significantly high levels of drug or toxin may be cautiously interpreted as being consistent with antemortem overdose, whatever was the circumstance of administration. In some cases where there are suspicions of the involvement of criminal poisoning, resulting in analysis of exhumed remains, the qualitative detection of a poison or of an agent not known to have been used by or prescribed to the deceased may be deemed sufficient to corroborate such allegations. This assumes, of course, that the idea may reasonably be eliminated that the source of any such agent was external contamination of exposed, buried remains or any of the other processes that may produce analytes, as discussed above. When dealing with buried but exposed remains or those in deteriorated coffins (often containing groundwater, soil, and other debris), appropriate controls from the areas in proximity to the remains should always be taken by the forensic investigator. An important example is when such remains are to be examined for metal content, where it is essential that control samples of soil and groundwater be tested.

13.6.5 Effects of Embalming Fluid

The effects of embalming fluid, which contains formaldehyde and small amounts of formic acid, on different classes and types of drugs is varied and complex (Rohrig, 1998; Cingolani et al., 2001–2005; Skopp, 2004). Many basic drugs (i.e., primary and secondary amines) are known to undergo varying degrees of methylation when the tissue containing them has been treated with formalin-containing solutions. Methylation will result in variable loss of the original compound as well as the appearance of a new methylated compound. In some cases the methylated compound may be the nonprescribed parent drug of the actual antemortem drug that has the demethylated structure. Primary amine drugs can be converted to secondary amines, and drugs containing a secondary amine can be methylated by formaldehyde to form N-methyl derivatives. This has been reported for the conversion of nortriptyline to amitriptyline, desipramine to imipramine, and fenfluramine to N-methylfenfluramine (Gannett et al., 2001). Sertraline was converted to N-methylsertraline (Suma and Sai Prakash, 2006). Methamphetamine converts to N,N-dimethylamphetamine. It may also be noted here that the analysis of formalin-fixed tissue should always be accompanied by the appropriate use of formalin-spiked calibrators. Caution must be exercised in general when interpreting findings from any formalin-fixed tissue matrix as opposed to data derived from "fresh" tissue. These chemical conversions may make it difficult to determine the original concentration of the parent drug. Nonetheless, a wide variety of drugs and toxins have been detected in exhumed bodies, and this is often a crucial factor in the investigation. Exhumed bodies often present a combination of problems, due to decomposition and the effects of embalming fluid. Occasionally, the toxicologist must analyze stored formalin-fixed specimens. One must consider that drugs from the fixed tissue will diffuse into the formalin, thereby decreasing the tissue concentration of drug.

CONCLUSION

The concentration of drug found at the time of sampling from postmortem specimens is the result of many factors, some relating to the host, some to postmortem changes, and some to the physicochemical properties of the analyte, and may not represent the concentration of drug at the time of death. Thus, a negative finding does not always mean that drugs were not involved; and a finding of high levels of drug does not always mean that an overdose occurred. The decedent may have been tolerant of the drug in question and may have required or been using higher doses. Analysis and comparison of drug levels in both blood and tissue (e.g., liver or brain) may be necessary and should be done whenever possible for accurate interpretation of quantitative findings. Finding therapeutic levels of drug may be the result of drug loss due to the factors listed above. Where postmortem exposure to formalin-containing solutions has occurred, finding parent drug may be the result of metabolites having been converted back to the parent drug. Estimating the dose of drug consumed from postmortem data is often unreliable, as discussed above. Also, finding several drugs raises the issue of additive or synergistic effects or drug interactions in causing death. Clearly, the longer the delay between the time of death and the taking of samples, the less accurate may be the interpretation of the role the drugs played in causing the death.

In summary, the forensic toxicologist is presented with a broad spectrum of case findings, ranging from those that point readily toward a particular opinion concerning whether or not one or more substances had a role in the cause of death, to those where the complexity or circumstances do not allow for a definitive interpretation or conclusion. In the latter case, the toxicologist must always resist pressure from those seeking to obtain an opinion not supported by the facts. Where there is uncertainty in interpretation, it is the obligation of the toxicologist to make known to the inquiring party the scientific basis for the uncertainty and all possible conclusions. The toxicologist must never forget that the role of the forensic scientist is first to obtain the most reliable data possible and then, accurately and impartially, interpret and educate the participants in the medicolegal process.

REFERENCES

Abusada, G. M., Abukhalaf, I. K., Alford, D. D., Vinzon-Bautista, I., Pramanik, A. K., Ansari, N. A., et al. (1993). Solid-phase extraction and GC/MS quantitation of cocaine, ecgonine methyl ester, benzoylecgonine, and cocaethylene from meconium, whole blood, and plasma. *J. Anal. Toxicol.*, **17**:353–358.

Anderson, D. T., and Muto, J. J. (2000). Duragesic® transdermal patch: postmortem tissue distribution of fentanyl in 25 cases. *J. Anal. Toxicol.*, **24**:627–634.

Ang-Lee, M. K., Moss, J., and Yuan, C.-S. (2001). Herbal medicines and perioperative care. *JAMA*, **286**:208–216.

Backer, R. C., Pisano, R. V., and Sopher, I. M. (1980). The comparison of alcohol concentrations in postmortem fluids and tissues. *J. Forensic Sci.*, **25**:327–331.

Beránková, K., Mutňanská, K., and Baliková, M. (2006). Gamma-hydroxybutyric acid stability and formation in blood and urine. *Forensic Sci. Int.*, **161**:158–162.

Berg, J. M. (April 24, 2007). Pharmacogenomics: a powerful and challenging approach to personalized medicine. http://www.ostp.gov/galleries/default-file/Berg_PCAST_Apr07.pdf. Accessed Aug. 17, 2008.

Bertol, E., Trignano, C., Grazia Di Milia, M., Di Padua, M., and Mari, F. (2008). Cocaine-related deaths: an enigma still under investigation. *Forensic Sci. Int.*, **176**:121–123.

Brunton, L., Lazo, J., and Parker, K. (Eds.) (2005). *Goodman & Gilman's The Pharmacological Basis of Therapeutics*, 11th ed. New York: McGraw-Hill.

Callahan, C. M., Grant, T. M., Phipps, P., Clark, G., Novack, A. H., Streissguth, A. P., et al. (1992). Measurement of gestational cocaine exposure: sensitivity of infant's hair, meconium and urine. *J. Pediatr.*, **120**:763–768.

Chamberlain, J. (1995). *The Analysis of Drugs in Biological Fluids*, 2nd ed. Boca Raton, FL: CRC Press.

Cingolani, M., Froldi, R., Mencarelli, R., Mirtella, D., and Rodriguez, D. (2001). Detection and quantitation of morphine in fixed tissues and formalin solutions. *J. Anal. Toxicol.*, **25**:31–34.

Cingolani, M., Cippitelli, M., Froldi, R., Gambaro, V., and Tassoni, G. (2004). Detection and quantitation analysis of cocaine and metabolites in fixed liver tissue and formalin solutions. *J. Anal. Toxicol.*, **28**:16–19.

Cingolani, M., Cippitelli, M., Froldi, R., Tassoni, G., and Mirtella, D. (2005). Stability of barbiturates in fixed tissues and formalin solutions. *J. Anal. Toxicol.*, **29**:205–208.

Coe, J. I. (1972). Use of chemical determinations on vitreous humor in forensic pathology. *J. Forensic Sci.*, **17**:541–546.

Coe, J. I. (1974). Post-mortem chemistry: Practical considerations and a review of the literature. *J. Forensic Sci.*, **19**:13–32.

Coe, J. I. (1976). Comparative postmortem chemistries of vitreous humor before and after embalming. *J. Forensic Sci.*, **21**:583–586.

Coles, R., Clements, T. T., Nelson, G. J., and Urry, F. M. (2005). Simultaneous analysis of the Δ^9-THC metabolites 11-nor-9-carboxy-Δ^9-THC and 11-hydroxy-Δ^9-THC in meconium by GS-MS. *J. Anal. Toxicol.*, **29**:522–527.

Coles, R., Kushnir, M. M., Nelson, G. J., McMillin, G. A., and Urry, F. M. (2007). Simultaneous determination of codeine, morphine, hydrocodone, hydromorphone, oxycodone and 6-acetylmorphine in urine, serum, plasma, whole blood, and meconium by LC-MS-MS. *J. Anal. Toxicol.*, **31**:1–14.

Corkery, J. M., Schifano, F., Hamid Ghodse, A., and Oyefeso, A. (2004). The effects of methadone and its role in fatalities. *Hum. Psychopharmacol. Clin. Exp.*, **19**:565–576.

Dalpe-Scott, M., Degouffe, M., Garbutt, D., and Drost, M. (1995). A comparison of drug concentrations in postmortem cardiac and peripheral blood in 320 cases. *Can. Soc. Forens. Sci. J.*, **28**:113–121.

Dasgupta, A., and Bernard, D. W. (2006). Herbal remedies: effects on clinical laboratory tests. *Arch. Pathol. Lab. Med.*, **130**:521–528.

Dolinak, D. (2005). Toxicology. In: Dolinak, D., Matshes, E. W., and Lew, E. O. (Eds.), *Forensic Pathology: Principles and Practice*. London: Elsevier Academic Press, pp. 487–502.

Drummer, O. H. (2004). Postmortem toxicology of drugs of abuse. *Forensic Sci. Int.*, **142**:101–113.

Drummer, O. H. (2007). Post-mortem toxicology. *Forensic Sci. Int.*, **165**:199–203.

Elliott, S. (2001). The presence of gamma-hydroxybutyric acid (GHB) in postmortem biological fluids [*letter to the editor*], *J. Anal. Toxicol.*, **25**:152.

Elliott, S. P. (2004). Further evidence for the presence of GHB in postmortem biological fluid: implications for the interpretation of findings. *J. Anal. Toxicol.*, **28**:20–26.

ElSohly, M. A., Kopycki, W., Feng, S., and Murphy, T. P. (1999a). Identification and analysis of the major metabolites of cocaine in meconium. *J. Anal. Toxicol.*, **23**:446–451.

ElSohly, M. A., Stanford, D. F., Murphy, T. P., Lester, B. M., Wright, L. L., Smeriglio, V. L., et al. (1999b). Immunoassay and GC-MS procedures for the analysis of drugs of abuse in meconium. *J. Anal. Toxicol.*, **23**:436–445.

Engelhart, D. A., and Jenkins, A. J. (2007). Comparison of drug concentrations in postmortem cerebrospinal fluid and blood specimens. *J. Anal. Toxicol.*, **31**:581–587.

Felscher, D., and Wulfmeyer, M. (1998). A new specific method to detect cyanide in body fluids, especially whole blood, by fluorimetry. *J. Anal. Toxicol.*, **22**:363–366.

Flockhart, D. A. (2007). Drug interactions: cytochrome P450 drug interaction table. http://medicine.iupui.edu/flockhart/table.htm. Accessed Aug. 1, 2008.

Gambaro, V., Arnoldi, S., Casagni, E., Dell'Acqua, L., Pecoraro, C., and Froldi, R. (2007). Blood cyanide determination in two cases of fatal intoxication: comparison between headspace gas chromatography and a spectrophotometric method. *J. Forensic Sci.*, **52**:1401–1404.

Gannett, P. M., Hailu, S., Daft, J., James, D., Rybeck, B., and Tracy, T. S. (2001). In vitro reaction of formaldehyde with fenfluramine: conversion to *N*-methyl fenfluramine. *J. Anal. Toxicol.*, **25**:88–92.

Garriott, J. C. (Ed.) (2008). *Medicolegal Aspects of Alcohol*, 5th ed. Tucson, AZ: Lawyers & Judges Publishing Company.

Goeringer, K. E., Raymon, L., Christian, G. D., and Logan, B. K. (2000). Postmortem forensic toxicology of selective serotonin reuptake inhibitors: a review of pharmacology and report of 168 cases. *J. Forensic Sci.*, **45**:633–648.

Grellner, W., and Glenewinkel, F. (1997). Exhumations: synopsis of morphological and toxicological findings in relation to the postmortem interval: survey on a 20-year period and review of the literature. *Forensic Sci. Int.*, **90**:139–159.

Hepler, B. R., and Isenschmid, D. S. (1998). Specimen selection, collection, preservation, and security. In: Karch, S. B. (Ed.), *Drug Abuse Handbook*. Boca Raton, FL: CRC Press, pp. 873–889.

Hilberg, T., Ripel, A., Slørdal, L., Bjørneboe, A., and Mørland, J. (1999a). The extent of postmortem drug redistribution in a rat model. *J. Forensic Sci*., **44**:956–962.

Hilberg, T., Rogde, S., and Mørland, J. (1999b). Postmortem drug redistribution: human cases related to results in experimental animals. *J. Forensic Sci*., **44**:3–9.

Hu, Z., Yang, X., Ho, P. C., Chan, S. Y., Heng, P. W., Chan, E., et al. (2005). Herb–drug interactions: a literature review. *Drugs*, **65**:1239–1282.

Jannetto, P. J., Wong, S. H., Gock, S. B., Laleli-Sahin, E., Schur, B. C., and Jentzen, J. M. (2002). Pharmacogenomics as molecular autopsy for postmortem forensic toxicology: genotyping cytochrome P450 2D6 for oxycodone cases. *J. Anal. Toxicol*., **26**:438–447.

Jenkins, A. J., and Lavins, E. S. (1998). 6-Acetylmorphine detection in postmortem cerebrospinal fluid. *J. Anal. Toxicol*., **22**:173–175.

Jin, M., Gock, S. B., Jannetto, P. J., Jentzen, J. M., and Wong, S. H. (2005). Genotyping cytochrome P450 3A4*1B and 3A5*3 for 25 fentanyl cases. *J. Anal. Toxicol*., **29**:590–598.

Jones, G. R. (1998). Interpretation of postmortem drug levels. In: Karch, S. B. (Ed.), *Drug Abuse Handbook*. Boca Raton, FL: CRC Press, pp. 970–985.

Jönsson, A. K., Holmgren, P., Druid, H., and Ahlner, J. (2007). Cause of death and drug use pattern in deceased drug addicts in Sweden, 2002–2003. *Forensic Sci. Int*., **169**:101–107.

Karch, D. B. (Ed.) (2007). *Drug Abuse Handbook*, 2nd ed. Boca Raton, FL: CRC Press.

Klaassen, C. D. (Ed.) (2008). *Casarett & Doull's Toxicology*, 7th ed. New York: McGraw-Hill.

Kugelberg, F. C., and Jones, A. W. (2007). Interpreting results of ethanol analysis in postmortem specimens: a review of the literature. *Forensic Sci. Int*., **165**:10–29.

Kupiec, T. C., Raj, V., and Vu, N. (2006). Pharmacogenomics for the forensic toxicologist. *J. Anal. Toxicol*., **30**:65–72.

Lee, C.-W., Yim, L.-K., Chan, D. T. W., and Tam, J. C. N. (2002). Sample pre-treatment for CO-oximetric determination of carboxyhaemoglobin in putrefied blood and cavity fluid. *Forensic Sci. Int*., **126**:162–166.

Lee, C.-W., Tam, J. C. N., Kung, L.-K., and Yim, L.-K. (2003). Validity of CO-oximetric determination of carboxyhaemoglobin in putrefying blood and body cavity fluid. *Forensic Sci. Int*., **132**:153–156.

Levine, B. (Ed.) (2006). *Principles of Forensic Toxicology*, rev. 2nd ed. Washington, DC: AACC Press.

Lewis, D. E., Moore, C. M., Leikin, J. B., and Koller, A. (1995). Meconium analysis for cocaine: a validation study and comparison with paired urine analysis. *J. Anal. Toxicol*., **19**:148–150.

Lewis, R. J., Johnson, R. D., and Canfield, D. V. (2004). An accurate method for the determination of carboxyhemoglobin in postmortem blood using GC-TCD. *J. Anal. Toxicol*., **28**:59–62.

Madea, B., and Rödig, A. (2006). Time of death dependent criteria in vitreous humor: accuracy of estimating the time since death. *Forensic Sci. Int*., **164**:87–92.

Marin, S. J., Coles, R., Merrell, M., and McMillin, G. A. (2008). Quantitation of benzodiazepines in urine, serum, plasma, and meconium by LC–MS–MS. *J. Anal. Toxicol*., **32**:491–498.

Marquardt, H., Schäfer, S. G., McClellan, R. D., and Welsch, F. (Eds.) (1999). *Toxicology*. San Diego, CA: Academic Press.

Milroy, C. M., and Forrest, A. R. W. (2000). Methadone deaths: a toxicological analysis. *J. Clin. Pathol.*, **53**:277–281.

Miranda- G. E., Sordo, M., Salazar, A. M., Contreras, C., Bautista, L., Rojas García, A. E., and Ostrosky-Wegman, P. (2007). Determination of amphetamine, methamphetamine, and hydroxyamphetamine derivatives in urine by gas chromatography–mass spectrometry and its relation to CYP2D6 phenotype of drug users. *J. Anal. Toxicol.*, **31**:31–36.

Moore, C., and Negrusz, A. (1995). Drugs of abuse in meconium. *Forensic Sci. Rev.*, **7**:103–118.

Moore, C., Deitermann, D., Lewis, D., and Leikin, J. (1995). The detection of hydrocodone in meconium: two case studies. *J. Anal. Toxicol.*, **19**:514–518.

Moriya, F., and Hashimoto, Y. (1999). Redistribution of basic drugs into cardiac blood from surrounding tissues during early-stages postmortem. *J. Forensic Sci.*, **44**:10–16.

Moriya, F., and Hashimoto, Y. (2005). Endogenous gamma-hydroxybutyric acid levels in postmortem blood. *Bull. Int. Assoc. Forensic Toxicol.*, **XXXV**:51.

Moriya, F., Chan, K. M., Noguchi, T. T., and Wu, P. Y. K. (1994). Testing for drugs of abuse in meconium of newborn infants. *J. Anal. Toxicol.*, **18**:41–45.

Muñoz, J. I., Suárez-Peñaranda, J. M., Otero, X. L., Rodríguez-Calvo, M. S., Costas, E., Miguéns, X., et al. (2001). A new perspective in the estimation of postmortem interval (PMI) based on vitreous [K^+]. *J. Forensic Sci.*, **46**:209–214.

Muñoz, J. I., Suárez-Peñaranda, J. M., Otero, X. L., Rodríguez-Calvo, M. S., Costas, E., Miguéns, X., et al. (2002). Improved estimation of postmortem interval based on differential behavior of vitreous potassium and hypoxanthine in death by hanging. *Forensic Sci. Int.*, **125**:67–74.

Omiecinski, C. J., Remmel, R. P., and Hosagrahara, V. P. (1999). Concise review of the cytochrome P450s and their roles in toxicology. *Toxicol. Sci.*, **48**:151–156.

Ortega García, J. A., Carrizo Gallardo, D., Ferris i Tortajada, J., García, M. M. P., and Grimalt, J. O. (2006). Meconium and neurotoxicants: searching for a prenatal exposure timing. *Arch. Disease Child.*, **91**:642–646.

Ostrea, E. M. (2001). Understanding drug testing in the neonate and the role of meconium analysis. *J. Perinat. Neonat. Nurs.*, **14**:61–82.

Ostrea, E. M., Brady, M. J., Parks, P. M., Asensio, D. C., and Naluz, A. (1989). Drug screening of meconium in infants of drug dependent mothers: an alternative to urine screening. *J. Pediatr,*, **115**:474–477.

Ostrea, E. M., Jr., Brady, M., Gause, S., Raymundo, A. L., and Stevens, M. (1992). Drug screening of newborns by meconium analysis: a large-scale, prospective, epidemiologic study. *Pediatrics*, **89**:107–113.

Ostrea, E. M., Jr., Romero, A., Knapp, D. K., Ostrea, A. R., Lucena, J. E., and Utarnachitt, R. B. (1994). Postmortem drug analysis of meconium in early-gestation human fetuses exposed to cocaine: clinical implications. *J. Pediatr.*, **124**:477–479.

Pélissier-Alicot, A.-L., Gaulier, J.-M., Champsaur, P., and Marquet, P. (2003). Mechanisms underlying postmortem redistribution of drugs: a review. *J. Anal. Toxicol.*, **27**:533–544.

Perrigo, B. J., and Joynt, B. P. (1989). Evaluation of current derivative spectrophotometric methodology for the determination of percent carboxyhemoglobin saturation in postmortem blood samples. *J. Anal. Toxicol.*, **13**:37–46.

Plueckhahn, V. D. (1967). The significance of blood alcohol levels at autopsy. *Med. J. Aus.*, July 15, pp. 118–124.

Prouty, R. W., and Anderson, W. H. (1990). The forensic science implications of site and temporal influences on postmortem blood–drug concentrations. *J. Forensic Sci.*, **35**:243–270.

Rodda, K. E., and Drummer, O. H. (2006). The redistribution of selected psychiatric drugs in post-mortem cases. *Forensic Sci. Int.*, **164**:235–239.

Roden, D. M., Altman, R. B., Benowitz, N. L., Flockhart, D. A., Giacomini, K. M., Johnson, J. A., Krauss, R. M., McLeod, H. L., Ratain, M. J., Relling, M. V., Ring, H. Z., Shuldiner, A. R., Weinshilboum, R. M., and Weiss, S. T. (2006). Pharmacogenomics: challenges and opportunities. *Ann. Intern. Med.*, **145**:749–757.

Rohrig, T. P. (1998). Comparison of fentanyl concentrations in unembalmed and embalmed liver samples. *J. Anal. Toxicol.*, **22**:253.

Saper, R. B., Kales, S. N., Paquin, J., Burns, M. J., Eisenberg, D. M., Davis, R. B., et al. (2004). Heavy metal content of ayurvedic herbal medicine products. *JAMA*, **292**:2868–2873.

Scott, W., Root, I., and Sanborn, B. (1974). The use of vitreous humor for determination of ethyl alcohol in previously embalmed bodies. *J. Forensic Sci.*, **19**:913–916.

Skopp, G. (2004). Preanalytic aspects in postmortem toxicology. *Forensic Sci. Int.*, **142**:75–100.

Smith, M. L., Vorce, S. P., Holler, J. M., Shimomura, E., Magluilo, J., Jacobs, A. J., et al. (2007). Modern instrumental methods in forensic toxicology. *J. Anal. Toxicol.*, **31**:237–253.

Spiehler, V. R., and Reed, D. (1985). Brain concentrations of cocaine and benzoylecgonine in fatal cases. *J. Forensic Sci.*, **30**:1003–1011.

Spitz, W. U., and Spitz, D. J. (Eds.) (2006). *Spitz and Fisher's Medicolegal Investigation of Death*, 4th ed. Springfield, IL: Charles C Thomas.

Steele, B. W., Bandstra, E. S., Wu, N.-C., Hime, G. W., and Hearn, W. L. (1993). m-Hydroxybenzoylecgonine: an important contributor to the immunoreactivity in assays for benzoylecgonine in meconium. *J. Anal. Toxicol.*, **17**:348–352.

Stephens, R. J., and Richards, R. G. (1987). Vitreous humor chemistry: the use of potassium concentration for the prediction of the postmortem interval. *J. Forensic Sci.*, **32**:503–509.

Stout, P. R., Horn, C. K., and Lesser, D. R. (2000). Loss of THCCOOH from urine specimens stored in polypropylene and polyethylene containers at different temperatures. *J. Anal. Toxicol.*, **24**:567–571.

Suma, R., and Sai Prakash, P. K. (2006). Conversion of sertraline to N-methyl sertraline in embalming fluid: a forensic implication. *J. Anal. Toxicol.*, **30**:395–399.

U.S. Food and Drug Administration (2007). FDA issues dietary supplements final rule. *FDA News*, June 22. http://www.fda.gov/bbs/topics/NEWS/2007/NEW01657.html. Accessed Aug. 9, 2008.

Watterson, J., Blackmore, V., and Bagby, D. (2006). Considerations for the analysis of forensic samples following extended exposure to the environment. *Forensic Examiner*, **15(4)**:19–25.

Wong, S. H. Y., and Sunshine, I. (Eds.) (1997). *Handbook of Analytical Therapeutic Drug Monitoring and Toxicology*. Boca Raton, FL: CRC Press.

Wong, S. H., Wagner, M. A., Jentzen, J. M., Schur, C., Bjerke, J., Gock, S. B., and Chang, C. C. (2003). Pharmacogenomics as an aspect of molecular autopsy for forensic pathology/toxicology: Does genotyping CYP 2D6 serve as an adjunct for certifying methadone toxicity? *J. Forensic Sci.*, **48**:1406–1415.

Woo, T. M. (2008). When nature and pharmacy collide. *Adv. Nurse Pract.*, July, pp. 69–72.

CHAPTER 14

Entomotoxicology: Drugs, Toxins, and Insects

JASON H. BYRD

University of Florida, College of Medicine, Gainesville, Florida

MICHELLE R. PEACE

Department of Forensic Science, Virginia Commonwealth University, Richmond, Virginia

Summary Forensic entomology is gaining widespread acceptance within the forensic sciences as one method to estimate a portion of the postmortem interval by utilizing the time of insect colonization of the body, also known as the period of insect activity. Additionally, insect evidence can be utilized as alternative toxicology samples in cases where no other viable specimens exist. This subfield, known as entomotoxicology, can provide useful qualitative information to investigators as to the presence of drugs in the tissues at the time of larval feeding. The presence of drugs can also alter the developmental period of the insects and should always be taken into consideration by the forensic entomologist. In this chapter we examine the relationship between toxicology and forensic entomology.

14.1	Introduction		484
14.2	The fly and forensic science		484
	14.2.1	History of forensic entomology, toxicology, and the rise of entomotoxicology	485
	14.2.2	Drugs and the fly life cycle	488
	14.2.3	Why use insects as a toxicological specimen?	490
	14.2.4	Drug extraction methods	492
	14.2.5	Qualitative versus quantitative	493
	14.2.6	Changes in insect development: toxins and drugs	494
	14.2.7	The future of entomotoxicology	494
	References		495

Forensic Chemistry Handbook, First Edition. Edited by Lawrence Kobilinsky.
© 2012 John Wiley & Sons, Inc. Published 2012 by John Wiley & Sons, Inc.

14.1 INTRODUCTION

The use of entomological evidence in a legal investigation can provide extraordinarily valuable information that may otherwise be overlooked. Most often, the utility of an insect in a forensic investigation is to assist investigators in establishing the postmortem interval. More specifically, the period of the postmortem interval is often termed the *period of insect activity* (PIA) or the *time of colonization* (TOC) and can be thought of as the minimum postmortem interval. It is often expressed as the minimum time period necessary to allow the insects collected to develop to the life stage in which they were discovered.

Several different approaches exist in making this determination; they range from insect succession to estimations based on larval length, larval weight, and the accumulated degree hour (or degree day) methodology. Each of these techniques has advantages and disadvantages. One of the most common issues in forensic entomology is that the investigator is not able to determine exactly what the climate conditions were at the death scene during the entire postmortem period. Since insect growth and development are heavily influenced by ambient temperatures, the difficulty in accounting for the conditions at the scene prior to the discovery introduces some uncertainty in application. Thus, the forensic entomologist must account for this variability when providing estimations of the time since colonization. If such variability is addressed properly, forensic entomology can be applied with an impressive degree of accuracy and precision.

Generally, this variability is accounted for by the extrapolation of meteorological recording station data from nearby locations. However, other factors can play a role in the influence and alteration of larval insect growth. One of these factors, often unaccounted for by the forensic entomologist, is the presence of various metabolites of illicit drugs and poisons within the insect that can alter its expected time of development. Such chemical compounds, if present in the host tissue and consumed by the larvae, may alter insect development by hours or days.

The term *Entomotoxicology* was first used in 1994 to describe the interface of entomology, forensic science, and toxicology, and we are only now beginning to understand how various drugs and toxins interact with a developing insect to alter the rate of development of various life stages. In its relatively short history, the discipline of entomotoxicology has been very broadly applied. The deposition of metals within the larvae has assisted in pinpointing human remains to a possible point of origin, or to corroborate or refute suspect and witness statements as to the activities of the victim prior to death. However, the most common utilization has been the detection of various chemical compounds from the larvae long after suitable soft tissues, blood, and urine from the body have decomposed or been consumed by the insects present (Nolte et al., 1992) (Table 14.1).

14.2 THE FLY AND FORENSIC SCIENCE

For many millennia, humans have been aware of, known, and understood the role of flies in the ecosystem. The first mention of the flies most commonly used in

TABLE 14.1 Drug Compounds Recovered from Calliphoridae and Sarcophagidae Larvae

Calliphoridae	Sarcophagidae
Alimenmazine	Amitrptyline
Amitrptyline	Barbiturates
Arsenic	Cocaine (benzoylecgonine)
Barbiturates	Heroin
Bromazepam	Malathion
Clomipramine	Methamphetamines
Cocaine (benzoylecgonine)	Mercury
Laudanum	phencyclidine
Levomepromazine	3,4-Methylenedioxymethamphetamine
Mercury	
Morphine	
Oxazepam	
Phenobarbital	
Triazolam	

present-day criminal investigations was in the fourteenth tablet of Harra-Hubulla, a list of wild animals from the time of Hammurabi 3600 years ago. The Egyptians also keenly understood the metamorphosis of flies, as indicated by the Papyrus Gizeh No. 18026:4:14 (Cairo, Egyptian Museum CG 58009), found in the mouth of a mummy, which bore the inscription: "The maggots will not turn into flies within you (Golénischeff 1927)."

The blowflies (Calliphoridae), flesh flies (Sarcophagidae), and house flies (Muscidae) are the most forensically important flies. These flies are most often the first insects to find and colonize remains. The blowfly can travel up to 20 km a day in search of an appropriate food source. A female can lay several batches of about 250 eggs each, depending on her size and health. Once a female has begun oviposition, other females will also lay in the same area, resulting in a mass of several thousand eggs. The eggs usually hatch within 24 hours and immature larvae penetrate the interior of the remains to avoid desiccation and predation. As the larvae feed and mature, they pass through three stadia: first, second, and third instars. In doing so, their feeding mass can reach temperatures of 50°C. Once the larvae stop feeding, they wander from the food source to find a suitable place to pupate, during which the soft-bodied larvae become shorter, wider, rigid, and dark. Once metamorphosis is complete, the adult flies will emerge from the casing, inflate their wings, and become sclerotized (Figure 14.1). (Kamal, 1958; Greenberg, 1991).

14.2.1 History of Forensic Entomology, Toxicology, and the Rise of Entomotoxicology

The *Washing Away of Wrongs*, a thirteenth-century Chinese manual on forensic and legal medicine, considered the oldest in existence from any civilization, detailed

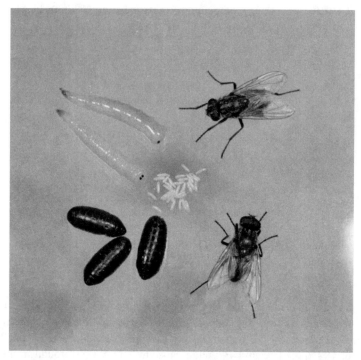

Figure 14.1 Life cycle of *Musca domestica*, showing complete metamorphosis. These life stages are typical of all Muscidae and Calliphoridae. (Photo courtesy of Clemson University; U.S. Department of Agriculture Cooperative Extension Slide Series.) *(See insert for color representation.)*

the death investigation of a man with a wound apparently inflicted by a sickle. The investigator gathered the local farmers with their sickles. In the hot afternoon sun, flies gathered on one sickle, indicating blood residue and incriminating the owner of that sickle. After interrogation and in light of the evidence, the owner of the sickle "knocked his head on the ground and confessed" to the murder (McKnight, 1981). In the late nineteenth century, scientists began to study and characterize the activities of arthropods and their application to death investigations. The first observation of insects on human remains in which the death was attributed to arsenic poisoning was in 1890. F. M. Webster noted that *Conicera* sp. were noted to have developed on remains buried for over two years. Analysis of the stomach contents proved that the poisoning was via arsenic. He noted that it was not surprising that these flies could develop on tissues laced with arsenic because "they are doubtless tenacious of life." Unfortunately, it is not apparent that he foresaw the forensic science application of their presence (Webster, 1890). In 1894, the French entomologist Mégnin pioneered the use of forensic entomology by painstakingly outlining and describing the eight successive waves of arthropod invasions on a corpse (Nuorteva, 1977; Keh, 1985; Smith, 1986). Those eight stages are still commonly recognized today.

Nonetheless, major issues still arose for forensic entomologists. Wide variability in geographical, topological, and environmental conditions led to even more widely variable insect species and behaviors for any given situation. These variable conditions made a well-characterized foundation of workable information difficult to attain. Differences between the entomologists' experimental conditions and actual forensic cases also compounded the problem of making useful correlations. Ultimately, a more thorough understanding of the basic principles as well as the limitations of using insects in forensic investigations was the result of the corroborative work of pathologists, biologists, detectives, and entomologists. These collaborative works had a significant impact in a variety of cases:

- Estimation of postmortem interval (PMI) by determining the life-cycle stage of the insects present on a body in conjunction with the stage of faunal procession (Bergeret, 1955; Nuorteva, 1977; Hall et al., 1986; Smith, 1986; Catts and Goff, 1992; Catts, 1992; Hall and Doisy, 1993).
- Estimation of PMI from the degree of insect colonization resulting from the degree of body exposure (totally exposed, partially covered, or completely buried) (Nuorteva, 1977; Smith, 1986).
- Acquittal of a person based on larval migration data (Nuorteva, 1977).
- Determination that a body had been moved from the original crime scene by identifying insects on that body not indigenous to the discovery area, thereby determining the area from which it was moved (Benecke, 1998).
- Linkage of suspects to the scene of a crime due to insect bites found on their bodies from insects specific to the vicinity where the body was discovered (Prichard et al., 1986).
- Determination of child abuse, neglect of the elderly, or disease or infection state from insect colonization prior to death (Goff, 1991; Benecke, 1998; Benecke and Lessig, 2001).
- Individualization of human DNA from insect gut contents (Lord et al., 1998).

Interestingly, Mathiew Orfila, a pathologist who is often credited as the father of modern toxicology, can potentially be credited with the first systematic observation of insect succession in a human cadaver (Greenberg, 1991). He documented 30 arthropods from a corpse and noted that live insects were recovered from bodies that were exhumed for several years. Bergeret (1955) is credited with the first case in which entomological evidence was utilized to create a time line. However, it turns out his assessment of the entomological evidence may have been incorrect; the period of insect activity could have been much shorter. Citations for the application of entomology to toxicological analysis did not begin to appear until the early 1970s. Nuorteva described a case in which the mercury concentration in larvae feeding on a corpse was used to determine if the deceased had lived in an area with high mercury pollution (Nuorteva, 1977). Subsequently, the deposition of copper, iron, and zinc and their effects on adult flies and larvae was demonstrated, and the bioaccumulation and effect of mercury on larvae was fully characterized (Sohal and Lamb, 1977; Nuorteva and Nuorteva, 1982).

The first examination of fly larvae from human remains for qualitative identification of drugs was proposed by Beyer et al. in 1980. In this case, Calliphoridae larvae were used to assist investigators in proving a suicide via an overdose of pentobarbital (Beyer et al., 1980). In this case fly larvae were used to determine the source of poisoning in a severely decomposed body that had no other source for a toxicological specimen (Beyer et al., 1980). Gunatilake and Goff (1989) utilized larvae collected from a body to aid in proving that a case involved poisoning by ingestion of malathion. This chemical was detected in the larvae as well as from body fat and gastric contents.

In 1990, Pascal Kintz published a landmark study from a 1987 case illustrating a "new" toxicological method of investigation which demonstrated the usefulness of entomological evidence. In this case example, larval concentrations of five drugs (triazolam, oxazepam, phenobarbital, alimenmazine, and clomipramine) were compared to concentrations in human heart, liver, lung, spleen, and kidney tissue as well as bile. In all cases, the drugs known to be in the human tissue were recovered form the larvae. One year later, in 1988, Kintz reported a case in which a severely decomposed corpse was found and insects were utilized as an alternative toxicological specimen because no other suitable tissue was available (Kintz et al., 1990).

Introna et al. (1990) collected livers from 40 decedents known to be positive for opiates (morphine) and colonized the tissues with eggs of *Calliphora vicina*. Upon analysis of the larvae and tissue via radioimmunoassay, the opiate concentration in the larvae correlated with that of the tissues on which they fed. Subsequently, investigations in entomology and toxicology illustrated the two important areas in forensic entomotoxicology: the effects of drugs on insect life cycles and the use of insects as an alternative toxicological sample (Goff et al., 1989, 1994).

The potential growth of entomotoxicology in casework can be moderately predicted based on the statistics that illustrate the increase in the abuse of illicit and prescription drugs in the past decade. The past decade has seen an increase in the abuse of illicit and prescription drugs. In a three-year time span alone (2000–2002), use of illicit drugs has increased from 6.3% to 8.3% in persons older than 12 years (NHS Survey, 2002). In 2000, 43% of emergency room visits were due to the misuse of prescription drugs (SAMHA, 2003). Not only have drug-related emergency room episodes and subsequent deaths increased (SAMHA 2003; DAWN, 2003), but a clear correlation exists among crime, violence, and drug usage. Of the approximate 14,000 homicides in the United States in 1998, 4.8% of them were murders committed during a narcotics felony (ONDCP, 2000). In the end, with this growth in potential must come research to support widespread acceptance and understanding of the expansive scope of applications and utility.

14.2.2 Drugs and the Fly Life Cycle

Sometimes, entomological evidence is used in death investigations as a tool to better understand the circumstances leading to death. Insects can be used to date a crime scene or estimate a PMI by applying the known developmental rates of insects and the successional patterns of different insect species on the remains (Smith, 1986;

Goff et al. 1988; Catts and Goff, 1992; Goff, 1993; Catts, 1992; Goff and Lord, 1994). As mentioned previously, the PMI estimation is an approximation of the time of colonization of the cadaver by insects. The succession of insects on carrion can be complicated by many geographical and environmental conditions, such as plant cover, water covering the body, inclement weather, and/or temperature (Nuorteva, 1977; Smith, 1986; Goff et al., 1988; Catts and Goff, 1992; Schoenly et al., 1992, 1996; Goff, 1993; Hall and Doisy, 1993; Catts, 1992; LaMotte and Wells, 2000). The time required for an insect to reach any developmental stage at certain environmental conditions has been well characterized, particularly for forensically important insects (Kamal, 1958; Greenberg, 1990a,b; Goff, 1991; Greenberg, 1991; Byrd and Butler, 1997, 1998; Anderson, 2000; Byrd and Allen, 2001). The time of colonization can be estimated using either of these techniques singularly or in combination. However, it is most common to use the stage of development in conjunction with meteorological data. This technique is most frequently referred to as the *accumulated degree method*. This method compares the temperature with the time of development and is often used to estimate the time of colonization and subsequent minimum postmortem interval.

Additionally, the stage of development or age can be determined by measuring the insect larvae's total length, crop length, or weight. Weight and crop length are the least useful estimates of age because broad variations result from the amount of food available, changes in the physiology of the developing larvae, and the lack of a well-defined informational database (Greenberg 1990a, 1991; Wells and LaMotte, 1995). The methods that utilize larval length and/or accumulated degree hours (which can incorporate larval length) are the most commonly used, particularly since an appreciable database has been developed (Greenberg, 1991; Goff, 1993; Byrd and Allen, 2001; Byrd and Castner, 2001).

Before Sohal and Nuorteva ushered in studies to investigate the effects of drugs on insects, Introna reported a solitary investigation in Japan in 1958 which found that flies were attracted to carrion differently depending on the poison ingested (Introna et al., 2001). Since then, Goff and Bourel have made strides in understanding the deposition and affects of drugs and toxins on forensically important insects. In 1989, Goff worked a case with Gunatilake in which a man expired from malathion intoxication. The man had been missing eight days, but the entomology-based PMI estimation was five days. Goff and Gunatilake concluded that the malathion had shortened the duration of the larval stage and that the pesticide retarded the succession of insects on the body (Gunatilake and Goff, 1989). Goff has also explored the effects of amitriptyline, cocaine, heroin, phencyclidine, methamphetamine, and 3,4-methylenedioxymethamphetamine (MDMA) on fly larvae. He found that amitriptyline had no effect on the rate of development of *Parasarcophaga ruficornis* larvae but prolonged the migratory stage. Amitriptyline also caused the body lengths and weights to be greater than those of the control larvae (Goff et al., 1993). Cocaine shortened total developmental times for *Boettcherisca peregrine* larvae and puparia. He found that heroin accelerated the growth rate of the *B. peregrine* larvae and that phencyclidine did not affect the rate of development for *P. ruficornis* larvae but shortened the migratory phase. Finally, Goff concluded that methamphetamine and MDMA generally shortened the duration of the feeding larval stage for *P. ruficornis*, and methamphetamine caused the larvae to grow significantly smaller than the control (Goff et al., 1989, 1992, 1994, 1997). Interestingly, Bourel found that morphine, a drug structurally similar to heroin, slowed the developmental rate of larvae for *Lucilia sericata* (Bourel et al., 1999). Ultimately, these results indicated three potentially significant points: (1) that drugs can affect developmental rates, contrary to expectations; (2) that the same drugs may affect different fly species differently; and (3) that PMI estimations could potentially be miscalculated if drug effects are not taken into account.

14.2.3 Why Use Insects as a Toxicological Specimen?

Invariably, some death investigations involve human remains that are severely decomposed or need to be exhumed after embalming and burial. Typical postmortem specimens for a toxicological analysis include brain, liver, kidney, heart blood, peripheral blood, vitreous humor, bile, urine, and gastric contents (Poklis et al., 1998; Levine, 1999). The tissue alteration and breakdown that results during the process of decomposition can make some toxicological testing methods more challenging. In some cases, the tissue samples are rendered unsuitable for analysis, or they are simply no longer present in the body. The most common problem in warm and humid climates is that no soft tissue remains on which to perform a toxicological analysis (Levine et al., 2000). Additionally, when a body is embalmed, the tissues are fixed, and blood and urine are typically no longer available. Toxicological analyses on embalmed tissue can and will result in drug concentrations that are difficult to interpret (Hanzlick, 1994; Alunni-Perret et al., 2003). Therefore, in these extreme cases, alternative specimens often need to be examined.

Properly preserved or freshly collected larvae can be treated as any other tissue for toxicological analysis (Beyer et al., 1980; Poklis et al., 1998; Levine et al., 2000). Two important premises underscore the utility of entomotoxicology.

1. When larvae feed, they shred the food source with mouth hooks and deposit the meal into a crop, which generally "reflects the adaptive imperative to eat now and digest later" (Greenberg, 1991).
2. Any drug detected in larvae could only have come from the body on which it was feeding (Beyer et al., 1980).

Beyer's case was a landmark in forensic entomotoxicology as the first case that used larvae as the sole source for a toxicological analysis. Toxicological analyses were performed on larvae found on a skeletonized body after the badly decomposed soft tissue was deemed unsuitable for analyses. Phenobarbital was identified in the larvae using thin-layer chromatography and gas chromatography–mass spectrometry (Beyer et al., 1980). In 1987, Kintz also used larvae as a toxicological specimen in a case of completely putrified remains. Despite the severe decomposition, organs were identified and analyzed for drugs. The concentrations of drugs (triazolam, oxazepam, phenobarbital, alimemazine, and clomipramine) in the larvae were compared to the concentrations in the body's bile, heart, liver, lung, and spleen. Kintz stated that they could not establish a correlation between the larval drug concentrations and organ concentrations. However, he concluded that "the application of the toxicological investigations in maggots will surely increase" and that the performance of these analyses on living material "is always more suitable" (Kintz et al., 1990). In 1989, larvae were used to detect the organophosphate suspected as the cause of death of a severely decomposed man (Gunatilake and Goff, 1989). In 1990, Introna reported results from the first systematic study in entomotoxicology in which he found a direct and significant correlation between the concentration of morphine in the food source and the larvae (Introna et al., 1990). In 1992, Nolte investigated a case in which he collected fly larvae and skeletal muscles from an almost completely skeletonized corpse for toxicological analysis. This was the first reported case in which cocaine and benzoylecgonine were recovered from fly larvae. Nolte concluded that the larvae were better specimens for toxicological analyses because the skeletal muscle contained tissue decomposition products which interfered with the analyses (Nolte et al., 1992). It is important to understand that the half-life of cocaine in human tissue is 0.7 to 1.5 h, depending on the route of administration (Baselt, 2004). Therefore, Nolte concluded that the detection of cocaine and its metabolites in larval tissue suggests the use of cocaine in the immediate hours before death. However, while finding that both the parent drug and its metabolite in the larvae could indicate drug usage immediately prior to death, other contributing factors, such as route of administration, dosage, and postmortem redistribution, will impinge on this conclusion, making it difficult or impossible to interpret the findings of drugs in larvae. During Goff's studies to better understand the effects of drugs on the insect life cycle, he found a direct correlation between the concentration of the drugs heroin, methamphetamine, and MDMA in the food source and the larvae. On the other hand, he observed no correlation between the two with phencyclidine (Goff et al., 1991, 1992, 1994, 1997).

Since the late 1990s, several attempts have been made to understand the implications of larval drug concentrations. Wilson and Sadler found that the larvae eliminated amitriptyline, nortriptyline, trimipramine, and trazodone and that the ratios of larvae to food source drug concentrations were widely variable (Wilson et al., 1993; Sadler et al., 1995, 1997a). Sadler also investigated barbiturates and other analgesics with variable results. He found that acetaminophen was eliminated from the larvae and that the ratios of larvae to food source drug concentrations were also widely variable for the barbiturates amphetamine and sodium salicylate. Ultimately, he concluded that his solid-phase drug extraction procedure was not rigorous enough (Sadler et al., 1997b). Most recently, Hédouin performed a whole animal study and found a correlation between mature larval morphine concentrations and the tissues (Hédouin et al., 1999, 2001).

A case in Maryland reported in 2000 underscores the growing importance of more thoroughly understanding the utility of fly larvae as an alternative toxicological specimen. In this case, larvae and the only soft tissue (a portion of calf muscle) from mostly skeletonized remains were analyzed for drugs. Secobarbital was found in the larvae but not in the skeletal muscle. If a large amount of drug was ingested which led immediately to death, the drug had insufficient time to distribute to the calf muscle in any appreciable concentration. Second, the presence of decompositional fluids in the muscle potentially complicated the analysis. Had the larvae not been available for analysis, the prescription bottle found near the deceased would have only served as circumstantial evidence (Levine et al., 2000).

14.2.4 Drug Extraction Methods

If using entomological evidence as a toxicological specimen, larvae and pupae can be treated similarly to human specimens. Standard operating procedures for drug extractions from human tissue and fluid in a toxicology laboratory are a reasonable starting point for elucidating an optimum drug extraction procedure for what can be considered a fairly novel and certainly different biological matrix, insects. However, an extraction procedure for any given drug will probably have to be modified to compensate for the entomological matrix. For example, Sadler admits that he applied methods routinely used in his laboratory to extract barbiturates from urine to extract the same from larvae, but conceded that his inconclusive results were due to needing a more rigorous drug extraction technique.

When extracting drugs from larvae, the first consideration should be the large fat body of the larvae (Downer, 1981; Gullan and Cranston, 1994; Elzinga, 1999). Lipophilic drugs can potentially be sequestered in this fat body, requiring extraction solvents that are more lipophilic and back-extractions to adequately separate the matrix from the drug. Additionally, in order to actually extract the drugs from the larval matrix (the fat body, the hemocoel, and hemolymph), the pK_a of the drug must be considered carefully using the Henderson–Hasselbach equation for acidic drugs,

$$pH = pK_a + \log \frac{[B]}{[BH^+]}$$

and for basic drugs (Levine, 1999),

$$pH = pK_a + \log \frac{[A^-]}{[HA]}$$

Assuming that a drug will be mostly ionized in an aqueous-based biological fluid, the initial pH of the aqueous fluid should be adjusted so that the drug is no longer ionized. Once the drug is deionized, it can be sequestered into a hydrophobic solvent, thereby effectively extracting it from the biological matrix. Some cellular components in the matrix may also become sequestered in the hydrophobic extraction solvent (particularly the fat body), requiring a back extraction or a "cleanup" extraction. The cleanup extraction can be executed by changing the pH to drive the drug back into an aqueous layer, leaving hydrophobic cellular components behind in the solvent. Since the drugs need to be in an organic solvent for instrumental analysis, a final pH adjustment will deionize the drugs, allowing their final sequestering in a hydrophobic solvent.

14.2.5 Qualitative Versus Quantitative

Several studies have demonstrated the correlation between the dose and the concentration of drug within the larval body. However, researchers have differed greatly in their interpretation of the meaning of this correlation. In application, the quantitative evaluation of drug concentration in larvae in actual forensic casework is quite limited. One of the primary reasons for this limitation is because of the difficulty in determining exactly where the larvae have fed, since the absorption and distribution of a drug will vary among skeletal muscle and organ tissues. Thus, the location of larval feeding is of critical importance. However, due to the larval movement on the body while feeding, it is difficult (if not impossible) to determine on which tissues the larvae have fed. Although it is possible for an investigator to document the location of larvae as they are collected, it is not known which tissue or tissues were utilized as a food source prior to collection.

Although the rates of drug deposition in the larval body are not known for most species, most commonly, homogenates of the entire larval body are used. It is not known if the drugs and their metabolites are localized in the fat bodies, puparium, or chitinous exoskeleton of the adult (Nolte, 1992). However, much of the published research on entomotoxicology indicates that ingested chemicals can be recovered from the soft tissues of the larvae, pupae, and chitinous exterior of the pupal case. As a result, initial studies on the application of entomotoxicology to casework were promising. The window of opportunity for the detection of chemical compounds in human remains could be extended for days or even weeks by utilizing larvae and puparia. If the chitinous pupa stage had formed, chemicals ingested by the feeding larvae could be detected in the puparial casings years after the insect itself had completed development. In one particular death investigation case, the authors detected the metabolites of cocaine from the puparium two years after the death of the victim.

However, the quantitative use of insects as toxicological specimens in actual casework is problematic. Tracqui et al. (2004) analyzed 29 cases in which organic compounds were known to be present. These compounds were detected in both the human tissue and the insect body. However, no correlation between the drug concentration in the human versus larval samples were noted. Additionally, it was found that interlarvae and intersite variations in drug concentrations were not reproduced between samples. Therefore, arthropod larvae are, at best, most appropriate for qualitative analyses in a toxicological investigation and unreliable and unsuitable as a toxicological sample used to interpret the episode of intoxication prior to death based on a quantitative analysis.

The method of calculating the accumulated degree hours to estimate the time since colonization is as follows (Byrd and Castner, 2001):

1. Accumulated degree hours are calculated for the crime scene using temperature data gathered for a specified period of time prior to the collection of larval evidence.

$$\text{accumulated degree hours} = \text{ADH} = (T - B)(t)$$

where T is the mean temperature (°C)[†] B the minimum threshold development temperature and t the time in hours.

2. Larval evidence is measured (mm) and compared to species developmental charts (larval length vs. time, or instar molt vs. time) generated at specific temperatures to obtain the ADH required for that size.
3. The ADH required, which is given in the developmental chart, is compared to the ADH calculated at the scene to age the larvae and "date" the scene.

14.2.6 Changes in Insect Development: Toxins and Drugs

Nourteva and Nourteva (1982) were the first to demonstrate a negative effect on larvae from toxic compounds with the larval food source. However, Goff et al. (1989) was the first to show an altered developmental rate for larvae feeding on tissues containing illicit drugs. Goff's study was the first to show that cocaine, metabolized in tissue and ingested by larvae, would shorten the developmental time expected. Subsequent studies by Goff et al. (1993, 1994) have demonstrated significant differences in rates of development for *P. ruficornis* and *B. peregrina* (Diptera: Sarcophagidae). These developmental changes potentially can be of critical importance when estimating the time since colonization. Few studies have been conducted on this effect for the Calliphoridae, which are typically the predominate flies on human remains in the early postmortem period.

With this said, the alteration of insect development by drugs and toxins is problematic on several levels. Often, the forensic entomologist is not aware of the postmortem toxicology results and does not know to take drug-induced alterations into consideration when developing a window for time of colonization. Second, a thorough investigation of all major drugs has not been undertaken with every species of forensically important fly. Therefore, the entomologist has no foundation on which to make adjustments in the time of colonization range. Most important, while some studies have suggested that the developmental rate of some larvae is affected by some drugs, the interpretation of the data must be considered carefully. It is possible that alteration in the life cycle due to drugs or toxins on board is well within the standard error in life-cycle development of drug-free larvae.

14.2.7 The Future of Entomotoxicology

Considered generally, the published literature is only a preliminary treatise of the interface between entomology, toxicology, and medicolegal death investigation. More research is needed on drug accumulation and retention in insect physiology. However, it is clear that the use of entomotoxicology as a quantitative measure is not yet suitable for application in forensic casework. Such a circumstance is unfortunate because the technical application of the premise works well, as the amount

---

[†]Collected from a nearby weather station. If provided, hourly temperatures may be gathered to calculate ADH and provide a defined window for age assessment of the larvae.

of drug compound within the larval body can be determined with great accuracy and precision. Using larvae as a toxicological specimen is readily put into practice because the analysis takes no specialized equipment beyond what is necessary for a standard toxicology analysis for human or animal tissue. Therefore, most any postmortem forensic toxicology laboratory possesses the equipment necessary for the testing. From the technical viewpoint, entomotoxicology looks promising.

However, it is the biology and behavior of the insect itself that may limit its application in forensic investigations. Entomologists must first conduct fundamental research on larval feeding and aggregation behavior. Currently, entomologists do not fully understand larval movement and distribution within the maggot mass, which often forms during the actively feeding third instar. After simple visual observation of the larval feeding and aggregation behavior, it is obvious that the larvae are in constant motion. Individual larvae seem to spend time in the center as well as the periphery of the mass. However, the movement of a single larva within the mass has not been tracked with great accuracy, and no studies have been published on the positional effect on the intensity of larval feeding. With standard entomological procedures currently in practice, it is unknown on what tissues the collected larvae have fed.

The absorption, distribution, metabolism, and elimination of drugs are well documented for most drugs and provides for the understanding of postmortem tissue distributions. However, since larval feeding locations are unknown in applied casework, the resulting larval concentration proves meaningless under current collection and analytical protocols. Insect larvae do accumulate drugs and toxins within bodies as a result of their feeding. Therefore, the qualitative analysis of these compounds may provide valuable information to forensic investigators.

Forensic entomologists are able to provide vast amounts of information to other forensic investigators based on the comparatively limited nature of currently published data sets. Biological variation may lead to cautious, conservative, and "incomplete" answers to many commonly posed questions from investigators. A tremendous amount of research must be conducted to better understand the patterns and implications of natural variables, particularly as they affect forensic investigations in the entire scope of forensic entomology, but specifically with regard to toxicology. As the knowledge base expands and more becomes known about larval behavior, entomotoxicology may one day become a valuable quantitative tool.

REFERENCES

Alunni-Perret, V., Kintz, P., Ludes, B., Ohayon, P., and Quatrehomme, G. (2003). Determination of heroin after embalmment. *Forensic Sci. Int.*, **134(1)**:36–39.

Anderson, G. S. (2000). Minimum and maximum development rates of some forensically important Calliphoridae (Diptera). *J. Forensic Sci.*, **45(4)**:824–832.

Baselt, R.C. (2004). *Disposition of Toxic Drugs and Chemicals*. 7th edition, Foster City, CA: Biomedical Publications.

Benecke, M. (1998). Six forensic entomology cases: description and commentary. *J. Forensic Sci.*, **43**:797–805.

Benecke, M., and Lessig, R. (2001). Child neglect and forensic entomology. *Forensic Sci. Int.*, **120(1–2)**:155–159.

Bergeret, M. (1955). Infanticide: Momification naturelle du cadavre. *Ann. Hyg. Legal Med.*, **4**:442–452.

Beyer, J. C., Enos W. F., and Stajic, M. (1980). Drug identification through analysis of maggots. *J. Forensic Sci.*, **25**:411–412.

Bourel, B., Hédouin, V., Martin-Bouyer, L., Bécart, A., Tournel, G., Deveaux, M., and Gosset, D. (1999). Effects of morphine in decomposing bodies on the development of *Lucilia sericata* (Diptera: Calliphoridae). *J. Forensic Sci.*, **44**:354–358.

Byrd, J. H., and Allen, J. C. (2001). The development of the black blow fly, *Phormia regina* (Meigen). *Forensic Sci. Int.*, **120**:79–88.

Byrd, J. H., and Butler, J. F. (1997). Effects of temperature on *Chrysomya rufifacies* (Diptera: Calliphoridae) development. *J. Med. Entomol.*, **34(3)**:353–358.

Byrd, J. H., and Butler, J. F. (1998). Effects of temperature on *Sarcophaga haemorrhoidalis* (Diptera: Sarcophagidae) development. *J. Med. Entomol.*, **35(5)**:694–698.

Byrd, J. H., and Castner, J. L. (2001). *Forensic Entomology: The Utility of Arthropods In Legal Investigations*. Boca Raton, FL: CRC Press.

Byrne, A. L., Camann, M. A, Cyr, T. L., Catts, E. P., and Espelie, K. E. (1992). Forensic implications of biochemical differences among geographic populations of the black blow fly, *Phormia regina* (Meigen). *J. Forensic Sci.*, **40(3)**:372–377.

Catts, E. P. (1994). Problems in estimating the postmortem interval in death investigations. *J. Agric. Entomol.*, **9(4)**:245–255.

Catts, E. P., and Goff, M. L. (1992). Forensic entomology in criminal investigations. *Anu. Rev. Entomol.*, **37**:253–272.

Connolly, H. M., Crary, J. L., McGoon, M. D., Hensrud, D. D., Edwards, B. S., and Edwards, W. D. (1997). Valvular heart disease associated with fenfluramine–phentermine. *N. Engl. J. Med.*, **337(9)**:581–588.

DAWN (2003). *Emergency Department Trends from the Drug Abuse Warning Network, Final Estimates 1995–2002*. DAWN Series. Washington, DC: Substance Abuse and Mental Health Services Administration, Office of Applied Studies.

DAWN (2004). *Mortality Data from the Drug Abuse Warning Network, 2002*. DAWN Series. Washington, DC: Substance Abuse and Mental Health Services Administration, Office of Applied Studies.

Dodd, M. (2003). Coroner: Ephedra helped kill pitcher. *USA Today*, Mar. 14, 2003.

Downer, R. G. H. (1981). *Energy Metabolism in Insects*. New York: Plenum Press, pp. 56–57.

Elzinga, R. J. (1999). *Fundamentals of Entomology*, 5th ed. Upper Saddle River, NJ: Prentice Hall, pp. 89, 107–115, 436–450.

Goff, M. L. (1991). Feast of clues: insects in the service of forensics. *The Sciences*, July 31, pp. 30–35.

Goff, M. L. (1993). Estimation of postmortem interval using arthropod development and successional patterns. *Forensic Sci. Rev.*, **5(2)**:81–94.

Goff, M. L., and Lord, W. D. (1994). Entomotoxicology, a new area for forensic investigation. *Am. J. Forensic Med. Pathol.*, **15**:51–57.

Goff, M. L., Omori, A. L., and Gunatilake, K. (1988). Estimation of postmortem interval by arthropod succession. *Am. J. Forensic Med. Pathol.*, **9(3)**:220–225.

Goff, M. L., Omori, A. I., and Goodbrod, J. R. (1989). Effect of cocaine in the tissues on the rate of development of *Boettcherisca peregrina* (Diptera: Sarcophagidae). *J. Med. Entomol.*, **26**:91–93.

Goff, M. L., Brown, W. A., Hewadikaram, K. A., and Omori, A. I. (1991). Effect of heroin in decomposing tissues on the development rate of *Boettcherisca peregrina* (Diptera: Sarcophagidae) and implications of this effect on estimation of postmortem intervals using arthropod development patterns. *J. Forensic Sci.*, **36(2)**:537–542.

Goff, M. L., Brown, W. A., and Omori, A. I. (1992). Preliminary observations of the effect of methamphetamine in decomposing tissues on the development rate of *Parasarcophaga ruficornis* (Diptera: Sarcophagidae) and the implications of this effect on the estimations of postmortem intervals. *J. Forensic Sci.*, **37(3)**:867–872.

Goff, M. L., Brown, W. A., and Omori, A. I., and LaPointe, D. A. (1993). Preliminary observations of the effects of amitriptyline in decomposing tissues on the development of *Parasarcophaga ruficornis* (Diptera: Sarcophagidae) and implications of this effect to estimation of postmortem interval. *J. Forensic Sci.*, **38(2)**:316–322.

Goff, M. L., Brown, W. A., Omori, A. I., and LaPointe, D. A. (1994). Preliminary observations of the effects of phencyclidine in decomposing tissues on the development of *Parasarcophaga ruficornis* (Diptera: Sarcophagidae). *J. Forensic Sci.*, **39**:123–128.

Goff, M. L., Miller, M. L., Paulson, J. D., Lord, W. D., Richards, E., and Omori, A. I. (1997). Effects of 3,4-methylenedioxymethamphetamine in decomposing tissues on the development of *Parasarcophaga ruficornis* (Diptera: Sarcophagidae) and detection of the drug in postmortem blood, liver tissue, larvae, and puparia. *J. Forensic Sci.*, **42**:276–280.

Golénischeff, Wladimir. (1927). Papyrus hiératiques (CGC) 58009 p. 45–54 [= Lieblein, Que mon nom fleurisse p. 12–16, pl. 17–27 (Lieblein, Jens)]

Goodman, L. S., and Gilman, A. G. (Eds.) (1995). *Goodman & Gilman's The Pharmacological Basis of Therapeutics*, 9th ed. New York: McGraw-Hill, p. 4.

Greenberg, B. (1990a). Behavior of postfeeding larvae of some Calliphoridae and a muscid (Diptera). *Ann. Entomol. Soc. Am.*, **83(6)**:1210–1214.

Greenberg, B. (1990b). Nocturnal oviposition behavior of blow flies (Diptera: Calliphoridae*). J. Med. Entomol.*, **27(5)**:807–810.

Greenberg, B. (1991). Flies as forensic indicators. *J. Med. Entomol.*, **28(5)**:565–577.

Gullan, P. J., and Cranston, P. S. (1994). *The Insects: An Outline of Entomology*. London: Chapman & Hall, pp. 82–84.

Gunatilake, K., and Goff, M. L. (1989). Detection of organophosphate poisoning in a putrefying body by analyzing arthropod larvae. *J. Forensic Sci.*, **34**:714–716.

Hall, R. D., and Doisy, K. E. (1993). Length of time after death: effect on attraction and oviposition or larviposition of midsummer blow flies (Diptera: Calliphoridae) and flesh flies (Diptera: Sarcophagidae) of medicolegal importance in Missouri. *Ann. Entomol. Soc. Am.*, **86**:589–593.

Hall, R. D., Anderson, P. C., and Clark, D. P. (1986). A case of myiasis caused by *Phormia regina* (Diptera: Calliphoridae) in Missouri, USA. *J. Med. Entomol.*, **23(5)**:578–579.

Hanzlick, R. (1994). Embalming, body preparation, burial and disinterment: an overview for forensic pathologists. *Am. J. Forensic Med. Pathol.*, **15(2)**:122–131.

Hédouin, V., Bourel, B., Martin-Bouyer, L., Bécart, A., Tournel, G., Deveaux, M., and Gosset, D. (1999). Determination of drug levels in larvae of *Lucilia sericata* (Diptera: Calliphoridae) reared on rabbit carcasses containing morphine. *J. Forensic Sci.*, **44(2)**:351–353.

Hédouin, V., Bourel, B., Martin-Bouyer, L., Bécart, A., Tournelm G., Deveauxm M., and Gosset, D. (2001). Determination of drug levels in larvae of *Protophormia terraenovae* and *Calliphora vicina* (Diptera: Calliphoridae) reared on rabbit carcasses containing morphine. *J. Forensic Sci.*, **46(1)**:12–14.

Introna, F., Lo Dico, C., Caplan, Y. H., and Smialek, J. E. (1990). Opiate analysis in cadaveric blowfly larvae as an indicator of narcotic intoxication. *J. Forensic Sci.*, **35**:118–122.

Introna, F., Campobasso, C. P., and Goff, M. L. (2001). Entomotoxicology. *Forensic Sci. Int.*, **120**:42–47.

James, M. T. (1947). *The Flies That Cause Myiasis in Man*. USDA Miscellaneous Publication 63. Washington, DC: U.S. Government Printing Office.

Kamal, A. S. (1958). Comparative study of thirteen species of sarcosaprophagous *Calliphoridae* and *Sarcophagidae* (Diptera). *Ann. Entomol. Soc. Am.*, **51**:261–271.

Keh, B. (1985). Scope and applications of forensic entomology. *Anu. Rev. Entomol.*, **30**:37–54.

Kintz, P., Godelar, B., Tracqui, A., Mangin, P., Lugnier, A. A., and Chaumont, A. J. (1990). Fly larvae: a new toxicological method of investigation in forensic medicine. *J. Forensic Sci.*, **35**:204–207.

LaMotte, L. R., and Wells, J. D. (2000). *p*-Values for postmortem intervals from arthropod succession data. *J. Agric. Biol. Environ. Stat.*, **5(1)**:58–68.

Levine, B. (1999). *Principles of Forensic Toxicology*. Washington, DC: AACC Press, pp. 3–5.

Levine, B., Golle, M., and Smialek, J. (2000). An unusual drug death involving maggots. *Am. J. Forensic Med. Pathol.*, **21**:59–61.

Lord, W. D., DiZinno, J. A., Wilson, M. R., Budowle, B., Taplin, D., and Meinking, T. L. (1998). Isolation, amplification, and sequencing of human mitochondrial DNA obtained from human crab louse, *Pthirus pubis* (L.), blood meals. *J. Forensic Sci.*, **43(5)**:1097–1100.

McKnight, B.E. (Transl.) (1981). *The washing away of wrongs: Forensic Medicine in thirteenth-Century China.* pp. 181, University of Michigan, Ann Arbor.

NHS Survey (2002). *Results from the 2001 National Household Survey on Drug Abuse*, Vol. I, *Summary of National Findings*. Washington, DC: Substance Abuse and Mental Health Services Administration, Office of Applied Studies.

Nolte, K. B., Pinder, R. D., and Lord, W. D. (1992). Insect larvae used to detect cocaine poisoning in a decomposed body. *J. Forensic Sci.*, **37**:1179–1185.

Nuorteva, P. (1977). Sarcosaprophagous insects as forensic indicators. In: Tedeschi, C. G. (Ed.), *Forensic Medicine: A Study in Trauma and Environmental Hazards*, Vol. V, *Death*. Philadelphia,W. B. Saunders, pp. 1072–1095.

Nuorteva, P., and Nuorteva, S. L. (1982). The fate of mercury in sarcosaprophagous flies and in insects eating them. *Ambio*, **11**:34–37.

ONDCP (2000) Drug-Related Crime. Walters, J. P., director. ONDCP Drug Policy Information Clearinghouse Fact Sheet. Washington, DC: Executive Office of the President, Mar.

Poklis, A., Sunshine, I., and Jentzen, J. (1998). Toxicology. In: Fierro, M. (Ed.) *Handbook for Postmortem Examination of Unidentified Remains*. Northfield, IL: College of American Pathologists, pp. 241–247.

Pounder, D. J. (1991). Forensic entomotoxicology. *J. Forensic Sci., Society*. **31**:469–472.

Prichard, J. G., Kossoris, P. D., Leibovitch, R. A., Robertson, L. D., and Lovell, F. W. (1986). Implications of trombiculid mite bites: report of case and submission of evidence in a murder trial. *J. Forensic Sci.*, **31**:301–306.

Sadler, D. W., Fuke, C., Court, F., and Pounder, D. J. (1995). Drug accumulation and elimination in *Calliphora vicina* larvae. *Forensic Sci. Int.*, **71**:191–197.

Sadler, D. W., Richardson, J., Haigh, S., Bruce, G., and Pounder, D. J. (1997a). Amitriptyline accumulation and elimination in *Calliphora vicina* larvae. *Am. J. Forensic Med. Pathol.*, **18**:397–403.

Sadler, D. W., Robertson, L., Brown, G., Fuke, C., and Pounder, D. J. (1997b). Barbiturates and analgesics in *Calliphora vicina* larvae. *J. Forensic Sci.*, **42**:481–485.

SAMHA (2003). Prevention Alert: Trouble in the Medicine Chest (I): *Rx Drug Abuse Growing*. Washington, DC: Center for Substance Abuse Prevention. Substance Abuse and Mental Health Administration.

Schoenly, K., Goff, M. L., and Early, M. A. (1992). BASIC algorithm for calculating the postmortem interval from arthropod successional data. *J. Forensic Sci.*, **37(3)**:808–823.

Schoenly, K., Goff, M. L., Wells, J. D., and Lord, W. D. (1996). Quantifying statistical uncertainty in succession-based entomological estimates of the postmortem interval in death scene investigations: a simulation study. *Am. Entomol.*, Summer, pp. 106–112.

Smith, K. G. V. (1986). *A Manual of Forensic Entomology*. London: Trustees of the British Museum.

Sohal, R. S., and Lamb, R. E. (1977). Intracellular deposition of metals in the midgut of the adult housefly, *Musca domestica*. *J. Insect Physiol.*, **23**:1349–1354.

Tracqui, A., Keyser-Tracqui, C., Kintz, P., Ludes, B. (2004). Entomotoxicology for the forensic toxicologist: much ado about nothing? *Institut de Médecine Légale*. **118(4)**:194–6. Epub 2004 May 26.

Tomita, T., and Zhao, Q. (2002). Autopsy findings of heart and lungs in a patient with primary pulmonary hypertension associated with use of fenfluramine and phentermine. *Chest*, **121(2)**:649–652.

Tz'u, S. (1981). *The Washing Away of Wrongs*. Translated by McKnight, B. Ann Arbor, MI: University of Michigan.

Webster, F. M. (1890). *Insect Life*, **5**: 356–358, 370–372.

Wells, J. D., and LaMotte, L. R. (1995). Estimating maggot age from weight using inverse prediction. *J. Forensic Sci.*, **40(4)**:585–590.

Wells, J. D., and Sperling, F. A. H. (2001). DNA-based identification of forensically important *Chrysomyinae* (Diptera: Calliphoridae). *Forensic Sci. Int.*, **120(1–2)**:110–115.

Wilson, Z., Hubbard, S., and Pounder, D. J. (1993). Drug analysis in fly larvae. *Am. J. Forensic Med. Pathol.*, **14**:118–120.

Wood, M., Laloup, M., Pien, K., Samyn, N., Morris, M., Maes, R. A. A., de Bruijn, E. A., Maes, V., and De Boeck, G. (2003). Development of a rapid and sensitive method for the quantitation of benzodiazepines in *Calliphora vicina* larvae and puparia by LC–MS–MS. *J. Anal. Toxicol.*, **27**:505–512.

INDEX

ABAcard HemaTrace, 279
Accelerants, 42
Acid phosphatase, 283
Accumulated degree hour, 484, 489, 493
Acrylic melamine enamel, 136
Adenine, 294
Alcohol, 436
Alcohol absorption, 440
Alcohol excretion, 440
Alcohol dehydrogenase, 377
Alcohol impairment, 441
Alcohol in urine and saliva, 450
Aldehyde dehydrogenase, 441
Allele frequency, 306
Allelic dropout, 311
Allelic ladder, 302
Alternate light source, 292
Amelogenin (Amel), 302
Amylase, 286, 287
Analyte, 13, 459. 467
Anderson reference sequence, 318
Antiparallel, 294
Architectural coatings, 140
Arson, 41, 49
Arthropod, 486
Atomic absorption spectroscopy (AAS), 469
Attenuated total reflectance (ATR) spectroscopy, 159, 162, 163, 256
Autolysis, 473
Automotive finish systems, 139, 144

Bile, 464
Binders, 135, 164
Biomarkers, 11
Bleed-through, 313
Blood, 271, 460
Blood alcohol concentration (BAC), 437
Blowflies, 484
Brain, 464

Breath alcohol testing, 446
Breathalyzer, 448

Calliphoridae, 485
Cambridge reference sequence (CRS), 318
Capillary electrophoresis, 301, 302
C-DNA microarray, 329
Cathodoluminescence, 124
Central nervous system (CNS), 435, 436, 442
Cerebrospinal fluid (CSF), 461
Chelex, 297
Chemical fingerprinting, 4
CODIS, 305, 343
Coherence anti-Stokes Raman spectroscopy, 263
Confocal Raman spectroscopy, 263
Colorant, 227
Color science, 134
Combustible liquids, 42
Complementary base pairing, 295
Complementary strand, 294
Confirmatory test, 270, 271
Control region (D-Loop), 296
Crossed-over immunoelectrophoresis, 281
Cyanide, 469
Cytochrome P450 (CYP450), 471
Cytosine, 294

Date rape drugs, 355, 361
Decomposition, 473
Degraded DNA, 308
Deoxynucleotide triphosphate, 298
Differential scanning calorimetry, 37, 38
Diffuse reflectance spectroscopy, 258
Dispersant, 244
Displacement loop (D-Loop), 295
Distillates, 76
DNA extraction, 297
DNA fingerprinting, 335, 336
DNA microarray, 328

Forensic Chemistry Handbook, First Edition. Edited by Lawrence Kobilinsky.
© 2012 John Wiley & Sons, Inc. Published 2012 by John Wiley & Sons, Inc.

DNA quantification, 297, 298
DNA structure, 292–294
DNA typing, 292
DUI, 446
Dyes, 136
Dynamic headspace concentration, 45

Electron capture detector (ECD), 35
Electropherogram, 303
Elemental analysis, 188–193, 204
Embalming fluid, 474
Energy dispersive x-ray spectrometry, 193
Entomology, 485
Entomotoxicology, 483, 484
Environmental forensics, 2
Explosives, 24
Ethanol, 389, 435, 473
Ethyl glucuronide, 450
Evaporation, 95–98
Exons, 299
Explosives, 23, 24
Expression profiling, 330
External reflection spectroscopy, 255
Extraction of DNA, 296

Filtered light examination, 236
Flame ionization detection (FID), 445
Flesh flies, 485
Flunitrazepam, 355
Forgery, 134
Formalin, 474
FTIR spectroscopy, 128, 159, 239, 246
Functional tolerance, 444

Gasoline, 66
Gastric contents, 462
GC, 35, 47, 48, 246, 468
GC-MS, 239, 240, 463
GC-FID, 49
Gel electrophoresis, 301
GHB (gamma hydroxybutyrate), 355, 459
Guanine, 294

Headspace gas chromatography, 445
Heme, 272
Heteroplasmy, 320
HPLC, 239, 240, 468
House flies, 484
Hydrocarbon Mixtures, 4
Hypervariable regions, 296, 318, 319

Ignitable liquid residues (ILRs), 41, 51
Immunochromatographic assays, 278, 285

Inductively coupled plasma emission spectroscopy (ICP-ES), 204
Infrared luminescence, 236
Infrared microspectroscopy, 259
Infrared spectroscopy, 47, 158, 253
Injection site, 466
Ink, 226
Ink analysis, 230
Ink composition, 227
Ink dating, 240
Inkjet ink, 242
Inorganic pigment, 171
Insect colonization, 483
Insect succession, 487
Intoxication, 436, 438
Introns, 299
Isotopes, 12
Isotope ratio mass spectroscopy, 35
Isozymes, 284, 364

Ketamine, 355

Lacquer, 135
Larva, 466, 484, 487–489, 491, 496
Laser desorption mass spectrometry, 207
Laser printers, 245
LC-MS-MS, 463
Leuco malachite green, 273
Liver, 464
Low copy number, 307
Luminol, 274

Maggots, 466, 491
Matrix-assisted laser desorption ionization time-of-flight mass spectrometry (MALDI-TOF), 309
Mass spectrometry, 178
MDMA, 358, 390, 491
Meconium, 463
Metabolic tolerance, 444
Metamorphosis, 486
Microarrays, 327, 328
Microarray probes, 329
Microcrystal assays, 275
Microspectrophotometry, 152–154
Microvariant allele, 304
Microscopy, 149
Minisatelllite DNA, 299, 340
Minisequencing, 314, 339
Mini STSR primers, 308, 309
Mitochondria, 377
Mitochondrial DNA, 295, 314–316, 318–320
Mitochondrial DNA sequencing, 316

Mixture interpretation, 310
Molecular autopsy, 471
Monoclonal antibodies, 278
Multiplex, 341

Nasal swabs, 466
Neutron activation analysis (NAA), 204
Nicotinamide-adenine dinucleotide (NAD+), 445
Nontemplate addition, 313
Nuclear DNA, 295
Null alleles, 311

Off-ladder allele, 304
Organic pigment, 172, 173
Ouchterlony assay, 280
Oxidation-reduction reactions, 272, 381

Paint coating, 134
Paint binder, 134
Paint additive, 134
Particle size distribution, 117
Period of insect activity (PIA), 484
PETN, 25
Petrographic microscope, 122
Phadebas assay, 287
Pharmacogenomics, 471
Phenolphthalin, 273
Photocopiers, 245
Pigments, 136
Plasticizer, 134
Polycyclic aromatic hydrocarbons (PAHs), 6
Polymerase chain reaction (PCR), 293, 298, 300
PCR inhibition, 310
Postmortem alcohol, 451
Postmortem interval, 483, 484
Postmortem redistribution, 460, 470
Postmortem toxicology, 457–459
Presumptive assay, 270, 271
Primers, 144, 298, 299, 312
Prostate specific antigen (P30), 285, 286
Pull-up artifact, 312, 313
Purine, 294
Putrefaction, 473
Pyrimidine, 294
Pyrolysis gas chromatography (PyGC), 178–188

Quality assurance, 36, 37, 105
Quantification of DNA, 297
Questioned document, 226

Raman spectroscopy, 128, 175, 239, 260
Reaction time, 442

Reference sequences, 317
Refractive index, 124
Resins, 228
Resonance Raman spectroscopy, 263
Restriction enzymes, 293
Restriction fragment length polymorphism (RFLP), 293
Reverse transcription, 330
RNA-based assays, 286, 289
RSID-blood, 279, 280

Saliva identification, 286
Sarcophagidae, 484
Satellite DNA, 340
Scanning electron microscope (SEM), 87, 125, 193, 246
Seat of explosion, 27
SEM–energy dispersive spectroscopy (SEM-EDS), 157, 193–201, 204, 246
Semen identification, 282
Serology, 270
Short tandem repeats (STRs), 298, 300, 340
Shroud of Turin, 213
Single nucleotide polymorphism (SNP), 313, 343
Sobriety tests, 443
Sodium fluoride, 466
Soil color, 115
Solid phase extraction, 46, 463, 467
Solvents, 54, 228
Spermatozoa, 284, 285
Starch-Iodine assay, 287
Statistics, 320
Steam distillation, 44, 56
Stereo binocular microscope, 120
Stutter artifact, 312
Surface enhanced Raman spectroscopy, 262

Taq polymerase, 298, 313
Takayama crystal assay, 275
TATP, 33
Teichmann crystal assay, 275
Tetramethylbenzidine, 273
THC (Tetrahydrocannabinol), 464
Thermal energy analyzer, 35
Thin layer chromatography (TLC), 238, 240
Thymine, 294
Time of colonization (TOC), 484
Tolerance, 444
Toner, 245, 246
Toxicogenomics, 332
Toxicology, 485
Tracers, 13
Transmission infrared spectroscopy, 255

Ultraviolet spectroscopy, 47
Urine, 460

Variable length probe array (VLPA), 342
Vibrational spectroscopy, 251, 252
Video spectral analysis (VSA), 236
Vinland map, 214

Vitreous humor, 461

Weathering, 18

x-ray diffraction, 126, 205
x-ray fluorescence spectrometry, 201–203, 205, 246